Chemical Reaction Engineering

Second Edition

Chemical Reaction Engineering

Parameter Estimation, Exercises and Examples

Second Edition

Martin Schmal & José Carlos Pinto

Department of Chemical Engineering, Federal University of Rio de Janeiro – COPPE/UFRJ, Brazil

CRC Press
Taylor & Francis Group
Boca Raton   London   New York   Leiden

CRC Press is an imprint of the
Taylor & Francis Group, an **informa** business

A BALKEMA BOOK

CRC Press/Balkema is an imprint of the Taylor & Francis Group, an informa business

First English edition © 2014 Taylor & Francis Group, London, UK

Originally published in Portuguese as "Cinética e Reatores: Aplicação à Engenharia Química"
  by Synergia Editora
© 2010 Martin Schmal

Translators:
Fabio Souza Toniolo
Carlos Alberto Castor Jr.
João Paulo Bassin
Martin Schmal

Second English edition © 2022 Taylor & Francis Group, London, UK
Authorised translation from the Portuguese language edition published by Synergia Editora
All rights reserved

Originally published in Portuguese as: "Cinética e Reatores – Aplicação na Engenharia
Química - teoria e exercícios – 3ª ed."
© 2017 Synergia Editora, Rio de Janeiro, Brazil

Typeset by MPS Limited, Chennai, India
Printed and Bound by CPI Group (UK) Ltd, Croydon, CR0 4YY.

Library of Congress Cataloging-in-Publication Data

A catalog record for this book has been requested

Published by: CRC Press/Balkema
                Schipholweg 107C, 2316 XC Leiden, The Netherlands
                e-mail: Pub.NL@taylorandfrancis.com
                www.routledge.com – www.taylorandfrancis.com

ISBN: 978-0-367-49446-9 (Pbk)
ISBN: 978-1-032-07060-5 (Hbk)
ISBN: 978-1-003-04660-8 (eBook)
DOI: 10.1201/9781003046608

Table of contents

Preface

The first English edition of this book was published in 2014. This book was originally intended for undergraduate and graduate students, its one major objective: to teach the basic concepts of kinetics and reactor design. The author's motivation was the fact that students frequently encounter great difficulty trying to explain the basic phenomena that occur in the practice of reactor design. So, instead of discussing specific projects of the industry, central concepts along with examples and many exercises are presented for each topic. The main objective then was to prompt students to observe and think about kinetic phenomena. Indeed, reactors cannot be designed and operated without knowledge of kinetics.

In the present edition, the empirical nature of kinetic studies is recognized; analyses related to how experimental errors affect kinetic studies are performed and illustrated with actual data. Particularly, analytical and numerical solutions are derived to represent the uncertainties of reactant conversions in distinct scenarios and are used to analyze the quality of the obtained parameter estimates. Consequently, the present edition includes new topics that focus on the development of analytical and numerical procedures for more accurate description of experimental errors in reaction systems and of estimates of kinetic parameters.

Finally, kinetics requires knowledge that must be complemented and tested in the laboratory. Therefore, practical examples of reactions performed in bench and semi-pilot scales are discussed in the final chapter.

We decided to organize this version of the book in two parts. In the first part, a thorough discussion regarding reaction kinetics is presented. In the second part, basic equations are derived and used to represent the performance of batch and continuous ideal reactors, isothermal and non-isothermal reaction systems and homogeneous and heterogeneous reactor vessels, as illustrated with several examples and exercises.

In closing, we would like to thank all of our undergraduate and graduate students for giving us the opportunity to discuss the subjects presented in the present book and to teach chemical kinetics in the classroom and in the lab. We also thank FAPERJ (Fundação Carlos Chagas Filho de Apoio à Pesquisa do Estado do Rio de Janeiro), UFRJ (Universidade Federal do Rio de Janeiro) and Synergia Editora for supporting the previous two editions of the book.

Martin Schmal
José Carlos Pinto
Rio de Janeiro, July 2021

ACKNOWLEDGEMENTS

We thank the many undergraduate and graduate students who attended our classes in the courses of Chemical Engineering at the Federal University of Rio de Janeiro, for providing us with the driving force and motivation to write this book. We thank, in particular, the colleagues of the Chemical Engineering Department of Escola de Química and Programa de Engenharia Química of COPPE for fruitful technical discussions and continuous support. Finally, we thank the colleagues and research team of NUCAT (Núcleo de Catálise) and EngePol (Laboratório de Engenharia de Polímeros) for partnership and sharing of opinions and points of view.

DEDICATION

This book is dedicated to my wife Vitoria for her patience and understanding, to my daughters Thaiz and Alice and especially to my grandchildren Camille, Sophie, Heitor and Catarina. I sincerely thank Prof. Martin Schmal for this opportunity and collaboration. I would also like to dedicate this book to my wife, Marcia, and my "children", Martina, Amon and Juninho. (JCP)

Nomenclature

Symbols	Meaning	Units
a, b, c	Stoichiometric coefficients	_____
a′, b′, c′	Reaction order	_____
A_i , B_i ...	Components, reactants and products	
C	Concentration	gmol/l
C_A	Concentration of reactant A	gmol/l
C_{AO}	Initial concentration of reactant A	gmol/l
C_P	Specific heat at constant pressure	cal/g°
E, E_D,E_R	Activation Energy, direct, reverse	cal/gmol
E	Distribution Function of residence time	
F, F_0	Molar flow, initial	moles/h
F_A	Molar flow of reactant A	moles/h
F_{Ao}	Initial molar flow of reactant A	moles/h
F(t)	Cumulative Distribution Function	
$\dot{G}_j$	Mass flow of component j	Kg/h
G	Gibbs free energy	Kcal/gmol
H_T	Enthalpy of reaction at T	Kcal/gmol
H^o	Enthalpy of reaction at 25°C	Kcal/gmol
K	Equilibrium constant	Kcal/gmol
Kc	Concentration equilibrium constant	Kcal/gmol
Kp	Equilibrium constant at pressure P	atm^{-1}
k, k′	Specific reaction rates, direct and reverse	eq.3.19
k_0	Frequency factor of Arrhenius equation	
L	Reactor length	cm (m)
M	Mass	g (Kg)
M	Molecular weight	
n	Global reaction order	
n	Number of moles	moles
$n_{A,B}$	Number of moles of reactants A, B ...	moles
n_0	Initial total number of moles	moles
n_T	Total number of moles	moles
P	Pressure	atm^{-1}
P_0	Initial pressure	atm

Symbols	Meaning	Units
Pé	Péclét Number	
q	Heat transfer rate	Kcal/h
$\dot{Q}$	Heat transfer rate	Kcal/h
$\dot{Q}_g$	Heat generated by reaction	Kcal/h
$\dot{Q}_c$	Convective heat transfer	Kcal/h
$\dot{Q}_s$	Sensible heat transfer	Kcal/h
$\dot{Q}_T$	Total heat transfer	Kcal/h
$\dot{Q}_r$	Removal heat transfer	Kcal/h
r	Reaction rate	mol /L.h
$(-r_A)$, $(-r_B)$	Rate of consumption of A, B	mol/L.h
r_j	Rate of formation of component j	mol/L.h
R	Gas constant	atm.L/mol.K
S	Selectivity	
s	Space velocity	s^{-1}
t	Time	s
t_f	Final time	s
$\bar{t}$	Mean residence time	s
T	Temperature	° C (°K)
U	Global heat transfer coefficient	cal/mol.K
v	Linear velocity	cm/s
v_0	Volumetric flow rate	cm^3/s
V_R	Reactor volume	cm^3 (L)
V	Reaction volume	cm^3
V_0	Initial volume	cm^3
x	Length	cm
X	Conversion	
X_A	Conversion of reactant A	
Y	Molar fraction	
z	Axial length	cm

Symbols

α	Degree of extension	
β	Energy parameter	
β	Kinetic parameter	
γ	Kinetic parameter	
ε	Porosity	
ε_A	Expansion or contraction parameter	
χ	Ratio of specific rate constants	
μ	Viscosity; chemical potential; Monod constant; reduced mass; average	
ν	Stoichiometric coefficient; Kinematic viscosity	
p	Density	g/cm^3
σ	Standard deviation	

τ	Space time	s
φ	Non-dimensional concentration	
ϕ	Local yield	
Φ	Global yield	
θ	Non-dimensional time; surface fraction	

Subscripts

A,R	Reactant or products
e	Equilibrium
f	Final
g	Gas
j	Component
i	Reaction number
r	Reaction
o	Initial

Other nomenclatures

MeO	–	Metal oxide
M^0	–	Metal
k_B	–	Boltzman constant
SSI	–	Structure sensitive reaction
SIS	–	Structure insensitive reaction
BET	–	Surface area (m^2/g)
TOF	–	Rate (frequency factor) (s^{-1})
DRX	–	x-ray diffraction
TPR	–	Temperature programmed reaction
MEV	–	Scanning electronic microscopy
TEM	–	Transmission Electronic microscopy
Φ_n	–	Thiele modulus
η	–	Effectiveness factor
Ω	–	Global effectiveness factor
L_{mf}	=	Height bed at minimum fluidization
ε_{mf}	=	Void fraction at minimum fluidization
ρ_s	=	Density of solid
ρ	=	Density of gas
f_b	–	Bubble fraction
f_e	–	Emulsion fraction of gas
k_i	–	Interface coefficient $(m^3/m^3$ bed x $h)$
$\bar{u}$	–	Mean velocity; collision velocity
u_b	–	Superficial velocity (m/s)
C_{Ab}	–	Concentration of gas in the bubble
C_{Ae}	–	Concentration of gas in the emulsion phase
Ke	=	Gas fraction in the emulsion phase
D_e	=	Effective diffusivity
u_e	=	Superficial velocity in the emulsion phase

U_{mf}	–	Minimum fluidization velocity (cm/s)
ρ_s	–	Particle density of solids
ρ_f	=	Fluid density
d_p	–	Particle diameter
g	–	Gravity
φs	–	Solid parameter
ε_{mf}	=	Void fraction
ρ_s	=	Solid density
ε	=	Porosity of the bed reactor
$\overline{d_p}$	–	Mean particle size
ρ_s	–	Solid density kg/m3
d_b	–	Bubble diameter (m)
d_t	=	External diameter of reactor (m)
k_i	=	Interface coefficient (cm^3/cm^3 bed $\times$ s)
C_{Ab}	=	Gas concentration in the bubble phase
C_{Ae}	=	Gas concentration in the emulsion phase
r''	=	Reaction rate (mole/g $\times$ s)
ε_D	–	Dispersion
d_t, dp	=	Diameter (tube and particles)
Sh	=	Sherwood number
Re	=	Reynolds number
Sc	=	Schmidt number
k_l	–	Mass transfer coefficient (cm/s)
a_l	–	External surface – liquid/solid – (cm^2/g)
m_p	=	Mass of particles (g)
H	–	Henry constant at the gas-liquid interface
pi	–	Partial pressure
k_d	–	Deactivation constant
φ	–	Yield
ϕ_A	–	Local yield
$\chi = k_2/k_1$	–	Relation between specific rate constants
S	–	Selectivity
V_j	–	Specific volume (m^3/mol)
E_j	–	Energy of component j
F_j	–	Molar flow rate of component j
W	–	Work
Q	–	Heat (J/mol)
k^*	–	Apparent specific constant
f_p	–	Partition function
ΔS^0	–	Entropy
V_{max}	–	Maximum reaction rate
$V_{ads}; V_{des}$	–	Adsorption or desorption volume
r_s	–	Substrate reaction rate
R	–	Transfer rate

About the authors

Martin Schmal graduated in Chemical Engineering at the Engineering Faculty of the Catholic University of S. Paulo (1964), he received a Master's degree in 1966 from the Federal University of Rio de Janeiro/COPPE, Brazil, and obtained his doctorate degree (Dr. Ing.) from the Technische Universität Berlin, Germany (1970). He became an associate professor at the Chemical Engineering Department of the Federal University of Rio de Janeiro in 1970, became Full Professor in 1985 and has been Emeritus since 2008. He specialized at the Institut du Recherche sur la Catalyse, Lyon, France (1981) and at the University of Karlsruhe, Germany (1983).

He has been teaching Kinetics and Reactors to undergraduate students at the Chemical Engineering School, Catalysis to graduate students at COPPE and a Postgraduate course in Engineering for over 40 years, since 1973.

The main research topics at the Nucleus of Catalysis centre at COPPE are catalysis, the catalytic process and nanoscience. Schmal maintains close contacts with the industry, in particular with Petrobras and various petrochemical industries: Oxiteno, Copene, Copesul, Petroquimica União, Braskem, Degussa and others, developing processes and catalysts, and has more than 20 patents to his name.

He is a member of the Brazilian Academy of Science (elected in 1999) and of the International Catalysis Society (since 2000). He has received several awards and in particular the Humboldt Research Award from the Humboldt Foundation – Germany (2002), the Premio Mexico for Science and Technology (Science Consulting Council, Mexico, 2002); the Senior Researcher award from the Ibero-American society (2010); the SCOPUS award – Elsevier-CAPES (2009) and finally the Distinguished Professor award from Coppe – Federal University of Rio de Janeiro (2013). He has published more than 220 articles in international journals; has presented and published 250 papers in Annals of International Congresses, Symposia and Meetings and published the following books: Chemical Kinetics and Reactor Design (1. Ed. 1982, 2. Ed. 2010), Heterogeneous Catalysis (2012) and Natural Gas Conversion VIII – Surface Science (2011).

His other activities include being peer reviewer for the following journals: Journals for Physical Chemistry, Journal of Catalysis, Surface Science, Applied Catalysis A and B, Catalysis Today, ACS Catalysis, Angewandte Chemie International Edition, Catalysis Letters, International Journal of Hydrogen Energy, Material Science, etc. He was on the editorial board of Applied Catalysis A (1992–1999) and of Catalysis Today (2000–2006). He was Founder and President of the Brazilian Catalysis Society (1998–2006).

His external collaborators include H.J. Freund (Max Planck Berlin), E. Lombardo (Argentina), Schlögl and Behrens (Max Planck Berlin), Albert Vannice (Penn State University, USA) and Ted Oyama (Virginia Tech, USA).

José Carlos Pinto graduated in Chemical Engineering from the Federal University of Bahia (1985) and obtained a DSc degree in Chemical Engineering from the Federal University of Rio de Janeiro (1991). He is currently a Full Professor of Programa de Engenharia Química/COPPE, Federal University of Rio de Janeiro, and a member of Programa de Pós-Graduação em Engenharia de Processos Químicos e Bioquímicos/Escola de Química, Federal University of Rio de Janeiro. Professor Pinto has been a member of the editorial committees of different scientific journals and a full member of the Brazilian Academy of Sciences since 2010 and of the National Engineering Academy since 2014. He has experience in chemical engineering, with emphasis on chemical reactors and with particular emphasis in modeling, simulation and control of polymerization reaction systems. He has published more than 450 papers in peer-reviewed scientific journals and conducted hundreds of joint projects with partners in both industrial and academic institutions.

Part I

Basic notions

Chapter 1

Definitions and stoichiometry

A chemical process involves not only chemical reactions but also involves surface and mass/energy transport phenomena. The chemical reactions are defined by the stoichiometry, in which reactants are directly related to the products of the reaction. Therefore, once defined the stoichiometry, the measurement of the composition of one of the components allows to relate it with the composition of other components. However, the order to the reaction rate does not always follow the stoichiometry. In this particular case, the kinetics of the reaction is not simply represented by a single step but involves several intermediate steps. In order to differentiate them, the reactions are divided as follows:

Irreversible reactions: those which are carried out in a single direction $A + B \rightarrow R + S$.
Reversible reactions: those which are carried out in both directions (direct and reverse).
Elementary reactions: those which are carried out in a single step.
Non-elementary reactions: those which are carried out in several elementary steps, and whose resultant reaction may not be elementary.

$$\begin{array}{ll} A + B \rightarrow AB & \text{elementary} \\ \underline{AB \rightarrow R} & \\ A + B \rightarrow R & \text{non-elementary} \end{array}$$

Simple reactions: those which are carried out in a simple step or not. If the reaction order follows the stoichiometry, then the reaction is simple and elementary.
Complex reactions: those which correspond to several reactions carried out simultaneously, either in series or combined.

Example

1. Hydrolysis of acetic anhydride

    $$(CH_3CO)_2O + H_2O \rightarrow 2CH_3COOH$$

    Kinetics: irreversible second order reaction $\Rightarrow$ irreversible and elementary

2. Decomposition of the acetaldehyde

$$CH_3CHO \rightarrow CH_4 + CO$$

Kinetics: irreversible fractional order reaction $\Rightarrow$ rate $\sim C_A^{1.5} \Rightarrow$ non-elementary reaction, where

$$A = CH_3CHO$$

3. Synthesis of ammonia:

$$N_2 + 3H_2 \Leftrightarrow 2NH_3$$

Kinetics: irreversible reaction of fractional order $\Rightarrow$ rate $\sim \Rightarrow$ non-elementary reaction

4. Methanation or *Fischer–Tropsch* synthesis

$$CO + 3H_2 \rightarrow CH_4 + H_2O$$
$$CO + 2H_2 \rightarrow [C_nH_{2n}]_n + H_2O$$

Kinetics: irreversible complex reaction of fractional order (in parallel)

5. Hydrogenation of crotonaldehyde

$$CH_3-CH_2 = CH_2-HC = O + H_2 \rightarrow CH_3-CH_3-CH_3-HC = O + H_2$$
$$\text{Crotonaldehyde} \qquad\qquad\qquad \text{Butyraldehyde}$$
$$\rightarrow CH_3-CH_3-CH_3-HCOH$$
$$\text{Butanol}$$

Kinetics: irreversible complex reaction of fractional order (in series)

6. Coal gasification:

$$C + H_2O \rightarrow CO + H_2$$
$$CO + H_2O \rightarrow CO_2 + H_2 \quad \text{(shift)}$$

Kinetics: irreversible complex and mixed reaction of fractional order.

1.1 MEASUREMENT VARIABLES

The description of the reaction rate represents the variation of the consumption of reactants or the formation of products as the reaction takes place. It can be graphically represented by a kinetic curve. The tangent to this curve indicates how the reaction rate as well as the consumption of reactants or formation of products varies as the reaction takes place. The reaction rate is observed to be high in the beginning of the reaction and gradually decreases as the reaction takes place, tending to zero when the equilibrium is reached or when the reactants disappear completely.

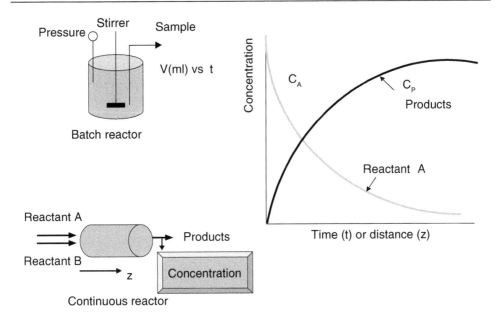

Figure 1.1   Kinetic curves for batch and continuous systems.

The kinetic curve of a reaction can be monitored by measuring the concentration, pressure, or other intensive variable, which is proportional to a characteristic property of the reaction (e.g., conductivity, wavelength, and binding energy). For a defined stoichiometric reaction, the variable of a component is followed with time or position in the reactor, depending on the system where the reaction is run. The concentrations of the other components can be calculated from this measured variable.

In a closed system, such as a batch reactor, the characteristic property varies with the reaction time. In an open system (such as continuous reactor), it varies with position or space time. In this case, the space time is defined as the ratio between the volume or mass of the reactor system and the inlet mixture flow. The schematic representation of the two systems is displayed in Figure 1.1.

The most important fact is that these are experimental measurements and depend on the type of reactor and on the analysis of reactants and products. In a batch reactor, several small samples are collected in different time intervals and the concentrations of reactants and products are measured. On the other hand, in a continuous reactor, the concentrations of reactants and products are measured along the reactor. The most used techniques for analysis are:

Titration: the samples are titrated with a neutralizing agent and with a chemical indi-
cator. The titration is conducted with an acid or a base which indicates the extent
of the reaction. This analysis is simple, but subject to errors and inaccuracy and
therefore should be repeated several times.
Gas or liquid chromatography: a measurement which is performed by thermal conductivity or by flame ionization. Each compound has its thermal or ionic properties

defined, which are detected and measured in the chromatograph after separation in appropriate columns. The selection of the separation chromatographic column depends on the components present. The conductivity and ionization are proportional to their concentrations and hence they can be related to each other.

Spectroscopy: detection of components through the measurement of their wavelengths in characteristic bands. The most frequently used instruments are infrared, ultraviolet, diffuse reflectance, etc. These properties are proportional to the concentrations of the components and may be related to their respective conversion factors.

Electrical conductivity: solutions containing H^+ and OH^- ions have electrical conductivity. In this solution, the substitution of an ion by another with different conductivities allows for determination of the reactant ion concentration.

Spectrophotometry: According to Beer's law, the absorbance, which is proportional to the concentration, is determined through spectrophotometry.

Dilatometry: in this technique, the mass variation is measured in a microbalance or through the dilatation of a quartz spring and the weight change is monitored as a function of the reaction time.

Total pressure: very simple method, in which the variation of the total pressure in a gas system kept at constant volume is measured. The pressure is monitored through a manometer.

1.2 CALCULATION OF MEASUREMENT VARIABLES

In general, it is recommended to determine a new variable which relates the concentrations of the reactants and products. Thus, it is possible to calculate the concentration of all components. We therefore define the extent of the reaction and the conversion of a particular component. If the reaction in a closed or open system is reversible, then the initial number of moles and the moles after a given time or position will be:

$$aA + bB \Leftrightarrow rR + sS \tag{1.1}$$

Initial moles: n_{A0} n_{B0} n_{R0} n_{S0}

Final moles: n_A n_B n_R n_S

where n is the number of moles; the indexes 0 and i represent the initial and instantaneous (or local) conditions, respectively, with A and B corresponding to the reactants and R and S corresponding to the products.

1.2.1 Extent of the reaction

α is defined as the extent of the reaction, indicating how much of reactant has been transformed or how much of product has been formed. Therefore, it can be generally represented by:

$$\alpha = \frac{n_{A0} - n_A}{a} = \frac{n_{B0} - n_B}{b} = \frac{n_R - n_{R0}}{r} = \frac{n_S - n_{S0}}{s} \tag{1.2}$$

where a, b, r, and s are the respective stoichiometric coefficients of the reaction.

Note that in this case the extent of the reaction is an extensive variable and is measured in moles. If α is known, the instantaneous or local number of moles of each component can be determined:

$$n_A = n_{A0} - a\alpha$$
$$n_B = n_{B0} - b\alpha$$
$$n_R = n_{R0} + r\alpha$$
$$n_S = n_{S0} + s\alpha \qquad (1.3)$$

Example

The reaction: $4PH_3 \rightarrow P_4 + 6H_2$
   From 1 mole of phosphine, after a certain time, we can find:

$$n_{PH_3} = 1 - \alpha$$
$$n_{P_4} = (1/4)\alpha$$
$$n_{H_2} = (6/4)\alpha$$

When the reaction is performed in a constant volume system, the extent of the reaction is directly determined as a function of the concentration, since:

$$C = \frac{n}{V} \left(\text{moles} / 1 \right)$$

Therefore,

$$C_A = C_{A0} - a\alpha$$
$$C_B = C_{B0} - b\alpha$$
$$C_R = C_{R0} + r\alpha$$
$$C_S = C_{S0} + s\alpha \qquad (1.4)$$

1.2.2   Conversion

The conversion is the most used variable. It is defined by the number of moles transformed or formed at a given time or local in relation to the initial number of moles. The conversion should always be defined for the limiting reactant of the reaction. The conversion has no unit, ranging from 0 to 1 for irreversible reactions or from 0 to X_{Ae} for reversible reactions. So, for irreversible reactions in which A is the limiting component, we have:

$$X_A = \frac{n_{A0} - n_A}{n_{A0}} \qquad (1.5)$$

For reversible reactions, we have:

$$X_{Ae} = \frac{n_{A0} - n_{Ae}}{n_{A0}} \tag{1.6}$$

where n_{Ae} is the number of moles in the equilibrium, and thus $X_{Ae} < 1.0$.

In a constant volume system, the conversion can be expressed as a function of concentration and thus,

$$X_A = \frac{C_{A0} - C_A}{C_{A0}} \tag{1.7}$$

If the conversion and reaction stoichiometry are known, the number of moles or the concentration of each component can be determined. Thus, in analogy to the extent of the reaction, we have for a reaction of the type like:

$$aA + bB \Leftrightarrow rR + sS$$

$$
\begin{aligned}
n_A &= n_{A0}(1 - X_A) \rightarrow V = \text{Const} \rightarrow C_A = C_{A0}(1 - X_A) \\
n_B &= n_{B0} - (b/a)X_A && C_B = C_{B0} - (b/a)X_A \\
n_R &= n_{R0} + (r/a)X_A && C_R = C_{R0} + (r/a)X_A \\
n_S &= n_{S0} + (s/a)X_A && C_S = C_{S0} + (s/a)X_A
\end{aligned} \tag{1.8}
$$

Note that for a variable volume system, the total change in the number of moles should be considered. Thus, for example, in a reaction of the type like $A + 3B \rightarrow 2R$, the total number of moles of reactants and products are 4 and 2, respectively. Therefore, there is a volume contraction. In this case, to express the concentrations in terms of the conversion, it is necessary to take into account the volume change. Using the ideal gas law, the total number of moles can be determined:

$$n = \frac{PV}{RT} \tag{1.9}$$

From the sum of the number of moles of each component, according to Equation 1.3, we can determine the total number of moles as a function of the extent of the reaction (α):

$$n = n_0 + (r + s - a - b)\alpha \tag{1.10}$$

Considering that $(r + s - a - b) = \Delta v$, it is possible to know whether there is volume contraction or expansion. Note that r, s, a, and b are the stoichiometric coefficients of the products and reactants, respectively, and $n_0 = n_{A0} + n_{B0} + n_{R0} + n_{S0}$ is the total number of initial moles. Thus,

$$n = n_0 + \Delta v \cdot \alpha \tag{1.11}$$

Substituting Equation 1.9 into Equation 1.11, we have:

$$\frac{PV}{RT} = \frac{P_0 V_0}{RT} + \Delta v \cdot \alpha \tag{1.12}$$

As a function of the conversion, replacing α by Equation 1.2 and rearranging, we obtain:

$$\Delta v \cdot \alpha = \Delta v \frac{n_{A0} - n_A}{a} \cdot \frac{n_0}{n_0} \cdot \frac{n_{A0}}{n_{A0}}$$

$$\Delta v \cdot \alpha = \frac{\Delta v}{a} \cdot X_A \cdot y_{A0} \cdot n_0$$

where X_A is the conversion (Equation 1.5) and y_{A0} is the initial molar fraction of A.
The term

$$\frac{\Delta v}{a} \cdot y_{A0}$$

is defined as the factor of contraction or expansion, known as ε_A. It indicates the volume change in the reaction system.
Substituting $\Delta v \alpha$ in Equation 1.12 and considering that

$$n_0 = \frac{P_0 V_0}{RT},$$

We obtain, for a system at constant pressure and temperature, the variation of the system volume as a function of the conversion, i.e.,

$$V = V_0(1 + \varepsilon_A X_A) \text{ at } P \text{ and } T = \text{constant} \tag{1.13}$$

Therefore, the factor ε_A will be defined as the ratio between the total change of the reaction volume and the initial volume. Thus,

$$\varepsilon_A = \frac{V_{X_A=1} - V_{X_A=0}}{V_{X_A=0}} \tag{1.14}$$

If in gas-phase reactions ε_A is positive, there will be expansion, otherwise there will be contraction. When ε_A is zero, no volume change will occur. In liquid-phase reactions, $\varepsilon_A = 0$.
The volume change in nonisothermal conditions must be corrected for temperature. For non-ideal conditions, the compressibility factor z is corrected. Thus, Equation 1.13 becomes:

$$V = V_0 \left(1 + \varepsilon_A X_A\right) \frac{T}{T_0} \frac{z}{z_0} \tag{1.15}$$

Note that the volume varies with the conversion of the limiting reactant A but may also vary with any other component. Only under equimolar conditions, we have that $\varepsilon_A = \varepsilon_B$. However, for any other condition, they are different. For the same total volume, it is known that:

$$V = V_0(1 + \varepsilon_A X_A) = V_0(1 + \varepsilon_B X_B)$$

Thus,

$$\varepsilon_A X_A = \varepsilon_B X_B \tag{1.16}$$

Example E1.1

$$N_2 + 3H_2 \leftrightarrow 2NH_3$$

Case 1	N_2	$3H_2$	$2NH_3$ (+ Inert)	Total	ε
Initial cond.	1	3	0	4	
Final cond.	0	0	2	2	$\varepsilon_A = -0.5$
Case 2					
Initial cond.	1	4	0	5	
Final cond.	0	1	2	3	$\varepsilon_A = -2/5$
Case 3					
Initial cond.	2	3	0	5	
Final cond.	1	0	2	3	$\varepsilon_B = -2/5$
Case 4					
Initial cond.	1	4	0 + 1 Inert	6	
Final cond.	0	1	2 + 1 Inert	4	$\varepsilon_A = -1/3$

Note that in case 3, the limiting component is B and in case 4, an inert is present.

1.3   CONTINUOUS SYSTEMS

Most of the reactions are carried out in continuous systems, for which other variable is used. In particular, we used the molar flow F (moles/time) in an open system:

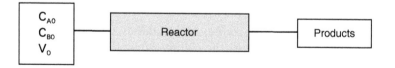

The local molar flow in relation to component A is defined as $F_A = C_A v$ (moles/h), where v is the volumetric flow rate (L/h). The molar inflow rate will be $F_{A0} = C_{A0} v_0$. For any other component, reactant, or product, we analogously obtain F_B, F_R. However, note that the local volumetric flow for a variable volume system is not constant and varies with the factor of contraction or expansion. In this case, for a system at constant pressure and temperature, we have an expression analogous to Equation 1.14 for the volumetric flow, i.e.,

$$v = v_0(1 + \varepsilon_A X_A) \tag{1.17}$$

Therefore, considering that the local concentration varies with ε_A, since, we have:

$$F_A = \frac{n_A}{V} \cdot v = \frac{n_{A0}(1 - X_A)}{v_0(1 + \varepsilon_A X_A)} \cdot v_0(1 + \varepsilon_A X_A) = F_{A0}(1 - X_A)$$

where $F_{A0} = C_{A0}v_0$.

Therefore, for a system at variable (ε_A) or constant volume $(\varepsilon_A = 0)$, we have the general definition:

$$F_A = F_{A0}(1 - X_A). \tag{1.18}$$

The conversion can therefore be defined as a function of the molar flow, i.e.,

$$X_A = \frac{F_{A0} - F_A}{F_{A0}} \tag{1.19}$$

This is similar to the definition given by Equation 1.5.

Therefore, for the extent of the reaction, we have an analogous relation to Equation 1.2:

$$\alpha = \frac{F_{A0} - F_A}{a} = \frac{F_{B0} - F_B}{b} = \frac{F_R - F_{R0}}{r} \tag{1.20}$$

The molar flows of the other components can be determined as a function of the conversion. We then obtain the following relations, which are valid for the reactions at variable or constant volume:

$$F_A = F_{A0}(1 - X_A)$$
$$F_B = F_{B0} - (b/a)F_{A0}X_A$$
$$F_R = F_{R0} + (r/a)F_{A0}X_A \tag{1.21}$$

1.4 PARTIAL PRESSURES

From the gas law, it is known that the partial pressure is a function of the total pressure, $p_A = y_A P$.

Therefore, if the molar fractions of an open or closed system are known, the partial pressures of each component can be determined. Considering again the chemical reaction:

$$aA + bB \rightarrow rR$$

The balance of the number of moles will be:

Initial moles:	$n_{A0}\ n_{B0}\ n_{R0}$	n_0 (initial total)
Moles at any t:	$n_A\ n_B\ n_R$	$n = n_0 + \Delta v\alpha$
Reacted moles:	$n_{A0} - a\alpha\ n_{B0} - (b/a)\alpha\ n_{R0} + (r/a)\alpha$	

We have the molar fractions:

$$y_i \Rightarrow \frac{n_{A0} - a\alpha}{n_0 + \Delta v \cdot \alpha}\ \frac{n_{B0} - (b/a)\alpha}{n_0 + \Delta v \cdot \alpha}\ \frac{n_{R0} + (r/a)\alpha}{n_0 + \Delta v \cdot \alpha} \tag{1.22}$$

We can also determine the molar fractions as a function of the conversion by using Equation 1.2, and the molar flow, knowing that:

$$\frac{\alpha}{n_{A0}} = \frac{n_{A0} - n_A}{a \cdot n_{A0}} = \frac{X_A}{a} \tag{1.23}$$

or

$$\frac{\alpha}{F_{A0}} = \frac{F_{A0} - F_A}{a \cdot F_{A0}} = \frac{X_A}{a} \tag{1.24}$$

Consequently, the partial pressures of each component can be calculated.

1.5 METHOD OF TOTAL PRESSURE

In a gas-phase system at constant volume, the reaction is monitored through the formation of products and change of total pressure. The pressure is a direct measure in a closed system. If the partial molar fractions are known, the partial pressure can also be determined as a function of the total pressure. For a gas-phase reaction of the type like:

$$aA + bB \rightarrow rR$$

The partial pressure of A will be:

$$p_A = y_A P$$

From the ideal gas law,

$$p_A = \frac{n_A}{V} RT = \frac{n_{A0} - a\alpha}{V} RT$$

Using Equation 1.1 and substituting $\alpha \Rightarrow n = n_0 + \Delta v\alpha$ and knowing that $p_{A0} = \frac{n_{A0}}{V} RT$, the partial pressure of A can be calculated:

$$p_A = p_{A0} - \frac{a}{\Delta v} \frac{(n - n_0)}{V} RT$$

Since the total initial pressure and pressure at time t are P_0 and P, respectively, we have:

$$p_A = p_{A0} - \frac{a}{\Delta\nu}(P - P_0) \tag{1.25}$$

Regarding the product, only the sign is changed from $(-)$ to $(+)$, i.e.,

$$p_R = p_{R0} + \frac{r}{\Delta\nu}(P - P_0) \tag{1.26}$$

1.6 GENERAL PROPERTIES

As seen, the reactants have specific properties, such as conductivity, wavelength, binding energy, resistivity, polarized light, and others. These properties are directly measured by chromatography, spectroscopy, and other methods and must be related with the usual measure, concentrations, or conversion.

Consider again a generic reaction of the type:

$$aA + bB \rightarrow rR$$

and, considering G_I as a characteristic property of each component. Thus,

$$G = G_A n_A + G_B n_B + G_R n_R \tag{1.27}$$

But considering the number of moles as a function of the extent of the reaction (Equation 1.3), we have:

$$G = G_A(n_{A0} - a\alpha) + G_B(n_{B0} - (b/a)\alpha) + G_R(n_{R0} + (r/a)\alpha)$$

We have:

$$G = G_0 + \Delta G\alpha \tag{1.28}$$

Considering that:

$$G_0 = G_{A0} n_{A0} + G_{B0} n_{B0} + G_{R0} n_{R0}$$

and

$$\Delta G = G_R(r/a) - [G_A a + G_B(b/a)]$$

In a reversible reaction in equilibrium, we have that $\alpha = \alpha_e$ and, thus, $G = G_e$. So, by replacing this term into Equation 1.28, we obtain ΔG:

$$\Delta G = \frac{G_e - G_0}{\alpha_e}$$

By replacing ΔG in Equation 1.28, we determine α as a function of the property G, as follows:

$$\frac{\alpha}{\alpha_e} = \frac{G - G_0}{G_e - G_0} \tag{1.29}$$

or as a function of the conversion:

$$\frac{X_A}{X_{Ae}} = \frac{G - G_0}{G_e - G_0} \tag{1.30}$$

If the reaction is irreversible and the property is related to the reactant, it is known that $X_{Ae} = 1$ and $G_e = G_\infty = 0$. Therefore, for irreversible reactions, we simplify to get:

$$X_A = \frac{G_0 - G}{G_0} \tag{1.31}$$

or $G_e = G_\infty$ and $G_0 = 0$ when it is related to the product:

$$X_A = \frac{G}{G_\infty} \tag{1.32}$$

1.7 SOLVED PROBLEMS

E1.1 A reaction of the type $A \to nR$ is carried out in gas phase. The reactant A is introduced with an inert I in a batch reactor at constant volume. The initial pressures of the reactant and inert are 7.5 and 1.5 mmHg, respectively.

The reaction is irreversible and the total pressure reaches 31.5 mm Hg after a long time.

(a) Determine the stoichiometry n and calculate the conversion after 20 min of reaction, knowing that the pressure was 19 mmHg.
(b) What would be the final volume considering that we have plug-flow reactor at constant pressure (equal to the initial pressure) and the reactor reached the same conversion of the previous system? It is known that the initial volume is 0.5 L.

Solution

(a) For a system at constant volume, the partial pressure of the reactant can be calculated as a function of the total pressure P by means of Equation 1.25:

$$p_A = p_{A0} - \frac{a}{\Delta \nu}(P - P_0) \tag{1.33}$$

The initial pressure is $P_0 = p_{A0} + p_I = 9\,\text{mmHg}$.

When the time is long enough, all the reactant is converted. The initial pressure of the reactant A is zero when the final pressure of the system is equal to 31.5 mmHg. Therefore,

$$0 = 7.5 - \frac{1}{\Delta \nu}(31.5 - 9)$$

thus,

$$\Delta \nu = 3 \text{ and } n = 4 \tag{1.34}$$

Therefore,

$$A \rightarrow 4R \tag{1.35}$$

Considering that both the volume and temperature are constant

$$p_A = C_A RT,$$

thus, from Equation 1.7, we obtain:

$$X_A = \frac{C_{A0} - C_A}{C_{A0}} = \frac{p_{A0} - p_A}{p_{A0}} \tag{1.36}$$

Substituting p_A from Equation 1.33,

$$X_A = \frac{1}{3}\frac{P - P_0}{p_{A0}} \tag{1.37}$$

Substituting the values of P_0 and P after 20 min, we obtain:

$$X_A = 0.44 \tag{1.38}$$

(b) Considering $P = 9 \text{ mmHg}$ and the conversion $X_A = 0.44$, the volume change can be calculated according to Equation 1.13,

$$V = V_0(1 + \varepsilon_A X_A) \tag{1.39}$$

therefore,

$$\Delta V = V - V_0 = V_0 \varepsilon_A X_A \tag{1.40}$$

The molar fractions can be calculated:

$$P_A = y_A P \Rightarrow y_A = 0.83$$
$$\Rightarrow y_I = 0.16$$

Thus, the value of ε_A can be calculated since,

$$A \rightarrow 4R \quad Inert$$

	A	R	I	Total
Initial	0.83	0	0.16	1.0
Final	0	3.3	0.16	3.5

$$\varepsilon_A = 2.5$$

Thus, according to Equation 1.38, the volume change is:

$$\Delta V = V_{0\varepsilon_A} X_A = 0.55$$

Therefore, the final volume will be:

$$V = 1.05\,\text{L}$$

E1.2 A gas-phase reaction of the type like $2A + 4B \rightarrow 2R$ is conducted in a continuous system at a constant pressure of 10 atm and temperature of 727°C. The reactants are introduced with equal initial concentrations and volumetric flow of 2.5 L/min. Calculate the molar flow of each component when the conversion reaches 25%.

P = 10 atm, T = 727°C

v_0 = 2.5 L/min   X = 0.25

Solution

Calculation of the initial concentration: for $R = 0.082$, $P = 10$ atm and $T = 1.000$ K

$$C_A = p_{A0}/RT = y_{A0}P/RT = 0.0609\,\text{moles/L}$$
$$C_B = p_{B0}/RT = y_{B0}P/RT = 0.0609\,\text{moles/L}$$

And considering that, $y_{A0} = y_{B0} = 0.5$,
  since the initial concentrations of the reactants are equal.
  The reactant B is the limiting reactant and, therefore, we need to calculate X_B.
  The initial molar flows are:

$$F_{A0} = C_{A0}v_0 = F_{B0} = C_{B0}v_0 = 0.0609 \times 2.5 = 0.152\,\text{moles/min}$$

The molar flows of the other components are calculated according to the following equations:

$$F_B = F_{B0}(1 - X_B) = 0.114\,\text{moles/min.}$$
$$F_A = F_{A0} - F_{B0}(b/a)X_B = F_{B0}(1 - X_B/2) = 0.133\,\text{moles/min.}$$
$$F_R = F_{R0} + F_{B0}(r/a)X_B = F_{B0}(0 + X_B/2) = 0.019\,\text{moles/min.}$$

E1.3 A reversible reaction $A + B \Leftrightarrow R + S$ was performed in a liquid-phase batch reactor. Several samples were taken as the reaction was run and titrated with a normal solution of 0.675 N in an ampoule containing 52.5 mlL of sample. The data in the table indicate the titrated volume in different reaction times.

t (min)	0	10	20	30	50	∞
V (mlL)	52.5	32.1	23.5	18.9	14.4	10.5

Determine the conversion as a function of the reaction time.

Solution

Calculation of the number of moles:

$$n = \frac{V \cdot N}{1000} = \frac{V}{1000} 0.675 \qquad (1.41)$$

Calculation of the conversions by means of Equation 1.7:

$$X_A = \frac{n_{A0} - n_A}{n_{A0}} \qquad (1.42)$$

Calculating according to the following table, we have:

t (min)	0	10	20	30	50	∞
V (mlL)	52.5	32.1	23.5	18.9	14.4	10.5
n_A (moles/L)	0.0354	0.02166	0.01586	0.0127	0.00972	0.00708
X_A	0	0.388	0.551	0.639	0.725	0.799

Note that the equilibrium conversion X_{Ae} is equal to 0.799.

1.8   ACCURACY AND PRECISION

An important point everyone involved with measurements must consider is that measurements are subject to errors and disturbances. This is particularly important in the field of reaction kinetics, given the empirical nature of activities performed in this field. For the sake of clarity, 'empirical' is used here to emphasize that the kinetic behavior of chemical systems cannot be unveiled without the support of experiments and measurements. As a matter of fact, kinetic behavior of real chemical systems is normally observed first through experimentation, followed by the proposal of reaction mechanisms and afterwards by the development of kinetic rate expressions, such as those the reader will see in the following chapters. Although it may be true that the current development of theoretical chemistry and simulation tools allow the analysis of certain relatively simple chemical systems *a priori*, in fact the field of chemical kinetics cannot dispense with the use of experiments for understanding real chemical reaction systems yet.

Many various factors can disturb the measurements used to characterize the behavior of an experimental system. For instance, all measuring devices possess finite capacity for describing the measured variable. For example, the ordinary ruler most students use can measure the length of objects in millimeters but is unable to determine any fraction of length smaller than one millimeter. Indeed, regardless of the measurement being analyzed, the real property of the measured entity can never be known, with its infinite number of decimal places. Besides, it is often necessary to infer the value of a certain variable based on another variable's available measurements. For example, temperature measurements made with a mercury thermometer rely on the variation of the mercury volume, as measured in a graduated cylinder. Since the volume of mercury changes with temperature, its measured volume can be related to system temperature. So a function that relates volume to temperature must be built, named as the calibration model. However, this leads to the problem of determining the best calibration model and guaranteeing that it be valid in all conditions of interest. These facts introduce additional uncertainties to the measurement process and measured values, since calibration models are not perfect and can perform differently depending on experimental conditions. Additionally, measuring systems combine equipment and experimental procedure and are, therefore, subject to failure. For example, a plastic ruler can deform when it is poorly packed in students' backpacks, introducing additional inaccuracies in the measurement process. Similarly, the existence of a bad electric contact in an electrical circuit can generate noise and deviations in the measurements provided by a piece of equipment. The significant problem is that such deviations and deformations are not always noticed by the experimenter. Finally, the analyst usually does not know all the variables that describe a given problem, so fluctuations can be expected due to the unknown variations of all the uncontrolled variables that affect a certain measurement. For example, due to thermal expansion, the millimeter of the above mentioned ordinary ruler changes with temperature.

As a consequence of the previously discussed facts (and many more!), some degree of uncertainty always exists in all real measurements, so it is not possible to guarantee with absolute certainty what the value of a real variable is. In addition, measurements performed with different devices and by different analysts can lead to different values for the same measured variable. Obviously, some systems are more liable to process disturbances than others. For instance, it is easier and simpler to measure the length of a sheet of paper than to measure the velocity of a seagull that flies in the sky. Hence, it is necessary to introduce some concepts related to the *quality* of the available measurements.

First of all, let us introduce the concept of **accuracy**, which is usually related to the ideas of correctness and exactness. The accuracy of a measurement is related to its distance from the true (usually unknown) value. The more accurate a measurement, the closer it will be to the unknown value. It is difficult to characterize the accuracy of a measurement because to do so requires knowledge of its true value, which is unknown in a real experiment – otherwise, measuring would not be necessary! For this reason, accuracy is usually associated with the instrument or procedure used to make the measurements. In this case, a standard (or entity with known properties) can be used to evaluate the accuracy of the measuring system. Standards can be obtained from respectable organizations and providers (for example, mass standards) or can be prepared by the analyst (for example, by mixing known quantities of pure components). In order to guarantee the accuracy of obtained measurements, companies and laboratories are obliged to comply with standardized procedures that establish the periodic evaluation of standard properties (often a blank or reference experiment) with available measuring systems; whenever unacceptable deviations from reference values are detected, the available measuring system must be replaced or sent to maintenance.

Now, let us introduce the concept of **precision**, which is usually related to the ideas of reproducibility and variability. The precision of a measurement is related to its distance from other measurements performed for the same entity and using the same experimental procedures. The more precise a measurement, the closer it will be to previously measured values. Therefore, while accuracy is a concept that can be used to characterize a single value, precision is a concept that necessarily involves a set of multiple measurements. As with accuracy, to guarantee good precision of obtained measurements, companies and laboratories are obliged to comply with standardized procedures that establish the periodic evaluation of standard properties (often a blank or reference experiment) with available measuring systems; whenever unacceptable deviations among the obtained values are detected, the available measuring system must be replaced or sent to maintenance.

Figure 1.2 explains graphically the meanings of accuracy and precision, where one can see that the two concepts are independent from each other, although in the ideal experimental scenario measurements must be simultaneously accurate and precise. Despite that, one should see that precision imposes a limit on accuracy, as the accuracy of available measurements cannot be better than the precision of these very same measurements (this is apparent when one of the experimental values is replaced by the standard value). For this reason, the two are often confused, which can be dangerous, as precise measurements can be quite inaccurate.

As discussed in the following sections, both the accuracy and the precision of experimental measurements used to carry out quantitative evaluations of system performances affect the quality of the proposed analyses. For this reason, accuracy and precision of experimental data must be carefully characterized by the analyst, as described throughout this book.

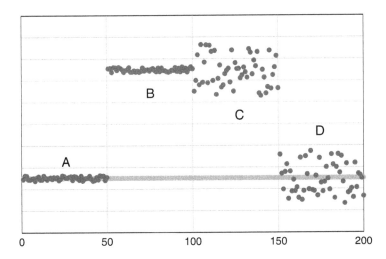

Figure 1.2 Graphical illustration of accuracy and precision. Line is the unknown true value, and dots are the experimental measurements. (A) Accurate and precise (best case); (B) inaccurate and precise; (C) inaccurate and imprecise (worst case); (D) accurate and imprecise..

1.9 MEASUREMENT ERRORS AND PRECISION

As variable measurements are contaminated by disturbances, as described in Section 1.8, it becomes necessary to evaluate simultaneously the variables that affect a problem and the precision and accuracy of the measured values. In order to convey the importance of this point, Figure 1.3 illustrates graphically the effect of measurement variability on the information content of experimental data. As one can observe in Figure 1.3, as the variability increases it becomes more difficult to observe the underlying linear relationship between variables X and Y; therefore, the occurrence of measurement disturbances leads to deterioration of the information content of available data, which is why measurement variability must be properly characterized by the investigator.

In order to characterize the **precision** of actual experimental measurements, the best strategy consists in replicating the experiment as many times as possible (Schwaab and Pinto, 2007). Assuming that a sufficiently large number of experimental measurements are available through replication $(x_1, x_2, ..., x_N)$ the most popular quantitative characterization of precision is the sample variance s_X^2, defined as:

$$s_X^2 = \frac{\sum_{i=1}^{N} \left(x_i - \overline{X}\right)^2}{N-1} \tag{1.9.1}$$

where $\overline{X}$ is the sample average, defined as:

$$\overline{X} = \frac{\sum_{i=1}^{N} x_i}{N} \tag{1.9.2}$$

and N is the number of replicates.

Some questions related to the calculation of s_X^2 must be considered at this time. First of all, the term 'replication' here refers to repetition of the whole experimental procedure being analyzed. For instance, if an experiment involves many steps (such as the manufacture of a catalyst; the preparation of a reactor bed; the conduction of the experiment through manipulation of pressures, temperatures and flowrates; the withdrawal of samples; and the characterization of samples with the appropriate experimental technique), then replication requires the repetition of the entire experimental procedure. The reader must observe that the withdrawal of N samples and analysis of the N resulting measurements cannot evaluate the precision of the whole experimental procedure, but only of the combined sampling and characterization techniques. If process variability is concentrated in one step, such as the catalyst manufacture process step (which it frequently is!), replication imposes the preparation of distinct catalyst loads and the use of different catalyst samples for conducting of the experiments; otherwise, the analyst will underestimate the level of uncertainty of the available experimental values and overestimate the information content of the data

The second point regards the random nature of both s_X^2 and $\overline{X}$ As experimental measurements are uncertain to some extent due to the measurement disturbances, both s_X^2 and $\overline{X}$ are also contaminated by these disturbances, as they are calculated

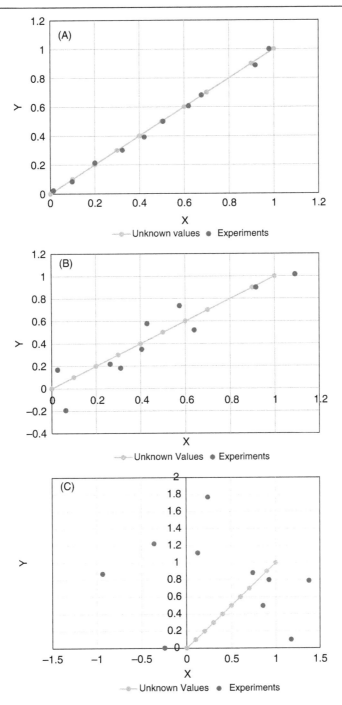

Figure 1.3 Effect of increasing variability on information content of experimental data. Measurements were generated through addition of random signals to both X and Y values with maximum amplitudes of: (A) $\pm$ 0.025; (B) $\pm$ 0.25; (C) $\pm$ 1.00..

with uncertain measurements. In other words, if N new experiments are performed, both s_X^2 and $\overline{X}$ change. Example E1.4 illustrates this particular point.

Example

E1.4 The function **random** of the Excel spreadsheet provides random numbers uniformly distributed in the interval [0,1]. This function is used here to simulate the disturbances that affect a measurement. In order to do that, numbers were generated in the form:

$$x_i = 0.5 + 0.2 * (random - 0.5) \tag{1.9.3}$$

so that measured values fall in the interval [0.4, 0.6]. Ten sequences of $N = 10$ values were then generated and used for calculation of s_X^2 and $\overline{X}$, as shown in Table E1.4. One can see in Table E1.4 that both s_X^2 and $\overline{X}$ were different in the distinct sequences, illustrating the random nature of these variables.

Table E1.4 Simulation of measurements with the Excel spreadsheet.

N	Set I	Set 3	Set 3	Set 4	Set 5	Set 6	Set 7	Set 8	Set 9	Set 10
1	0.487075	0.560035	0.560069	0.501688	0.542396	0.560716	0.414854	0.450284	0.476407	0.422055
2	0.584286	0.413638	0.557084	0.42373	0.580721	0.430117	0.402575	0.527913	0.500483	0.40053
3	0.505908	0.46383	0.470057	0.554682	0.592128	0.487544	0.525618	0.555914	0.582956	0.42903
4	0.54224	0.578394	0.519186	0.44165	0.511674	0.419018	0.532482	0.493348	0.563406	0.524839
5	0.595285	0.560245	0.524326	0.411778	0.420495	0.595182	0.468839	0.466339	0.437386	0.479597
6	0.419107	0.458907	0.573627	0.563084	0.46288	0.524981	0.528729	0.561878	0.434244	0.425517
7	0.513149	0.555218	0.465596	0.400113	0.466827	0.569699	0.549183	0.425047	0.596886	0.412573
8	0.441502	0.424353	0.545163	0.589763	0.542669	0.422123	0.595255	0.453755	0.592232	0.462876
9	0.50472	0.514275	0.518946	0.553456	0.592263	0.512453	0.564784	0.496776	0.535349	0.426062
10	0.497913	0.574934	0.542915	0.488239	0.418174	0.454582	0.559163	0.513593	0.571652	0.494425
$\overline{X}$	0.509118	0.510383	0.527697	0.492818	0.513023	0.497642	0.514148	0.494485	0.5291	0.44775
s_x^2	0.003056	0.00415	0.001322	0.004944	0.004563	0.004228	0.00416	0.002115	0.003946	0.001645

The third point regards the levels of uncertainty of the calculated s_X^2 and $\overline{X}$ values, as these variables are subject to unknown disturbances. In order to solve this problem rigorously, it is necessary to know exactly how disturbances affect the measured values, which is usually impossible, as this would require the execution of infinitely many (or a sufficiently high number of) experiments or the derivation of a theoretical expression that links measurements and possible disturbances, which is rare. For this reason, the commonest approach makes use of the gaussian assumption, which allows the development of analytical solutions for the proposed problem (Schwaab and Pinto, 2007). According to the gaussian approach, it is assumed that the probability of observing a value of x in the interval $[x_1 x_2]$ $P(x_1 \leq x \leq x_2)$, follows the function:

$$P(x_1 \leq x \leq x_2) = \int_{x_1}^{x_2} \wp(x)\, dx \tag{1.9.4}$$

$$\wp(x) = \frac{1}{\sigma_x\sqrt{2\pi}}\exp\left(-\frac{1}{2}\left(\frac{x-\mu_x}{\sigma_x}\right)^2\right), \quad -\infty < x < \infty \tag{1.9.5}$$

where μ_x is the unknown mean value (or unknown true value) of the variable x, around which the measured values fluctuate, and σ_x^2 is the unknown variance of the fluctuations. If $s_{\overline{X}}^2$ and $\overline{X}$ were determined with perfect accuracy, they would be equal to $\sigma_{\overline{X}}^2$ and μ. The difference ε, given by:

$$\varepsilon = x - \mu_x \tag{1.9.6}$$

is a measure of the experimental disturbance. In this case, the uncertainties can be described in the form (Schwaab and Pinto, 2007):

$$\Delta\overline{X} = \pm\frac{ts_x}{\sqrt{N}} \tag{1.9.7}$$

$$s_{x,\min}^2 = \frac{(N-1)s_x^2}{\chi_{\max}^2}; \quad s_{x,\max}^2 = \frac{(N-1)s_x^2}{\chi_{\min}^2} \tag{1.9.8}$$

$$\Delta x = \pm cs_x \tag{1.9.9}$$

where $\Delta\overline{X}$ is the uncertainty of the calculated sample average; Δx is the uncertainty of individual measurements; $s_{x,\min}^2$ and $s_{x,\max}^2$ are the limits of the confidence interval of s_x^2; and t, $\chi_{\max}^2$, $\chi_{\min}^2$ and c are reference numbers that depend on the number of samples N and on the confidence level. In short, the confidence level describes the fraction of times that the disturbances will fall inside the confidence interval. When one says that the confidence level is equal to 95%, one simultaneously says that accepts to be mistaken one time out of 20. As the gaussian reference is defined in the infinite real interval, the only possibility to be correct 100% of the times is to admit that disturbances can be infinitely large, which is useless for practical use.

t values in Equation (1.9.7) can be obtained with help of the Student's t-distribution. Table 1.9.1 presents t values for different sample sizes and degrees of confidence. Values of $\chi_{\min}^2$ and $\chi_{\min}^2$ in Equation (1.9.8) can be obtained with help of the χ^2 distribution. Tables 1.9.2 and 1.9.3 present values of $\chi_{\min}^2$ and $\chi_{\min}^2$ for different sample sizes and degrees of confidence. c values in Equation (1.9.9) can also be obtained with help of the Student's t distribution of Table 1.9.1, by assuming that c values are approximately equal to t values in most practical situations. Example 2 presents an illustrative explanation about the use of Equations (1.9.7–9) and Tables 1.9.1–3 for characterization of confidence intervals.

Table 1.9.1 t values for distinct sample sizes and confidence levels.

N	2	3	4	5	6	7	8	9	10	11
90%	6.313752	2.919986	2.353363	2.131847	2.015048	1.94318	1.894579	1.859548	1.833113	1.812461
95%	12.7062	4.302653	3.182446	2.776445	2.570582	2.446912	2.364624	2.306004	2.262157	2.228139
98%	31.82052	6.964557	4.540703	3.746947	3.36493	3.142668	2.997952	2.896459	2.821438	2.763769
99%	63.65674	9.924843	5.840909	4.604095	4.032143	3.707428	3.499483	3.355387	3.249836	3.169273

Table 1.9.2 χ^2_{min} values for distinct sample sizes and confidence levels.

N	2	3	4	5	6	7	8	9	10	11
90%	0.00393214	0.102587	0.351846	0.710723	1.145476	1.635383	2.16735	2.732637	3.325113	3.940299
95%	0.000982069	0.050636	0.215795	0.484419	0.831212	1.237344	1.689869	2.179731	2.700389	3.246973
98%	0.000157088	0.020101	0.114832	0.297109	0.554298	0.87209	1.239042	1.646497	2.087901	2.558212
99%	3.92704E-05	0.010025	0.071722	0.206989	0.411742	0.675727	0.989256	1.344413	1.734933	2.155856

Table 1.9.3 χ^2_{max} values for distinct sample sizes and confidence levels.

N	2	3	4	5	6	7	8	9	10	11
90%	3.841458821	5.991465	7.814728	9.487729	11.0705	12.59159	14.06714	15.50731	16.91898	18.30704
95%	5.023886187	7.377759	9.348404	11.14329	12.8325	14.44938	16.01276	17.53455	19.02277	20.48318
98%	6.634896601	9.21034	11.34487	13.2767	15.08627	16.81189	18.47531	20.09024	21.66599	23.20925
99%	7.879438577	10.59663	12.83816	14.86026	16.7496	18.54758	20.27774	21.95495	23.58935	25.18818

Example

E1.9.2 It is assumed that triplicate experiments were carried out to characterize the activity of a certain catalyst, leading to the following conversion values of the main reactant: $x_1 = 0.90$, $x_2 = 0.95$ and $x_3 = 0.98$. In this case:

$$\overline{X} = \frac{0.90 + 0.95 + 0.98}{3} = 0.94333 \tag{1.9.10}$$

$$s_x^2 = \frac{(0.90 - 0.94333)^2 + (0.95 - 0.94333)^2 + (0.98 - 0.94333)^2}{3 - 1} \tag{1.9.11}$$

$$= 0.001633$$

$$s_x = \sqrt{s_x^2} = 0.040415 \tag{1.9.12}$$

Assuming a confidence level of 95% (and accepting one wrong conclusion out of 20), then:

$$\Delta\overline{X} = \pm\frac{4.302653 * 0.040415}{\sqrt{3}} = 0.100395 \tag{1.9.13}$$

$$s_{x,min}^2 = \frac{(3-1) * 001633}{7.377759} = 0.000433; \tag{1.9.14}$$

$$s_{x,max}^2 = \frac{(3-1) * 0.001633}{0.050636} = 0.064513$$

$$\Delta x = \pm(4.302653 * 0.040415) = 0.17389 \tag{1.9.15}$$

Therefore, the calculated values are:

$$0.73 < x_1 = 0.90 < 1.07$$
$$0.78 < x_2 = 0.95 < 1.12$$
$$0.81 < x_3 = 0.98 < 1.15$$
$$0.84 < \overline{X} = 0.94 < 1.04$$
$$4.33 \times 10^{-3} < s_x^2 = 1.63x10^{-3} < 64.51 \times 10^{-3}$$

As one can see, the confidence intervals can be too large (and precision too low) when only few samples are available. Besides, one can also see that the gaussian approach frequently leads to calculation of unfeasible confidence intervals (for instance, conversions cannot be larger than 1).

The gaussian distribution frequently leads to calculation of highly imprecise confidence intervals, particularly when the number of samples is small, as shown in Example 1.9.2. This happens for many reasons, including: (i) the actual disturbances do not necessarily follow the gaussian distribution of Equations (1.9.4–5); (ii) the gaussian distribution is symmetrical and actual phenomena can present skewed behavior, with unequal numbers of positive and negative deviations; and (iii) the gaussian distribution is defined in the infinite real interval, while real measurements are normally constrained within well-defined limits (for instance, conversions must fall within the [0,1] interval). Consequently, one might wonder why one should use the gaussian approach for calculation of confidence intervals and experimental precision; possible answers are: (i) the behavior of actual disturbances is usually unknown; (ii) the gaussian distribution allows the analytical computation of confidence intervals; (iii) the gaussian distribution is simple and depends on only two parameters, σ_x^2 and μ_x; (iv) the very important central limit theorem states the sum of infinitely many disturbances leads to a signal that follows the gaussian distribution (in other words, if a measurement is subject to an infinite number of individual disturbances, the effective resulting disturbance will follow the gaussian distribution).

Based on the previous paragraphs and on Example 1.9.2, it is possible to conclude that the gaussian distribution should not be regarded as the formal theoretical description of the experimental variability problem, but as an approximation of experimental fluctuations that can be useful for characterization of the experimental precision, particularly when the number of samples is large. For this reason, there are reasons for more involving and precise characterization of experimental errors in reaction kinetic problems, as described by Pacheco *et al.* (2018). The interested reader is encouraged to download and read the proposed manuscript.

Finally, it is important to emphasize that it is very difficult to evaluate the **accuracy** of a measurement, as the true value is normally unknown (with exception of standards used for evaluation of instruments and measuring procedures). For this reason, Equations (1.9.7) and (1.9.9) can be regarded as inferior limits for the accuracy of sample averages and individual measurements, respectively. Therefore, if the difference observed between the reference standardized value and the sample average falls within the confidence interval defined by Equation (1.9.7), the sample average can be regarded as accurate. Similarly, if the difference observed between the reference standardized value and the individual measurements fall within the confidence interval defined by Equation (1.9.9), the individual measurements can be regarded as accurate. On the other hand, if Equations (1.9.7) and (1.9.9) are not satisfied by the differences observed between the reference value and the analyzed experimental measurement, the measurement is assumed to be biased and corrective actions should be considered by the analyst to enhance the performance of the measuring system.

Chapter 2

Chemical equilibrium

For choosing the conditions of a chemical process, we should know the thermodynamic properties and particularly the conditions of the chemical equilibrium. Before determining the reaction kinetics, it is necessary to verify if the reaction is thermodynamically possible. The pressure and temperature conditions are important to calculate the conversion of a reversible or irreversible reaction. For reversible reactions, we need to determine the chemical equilibrium constant, which is temperature dependent. With this constant, it is possible to predict what is the maximum equilibrium conversion of a reversible reaction. Therefore, the reversibility of the reaction imposes serious limitations.

By means of thermodynamics, it is possible to predict whether a chemical reaction occurs and determine its composition. The thermodynamic equilibrium conversion represents the maximum conversion that can be achieved, regardless of the catalyst and reaction rates. However, the rates and conversions depend only on temperature, pressure, and inflow composition.

When a reaction occurs at constant temperature and pressure, it proceeds spontaneously varying in the direction of the increase in entropy. Once the equilibrium is reached, this entropy does not increase further. Consequently, from the first law of thermodynamics, the total change of the free energy of the system is always negative for any spontaneous reaction and zero at the equilibrium.

In a reaction of the following type:

$$aA + bB \Leftrightarrow rR + sS$$

The variation of free energy with temperature and pressure for an open system will be:

$$dG = -S\,dT + V\,dP + \sum \mu_j\,dn_j \tag{2.1}$$

where μ_j is the chemical potential j and n_j is the number of moles of the component. At constant temperature and pressure, we have:

$$dG = 0$$

Therefore:

$$\sum \mu_j\,dn_j = 0 \tag{2.2}$$

The chemical potential for ideal gases expressed as a function of the partial pressure is defined by:

$$\mu_j = \mu_j^0 + RT \ln p_j$$

We then have:

$$RT \sum a_j \ln p_j = -\sum a_j \mu_j^0 \qquad (2.3)$$

By definition, the total Gibbs free energy is:

$$G^0 = \sum n_j \mu_j^0 = \mu_j^0 (n_0 + a_j \alpha)$$

with the extent of the reaction (α) defined by Equation 1.2:

$$\alpha = \frac{n_{j0} - n_j}{a_j}$$

Differentiating with respect to α, we obtain:

$$\frac{\partial G^0}{\partial \alpha} = \sum \mu_j a_j = \Delta G^0 \qquad (2.4)$$

Therefore, by substituting Equation 2.4 into Equation 2.3, we have:

$$-\frac{\Delta G^0}{RT} = \sum_{\text{products}} a_j \ln p_j - \sum_{\text{reactants}} a_j \ln p_j \qquad (2.5)$$

For a reaction of the type, we have:

$$\frac{p_R^r p_S^s}{p_A^a p_B^b} = \exp\left(-\frac{\Delta G^0}{RT}\right) = K \qquad (2.6)$$

Thus:

$$\Delta G^0 = -RT \ln K \qquad (2.7)$$

Considering that:

$$\frac{p_R^r p_S^s}{p_A^a p_B^b} = K \qquad (2.8)$$

If the entropy of the system is known, we can determine the equilibrium constant by calculating the Gibbs free energy DG^0 from the enthalpy of reaction, according to the following relation:

$$\Delta G^0 = \Delta H^0 - T \Delta S^0 \qquad (2.9)$$

By substituting Equation 2.9 into Equation 2.7, we obtain:

$$K = \exp\left(\frac{\Delta S^0}{R}\right) \exp\left(-\frac{\Delta H^0}{RT}\right) \tag{2.10}$$

This is the Van't Hoff equation, in which the term $\exp(\Delta S^0/R)$ is independent of temperature, but depends on the entropy of the system. Generically, the equilibrium constant depends on the temperature, according to equation:

$$K = K_0 \exp\left(-\frac{\Delta H^0}{RT}\right) \tag{2.11}$$

The enthalpy change ΔH^0 is known in standard conditions. The variation of the equilibrium constant with temperature can also be determined by differentiating Equation 2.11:

$$\frac{d\ln K}{dT} = \frac{\Delta H^0}{RT} \tag{2.12}$$

For reversible reactions, an increase or decrease in temperature tends to directly influence the equilibrium constant, but depends on whether the reaction is exothermic or endothermic. When the reaction is exothermic ($\Delta H^0 < 0$), an increase in temperature favours the reverse reaction, since $K << 1$. For endothermic reactions, the opposite is valid.

The enthalpy of the reaction varies with temperature and depends on the specific heat of each component. For ideal gases, the specific heat varies according to a polynomial function, namely:

$$c_{pj} = \alpha_j + \beta_j T + \gamma_j T^2 \tag{2.13}$$

where α_j, β_j, and γ_j, are constants of each gas component under ideal conditions.

The enthalpy of each component varying with the temperature will be:

$$H_j = \int_{T_0}^{T} c_p \cdot dT$$

$$H_j = \int_{T_0}^{T} (\alpha_j + \beta_j T + \gamma_j T^2) \cdot dT \tag{2.14}$$

and the change in the reaction enthalpy will be given by:

$$\Delta H_r^0 = \text{Total enthalpy of the reaction: } \left(\sum H_{\text{products}} - \sum H_{\text{reactants}}\right)$$

i.e.:

$$\Delta H_T = \Delta H_T^0 + \int_{T_0}^{T} \left(\Delta\alpha_j + \Delta\beta_j T + \Delta\gamma_j T^2 \right) \cdot dT \tag{2.15}$$

where $\Delta\alpha_j$, $\Delta\beta_j$, and $\Delta\gamma_j$ are constants (products minus reactants).

Finally, the variation of the equilibrium constant with temperature is determined by integrating Equation 2.12, i.e.:

$$\ln \frac{K_T}{K_{T_0}} = \int_{T_0}^{T} \frac{\Delta H_r^0}{RT} dT \tag{2.16}$$

In order to obtain the values of the conversion and the composition of species j in the thermodynamic equilibrium for the reaction, we should perform a simulation which takes into account the possible existence of products of the reaction A_j $_{products}$.

In the thermodynamic study of the reaction, one can also use the HYSYS ® software version 3.1, particularly the Gibbs reactor module, with the thermodynamic package Peng–Robinson and the method of minimization of the Gibbs free energy (G) of the system, given by the following equation:

$$G = \sum_{1}^{n} n_j G_j(T, P, C_j) \tag{2.17}$$

where n_j is the number of species and G_j is the partial molar Gibbs energy of the species j, which is dependent on temperature, pressure, and composition. The mass balance for the species j is represented by Equation 2.17, in which k symbolizes the elements (H, O, C), n_j is the number of molecules of the species j present in the system, a_{jk} is the number of atoms of the element k present in the molecule of the species j, and, finally, A_k is the total numbers of atoms of the element k:

$$\sum_{j} n_j a_{jk} - A_k = 0$$

Example

E2.1 Determine the equilibrium concentrations of the conversion of the hydrobutyric acid in dilute solution, i.e.:

$$CH_3 - CH_2 - CH_2 - COOH \Leftrightarrow CH_2 - CH_2 - \underset{\underset{\displaystyle O}{|\quad\quad|}}{\overset{\displaystyle \overset{CH_2}{|}}{C}} = O + H_2$$

The initial concentration of the acid at 25°C is 0.182 mol/L and the equilibrium constant is 2.68.

Solution
Since the solution is diluted with excess of water, we can assume a direct first order reversible reaction, i.e.:
By making $CH_3-CH_2-CH_2-COOH = A$

$$CH_2 - \underset{\underset{O}{|}}{\overset{\overset{CH_2}{|}}{CH_2}} - C = O \; = R$$

$$A \Leftrightarrow R$$

From the definition of the equilibrium constant, as a function of the concentration, we have:

$$K = \frac{C_{Re}}{C_{Ae}} = \frac{X_{Ae}}{1 - X_{Ae}}$$

Therefore, the equilibrium conversion can be calculated:

$$X_{Ae} = \frac{K}{1 + K} = 0.73$$

Thus, the equilibrium concentrations are:

$$C_{Re} = 0.133 \, mol/L$$
$$C_{Ae} = 0.049 \, mol/L$$

Example

E2.2 In the previous example, determine the equilibrium conversion as a function of the temperature, knowing the equilibrium constant as a function of the temperature, i.e.:

$$\ln K = \frac{9060}{T} - 27.4$$

Solution
Starting from the equation of the previous problem:

$$K = \frac{X_{Ae}}{1 - X_{Ae}} = K_0 \, e^{-\theta}$$

in which:

$$\theta = \frac{\Delta H}{RT} = -\frac{9060}{T}$$

and

$$K_0 = e^{-27.4} = 1.186 \times 10^{-12}$$

Thus, the solution will be:

$$X_{Ae} = \frac{1186 \times 10^{-12} \, e^{9060/T}}{1 + 1186 \times 10^{-12} \, e^{9060/T}}$$

The results are given in the following table below and in Figure E2.1.

T (°C)	T (K)	X_{Ae}
10	283	0.989582
20	293	0.969559
30	303	0.919868
40	313	0.815365
50	323	0.643159
60	333	0.437029
70	343	0.259909
80	353	0.142496
90	363	0.075735
100	373	0.040276

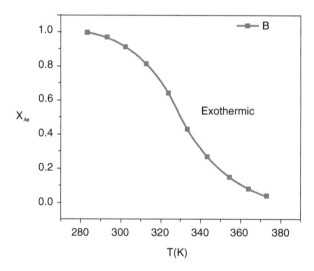

Figure E2.1   Equilibrium conversion as a function of the temperature.

Example

E2.3 To exemplify the use of the software HYSYS®, we will do the analysis for the partial oxidation of methane, which generates the following main products: CH_4, O_2, CO_2, H_2O, CO, and H_2. This reaction is conducted at atmospheric pressure, since for higher values of temperature, the conversion of CH_4 and selectivity to CO and H_2 decrease, as suggested by York et al. (2003). Moreover, higher hydrocarbons such as ethane, ethylene, and others were not taken into account because their concentrations are very low in the wide temperature range studied (200–1000°C).

Different gas compositions in the inflow were evaluated and the results for the conversion of CH_4 as a function of the temperature are shown in Figure E2.2. Between 500°C and 700°C, there is a considerable increase in the conversion of CH_4. At temperatures higher than 800°C, more than 90% of conversion is reached. If an inert gas (e.g., He) is assumed as the diluent of the system in a $CH_4/O_2/He$ ratio of 2/1/8, the conversion of CH_4 increases in the temperature range of 400 to −800°C in comparison to the CH_4/O_2 system without inert gas.

For ideal gases, the presence of any inert does not affect the thermodynamic equilibrium. However, for situations involving real gases, investigations concerning the effects of variations in pressure and temperature revealed that the disturbance results are not predicted by the expression of ideal gases and by the Le Chatelier's principle. When taking into account the deviations from ideality in real systems, in which inert gases are present, the equilibrium is affected, although the changes become significant only at moderate pressures. Whether the equilibrium is shifted in the direction of the products or reactants depends on the system constituents and the inert gas present (e.g., He, Ne, Ar, Kr, and Xe).

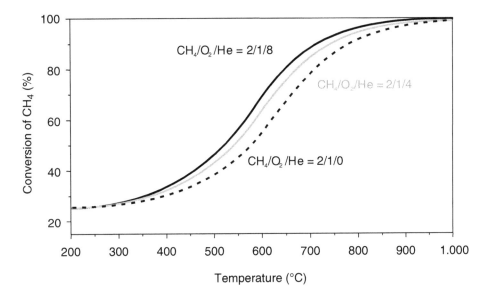

Figure E2.2 Conversion of CH_4 as a function of the temperature in the thermodynamic equilibrium for different feeding compositions.

The mole fraction of the species involved in the partial oxidation of CH_4 for the inflow composition $CH_4/O_2/He = 2/1/8$ was calculated and shown in Figure E2.2. At temperatures below 400°C, CH_4 is predominantly converted to CO_2 and H_2O, which indicates that the reaction most likely to occur is the highly exothermic total oxida-tion of methane. However, other substoichiometric oxidation reactions can occur. It is also important to note that the mole fraction of O_2 is zero throughout the temperature range studied, since the conversion of this reactant in the equilibrium is 100%.

Example

E2.4 Determine the equilibrium conversion of the following reaction:

$$C_2H_4 + H_2O\,(g) \rightarrow C_2H_5OH$$

As a function of the temperature, at 30 atm, and steam/ethylene molar ratio equals to 10.

As a function of the temperature, varying the total pressure varies.

Given that

$$\Delta H^0_{298} = -10,940\,\text{cal/mol}$$

$$K = 6.8 \times 10^{-2}\,\text{atm}^{-1}\,\text{at }145°C$$

$$c_{p_{C_2H_4}} = 2.83 + 28.6 \times 10^{-3}T - 8.73 \times 10^{-6}T^2$$

$$c_{p_{H_2O}} = 7.26 + 2.3 \times 10^{-3}T + 0.28 \times 10^{-6}T^2$$

$$c_{p_{C_2H_4O}} = 6.99 + 39.7 \times 10^{-3}T - 11.93 \times 10^{-6}T^2$$

Calculation of ΔH_{418K}:

$$\Delta H_{418} = \Delta H^0_{298} + \int_{298}^{418} (-3.1 + 8.84 \times 10^{-3}T - 3.48 \times 10^{-6}T^2) \cdot dT$$

Thus:

$$\Delta H_{418K} = -10.978\,\text{cal/mol}$$

Calculation of the equilibrium constant from Equation 2.16:

$$\ln \frac{K_T}{K_{T_0}} = \int_{418}^{673} \frac{\Delta H^0_r}{RT} dT$$

The equilibrium constant and the equilibrium conversion are shown in Figures E2.3 and E2.4, respectively.

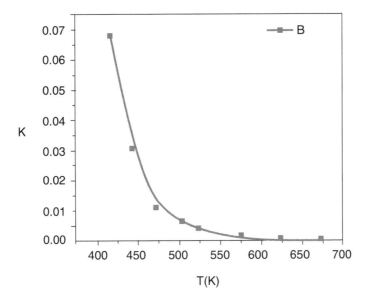

Figure E2.3 Equilibrium constant as a function of the temperature.

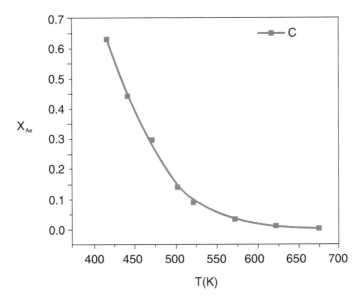

Figure E2.4 Equilibrium conversion as a function of the temperature.

Chapter 3

Kinetics of reactions

3.1 REACTION RATES—DEFINITIONS

When a mass or energy balance is carried out for a chemical reactor, one must consider the chemical reaction in which the reactants are transformed into products. These transformations are represented by the reaction rates.

In an open system, a molar balance is carried out for all the components of a chemical reaction of the type $A + B \Leftrightarrow R + S$. Generically, we can represent the system as follows:

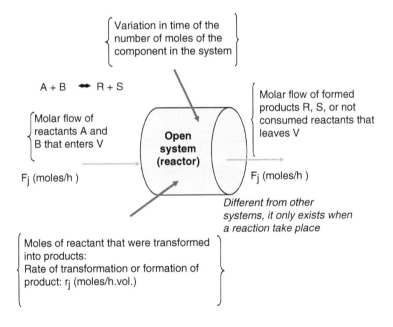

The reaction rate is defined as the rate of formation of product or disappearance of reactant per unit of volume and represents the variation of the number of moles formed or disappeared per unit of time and volume, respectively. The rate of formation of any component j is represented by r_j and the rate of disappearance of the component A is represented by $(-r_A)$.

Since the rates of formation or disappearance are intensive properties, it is convenient to relate them to a reference system. In a homogeneous phase system, the unit of volume (mol/L h) is used as reference. On the other hand, in heterogeneous systems, we used as reference the superficial or interfacial area (mol/m² h).

Homogeneous systems:

r_j = rate of formation of component j/unit of volume (mol/L h)

Heterogeneous systems:

r_j = rate of formation of component j/unit area (moles/m²h)

1   Reactions in homogeneous phase

The homogeneous reactions are very common in liquid and gas phases. For example, Gas phase—Cracking of ethane to obtain ethylene and hydrogen:

$$C_2H_6 \rightarrow C_2H_4 + H_2$$

2   Heterophase reactions

Several reactions are carried out in heterogeneous phase:

Gas/solid (fixed or fluidized bed).
Gas/liquid (fluid bed).
Liquid/liquid (fluid bed).
Gas/liquid/solid (three-phase slurry bed).

The oxidation of ethylene is conducted in gas phase with the presence of a catalyst. Therefore, it is a heterophase reaction.

Examples

1.   Ethylene oxidation

$$C_2H_4 + 1/2O_2 \rightarrow C_2H_4O$$
$$C_2H_4 + 3O_2 \rightarrow CO_2 + H_2O$$

2.   Coal gasification

The solid–gas reactions occur when a solid reacts in the presence of a gas, such as coal gasification:

$$C + H_2O \rightarrow CO + H_2$$

3.   Hydrogenation of oils

The reactions in gas–solid–liquid phase can be represented by the hydrogenation of oil, in which small particles of Raney–Nickel are used as a catalyst and hydrogen is bubbled:

$$\text{Soy oil } (HC_{unsaturated}) + H_2 \xrightarrow{\text{Ni Raney}} C_{saturated}$$

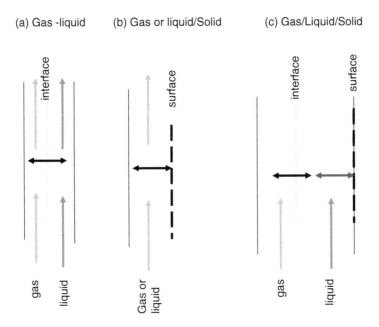

Figure 3.1 Homogeneous and heterogeneous models.

Reactions in homogeneous and heterogeneous phase are schematically shown in Figure 3.1.

Several times, it is difficult to know the interface area due to the presence of pores and irregularity of the external contour or modification of the interface, as in the case of a bubble through a liquid. Thus, in the case of solids, particularly catalysts, the mass of the solid is used as reference and consequently the rate is expressed as:

$r_j'' = $ rate of formation of component j/unit of mass (mol/g h)

The defined rates are intrinsic properties which depend on concentration, temperature, and pressure of the system. In a batch reactor, the concentration varies with the reaction time, and, in an open system, the concentration varies with the position.

A positive sign is conventionally used when the rate is related to the formation of products. When the rate is referred to the disappearance of a reactant, a negative sign is used.

In a system at constant volume, the rates can be expressed in terms of concentrations. In an open system, the rate depends on the variation of moles or on the molar flow F_j per volume, area, or mass. In general, we have:

$r_j = f(F_j)/V (\text{mol}/(\text{L h}))$

where $f(F_j)$ represents the flow of reactants and/or products in any position and V is the reaction volume.

Note that this is the reaction rate or activity. However, this definition takes into account the reaction medium, be it volume, surface, or interface, and not exactly the active sites. Not all mass or surface is active, but part of its outer surface has active sites, which are truly the sites where the chemical reaction occurs. Therefore, r_j in fact represents the apparent rate. An important example of reaction that allows to differentiate the apparent from the true rate is the hydrogenation of carbon monoxide to form methane, which is conducted with different catalysts. With iron and cobalt catalysts, the rate per unit of mass of catalyst, used as reference, has shown controversial values. The activity of the catalysts in the Fischer–Tropsch synthesis to form hydrocarbons would decrease according to the order Fe > Co > Ni. However, when the rate per active site was defined, the order of activity was different, i.e., Co > Fe > Ni. This controversy was resolved by Boudart, who defined the intrinsic activity, i.e., the rate per active site. To make it more clear, the turnover frequency (TOF) was defined. Thus, the intrinsic activity is determined, knowing the active sites, i.e.:

TOF = Rate of formation of componet j/active site.

$$\text{TOF} = \frac{F_j N_A}{S_i} (s^{-1})$$

where F_j = molar flow (mol/s) N_A = Avogadro's number = 6.023×10^{23} molecules S_i = active site = $[L] \times [S]$ $[L]$ = atoms density—(atoms/m^2) $[S]$ = superficial area (m^2).

This measurement gives us an idea of how many cycles a molecule reaches the surface and reacts to form the product or how many cycles per second the molecule collides with the active sites and reacts. In fact, not all molecules that collide with the active sites react. Some of them only collide but return without reacting. However, when they have enough energy to react, part of these molecules which collide with the surface reacts and are transformed into products, according to the theory of collisions.

In this case, active sites are understood as the surface which presents sites with specific characteristics. Thus, in a platinum supported catalyst, it is assumed that the outer surface has a collection of atoms, representing the density of the platinum atoms, i.e., the number of atoms that occupy an area of 1 m^2. Many solids have different types of sites, such as the acid sites in the surface of zeolites. For more complex systems such as liquid–gas and gas–liquid–solid interface, the rates per unit of surface or mass are conventionally used.

The difference between the two definitions is that the rates vary with time or position, while the TOF only depends on the availability of the sites present. The activity of the catalysts and the selectivity of the reaction depend on the characteristics of the material. Some important features of these reactions should be remembered. The first one is related to the sensitivity of the chemical reaction to the catalyst structure. According to Boudart's theory, the supported catalysts consist of metal particles of different and variable sizes. With the increase of particle diameter, the concentrations of metal atoms and exposed sites vary significantly, indicating that no changes in surface structure have happened. The structure sensitive reactions (SSR) are those in which the intrinsic reaction rate relative to the number of surface active sites, i.e., the frequency of the reaction varies with the particle sizes, which does not happen in structure insensitive reactions (SIR). This means that in the structure sensitive reactions, the frequency of

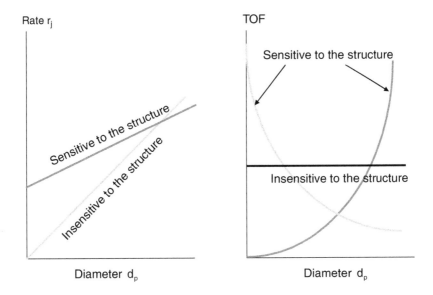

Figure 3.2 Activity— Structure sensitive and insensitive reactions.

the reaction significantly depends on particle size, dispersion of the metals, or structure of the catalyst active sites. On the other hand, in structure insensitive reactions , the activity is independent of these parameters, according to Figure 3.2.

In the dehydrogenation, the activity of the catalyst, TOF, is independent of the particle size of the metal or the catalyst used. Therefore, the reaction is insensitive to the structure.

In hydrogenations or hydrogenolysis, the intrinsic activity (TOF) depends on the structure. It depends on the particles size and the surface structure of the catalyst. The rate is not proportional to the type of surface site.

The intrinsic activity therefore depends on the type of site and the nature of the external surface. Generally, Figure 3.2 shows this behavior.

3.2 REACTION RATE

The reaction rate depends on concentration, pressure, and temperature of the reactants and products of the reaction. It is an intensive property because it has specific units and applies for any closed or open system. Since the concentration varies with time in a batch system or with position in a continuous system, the reaction rate also depends on these variables. This rate decreases with time or position, and tends to zero when all the reactant is consumed or in the equilibrium. The rate of the reaction is defined as a function of a component, and for a reversible reaction of the type $aA + bB \Leftrightarrow rR + sS$, the resulting reaction rate is expressed as follows:

$$r = k C_A^{a'} C_B^{b'} - k' C_R^{r'} C_S^{s'} \qquad (3.1)$$

where the first term on the right-hand side represents the direct rate of transformation of the reactants and the second term corresponds to the reverse rate of decomposition of the product in a reversible reaction.

Note that the direct rate is proportional to the concentration of the reactants, and the factor of proportionality is the rate constant of the direct reaction k. Analogously, we define the rate constant of the reverse reaction as k'.

The exponents of the concentrations a', b', r', and s' represent the orders of the reaction in relation to the respective components and are *different* from the stoichiometric coefficients of the reaction. They coincide if the reaction is elementary.

Summarizing:

$k, k' \Rightarrow$ specific rate constants of the direct and reverse reactions.
$a', b' \Rightarrow$ Reaction order in relation to the reactants A e B.
$r', s' \Rightarrow$ Reaction order in relation to the products R e S.

$$\text{If } a = a', b = b', r = r', s = s' \Rightarrow \text{it is an elementary reaction} \tag{3.2}$$

The specific rates have units and depend on the order of the reaction. When the order of the reaction is integer, we have some particular cases:

- Zero order $\Rightarrow r = kC_A^0 C_B^0 \Rightarrow k(\text{mol/L h})$
- First order $\Rightarrow r = kC_A \Rightarrow k(\text{h}^{-1})$
- Second order $\Rightarrow r = kC_A^2 \Rightarrow k(\text{L/mol h})$
  Or $\Rightarrow r = kC_A C_B \Rightarrow k(\text{L/mol h})$
- Third order $\Rightarrow r = kC_A^3 \Rightarrow k(\text{L/mol})^2 \text{ h}^{-1}$

If we have a fractional order reaction, the units of k and k' have the corresponding units.

At the equilibrium, the resulting rate of a reversible reaction is zero. Thus, for a constant temperature, we have $r = 0$ and, consequently:

$$\frac{k}{k'} = \frac{C_{Re}^{r'} C_{Se}^{s'}}{C_{Ae}^{a'} C_{Be}^{b'}} = K \rightarrow \text{equilibrium} \tag{3.3}$$

By substituting Equation 3.3 into Equation 3.1, we have:

$$r = k \left[C_A^{a'} C_B^{b'} - \frac{1}{K} C_R^{r'} C_S^{s'} \right] \tag{3.4}$$

Note that when the equilibrium constant is large, the reaction favorably moves to the right in the direction of products formation. The reaction is considered irreversible when $K \rightarrow \infty$. Therefore:

$$r = kC_A^{a'} C_B^{b'} \Rightarrow \text{Irreversible reaction} \tag{3.5}$$

The most common cases are:

- Irreversible reaction of 1st order $\Rightarrow a' = 1, b' = 0 \Rightarrow -r_A = kC_A$
- Irreversible reaction of 2nd order $\Rightarrow a' = 1, b' = 1 \Rightarrow -r_A = kC_A C_B$
$$\Rightarrow a' = 2, b' = 0 \Rightarrow -r_A = kC_A^2$$
- Irreversible reaction of n order (global) $a' + b' = n \Rightarrow -r_A = kC_A^n$

Examples

1. $CH_3 CHO \rightarrow CH_4 + CO$

   It is equivalent to a first or second order irreversible reaction:

   $$A \rightarrow R + S$$

   - Rate $\Rightarrow -r_A = kC_A$ or

     $$\Rightarrow -r_A = kC_A^2$$

2. $C_4H_{10} + NaOH \rightarrow C_2H_5ONa + C_2H_5OH$

   It is equivalent to a second order irreversible reaction:

   $$A + B \rightarrow R + S$$

   The corresponding rate will be:

   $$-r_A = -r_B = kC_A C_B$$

The rates of complex reactions involve the rates of the components which participate in the several reactions in series, parallel, or combination of both. For simplicity, we will consider the rates of elementary reactions with integer order, i.e., when the stoichiometric coefficients coincide with the order of the reaction. There are three classic cases:

1. *Reactions in parallel*: e.g., decomposition reactions, which are represented by:

   $$A \xrightarrow{k_1} P$$
   $$A \xrightarrow{k_2} R$$

   The rates of the corresponding components for the irreversible reaction are:

   $$r_P = k_1 C_A$$
   $$r_R = k_2 C_A$$
   $$-r_A = k_1 C_A + k_2 C_A \qquad (3.6)$$

2.   *Reactions in series*: e.g., decomposition reactions, which can be represented by:

$$A \xrightarrow{k_1} P \xrightarrow{k_2} R$$

The corresponding rates of each component are given by:

$$-r_A = k_1 C_A$$
$$r_P = k_1 C_A - k_2 C_P$$
$$r_R = k_2 C_P \tag{3.7}$$

3.   *Combined reactions*: when irreversible, they can be represented by:

$$A + B \xrightarrow{k_1} P$$
$$A + P \xrightarrow{k_2} R$$

The corresponding rates of each component, in an irreversible reaction, can be represented as follows:

$$r_P = k_1 C_A C_B - k_2 C_A C_P$$
$$r_R = k_2 C_A C_P$$
$$-r_A = k_1 C_A C_B + k_2 C_A C_P \tag{3.8}$$

In reversible reactions, the reverse rates should be considered, i.e., the decomposition of the products in their respective components, whose respective specific reverse rates would be k_i' of each reaction i.

Examples

1.   In the coal gasification, two main reactions take place:

$$C + H_2O \rightarrow CO + H_2 \quad \Delta H = 118.5 \, kJ/mol$$
$$CO + H_2O \leftrightarrow CO_2 + H_2 \quad \Delta H = -42.3 \, kJ/mol$$

This is series–parallel reaction, in which the carbon monoxide reacts with water, known as a shift reaction. It is reversible, depending on the pressure and temperature conditions. The first reaction is endothermic, and thus thermodynamically unfavorable, and only reacts when heat is supplied, while the second is exothermic, and thus thermodynamically favorable (spontaneous). The rates corresponding to each component in both reactions are indicated above. The reversible term should be added.

2. Utilization of the synthesis gas

Methanation:

$$CO + 3H_2 \rightarrow CH_4 + H_2O \Rightarrow \Delta H = -206.6 \, kJ/mol$$

Fischer–Tropsch synthesis:

$$CO + 2H_2 \rightarrow [C_nH_{2n}]_n + H_2O \Rightarrow \Delta H = -165.0 \, kJ/mol$$

Synthesis of methanol:

$$CO + 2H_2 \rightarrow CH_3OH \Rightarrow \Delta H = -90.8 \, kJ/mol$$

Shift reaction:

$$CO + H_2O \leftrightarrow CO_2 + H_2 \Rightarrow \Delta H = -39.8 \, kJ/mol$$

These reactions may occur simultaneously or not, depending on the thermodynamic conditions and especially on the catalyst. In the methanation Ni , catalysts are used, while in Fischer–Tropsch reactions, Fe or Co catalysts are employed. In the reactions for methanol synthesis, mixed oxide catalysts of CuO/ZnO are used, and for the shift reaction, Ni supported catalyst.

The reaction of methanation occurs around 300°C, whereas the Fischer–Tropsch synthesis occurs at lower temperatures, between 250°C and 280°C. Both reactions can occur simultaneously in this temperature range.

The corresponding rates can be written as Equation 3.8, considering the reversibility or not.

3. Hydrogenation of crotonaldehyde to butanol

This transformation can be represented by a reaction in series of the type:

$$CH_3 - CH_2 = CH_2 - HC = O + H_2 \rightarrow CH_3 - CH_3 - CH_3 - HC = O + H_2$$
$$\text{Crotonoaldehyde} \qquad\qquad\qquad \text{Butyraldehyde}$$

$$\rightarrow CH_3 - CH_3 - CH_3 - HCOH$$
$$\text{Butanol}$$

The corresponding rates are shown in Equation 3.7, considering all the limiting components, since the reaction occurs with excess of hydrogen, and, therefore, the rate is independent of the concentration of hydrogen.

3.2.1 Kinetic equations

The reaction rates are kinetic equations, written in terms of measurement variables, such as concentration, partial pressure, and particularly, conversion or extent of reaction. The rate of product formation or transformation of the reactant is expressed in relation to the concentration of the limiting reactant and is valid for any closed or open system, at variable or constant volume.

3.2.1.1 Irreversible reaction at constant volume

Consider the reaction, $A + B \rightarrow R + S$, in which A is the limiting reactant. Thus, the rate will be, according to Equation 3.5:

$$r = kC_A^{a'} C_B^{b'} \Rightarrow \text{irreversible reaction} \tag{3.9}$$

By defining the conversion in relation to the limiting component A, we have:

$$C_A = C_{A0}(1 - X_A)$$
$$C_B = B_{B0} - (b/a)X_A = C_{A0}(M - (b/a)X_A)$$

where

$$M = \frac{C_{B0}}{C_{A0}} \text{ is always} \geq 1,$$

relating the initial concentration of the reactants, with A as the limiting reactant. When B is the limiting reactant, the reaction is inverted, because M is always ≥ 1. When the initial concentrations are equal, we have $M = 1$.

By substituting the concentrations C_A and C_B into Equation 3.9, we have:

$$-r_A = C_{A0}^n (1 - X_A)^{a'} (M - (b/a)X_A)^{b'} \tag{3.10}$$

where

$$n = a' + b' \Rightarrow \text{global order}$$

We can define the rate of formation of the products r_R, but one must be careful when relating it with the rate of transformation of reactant. From the proportionality law, we always have the relationship:

$$\frac{(-r_A)}{a} = \frac{(-r_B)}{b} = \frac{r_R}{r} \tag{3.11}$$

Therefore, the specific rates are also defined in relation to each component. By choosing the reactant A we have k_A, and according to Equation 3.11, we can relate it with any other component, either a reactant or a product:

$$\frac{k_A}{a} = \frac{k_B}{b} = \frac{k_R}{r} \tag{3.12}$$

We will use the specific rate without subscript in order to implicitly correspond to the rate defined in relation to a particular component.

Particular cases

In most cases, the reactions are irreversible and of integer order, at maximum until the third order. One must be careful when the stoichiometry is different from the reaction order, as in the below example:

Reaction : $A + 3B \rightarrow$ products

Kinetics: Second order and first order in relation to each component. Therefore, the reaction rate will be:

$$-r_A = kC_{A0}^2(1 - X_A)(M - 3X_A) \tag{3.13}$$

where

$$(b/a) = 3$$

Pseudo-first-order
      The concentration of a particular reactant is much higher than the concentration of another component. This happens in liquid phase reactions, where one of the components, generally water, participates as a reactant and diluent simultaneously. In this case, the reaction of second order is simplified to pseudo-first-order:

$$-r_A = kC_{A0}^2 M(1 - X_A) \tag{3.14}$$

where:

$$M \gg 1 \Rightarrow C_{B0} \gg C_{A0}$$

The rate is represented as a function of the apparent constant $k^* = kC_{A0}^2 M$, i.e.:

$$-r_A = k^*(1 - X_A) \tag{3.15}$$

Generic order n
      When there is a stoichiometric proportionality in the reaction, we can simplify the global equation. So:

$$\frac{C_A}{a} = \frac{C_B}{b} = \cdots \tag{3.16}$$

By expressing the concentration C_B as a function of C_A in Equation 3.9:

$$-r_A = kC_A^{a'} C_B^{b'}$$

We obtain:

$$-r_A = k^* C_A^n \tag{3.17}$$

where:

$$k^* = k \left(\frac{b}{a}\right)^{b'}$$

$$n = a' + b'$$

We also can write this Equation as a function of the conversion (Equation 3.10), and then we obtain:

$$-r_A = k^* C_{A0}^n (1 - X_A)^n \tag{3.18}$$

3.2.1.2 Reversible reactions at constant value

The reversible reactions are generically represented by three types:

1. Reaction of generic order:

$$aA + bB \Leftrightarrow rR + sS$$

whose resultant rate will be:

$$r = k\left[C_A^{a'} C_B^{b'} - \frac{1}{K} C_R^{r'} C_S^{s'}\right]$$

2. Elementary reaction of direct and reverse first order:

$$A \Leftrightarrow R$$

whose rate is represented by:

$$r = k\left[C_A - \frac{1}{K} C_R\right]$$

3. Elementary reaction of direct second order and reverse first order, or vice versa, of the type:

$$A + B \Leftrightarrow R$$
$$A \Leftrightarrow R + S$$

whose rates are, respectively:

$$r = k\left[C_A C_B - \frac{1}{K} C_R\right]$$

$$r = k\left[C_A - \frac{1}{K} C_R C_S\right]$$

where: K is the chemical equilibrium constant.

By expressing the rates as a function of the conversion, the concentrations are replaced:

$$C_A = C_{A0}(1 - X_A)$$
$$C_B = C_{A0}(M - (b/a)X_A)$$

or products,

$$C_R = C_{A0}(R + (r/a)X_A)$$

By replacing for the case 2, we have:

$$r = kC_{A0}\left[(1 - X_A) - \frac{1}{K}(R + (r/a)X_A)\right] \tag{3.19}$$

However, at the equilibrium, the resulting rate is zero. Thus, from Equation 3.3 and considering the simplest stoichiometry $(a = r = 1)$, we have:

$$\frac{k}{k'} = \frac{(R + X_{Ae})}{(1 - X_{Ae})} = K \rightarrow \text{equilibrium} \tag{3.20}$$

By replacing K of Equation 3.20 into Equation 3.19, we obtain the rate as a function of the equilibrium conversion (X_{Ae}). The equilibrium conversion can be determined from the thermodynamic equilibrium constant (knowing ΔG^0), or from the experimental data of the kinetic curve, knowing that when $t \rightarrow \infty$, the conversion $X_A \rightarrow X_{Ae}$.
Thus,

$$r = \frac{kC_{A0}(R + 1)}{(R + X_{Ae})}(X_{Ae} - X_A) \tag{3.21}$$

where k (min^{-1}) is the direct specific rate and R is the relation between the initial concentrations of the product C_{R0} and of the reactant C_{A0}. Starting from a pure reactant, which is the most common case, we have that $R = 0$.
The rate varies with the conversions. Initially, it is maximum when $X_A = 0$, decreasing thereafter until the equilibrium is reached and zero when $X_A \rightarrow X_{Ae}$. However, by differentiating Equation 3.21, it is observed that the rate variation will be:

$$\frac{dr}{dX_A} = \frac{kC_{A0}(R + 1)}{(R + X_{Ae})}(-1) < 0 \tag{3.22}$$

Therefore, the rate decreases negatively and the curve is always concave, yielding a variation of the type:

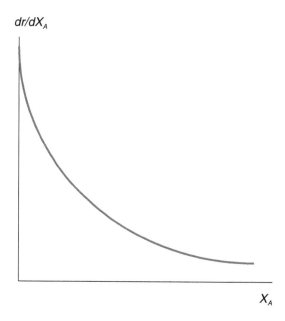

3.2.1.3   Irreversible or reversible reactions at variable volume

A reaction system at constant pressure can be open or closed and can occur in liquid, gas, or vapor phase. When the reactions are carried out in gas or vapor phase in an open system and there is variation of the number of moles, there will be volume expansion or contraction. In a closed system, consider a piston moving without friction, according to the schemes:

Open System:

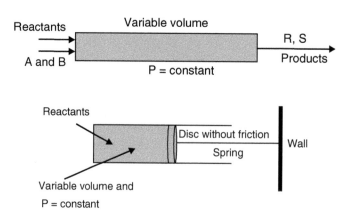

There are two ways to express the equation of the reaction rate:

1. As a function of the partial pressures for an irreversible reaction of second order, in which A is the limiting reactant:

$$-r_A = kC_A C_B \Rightarrow \text{irreversible reaction} \tag{3.23}$$

Since the concentrations in a system at constant volume are:

$$C_A = \frac{p_A}{RT}, \ C_B = \frac{p_B}{RT}$$

we have:

$$-r_A = \frac{k}{(RT)^2} p_A p_B \tag{3.24}$$

It is also possible to express it as a function of the molar fraction and total pressure of the system, because:

$$y_A = \frac{p_A}{P}, \ y_B = \frac{p_B}{P}$$

Thus:

$$-r_A = \frac{k}{(RT)^2} y_A y_B P^2 \tag{3.25}$$

In this case, the molar fraction can be expressed as a function of the extent of reaction (α), by using Equation 1.22, for a reaction of the type:

$$aA + bB \rightarrow rR$$

or

$$-r_A = \frac{k}{(RT)^2} \frac{(n_{A0} - a\alpha)(n_{B0} - (b/a)\alpha)}{(n_0 + \Delta v \cdot \alpha)^2} P^2 \tag{3.26}$$

2. As a function of the molar flows in an open system, for an irreversible reaction of second order, we start from the same Equation 3.23:
    It is known that the molar flows are, respectively:

$$F_{A0} = C_{A0} v_0$$
$$F_A = C_A v$$
$$F_B = C_B v$$

By replacing the concentration in Equation 3.23, we have:

$$-r_A = k \left(\frac{F_A}{v} \right) \left(\frac{F_B}{v} \right)$$

The molar flows and the volumetric flow v are known as a function of the conversion, for reactants or products, by means of Equations 1.21, displayed below:

$$F_A = F_{A0}(1 - X_A)$$
$$F_B = F_{B0} - (b/a)F_{A0}X_A$$

Therefore, the final rate will be:

$$-r_A = kC_{A0}^2 \frac{(1 - X_A)(M - (b/a)X_A)}{(1 + \varepsilon_A X_A)^2} \tag{3.27}$$

Generically, for an order n and partial orders a', b' and with temperature variation, we should take into account the volume variation with temperature, correcting the previous expression. Under these more general conditions, we have:

$$-r_A = kC_{A0}^n \frac{(1 - X_A)^{a'}(M - (b/a)X_A)^{b'}}{(1 + \varepsilon_A X_A)^n \left(\frac{T}{T_0}\right)^n} \tag{3.28}$$

Examples

E3.1 Consider a second order reversible reaction of decomposition $A \rightarrow rR$, carried out in gas phase. Initially, a test I was conducted in a batch reactor, introducing pure A, at 300 K. After 10 min, the pressure was 3 atm. After a sufficiently long time, the pressure reached 5 atm, remaining constant afterward.

Subsequently, a test was carried out in a closed system with a piston without friction, but at constant pressure of 1 atm. The final volume was doubled. Determine the equation of the rate and calculate both the rate for a conversion of 50% and the initial rate at such conditions. It is known that $k = 003$ L/(mol min).

Solution

The first part of the solution is equal to the Problem E1.1.
If the volume is constant, the partial pressure will be 1.25:

$$p_A = p_{A0} - \frac{a}{\Delta v}(P - P_0)$$

Since

$$a = 1$$

for $t = \infty \Rightarrow p_A = 0$, $p_{A0} = P_0 = 1$, and $P = 5$, we have:

$$\Delta v = 4 \Rightarrow r = 5$$

But

for $t = 10\,\text{min} \Rightarrow p_{A0} = 0$, $P = 1$, and $P = 3$

$$X_A = \frac{1}{3}\frac{P - P_0}{P_0} = 0.5$$

For the second condition, in a plug-flow, the conversion is the same at constant pressure. But the volume change is:

$$V = V_0(1 + \varepsilon_A X_A)$$

Since:

$$V = 2V_0$$

we obtain:

$$2V_0 = V_0 + V_0\varepsilon_A X_A$$

$$V_0\varepsilon_A = 2$$

With the same conversion, $X_A = 0.5 \Rightarrow \varepsilon_A = 4$.
The rate is of second order and irreversible, and for the batch system this will be:

$$-r_A = kC_{A0}^2(1 - X_A)^2$$

or for the plug-flow, at variable volume, this will be:

$$-r_A = kC_{A0}^2\frac{(1 - X_A)^2}{(1 + \varepsilon_A X_A)^2} \tag{3.29}$$

But:

$$C_{A0} = \frac{p_{A0}}{RT} = \frac{P_0}{RT} = \frac{1}{0.082 \times 300} = 4.06 \times 10^{-2}\,\text{mol/L}$$

By replacing the values in Equation 3.29, for $X_A = 0.5$, we have:

Batch $(-r_A) = 1.23 \times 10^{-5}\,\text{mol/L min}$

Plug-flow: $(-r_A) = 1.37 \times 10^{-6}\,\text{mol/L min}$

The initial rate is equal for both cases:

$$(-r_A)_0 = 4.95 \times 10^{-5}\,\text{mol/L min}$$

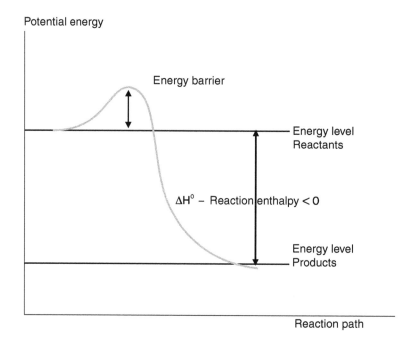

Figure 3.3 Potential energy over reaction time.

3.3   INFLUENCE OF TEMPERATURE ON THE REACTION RATE

It was observed experimentally that an increase of 10°C in temperature can double the rate in an exothermic reaction. This empirical observation greatly depends on the reaction. But the fact is that the reaction rate is very sensitive to temperature variation. A general law of dependence of the rate with the temperature was therefore needed.

It is known that the potential energy varies as the reaction occurs and presents an energy barrier which should be surpassed for the occurrence of the reaction. Starting from the different energy levels of the reactants and products, which characterizes the transformation reaction enthalpy of a chemical reaction, the potential energy curve varies according to Figure 3.3.

The energy barrier is related to the activation energy, inherent to each reaction. If the activation energy is higher than the energy barrier, the reaction will occur. Thus, Arrhenius defined a reaction rate constant which depends mainly on the temperature and is a function of activation energy E. The Arrhenius equation shows that the reaction rate constant varies exponentially with the temperature, according to Equation 3.30.

$$k = k_0 \cdot exp(-E/RT) \tag{3.30}$$

In this expression, the pre-exponential factor k_0 represents a factor proportional to the number of collisions of the molecules and can be considered constant.

Placing $\dfrac{k}{k_0}$ in a graph as a function of $\dfrac{RT}{E}$,

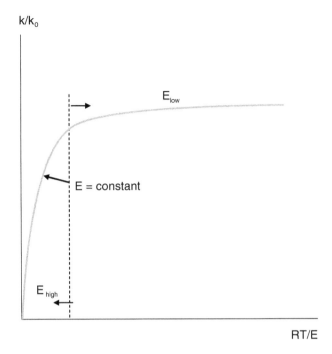

Figure 3.4 Influence of temperature on the rate constant.

According to Figure 3.4, it is possible to observe the influence of the activation energy on the reaction rate constant and the effect of the temperature.

If the activation energy is constant, the reaction rate constant k varies exponentially with temperature and is very sensitive in a small temperature range but depends on the order of magnitude of the activation energy. If the activation energy is high, small changes in temperature significantly affect the constant k. On the other hand, for low activation energies, the constant k is little sensitive in a wide temperature range.

The frequency factor k_0 is apparently constant. It depends on the probability of collisions of the molecules and on the temperature. The probability of collision of the molecules in a space depends on the system energy and reaction medium. In an empty space, the probability of collision of the molecules is relatively small compared to a space which is filled with particles.

In this case, the contact surface is much larger and the approach of the molecules with the surface increases the probability of collision between the molecules. Thus, for example, for different values of k_0 and the same activation energy $E = 60$ kcal/mol, at constant temperature of 600 K, very different values of k are obtained, which can reach up to 100 times higher.

$\Rightarrow$ for $k_0 = 10^{20} \Rightarrow k = 1.164 \times 10^{-2}$ min^{-1} $\Rightarrow$ (empty volume)
$\Rightarrow$ for $k_0 = 10^{22} \Rightarrow k = 1.164$ min^{-1} $\quad\quad \Rightarrow$ (volume with particles)

The effect of temperature on k_0 is small and, according to the theory of collisions, varies at the maximum with a power m. Admitting that:

$$k_0 = k_0' T^m \qquad (3.31)$$

The reaction rate constant will be:

$$k = k_0' T^m \cdot \exp\left(-E/RT\right) \qquad (3.32)$$

Taking the natural log and differentiating with respect to temperature, we have:

$$\frac{d\ln k}{dT} = \frac{m}{T} + \frac{E}{RT^2} = \frac{mRT + E}{RT^2} \qquad (3.33)$$

The numerator of the last term of this expression shows that the value of mRT is much smaller when compared to E. By assigning a maximum value of m to 2, $R = 1.98$ cal/mol and considering the average activation energy of about 25000 cal/mol, for a temperature of 300 K, we have:

$$mRT \approx 2370$$

This value is 20 times smaller than $E = 25{,}000$, and therefore, we can neglect the term mRT in Equation 3.33, and, consequently, the frequency factor k_0 can be considered constant. The variation of the reaction rate constant k with the temperature will be:

$$\frac{d\ln k}{dT} = \frac{E}{RT^2} \qquad (3.34)$$

3.3.1 Reversible reactions

In reversible reactions of the type, there are two reaction rate constants (direct and reverse), with the corresponding activation energies. Thus:

Direct reaction $\Rightarrow r_d = kf(C_A, C_B), k = k_0\, e^{-(E/RT)}$

Reverse reaction $\Rightarrow r_r = kf(C_R), k' = k_0'\, e^{-(E'/RT)}$

We have seen that the chemical equilibrium constant for the same temperature is given by eq. 3.20:

$$K = \frac{k}{k'}.$$

Taking the natural log of both sides of the equation, we have:

$$\ln K = \ln k - \ln k'$$

The variation of the constants with temperature will be:

$$\frac{d \ln K}{dT} = \frac{d \ln k}{dT} - \frac{d \ln k'}{dT}$$

By substituting the first term by the Van't Hoff equation (2.12)

$$\frac{d \ln K}{dT} = \frac{\Delta H^0}{RT^2}$$

and the other terms by the Arrhenius equation (Equation 3.34) for the two constants, we have:

$$\Delta H^0 = E - E' \tag{3.35}$$

This equation relates the enthalpy of the reaction, which is a thermodynamic property, with the direct (E) and reverse (E') activation energies. If the reaction is exothermic, the enthalpy $\Delta H^0 < 0$ and, therefore, $E < E'$. Therefore, the direct activation energy is always lower than the reverse, and, consequently, the direct reaction is facilitated because the lower the activation energy, the easier the reaction occurs. The opposite happens with endothermic reactions, in which the enthalpy $\Delta H^0 > 0$, and, therefore $E < E'$. In this context, the reverse reaction is easier.

The effect of temperature on the resulting rate is generally represented by:

$$r = k_0 \cdot \exp(-E/RT) \cdot f(C_A, C_B) - k'_0 \cdot \exp(-E'/RT) \cdot f(C_R)$$

Differentiating with respect to the temperature and considering that the first term corresponds to the direct rate r_d and the second to the reverse rate r_r, we have:

$$\frac{d \ln r}{dT} = \frac{E}{RT^2} r_d - \frac{E'}{RT^2} r_r \tag{3.36}$$

Note that the direct (r_d) and reverse (r_r) rate increase with the increase in temperature and are always positive. Moreover, the direct and reverse activation energies are always positive. Therefore, the variation of the rate with the increase in temperature is always positive. However, depending on whether the reaction is exothermic ($\Delta H^0 < 0$ and $E < E'$), the variation of the resulting rate will depend on the difference $Er_d - E'r_r$. Therefore, the resulting rate increases and is always positive, and when the reverse rate increases, the resulting rate reaches a maximum value and decreases positively (according to Figure 3.5). Thus:

$$\frac{d \ln r}{dT} > 0$$

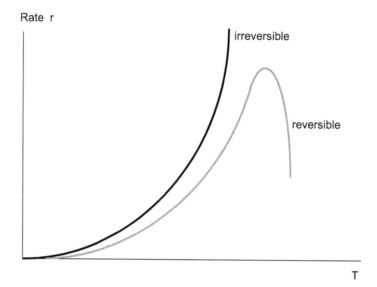

Figure 3.5 Influence of temperature on the reaction rate.

3.3.2 Interpretation remarks

1. Order of magnitude of the activation energy
    By expressing the Arrhenius equation in logarithmic form (Equation 3.37) and representing it graphically as shown in Figure 3.6, a straight line is obtained, whose angular coefficient is $\left(-\dfrac{E}{R}\right)$:

$$\ln k = \ln k_0 - \frac{E}{RT} \tag{3.37}$$

   The slopes are different and from the graph, we can note that the activation energies are constant, but $E_1 < E_2$. This means that depending on the reaction, the activation energy changes. Through this graph, it is possible to determine the activation energy of the reaction by measuring the rates or specific rates at different temperatures. It is recommended to carry out at least three experiments or three measurements at different temperatures.

2. Intermediate steps
    The nonelementary reactions consist of several steps of elementary reactions. They are recognized by the change in activation energy. Thus, if the form of Arrhenius plots is a curve, one can recognize two or more intermediate steps, as shown in Figure 3.7.
    The activation energy varies with the temperature during the reaction, indicating that there was a change of mechanism or compensatory effect.

3. Diffusive effects
    The activation energy at high temperatures is lower than at lower temperatures, according to Figure 3.8, which means that the kinetic regime changed.

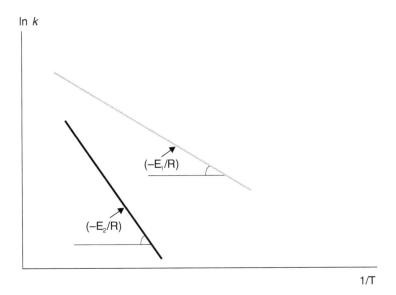

Figure 3.6 Arrhenius graph.

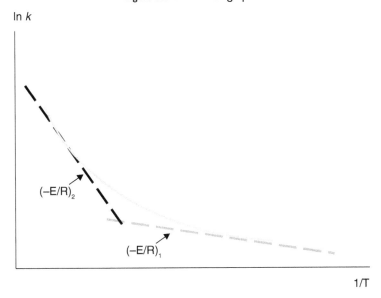

Figure 3.7 Intermediate steps.

The activation energy at lower temperatures represents the kinetic regime. The behavior at high temperatures is influenced by the diffusion or mass transfer. Therefore, there are diffusive effects hampering the determination of the "real" chemical reaction in the kinetic regime. An "apparent" activation energy is obtained. Therefore, it is essential to choose the temperature range for the determination of the activation energy, ensuring the kinetic regime. It is also a way to check if the experiment was conducted without diffusion effects and an indication of the validity of a kinetic experience.

Figure 3.8 shows a different behavior at high and low temperatures.

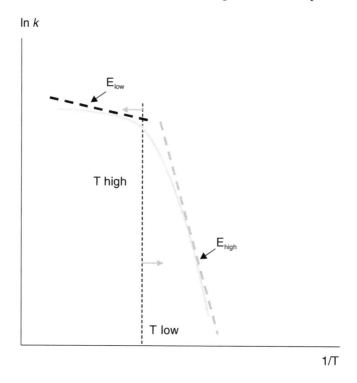

Figure 3.8 Diffusive effects.

3.3.3 Reparameterization of the Arrhenius equation

Commonly, mathematical models contain variables that cannot be measured directly. The values of these variables, called model parameters, can be estimated from experimental data with the help of parameter estimation procedures, which consist in minimizing some sort of objective function that takes into account the difference between model responses and experimental observations, as described in Chapter 11. A common model used to describe the temperature dependence of reaction rates in kinetic problems is the Arrhenius equation, described in previous sections:

$$k = k_0 \exp\left(-\frac{E}{RT}\right) \tag{3.38}$$

where k is the reaction rate constant (or the specific reaction rate), T is the absolute temperature, R is the ideal gas constant, k_0 is the frequency (or pre-exponential) factor and E is the activation energy. Both k_0 and E are the parameters of the Arrhenius equation, usually estimated from available experimental data. However, the intrinsic mathematical structure of the Arrhenius equation introduces a very strong dependence between the parameters k_0 and E (called 'parameter correlation'), making the estimation of the parameter values difficult. To overcome this difficulty, many authors (Box, 1960; Himmelblau, 1970; Pritchard and Bacon, 1975) suggested the reparameterization of the Arrhenius equation by introducing a reference temperature in the form:

$$k = k_{ref} \exp\left[-\frac{E}{R}\left(\frac{1}{T} - \frac{1}{T_{ref}}\right) \right] \tag{3.39}$$

$$k = \exp\left[A - \frac{E}{R}\left(\frac{1}{T} - \frac{1}{T_{ref}}\right) \right] \tag{3.40}$$

$$k = k_{ref} \exp\left[B\left(\frac{T - T_{ref}}{T}\right) \right] \tag{3.41}$$

$$k = \exp\left[A + B\left(\frac{T - T_{ref}}{T}\right) \right] \tag{3.42}$$

where the parameters of the reparameterized equations can be related to the parameters of the more traditional form of the Arrhenius equation as:

$$k_{ref} = k_0 \exp\left(-\frac{E}{RT_{ref}} \right) \tag{3.43}$$

$$A = \ln(k_{ref}) = \ln(k_0) - \frac{E}{RT_{ref}} \tag{3.44}$$

$$B = \frac{E}{RT_{ref}} \tag{3.45}$$

The introduction of reference temperature into the Arrhenius equation allows for reduction of the parameter correlation and, consequently, for reduction of the computational effort required for estimation of model parameters (Espie and Macchietto, 1988; Schwaab and Pinto, 2007). The reference temperature is usually defined as a suitable average temperature of the analyzed experimental data. For instance, Veglio et al. (2001) suggested the use of the inverse average:

$$\frac{1}{T_{ref}} = \frac{1}{NE} \sum_{i=1}^{NE} \frac{1}{T_i} \tag{3.46}$$

where NE is the number of experimental temperature values and T_i is the temperature of the individual experiments i. Particularly, Schwaab and Pinto (2007) showed that the proper definition of the reference temperature can eliminate the parameter correlation in kinetic models containing a single kinetic constant. It was shown both analytically and numerically that proper definition of reference temperature allows for elimination of the parameter correlation and simultaneous improvement of the precision of parameter estimates. For this reason, in the following sections the reparametrized form of the Arrhenius equation will be used frequently for quantitative analyses of kinetic data, as described in detail in Section 11.8.

Chapter 4

Molar balance in open and closed systems with chemical reaction

The kinetic experiments are carried out in reactors in closed or open systems. There are three common cases:

Batch.
Continuous.
Semicontinuous.

For kinetic purposes, the variation of the concentration or the pressure with reaction time is accompanied in a batch reactor. In a continuous or open system, the pressure is constant and the concentrations or molar flows of the reactants and products are accompanied in the course of the reaction or along the reactor. The time is substituted by an equivalent variable called space time. The space time takes into account the inlet volumetric flow of the reactants and the volume of the reactor, and, thus, has unit of time, here designated by:

$$\tau = \frac{V}{v_0}(h)$$

The inverse is the space velocity.

The molar balance in an open system and for any reaction of the type, initially at constant temperature, is shown in Scheme 4.1, in which j is a component, reactant, or product:

Molar flow of the component j which enters the reactor per unit of volume V		Molar flow of the component j which leaves the reactor per unit of volume V		Rate of generation or consumption of the component j through the chemical reaction per volume V		Rate of accumulation of the component j in the volume V
	−		+		=	
[1]		[2]		[3]		[4]

Scheme 4.1

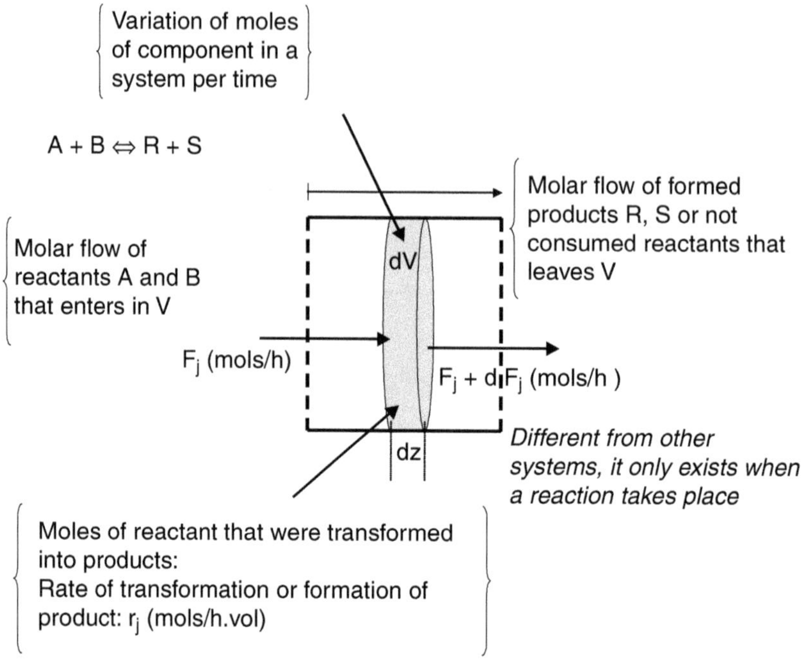

A + B ⇔ R + S

Figure 4.1

Considering F_j the molar flow of the component j, G_j the rate of generation of products or consumption of reactants, and n_j the number of moles of component j, we have the following molar balance:

$$F_{j0} - F_j + G_j = \frac{dn_j}{dt} \tag{4.1}$$

Note that the balance is carried out to any component, reactant, or product of the reaction, and has the unit of mol/time.

Schematically, it is shown in Figure 4.1.

The rate of generation of products or consumption of reactants in the system is represented per unit of volume, within each element of volume ΔV. So:

$$G_j = r_j V$$

where: r_j (rate) $\Rightarrow$ (mol/time·volume)

In the elements of volume ΔV_i, we have:

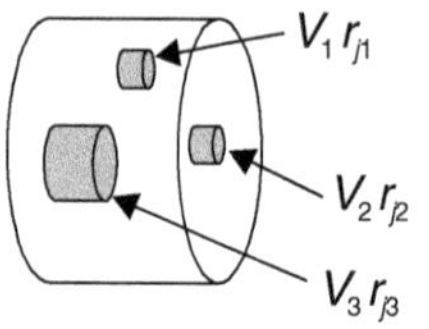

$$\{ \;\Rightarrow\; G_j = \sum \Delta G_{ji} = \sum r_{ji} \Delta V_i$$
$$\Downarrow$$
$$G_j = \int r_j dV$$

By substituting the expression of the rate G_j in Equation 4.1, we obtain:

$$F_{j0} - F_j + \int r_j \, dV = \frac{dn_j}{dt} \qquad (4.2)$$

This is the general equation of the molar balance for any component j.
Thereafter, one can make several simplifications.

4.1 BATCH

In this case, there is no flow, and, therefore, the terms referring to flow disappear. We then obtain the equation for the batch reactor:

$$\int r_j \, dV = \frac{dn_j}{dt} \qquad (4.3)$$

If the volume is constant, we have:

$$r_j V = \frac{dn_j}{dt} \qquad (4.4)$$

An expression equal to the rate already defined is obtained, representing the variation of the number of moles of a component, reactant, or product, per unit of volume, in a closed system. We should remember that if the rate corresponds to the formation of the product, it has positive sign (r_j). On the other hand, if the rate corresponds to the transformation of the reactant, it has negative sign $(-r_j)$.

Consider the reaction of the type:

$$A \rightarrow products$$

If the reaction is of first order, we have as a function of the conversion:

$$n_A = n_{A0}(1 - X_A)$$

From Equation 4.4, we have:

$$(-r_A) = -\frac{1}{V}\frac{dn_A}{dt} = \frac{n_{A0}}{V}\frac{dX_A}{dt} \qquad (4.5)$$

We then integrate in the interval $t \Rightarrow 0 \rightarrow t$ and $X_A \Rightarrow 0 \rightarrow X_A$, in order to obtain:

$$t = n_{A0} \int_{0}^{X_A} \frac{1}{V}\frac{dx_A}{(-r_A)} \qquad (4.6)$$

In a batch reactor, the volume is constant.

Therefore:

$$t = C_{A0} \int_0^{X_A} \frac{dx_A}{(-r_A)} \tag{4.7}$$

4.2 CONTINUOUS STIRRING TANK REACTOR

Consider as the starting point the general equation (Equation 4.2) for any component of the reaction. In general, the regime is considered to be permanent. When there is disturbance of the system, the term of the Scheme 4.1 is included. Otherwise, the respective term is not included. In most cases, this type of reactor is used for reactions in liquid phase, and, therefore, the volume is constant. If gas phase reactions occur, we should consider the variation of the volume, according to Equation 1.13. Thus:

$$F_{j0} - F_j + r_j V = 0 \tag{4.8}$$

For a reactant A or B of the reaction of the type:

$$aA + bB \rightarrow rR$$

we have:

$$F_{A0} - F_A + r_A V = 0 \tag{4.9}$$

Considering the molar flow as a function of the conversion, according to Equation 1.21, i.e.:

$$F_A = F_{A0}(1 - X_A)$$

And substituting F_A in Equation 4.9, we have:

$$F_{A0} X_A = (-r_A)V \tag{4.10}$$

Hence, we can determine the rate $(-r_A)$ with the conversion, reactor volume, and inlet molar flow.

If the molar balance is carried out in relation to the product R or S:

$$F_{R0} - F_R + r_R V = 0 \tag{4.11}$$

Thus, by using the relation shown in Equation 1.20:

$$\frac{F_{A0} - F_A}{a F_{A0}} = \frac{F_{B0} - F_B}{b F_{A0}} = \frac{F_R - F_{R0}}{r F_{A0}}$$

We obtain:

$$F_R - F_{R0} = (r/a) F_{A0} X_A$$

Substituting in Equation 4.11, we have:

$$F_{A0}X_A = \left(\frac{a}{r}\right)(r_R)V \tag{4.12}$$

Note the positive signal corresponding to the formation rate of R. This rate can be determined, knowing the conversions, the reactor volume, and inlet molar flux of the reactant.

4.3   CONTINUOUS TUBULAR REACTOR

The concentration and therefore the molar flows and reaction rate vary along the reactor. The differential balance is performed, starting from Equation 4. Considering a permanent regime for simplification, as in the previous case, we obtain Scheme 4.2:

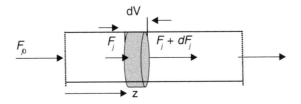

By performing the balance in the element dV, considering that the cross section of the tube is constant, we obtain:

$$F_j - (F_j + dF_j) + r_j\, dV = 0 \tag{4.13}$$

Therefore:

$$dF_j = r_j\, dV \tag{4.14}$$

Analogously, for a reaction of the type:

$$aA + bB \rightarrow rR$$

The molar balance in relation to the reactant A (or B), considering that:

$$F_A = F_{A0}(1 - X_A)$$
$$\Downarrow$$
$$dF_A = -F_{A0}\, dX_A$$

Thus:

$$F_{A0}\, dX_A = (-r_A)\, dV \tag{4.15}$$

or in relation to the product R, by using the relation 1.20:

$$F_{A0}\, dX_A = (r_R)dV \tag{4.16}$$

Again, note that in the transformation of the reactant, the rate has negative sign, and in the formation of the product, the rate has positive sign, in accordance with the definition initially adopted.

In any case, we can integrate with respect to the total volume of the reactor V and the final conversion reached X_A. If we use the definition of space time t, which relates the volume of the reactor with the volumetric flow:

$$\tau = \frac{V}{v_0}(b)$$

We obtain:

$$\tau = \frac{V}{v_0} = C_{A0} \int_0^{X_A} \frac{dX_A}{r_j} \tag{4.17}$$

where: $\Rightarrow \quad F_{A0} = C_{A0} v_0$.

In this case, it is important to note that the conversion is defined in relation to the limiting reactant A.

For kinetic purposes, when the objective is to determine the reaction rate as a function of the concentration, pressure, or conversion, we use these systems in ideal conditions at constant temperature. Diffusive or mass transfer effects should be minimized or eliminated.

Part II

Kinetics

Chapter 5

Determination of kinetic parameters

The kinetic parameters of the reaction rate are the rate constants k and the orders (a', b', n) of the reaction in relation to each component. The effect of the temperature is incorporated in the reaction rate and in order to determine it, the activation energy E and the frequency factor k_0 should be determined.

There are two methods: integral and differential. The integral method has an advantage of having an analytical solution. The differential method has an approximate or numerical solution. For all cases, experimental data obtained in laboratory are necessary, both for batch and continuous systems.

INTEGRAL METHOD

There are four successive steps:

1. *Selection of the kinetic model*: a reaction with defined order (integer or fractional), and the appropriate reaction rate is written:

$$r_j = kC_A^{a'} C_B^{b'} = kf(C_j^n) = kC_{A0}^n f(X_A)$$

where r_j is the rate of the component j of the reaction, reactant, or product; A is the limiting reactant, $f(X_A)$ is a function of the conversion X_A for irreversible and reversible reactions.

Generally, we assume an integer order. Therefore:

$$r_j = kC_{A0}^n f(X_A) \tag{5.1}$$

2. *Selection of the system*: **batch** or **continuous (tubular)**. For both cases, the expression of the rate is substituted, and we obtain a solution of the type:

Batch (Equation 4.7):

$$t = C_{A0} \int_0^{X_A} \frac{dX_A}{r_j} = C_{A0} \int_0^{X_A} \frac{dX_A}{kC_{A0}^n f(X_A)} \tag{5.2}$$

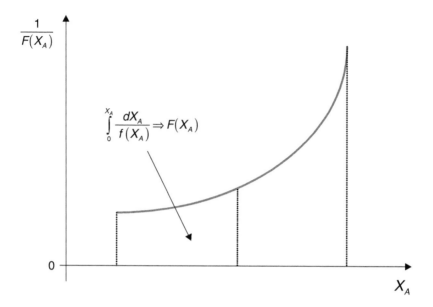

Figure 5.1 Representation of the integral function.

Tubular (Equation 4.17):

$$\tau = \frac{V}{v_0} = C_{A0} \int_0^{X_A} \frac{dX_A}{r_j} = C_{A0} \int_0^{X_A} \frac{dX_A}{k C_{A0}^n f(X_A)} \tag{5.3}$$

where: t is the measurement time and τ is the space time.

3. *Mathematical solution*: In general, the integral is solved analytically when the order of the reaction is an integer. When the order is fractional or the model is more complex, the integral is solved numerically. Figure 5.1 illustrated the integral solution.

    Note that the area under the curve is equal to the integral represented in Equations 5.2 and 5.3, according to Figure 5.1. If we represent the areas by $F(X_A)$, for each value of X_A, we have a linear equation of the type:

$$F(X_A) = C_{A0}^{n-1} k\tau \tag{5.4}$$

This equation is graphically presented in Figure 5.2, where k is a constant.

4. *Experimental verification*: The laboratory experiments provide the measurements of the concentration as a function of time or space time, for a constant temperature. The conversions, the function $f(X_A)$, and finally the specific rate (rate constant) k are calculated. If the experimental values of $F(X_A)$ *versus* t or τ are on the line, it can be concluded that the proposed model is correct. Otherwise, we would have to choose other model. In Figure 5.2, we can see the gray dots that follow the model. The black dots indicate that the model is not appropriate.

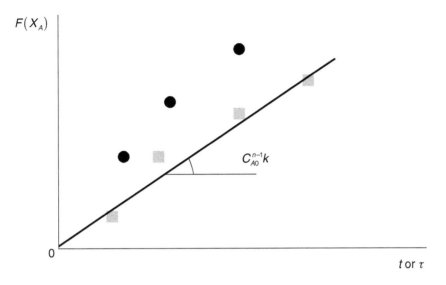

Figure 5.2 Representation of Equation 5.4.

SPECIFIC CASES

5.1 IRREVERSIBLE REACTION AT CONSTANT VOLUME

Reaction of the type: $A \rightarrow$ products

5.1.1 Kinetic model of first order

$$(-r_A) = kC_A = kC_{A0}(1 - X_A)$$

The continuous tubular system is chosen and substituting r_j by $(-r_A)$ in Equation 5.3, we have:

$$\tau = C_{A0} \int_0^{X_A} \frac{\mathrm{d}X_A}{kC_{A0}(1 - X_A)}$$

If the temperature is constant, we obtain:

$$-\ln(1 - X_A) = k\tau \tag{5.5}$$

For a batch system, the same solution is obtained:

$$-\ln(1 - X_A) = k\tau \tag{5.5a}$$

5.1.2 Kinetic model of second order (global)

This model represents the reaction of the type:

Case (a) $2A \rightarrow$ products
Case (b) $A + B \rightarrow$ products

The corresponding rates, according to Equation 1.5a:

For irreversible reactions of second order:

$$\Rightarrow a' = 1, b' = 1 \Rightarrow (-r_A) = kC_A C_B = kC_{A0}^2(1 - X_A)(M - X_A)$$
$$\Rightarrow a' = 2, b' = 0 \Rightarrow (-r_A) = kC_A^2 = kC_{A0}^2(1 - X_A)^2$$

In this case, we have a reaction which is first order in respect to each component and presents global second order. We assume that the stoichiometric coefficients are also equal, which characterizes an elementary reaction. However, the coefficients can be different, as for example:

Case (c) $A + 2B \rightarrow$ products

With respect to the component B, the reaction is first order ($b' = 1$). Thus:

$$\Rightarrow (-r_A) = kC_{A0}^2(1 - X_A)(M - 2X_A)$$

Substituting the expressions of the rates in Equation 5.3 for a continuous tubular system, we obtain the following solution (case (b)):

$$\frac{1}{(M - 1)} \ln \frac{(M - X_A)}{M(1 - X_A)} = kC_{A0}\tau \tag{5.6}$$

Or for a batch system (Equation 5.2):

$$\frac{1}{(M - 1)} \ln \frac{(M - X_A)}{M(1 - X_A)} = kC_{A0}t \tag{5.6a}$$

where:

$$M = \frac{C_{B0}}{C_{A0}}, \text{ when } A \text{ is the limiting reactant}$$

For a continuous stirred tank system, the solution will be different, because when substituting the rate $(-r_A)$ in Equation 4.10, we obtain:

$$\frac{X_A}{(1 - X_A)(M - X_A)} = \tau k C_{A0} \tag{5.7}$$

Therefore, the choice of the reactor is very important in order to determine the kinetic parameters and use the specific kinetic expressions.

If we have an irreversible reaction of second order in which the initial concentrations are equal (thus, $M = 1$), we cannot simplify Equation 5.6, because it is undetermined. We should then start from another kinetic model, i.e.:

$$(-r_A) = kC_A^2 = kC_{A0}^2(1 - X_A)^2$$

The solutions for the batch or tubular tank are, respectively:

(tubular or batch)

$$\frac{X_A}{(1 - X_A)} = kC_{A0}\tau \quad \text{or} \quad \frac{X_A}{(1 - X_A)} = kC_{A0}t \quad \text{or} \tag{5.8}$$

(tank)

$$\frac{X_A}{(1 - X_A)^2} = kC_{A0}\tau \tag{5.8a}$$

Example

***

E5.1 A reaction $A \rightarrow 4R$ is carried out in a batch reactor. A is introduced with 50% (vol) of inert.

The reaction is irreversible and the reaction order is integer. However, it is not known if the reaction is first or second order. Do a test and show the difference. The reaction is carried out at constant temperature at 27°C. The final pressure measured was 10 atm, which remained constant. It was observed that after 8 min the total pressure was 8 atm.

Solution

It is observed that the final pressure of 10 atm was reached when all the reactant was transformed into product. However, the initial total pressure is unknown, and 50% corresponds to the inert. Therefore, by applying the law of partial pressures, we have:

$$p_A = p_{A0} - \frac{a}{\Delta v}(P - P_0)$$

In this case, $p_A = 0$, $p_{A0} = 0.5$, and $P_0 = 1/3$. Thus, when $P = 10$ atm, we have $P_0 = 4$ atm.

With these values, we can calculate the conversion after 8 min, since:

$$X_A = \frac{p_{A0} - p_A}{p_{A0}} = \frac{1}{3}\frac{(P - P_0)}{0.5P_0} = 0.67$$

Considering a first-order reaction: Equation 5.5a:

$$-\ln(1 - X_A) = kt \tag{5.6a}$$

Thus:

$$k = 0.11 \text{ min}^{-1}$$

Considering a second-order reaction: Equation 5.8:

$$\frac{X_A}{(1 - X_A)} = kC_{A0}t$$

Since:

$$C_{A0} = \frac{p_{A0}}{RT} = \frac{0.5P_0}{0.082 \times 330} = 7.39 \times 10^{-2} \text{ mol/L}$$

we obtain:

$$k = 2.74 \text{ L/mol min}$$

Note that values of k are different. It is therefore necessary to verify experimentally and observe if the value of k is constant. Since in the present case there is only one experimental data, it is not possible to conclude what is the true order of the reaction and the reaction rate constant.

5.2 IRREVERSIBLE REACTIONS AT VARIABLE VOLUME

The typical reaction at variable volume can be represented as follows:

Case (a) $A \rightarrow R + S$ $\Rightarrow$ with volume expansion
Case (b) $A + B \rightarrow R$ $\Rightarrow$ with volume contraction
Case (c) $2A \rightarrow R$ $\Rightarrow$ with volume contraction

5.2.1 Irreversible of first order

Consider initially a *first-order reaction*. From the expression of the rate (3.9), when $a' = 1$, $b' = 0$ and $n = 1$, together with Equation 1.17. Thus, for a constant T, we have:

$$(-r_A) = kC_{A0} \frac{(1 - X_A)}{(1 + \varepsilon_A X_A)}$$

For a tubular reactor (Equation 5.3), the expression of the rate is substituted, and after the integration, we obtain the following equation:

$$-(1 + \varepsilon_A) \ln(1 - X_A) - \varepsilon_A X_A = k\tau \tag{5.9}$$

This solution is different from that of Equation 5.5 for a system at constant volume. The solution is the same when $\varepsilon_A = 0$.

For a batch reactor, the expression change, since we start from the expression at variable volume (Equation 4.6), i.e.:

$$t = n_{A0} \int_0^{X_A} \frac{dX_A}{V(-r_A)} \qquad (4.6)$$

For gas-phase reactions, such as in a plug-flow batch reactor at constant pressure, the volume varies according:

$$V = V_0(1 + \varepsilon_A X_A)$$

Substituting the rate and volume expressions, we obtain:

$$-\ln(1 - X_A) = kt \qquad (5.10)$$

This expression is equal to the batch or tubular systems at constant volume. Therefore, if the volume varies, this variation will be:

$$\Delta V = V - V_0 = V_0 \varepsilon_A X_A$$

Thus:

$$X_A = \frac{\Delta V}{V_0 \varepsilon_A}$$

Substituting in Equation 5.10, we obtain:

$$-\ln\left(1 - \frac{\Delta V}{V_0 \varepsilon_A}\right) = kt \qquad (5.11)$$

5.2.2 Irreversible reactions of second order

The reactions are represented by the examples *b* and *c*. According to Equation 3.27:

Case (a)

$$(-r_A) = kC_{A0}^2 \frac{(1 - X_A)(M - X_A)}{(1 + \varepsilon_A X_A)^2}$$

When $M = 1$, we can simplify for the case (b).

In isothermal conditions for an elementary reaction $(a = b = 1)$, we proceed in the same way, substituting the expression of rate in Equation 5.3 for a tubular reactor:

$$\tau = C_{A0} \int_0^{X_A} \frac{dX_A}{r} = \int_0^{X_A} \frac{(1 + \varepsilon_A X_A)^2 \, dX_A}{kC_{A0}(1 - X_A)(M - X_A)} dX_A$$

We obtain the following solutions:

$$\left[\frac{(1+\varepsilon_A M)^2}{(M-1)}\right]\ln\frac{(M-X_A)}{M} - \frac{(1+\varepsilon_A)^2}{(M-1)}\ln(1-X_A) + \varepsilon_A^2 X_A = \tau k C_{A0} \qquad (5.12)$$

and when $M=1$:

$$(1+\varepsilon_A)^2\frac{X_A}{(1-X_A)} + \varepsilon_A^2 X_A + 2\varepsilon_A(1+\varepsilon_A)\ln(1-X_A) = \tau k C_{A0} \qquad (5.13)$$

Example

E5.2 A reaction $A \rightarrow 2R + S$ is carried out in a tubular reactor, and the reactant A is introduced with 30% of inert. The reaction is irreversible and of second order. The reactor of 0.2 L is isothermal, and the reaction occurs at 800 K and pressure of 10 atm. It is known that the outflow of the product R is 0.034 mol/s and the conversion was 10%. Calculate the reaction rate constant. If this reaction would be further carried out in a batch reactor, calculate the reaction time for the previous conditions.

Solution

Since we have a second-order irreversible reaction, the rate is:

$$(-r_A) = k C_A^2$$

and consequently as a function of the conversion we have:

$$(-r_A) = k C_{A0}^2 \frac{(1-X_A)^2}{(1+\varepsilon_A X_A)^2}$$

Since the reactor is tubular, we have:

$$\tau C_{A0} = \int_0^{X_A} \frac{(1+\varepsilon_A X_A)^2\, \mathrm{d}X_A}{k(1-X_A)^2}\mathrm{d}X_A$$

whose solution was shown in Equation 5.13:

$$(1+\varepsilon_A)^2\frac{X_A}{(1-X_A)} + \varepsilon_A^2 X_A + 2\varepsilon_A(1+\varepsilon_A)\ln(1-X_A) = \tau k C_{A0}$$

We can calculate the conversion through the outflow of the product R:

$$\frac{F_{A0} - F_A}{F_{A0}} = \frac{F_R - F_{R0}}{2F_{A0}} = \frac{F_S - F_{S0}}{F_{A0}}$$

Where $F_{R0} = F_{S0} = 0$

Therefore:

$$F_R = 2F_{A0} \cdot X_A$$

$$0.034 = 2F_{A0} \times 0.1$$

$$F_{A0} = 0.170 \, \text{mol/s}$$

But:

$$\tau = \frac{V}{v_0} = \frac{VC_{A0}}{v_0 C_{A0}} = \frac{VC_{A0}}{F_{A0}} = \frac{0.2 \times 1.06 \times 10^1}{0.170} = 0.124 \, \text{s}^{-1}$$

Calculation of ε_A:

	A	2R	S	INERT	TOTAL
Initial	0.7	0	0	0.3	1.0
Final	0	1.4	0.7	0.3	2.4

$$\varepsilon_A = \frac{2.4 - 1}{1} = 1.4$$

Substituting in Equation (a) with $X_A = 0.1$, we have:

$$= 9.73 \text{L/mol} \cdot \text{s}$$

(b) If the reaction is conducted in a batch system at constant volume, we have that $\varepsilon_A = 0$.

Thus, we calculate the time necessary to reach the same previous conversions, by using the expression for the batch system, i.e., Equation 5.8:

$$\frac{X_A}{(1 - X_A)} = kC_{A0}t$$

Thus,

$$t = 0.107 \, \text{s}$$

When comparing with $\tau = 0.124 \, \text{s}^{-1}$, we can observe a difference due to the expansion of the gas-phase continuous system.

5.3 IRREVERSIBLE REACTIONS OF ORDER n–HALF-LIFE METHOD

There are processes with reactions that is not well defined and with several components. Several elementary or intermediate reactions can occur, which cannot be classified in the simple models of integer order. In the pyrolysis of naphtha or bituminous compounds, it is known that some major component is transformed, with the formation of several nonidentified products simultaneously. A main component is chosen as a reference for the kinetic study and we assume a global order n. Schematically, we have:

$$aA + bB \rightarrow \text{products}$$

or

$$aA \rightarrow R + S + T$$

The rate is generically represented in relation to the principal component A:

$$(-r_A) = kC_A^n \tag{3.17}$$

in which k is the apparent constant and n is the apparent global order.

In a batch reactor, the volume is constant; therefore, starting from Equation 5.2 as a function of the concentration:

$$t = C_{A0} \int_0^{X_A} \frac{dX_A}{(-r_A)} = -\int_{C_{A0}}^{C_A} \frac{dC_A}{kC_A^n}$$

Integrating, we obtain:

$$C_A^{1-n} - C_{A0}^{1-n} = (n-1)kt \tag{5.14}$$

Or as a function of the conversion of A, we obtain:

$$C_A^{1-n}[(1-X_A)^{1-n} - 1] = (n-1)kt \tag{5.15}$$

Note that there are two unknowns, k and n, and hence the solution must be iterative. In general, we simplify by choosing a concentration and the corresponding time. It is difficult to follow reactions which have too low or too high rates, since they take too long or too short to occur, respectively.

In this method, it is proposed to set a time and determine the concentration at that moment. However, the experiment also changes, since it is necessary to start it with a new initial concentration. The concentration is then measured as a function of time for different initial concentrations and for a certain time t.

Usually, we adopt the half-life criterion, considering the half-life time ($t_{1/2}$), which corresponds to a conversion of 50% or half of the initial concentration (Figure 5.3).

Note that it is possible to choose any conversion and determine the corresponding time. As an example, for $X_A = 0.7$, the corresponding time is $t_{0.7}$.

In the case of half-life, Equation 5.15 becomes:

$$C_{A0}^{1-n}[2^{(n-1)} - 1] = (n-1)kt_{1/2} \tag{5.16}$$

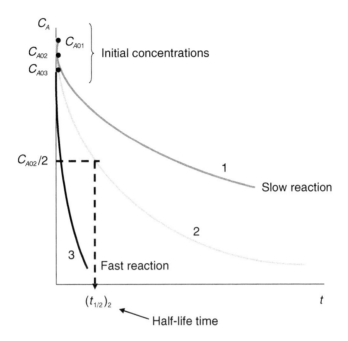

Figure 5.3 Half-life time.

There are special cases in which the specific rate can be determined directly through a single measurement, knowing the initial concentration and the half-time life. Exact solutions are obtained. However, when the reaction is of first order $(n=1)$, Equation 5.16 is undetermined. So, we start from Equation 5.5, making $X_A = 0.5$. For a reaction of zero order, in which the rate is independent of the concentration, the expression above is simplified.

Therefore:

$$\text{For } n=0: \qquad k = \frac{C_{A0}}{2t_{1/2}} \qquad\qquad (5.17)$$

$$\text{For } n=1: \qquad k = \frac{0.693}{t_{1/2}} \qquad\qquad (5.18)$$

For generic solutions, Equation 5.16 is rearranged and natural log is taken in both sides of the equation. Therefore:

$$t_{1/2} = \frac{[2^{(n-1)} - 1]}{(n-1)k} C_{A0}^{1-n} \qquad\qquad (5.19)$$

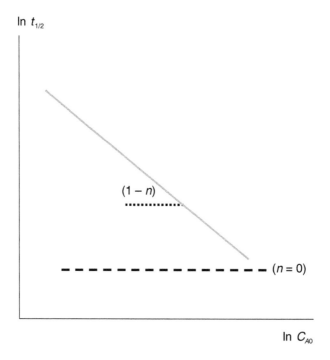

$Figure\ 5.4$ Kinetics of half-life.

Thus:

$$\ln t_{1/2} = \ln k^* + (n - 1)\ln C_{A0} \tag{5.20}$$

where:

$$k^* = \frac{[2^{(n-1)} - 1]}{(n - 1)k} \tag{5.21}$$

Graphically, we have the following representation as shown in Figure 5.4.

Example

E5.3 A reaction $A \rightarrow R + S$ is carried out in a batch reactor (*constant volume*). The experiment was started with several initial concentrations and when half of the concentration was reached, the corresponding time was measured. The following table gives the values for two different temperatures. Determine the order of the reaction, the specific rate, and the activation energy.

$$T = 100°C$$

100°C	C_{A0} (mol/L)	0.0250	0.0133	0.010	0.05	0.075
	$t_{1/2}$ (min)	4.1	7.7	9.8	1.96	1.30
110°C	C_{A0} (mol/L)	0.025				
	$t_{1/2}$ (min)	2.0				

Solution

Look at the table and note that the reaction occurs in the gas phase and an expansion occurs. However, in a batch reactor, the volume is constant. We can directly apply Equations 5.19 and 5.20. Note that only one value was provided for the temperature of 110°C. The order n is constant and it will be determined with the first data set. Therefore, we can obtain the order n and the constant k through Equation 5.20, i.e.:

$$\ln t_{1/2} = \ln k^* + (n-1) \ln C_{A0}$$

C_{A0} (mol/L)	$t_{1/2}$ (min)	$\ln C_{A0}$	$\ln t_{1/2}$
0.0250	4.1	−3.59	1.41
0.0133	7.7	−4.32	2.04
0.0100	9.8	−4.06	2.28
0.0500	1.96	−3.00	0.67
0.0750	1.30	−2.59	0.26

It can be solved graphically or we can find a correlation:

$$Y = -2.77 - 1.16X$$

Therefore:

$$(1 - n) = -1.16$$
$$n = 2.16$$

with a mean standard deviation of 0.99.
Thus:

$$(-r_A) = k^* C_A^{2.16}$$

The constant k^* can be calculated by substituting the value of n in the equation above and by using an experimental value, for example, the first value of the table: $C_{A0} = 0.025$ when $t_{1/2} = 4.1$.
We obtain from equation 5.21:

$$k = 16.5 \, \text{min}^{-1}$$

The solution is simplified to an exact solution (considering $n = 2$), and comparing with the previous case. In this case, we start from the Equation 5.8 (for a batch system):

$$\frac{X_A}{(1 - X_A)} = kC_{A0}t$$

with $X_A = 0.5$ and $t = t_{1/2}$, we obtain:

$$k = \frac{1}{C_{A0}t_{1/2}} \quad (A) \tag{5.22}$$

Thus:

$$k_{100} = 9.76 \, \text{L mol}^{-1} \, \text{min}^{-1}$$

error is 17%, and, therefore, not neglegible.

With this consideration, we may calculate the value of k, assuming a second-order reaction, and using the values of the table for 110°C in Equation 5.22. We then obtain for $C_{A0} = 0.025$ and $t_{1/2} = 2.0$:

$$k_{110} = 16.0 \, \text{L min}^{-1}$$

Note that the value of k is doubled. Moreover, we can see from the table that, or the same initial concentration C_{A0}, the half-life time at 110°C was exactly half of the value at 100°C. Consequently, the constant k is expected to double. Note that with a variation of 10°C, the constant doubled, indicating a significant effect of temperature on the reaction rate.

We can now calculate the constant k for any temperature, determining the activation energy E and the constant k_0 of the Arrhenius equation (Equation 3.30):

$$k = k_0 \exp(-E/RT)$$

Taking the natural log on both sides of the equation, we have (Equation 3.37):

$$\ln k = \ln k_0 - \frac{E}{RT}$$

Filling up the table with the correct data set:

k	T (K)	ln k	1/T
10	373	2.32	2.68×10^{-3}
20	383	3.00	2.61×10^{-3}

Thus:

$$Y = 28.2 - 9.657X$$

Therefore:

$$\frac{E}{R} = 9657$$

With $R = 1.98\, cal/mol$
$E = 19.120\, cal/mol$

5.4   REVERSIBLE REACTIONS AT CONSTANT VOLUME

In the reversible reactions at constant volume, we have to determine two reaction rate constants, i.e., the direct constant (k) and the reverse constant (k'), which are temperature dependent and the units depend on the order of reaction of the reactants or products. The determination of these constants follows the same procedure. We will use here the two simplest cases, i.e., the integer order and constant volume.

5.4.1   Direct and reverse first-order elementary reaction

$A \Leftrightarrow R$

The resulting rate is represented by:

$$r = k\left[C_A - \frac{1}{K}C_R\right]$$

Substituting the concentration by the conversion of the reactant A, according to the set of Equations 1.8:

$$C_A = C_{A0}(1 - X_A)$$

$$C_B = C_{A0}(M - X_A)$$

$$C_R = C_{A0}(R + X_A)$$

we have:

$$r = kC_{A0}\left[(1 - X_A) - \frac{1}{K}(R + X_A)\right] \qquad (3.19)$$

where:

$$\frac{k}{k'} = \frac{(R + X_{Ae})}{(1 - X_{Ae})} = K \rightarrow \text{equilibrium constant} \qquad (3.20)$$

and X_{Ae} is the equilibrium conversion.

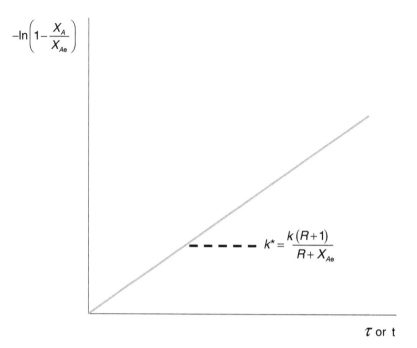

Figure 5.5 Kinetics of a reversible reaction.

Substituting the constant K of Equation 3.20 into Equation 3.19, we have an equation as a function of X_A, X_{Ae}:

$$r = \frac{kC_{A0}(R+1)}{(R+X_{Ae})}(X_{Ae} - X_A) \tag{3.21}$$

We choose the batch or continuous system, whose solution is according to Equation 5.2 or 5.3:

$$\tau(t) = C_{A0} \int_0^{X_A} \frac{dX_A}{r} \tag{5.23}$$

Substituting Equation 5.23 into Equation 5.3 and integrating, we obtain:

$$-\ln\left(1 - \frac{X_A}{X_{Ae}}\right) = \frac{k(1+R)}{(R+X_{Ae})}\tau \text{ or } (t) \tag{5.24}$$

Or graphically:

Note that $k^* = k + k'$ and can be determined through experimental results (graph), by measuring the conversion as a function of the time t or the space time space τ. Knowing the equilibrium constant or the equilibrium conversion, we can determine k and k' separately.

5.4.2 Direct and reverse second-order elementary reaction

$A + B \Leftrightarrow 2R$

The rate will be (Equation 3.4):

$$r = k\left[C_A C_B - \frac{1}{K}C_R^2\right] \tag{5.25}$$

It is assumed that the reactants A and B enter the system as pure components and therefore $C_{r0} = 0$. Furthermore, it is assumed that the reactants enter the system with equal initial concentrations. Therefore, $C_{A0} = C_{B0}$ $(M = 1)$.

By using the concentration of the reactants and products as a function of the conversion (Equation 5.25), this equation becomes:

$$r = kC_{A0}^2\left[(1 - X_A)^2 - \frac{4}{K}X_A^2\right] \tag{5.26}$$

Knowing that: $C_R = 2C_{A0}X_A$

In the equilibrium, the resulting rate r is zero; therefore:

$$(1 - X_{Ae})^2 \times K - 4X_{Ae}^2 = 0$$

Knowing that and solving the quadratic equation, i.e.:

$$(K - 4)X_{Ae}^2 - 2KX_{Ae} + K = 0$$

Two roots are obtained:

$$X_{Ae} + X'_{Ae} = \frac{2K}{K - 4} = -\frac{b}{a}$$

$$X_{Ae} \times X'_{Ae} = \frac{K}{K - 4} = \frac{c}{a}$$

The roots are X_{Ae} (real root) and X'_{Ae} (not real). There is only one real value of equilibrium X_{Ae} ranging from 0 to $X_{Ae} < 1$.

Transforming Equation 5.24, we get:

$$r = \frac{kC_{A0}^2(K - 4)}{K}\left[X_A^2 - \frac{2K}{(K - 4)}X_A + \frac{K}{(K - 4)}\right]$$

$$\Downarrow \qquad\qquad \Downarrow$$

$$(X_{Ae} + X'_{Ae})\ (X_{Ae} \times X'_{Ae})$$

Therefore, the resulting rate as a function of the conversions will be:

$$r = \frac{kC_{A0}^2(K - 4)}{K}\left[X_A^2 - \frac{2K}{(K - 4)}X_A + \frac{K}{(K - 4)}\right]$$

$$r = \frac{kC_{A0}^2 (K-4)}{K}[(X_{Ae} - X_A).(X'_{Ae} - X_A)] \tag{5.27}$$

Choosing the batch or continuous system and substituting Equation 5.27 into Equation 5.3, it is possible to analytically integrate to give the following final expression:

$$\ln\left[\left(\frac{1}{\kappa}\right)\frac{\left(\kappa - \frac{X_A}{X_{Ae}}\right)}{\left(1 - \frac{X_A}{X_{Ae}}\right)}\right] = k^*\tau \tag{5.28}$$

(or in the batch system (t))
where

$$\kappa = \frac{X'_{Ae}}{X_{Ae}}$$

and

$$k^* = k(\kappa - 1)\frac{(K-4)}{K}C_{A0}X_{Ae} \tag{5.29}$$

Similarly, the graphical solution of Equation 5.28 is identical to the previous one.

Knowing k^* from the experimental values of the conversions X_A as a function of the time τ (continuous) or t (batch) and values of the equilibrium conversion X_{Ae} or equilibrium constant K, we can separately determine the specific rates k and k'.

Note that Equation 5.28 is also valid for other reversible reactions and may have a different order in the direct and reverse direction. The specific rates and consequently, the value of the k^* from Equation 5.29 and the thermodynamic constant K are the parameters that change. Thus, for example, if the reaction is of the type:

$$2A \Leftrightarrow 2R$$

Consequently, the resulting rate will be:

$$r = kC_{A0}^2\left[(1 - X_A)^2 - \frac{1}{K}X_A^2\right]$$

With the following parameters:

$$K = \frac{k}{k'} = \frac{(1 + X_{Ae})^2}{(1 - X_{Ae})^2}$$

and

$$k^* = k\frac{(K-1)}{K}$$

Example

E5.4 The reaction $A\ R$ will be carried out in a tubular reactor (PFR) at 87°C. Pure reactant is introduced isothermally at 1.0 L/min and 10 atm. However, the specific rates are unknown. Therefore, experiments were separately conducted in batch reactor. Pure reactant was introduced at 1 atm and 27°C. It was observed that after 100 min, the conversion was 40%. When the conversion reached 90%, no change was observed anymore. In another experiment at 107°C, the same conversion of 40% was obtained, but in less time (20 min). After reaching a conversion of 60%, there was no more change. Determine the volume of the PFR under the specified conditions, knowing that the outflow of the product R was 0.20 mol/min.

Solution

(a) Determination of the direct (k) and reverse (k') specific rates.
   For a reversible reaction that is first order in both directions, we have:

$$r = kC_A - k'C_R$$

The specific rates are k and k'. The batch reactor was chosen in order to determine these constants. Therefore, substituting the rate in the equation of the batch reactor and integrating, we obtain the expression already deduces previously (Equation 5.24):

$$-\ln\left(1 - \frac{X_A}{X_{Ae}}\right) = \frac{k(1+R)}{(R+X_{Ae})}t \tag{5.30}$$

In this case, since the reactant is only present in the beginning, $R = 0$. Thus, if the equilibrium conversion $X_{Ae} = 0.90$, and the conversion $X_A = 0.40$ after 100 min, we can determine the constant k:

$$K_{27} = 5.29 \times 10^{-3} \text{ min}^{-1}$$

In the equilibrium, the resulting rate is zero and the constant K can be determined. Therefore, since the equilibrium conversion $X_{Ae} = 0.90$, we have:

$$K = \frac{k}{k'} = \frac{X_{Ae}^2}{(1 - X_{Ae})} = 9$$

Thus, the reverse specific rate will be:

$$k'_{27} = 5.87 \times 10^{-4} \text{ min}^{-1}$$

The specific rates were calculated for a temperature of 27°C (300 K). In order to know the specific rate at 87°C, we need to calculate the activation energy E. For this, we use the data at 107°C. By using the same equations described above, we have:

$$k_{107} = 3.29 \times 10^{-2} \text{ min}^{-1}$$

$$k'_{107} = 2.19 \times 10^{-2} \text{ min}^{-1}$$

We then calculate the activation energy, using the two values of k and of k':

$$E = R \frac{\ln k_{107}/k_{27}}{(1/T_{27} - 1/T_{107})}$$

and analogously the reverse activation energy E'. Note that the temperatures are given in kelvin.

Therefore:

$$E = 5.135 \, \text{cal/mol}$$

$$E' = 10.211 \, \text{cal/mol}$$

These data show that the direct rate is greater than the reverse rate, since K is higher. Furthermore, since the reverse activation energy is greater than the direct one, the barrier energy of the reverse reaction is higher than that of the direct reaction.

From the Arrhenius equation, we have to determine the frequency factor k_0, which are independent of the temperature. Thus:

$$k = k_0 \exp(-E/RT)$$

For $T = 300 \, \text{K}$, $k_{27} = 5.29 \times 10^{-3} \, \text{min}^{-1}$. Thus:

$$k_0 = 30.0 \, \text{min}^{-1}$$

The same procedure is valid for determining the constant $k'_0 = 1.71 \times 10^4 \, \text{min}^{-1}$. With these values, we have the following equations valid for any temperature:

$$k = 30 e^{-5135/RT}$$

$$k' = 1.71 \times 10^4 e^{-10211/RT}$$

We finally can calculate the specific rate at $87°C$. Substituting the values, we have:

$$k_{87} = 2.25 \times 10^{-2} \, \text{min}^{-1}$$

$$k'_{87} = 1.03 \times 10^{-2} \, \text{min}^{-1}$$

The equilibrium conversion can be determined:

$$K_{87} = \frac{k_{87}}{k'_{87}} = 2.16 = \frac{X_{Ae}}{(1 - X_{Ae})}$$

Thus:

$$X_{Ae} = 0.684$$

Calculation of the final conversion in the PFR.

If we have the molar flow in the reactor outlet, we can calculate the conversion:

$$F_R = 20\,\text{mol/min}$$

$$F_R = C_R v_0 = C_{A0} v_0 X_A$$

The initial concentration can be calculated through the general gas law:

$$C_{A0} = \frac{P}{RT} = \frac{10}{0.082 \times 360} = 3.38 \times 10^{-1}\,\text{moles/L}$$

Thus:

$$F_R = C_{A0} v_0 X_A = 3.38 \times 10^{-1} \times 1.0 \times X_A = 0.20$$

$$X_A = 0.59$$

Substituting these values in Equation 5.24, we obtain:

$$\tau = 60\ \text{min}$$

And the volume of the reactor will be:

$$V = 60\,\text{L}$$

Example

E5.5 The reversible second-order (direct and reverse) reaction $2C_2H_5NCO \rightarrow C_2H_5N_2$ $(C=O)C_2H_5$ is conducted in a 5 mL continuous reactor (PFR) at 25°C. Under these conditions, the equilibrium constant is 0.125. Pure reactant with a concentration of 0.2 mol/L is introduced at 0.36 L/h and the final conversion is 70% of the equilibrium conversion. Calculate the direct and reverse specific rate.

Solution

It is known that the reaction is of the type:

$$2A \leftrightarrow 2R$$

and consequently the resulting rate is:

$$r = kC_{A0}^2\left[(1 - X_A)^2 - \frac{1}{4K}X_A^2\right]$$

In the equilibrium, the resulting rate is zero. Therefore:

$$K = \frac{k}{k'} = \frac{X_{Ae}^2}{4(1 - X_{Ae})^2} = 0.125$$

or

$$(4K - 1)X_{Ae}^2 - 8KX_{Ae} + 4K = 0$$

$$X_{Ae} = 0.414$$

$$X'_{Ae} = -2.41$$

Thus, the resulting rate as a function as of the conversions will be:

$$r = \frac{kC_{A0}^2(4K - 1)}{K}\left[\left(1 - \frac{X_A}{X_{Ae}}\right)\left(1 - \frac{X_A}{X'_{Ae}}\right)\right] \tag{5.31}$$

Substituting the expression of the rate in the equation of PFR and integrating, we obtain Equation 5.28, i.e.:

$$\ln\left[\left(\frac{1}{\kappa}\right)\frac{\left(\kappa - \frac{X_A}{X_{Ae}}\right)}{\left(1 - \frac{X_A}{X_{Ae}}\right)}\right] = k^*\tau(t) \tag{5.32}$$

where:

$$k^* = k(\kappa - 1)\frac{(4K - 1)}{4K}C_{A0}X_{Ae} = 0.956$$

where:

$$\kappa = \frac{X'_{Ae}}{X_{Ae}} = -5.83$$

$$X_A = 0.7 \times X_{Ae}$$

From this, we get:

$$k\tau = 1.432$$

Since,

$$\tau = \frac{V}{v_0} = 0.83 \text{ min}$$

Thus:

$$k = 1.719 \text{ L/mol min}$$

$$k' = 13.7 \text{ L/mol min}$$

5.5 DETERMINATION OF THE KINETIC PARAMETERS BY THE DIFFERENTIAL METHOD

In the differential method, we have two approaches:

1. Using the concentration data as a function of time obtained experimentally, the rates are determined by approximation from the kinetic curves.
2. Determining the rates directly, using the data from a differential reactor.

Therefore, for a reaction of the type:

$$A \rightarrow R + S + T$$

The rate can also be generically represented as a function of the principal component A, with the general order:

$$(-r_A) = k^* C_A^n$$

where k^* is the apparent constant and n is the global order.

In a system at constant volume, the rate will be:

$$(-r_A) = -\frac{dC_A}{dt}$$

Thus, taking the natural log on both sides of the equation, we have:

$$\ln(-r_A) = \ln\left(-\frac{dC_A}{dt}\right) = \ln k^* + n \ln C_A \tag{5.33}$$

From the kinetic curve (Figure 5.6A) obtained from the experimental data, we have the corre3sponding rate for each concentration (tangent of the curve). We can then calculate the valued of the table and the graph corresponding to Equation 5.33.

Time (min)	Concentration (mol/L)	Rate (mol/L min)
$t = 0$	C_{A0}	
$t = t_1$	C_{A1}	
$t \neq 0$	C_A	

With the experimental data shown in Figure 5.6 and with the table, we can obtain the graph displayed in Figure 5.7 corresponding to Equation 5.33. Thus, we can directly determine the order of the reaction n and the apparent constant k^*.

When the order in relation to each component is different and not an integer, we use the same methodology. However, in this case, we would have to carry out different experiments.

When the reaction involves several components, we should determine the order of the reaction in relation to each component. This is only possible if the reaction can be

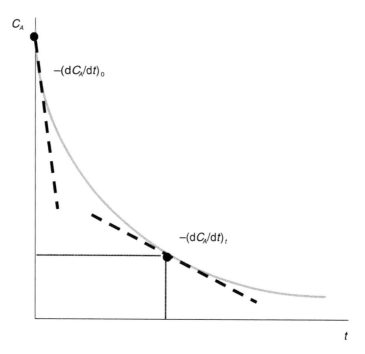

Figure 5.6a Kinetic curve.

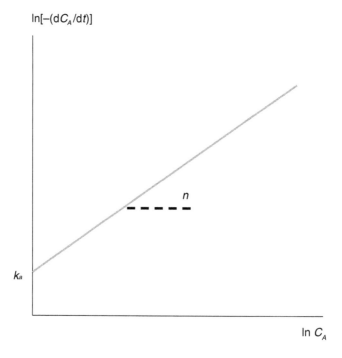

Figure 5.6b Differential method.

controlled in such a way that the concentration of only one component varies. There are three possibilities:

(a)   The concentration of the principal component varies, and the other components are not consumed during the reaction.
(b)   The concentration of the principal component is monitored, maintaining the concentrations of other components constant and adding reactant in a proportion that compensates its consumption during the reaction.
(c)   The concentration of the principal component is monitored, maintaining the concentrations of other components constant and introducing them in excess, relative to the concentration of the principal component. The variation in the concentration of the reactants in excess is negligible.
(d)   Both concentrations vary, and we should proceed in the same way, as shown in the following example. This is method of the partial rate.

For an irreversible reaction at constant volume of the type:

$$A + B \rightarrow \text{products}$$

The corresponding rate will be:

$$(-r_A) = k C_A^{a'} C_B^{b'}$$

Therefore, keeping the concentration of B constant, the expression can be simplified:

$$(-r_A) = k^* C_A^{a'}$$

where:

$$k^* = k C_B^{b'}$$

By proceeding analogously, we can calculate k^* and a'. Subsequently, we now keep the concentration of A constant. Thus, we obtain the new rate:

$$(-r_A) = k'' C_B^{b'}$$

where:

$$k'' = k^* C_A^{a'}$$

Analogously, we can determine the exponent b' and the constant k''. We have the following situations, as shown in Figure 5.7.
Finally, knowing the rate $(-r_A)$ for the defined concentrations and the orders a' and b', we can directly calculate the specific rate k^*, since:

$$k^* = \frac{(-r_A)}{C_A^{a'} C_B^{b'}} \tag{5.34}$$

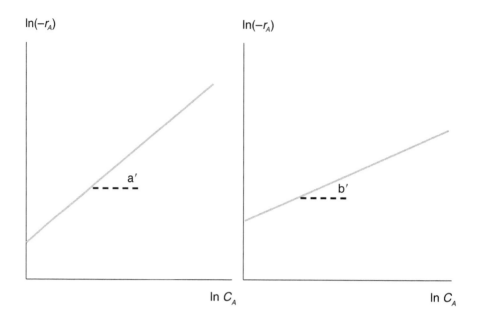

Figure 5.7 Determination of the partial orders.

The kinetic curves may have opposite behavior. When the reaction is fast, concentration drops sharply with time, whereas when the reaction is very slow the concentration drops slowly with the reaction time. In both cases, the rates obtained by the tangents to the curve are very imprecise.

Another way is to determine only the initial rates $(t = 0)$ and proceed in the same way. The downside is that several experiments should be carried out, starting from different initial concentrations. It is not necessary to stop the experiment. With the initial values, we can determine the kinetic curve and the reaction are interrupted. Furthermore, it is more accurate. Therefore, we use Equations 5.33 and 5.34, but with the initial values, i.e.:

$$\ln\left(-r_{A0}\right) = \ln\left(-\frac{dC_A}{dt}\right)_0 = \ln k^* + n \ln C_{A0} \tag{5.35}$$

and

$$k^* = \frac{(-r_{A0})}{C_{A0}^{a'} C_{B0}^{b'}} \tag{5.36}$$

These methods are not sufficiently accurate and thus the conventional numerical methods are used. From Equation 5.33 or 5.35, we have the polynomial equation of the type:

$$\ln(-r_A) = \ln\left(-\frac{dC_A}{dt}\right) = \ln k^* + n \ln C_A \tag{5.37}$$

or

$$Y = a_0 + a_1 x_1 + a_2 x_2$$

With an experiment j, we have therefore:

$$Y_j = a_{0j} + a_1 x_{1j} + a_2 x_{2j} \tag{5.38}$$

For a reaction system, we obtain:

$$\sum Y_j = N a_{0j} + a_1 \sum x_{1j} + a_2 \sum x_{2j}$$

$$\sum x_{1j} = a_0 \sum x_{1j} + a_1 \sum x_{1j}^2 + a_2 \sum x_{2j} x_{1j}$$

$$\sum x_{2j} = a_0 \sum x_{2j} + a_1 \sum x_{2j} x_{1j} + a_2 \sum x_{2j}^2$$

With three linear equations, the three unknowns a_0, a_1, and a_2 and consequently the specific rate k^* is solved.

Example

E5.6 In the formation of HBr, we obtain the following experimental results (adapted from Hill, 1977):

C_{H2O}	C_{Br2O}	$(-r_{HBr})_0 \times 10^3$ (mol/L min)
0.225	0.2250	1.76
0.90	0.90	10.9
0.675	0.675	8.19
0.450	0.450	4.465
0.5637	0.2947	4.82
0.2881	0.1517	1.65
0.3103	0.5064	3.28
0.1552	0.2554	1.267

Determine the order of the reaction in relation to each component (adapted from Fogler, 2000 and Hill, 1977).

Solution

We start from the rate of formation of HBr, i.e.:

$$r_{HBr} = k C_{H_2}^{a'} C_{Br}^{b'}$$

Considering $a' + b' = n$, we obtain:

$$(r_{HBr})_0 = k^* C_{H_2O}^n$$

where:

$$k^* = k \left(\frac{C_{Br0}}{C_{H_2O}} \right)^{b'}$$

Taking the natural log on both sides of the equation and constructing the graph (Figure 5.9), we obtain the reaction order n and the constant k_a.
Thus:

$$n = 1.28$$

$$k^* = 1.198 \times 10^{-2}$$

On the other hand, since

$$(r_{HBr})_0 = k \, C_{H2}^{a'} \, C_{Br}^{(n-a')}$$

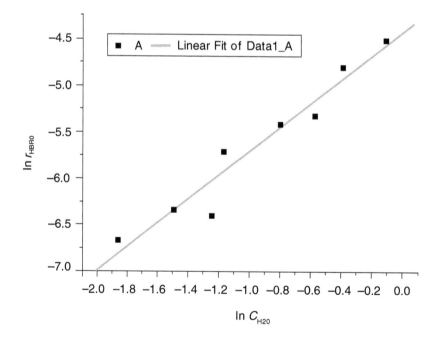

Figure 5.8 Differential method—initial rates.

or

$$(r_{\mathrm{HBr}})_0 = k C_{\mathrm{Br0}}^n \left[\frac{C_{\mathrm{H20}}}{C_{\mathrm{Br0}}} \right]^{a'}$$

Taking the natural log on both sides of the equation, we have:

$$\ln \frac{(r_{\mathrm{HBr}})_0}{C_{\mathrm{Br0}}^n} = \ln k + a' \ln \left[\frac{C_{\mathrm{H20}}}{C_{\mathrm{Br0}}} \right]$$

$\ln [C_{H20}/C_{Br20}]$	$C_{Br20}^{1.27}$	$\ln [(-r_{HBr})_0/C_{Br20}^{1.27}]$
0.0	0.150	−4.445
0.0	0.874	−4.38
0.0	0.6070	−4.30
0.0	0.362	−4.395
0.648	0.211	−3.779
0.641	0.0911	−4.011
−0.489	0.421	−4.85
−0499	0.1766	−4.937

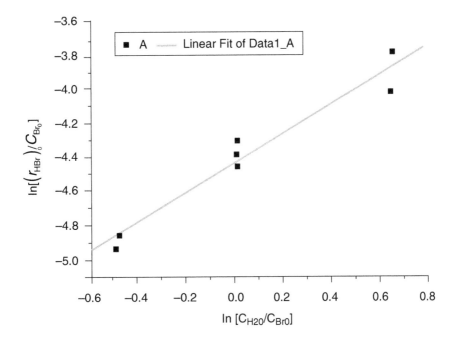

Figure 5.9 Determination of a' and b'.

Thus, from Figure 9 we determine the exponent $a' = 0.86$. Knowing n, we can determine $b' = 0.42$ and the constant:

$$k = 1.205 \times 10^{-2} \ [\text{L/mol}]^{1.27} \ \text{min}^{-1}$$

5.5.1 Differential reactor

To determine the kinetic parameters, we use the differential reactor. In this continuous flow system, the variation in concentration between the inlet and outlet of the reactor should be small and finite. The conversions should be around 5–10%. Under these conditions, the diffusive and mass transfer effects are avoided, assuring a kinetic regime for the determination of the kinetic parameters. Unlike the case of the batch system, the spatial time and consequently, the inlet flow and the mass or volume of the reactor are varied. Therefore, the reaction rate is directly determined.

In a continuous system, the differential molar balance in a finite volume $\cdot \Delta V$ will be as follows:

The molar flow in the reactor inlet and outlet will be F_{A0} and F_{As}, respectively. Therefore (Equation 4.13):

$$F_{A0} - F_{As} + r_A \Delta V = 0$$

Considering that $\Delta V = V$, we have:

$$(-r_A) = \frac{F_{A0} - F_{As}}{V} \qquad [\text{mol/L min}] \tag{5.39}$$

If the reaction takes place at constant volume, we can approximate the rate for concentration, i.e.:

$$(-r_A) = \frac{(C_{A0} - C_{As})v_0}{V} \tag{5.40}$$

where v_0 is the volumetric flow and V is the reactor volume.

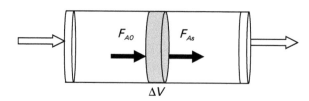

Figure 5.10 Differential reactor—initial rates.

Note that when the reactor volume is small, the space velocities are high, but the space time is very small, and consequently, the conversions are very low. Thus, the rate practically corresponds to the initial rate, i.e.:

$$(-r_A) = (-r_A)_0 \qquad (5.41)$$

This is the rate measured experimentally and to determine the reaction rate constants, we need a kinetic model, which may be of integer or fractional order, as has been done previously. The difference here is that the rate was obtained experimentally, thus avoiding problems of extrapolation and the use of approximation approaches in the determinations of the tangents to the kinetic curves seen in the differential method. Furthermore, complex integrations are avoided, as in the case of the integral method.

Taking the natural log on both sides of the equation, the parameters are determined according to Equation 5.35, i.e.:

$$\ln(-r_{A0}) = \ln k^* + n \ln C_{A0} \qquad (5.42)$$

Example

E5.7 The hydrogenation of octene to octane was performed in a reactor differential, by measuring the initial rates as a function of the respective concentrations of the reactant B. It is known that the reaction order of hydrogen is 1. However, the order in relation to octane is not known. The reaction is irreversible and the experiments were performed at 1 atm and 500°C. Under these conditions, the following experimental data was obtained:

$$C_8H_{16} + H_2 \rightarrow C_8H_{18}$$

$$C_8H_{16} \approx B$$

C_{B0}	mol/L	0.1	0.5	1.0	2.0	4.0
$(-r_B)_0 \times 10^2$	mol/L	0.073	0.70	1.84	4.86	12.84

Using equation:

$$\ln(-r_{B0}) = \ln k^* + n \ln C_{B0} \qquad (5.43)$$

and placing the values of the table above in the graph (Figure 5.11), we obtain:

The angular coefficient is $n = 1.4$, which is the reaction order with respect to octene. The linear coefficient would be the reaction rate constant. However, substituting $n = 1.4$ in the equation above and the values corresponding to a particular experimental data, the reaction rate constant is calculated. Therefore:

$$k = 1.844 \times 10^{-2} (\text{L/moles})^{2.4} \, \text{h}^{-1}$$

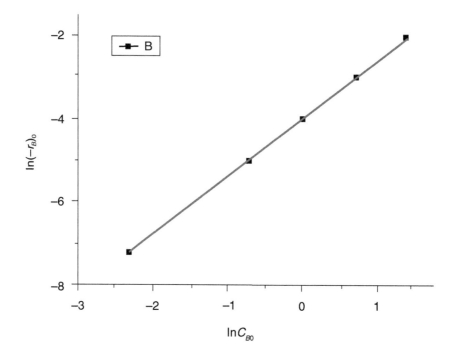

Figure 5.11 Determination of the specific rate.

Thus the rate will be:

$$(-r_A) = 1.844 \times 10^{-2} C_A^{1.4} C_{H_2}$$

5.6   UNCERTAINTIES OF KINETIC PARAMETERS

An important point to consider during the determination of kinetic parameters is the fact that experimental measurements can only be obtained with finite precision because of the unavoidable occurrence of experimental fluctuations, as discussed in Section 1.8 and Section 1.9. Consequently, as kinetic parameters must be obtained with help of uncertain experimental data, they are also subject to uncertainties and fluctuations.

For illustrative purposes, let us consider the first-order kinetic rate equation used to describe the batch reaction system presented in Example E5.1. Based on the proposed example:

$$k = -\frac{\ln (1 - X_A)}{t} \tag{5.44}$$

where

$$X_A = \frac{2}{3} \frac{P - P_0}{P_0} \tag{5.45}$$

Besides, it was assumed that experimental measurements were available, as $P_0 = 4$ atm and $P = 8$ atm after $t = 8$ min. However, the information presented in Example E5.1 was not complete, because the pressure and time measurement systems were certainly subject to uncertainties and the intrinsic precisions of the measuring devices were not reported. Thus, let us assume that the pressure and time measurements were subject to fluctuations that presented sample variances s_P^2 and s_t^2, respectively, as defined in Equation (1.9.1) of Section 1.9. Then, if one additionally assumes that measurement fluctuations were small (and they often are large!), it becomes possible to write:

$$dk = \frac{1}{(1 - X_A)t} dX_A + \frac{\ln(1 - X_A)}{t^2} dt = \frac{2}{(5P_0 - 2P)t}$$

$$(5.46)$$

$$dP - \frac{2P}{(5P_0 - 2P)t} \frac{dP_0}{P_0} + \frac{\ln\left(\frac{5P_0 - 2P}{3P_0}\right)}{t^2} dt$$

Equation (5.46) shows very clearly that uncertainties in k (dk) result from the intrinsic uncertainties of the measuring devices (dP, dP_0 and dt). The mathematical operation performed to obtain Equation (5.46) is the well-known error propagation procedure, valid when the fluctuations are small and can be somehow represented by the differential of the analyzed variable or expression. In other words, if k can be obtained as a complex mathematical transformation $f()$ of other variables $x_1, x_2, \ldots, x_{NX}$ in the form:

$$k = f(x_1 \cdots x_{NX})$$

$$(5.47)$$

then

$$dk = \sum_{i=1}^{NX} \frac{\partial f}{\partial x_i} dx_i$$

$$(5.48)$$

Finally, if it is assumed that the error measurements are independent and do not affect each other (and real measurements are frequently correlated!!!), then it becomes possible to write (Schwaab and Pinto, 2007):

$$s_k^2 = \sum_{i=1}^{NX} \left(\frac{\partial f}{\partial x_i}\right)^2 s_{x_i}^2$$

$$(5.49)$$

so that Equation (5.46) becomes:

$$s_k^2 = \left[\frac{2}{(5P_0 - 2P)t}\right]^2 s_P^2 + \left[\frac{2}{(5P_0 - 2P)t}\right]^2 s_{P_0}^2 + \left[\frac{\ln\left(\frac{5P_0 - 2P}{3P_0}\right)}{t^2}\right]^2 s_t^2$$

$$(5.50)$$

Now, using the available data and assuming that the sample standard deviations are equal to $s_P = 0.1$ atm and $s_t = 0.1$, Equation (5.50) becomes:

$$s_k^2 = \left[\frac{2}{(20-16)8}\right]^2 0.01 + \left[\frac{2}{(20-16)8} \cdot \frac{1}{4}\right]^2 0.01 + \left[\frac{\ln\left(\frac{20-16}{12}\right)}{8^2}\right]^2$$

(5.51)

$$0.01 = 1.98 \times 10^{-4} \text{ min}^{-2}$$

Therefore, in this particular case, $k = 0.11$ min^{-1} and $s_k^2 = 1.98 \times 10^{-4}$ min^{-2}, so that $s_k = 0.014$ min^{-1}. If it is assumed that the kinetic parameter follows the normal distribution (and model parameters mostly do not follow this particular distribution!), then it is possible to write Equation (1.9.9) in Section 1.9:

$$k = (0.137 \pm 2 * 0.014) \text{ min}^{-1} \text{ with 95\%confidence}$$
$$k = (0.137 \pm 2.5 * 0.014) \text{ min}^{-1} \text{ with 99\%confidence}$$

which shows that the uncertainty of k is in the second decimal place and is of order 0.03 min^{-1} with 95% confidence or 0.04 min^{-1} with 99% confidence. Therefore, it is more appropriate to describe the kinetic parameter in the form:

$$k = (0.14 \pm 0.03) \text{ with 95\%confidence}$$
$$k = (0.14 \pm 0.04) \text{ with 99\%confidence}$$

As one can see in Equation (5.49), the estimation of parameter uncertainties can be quite laborious, although very important for practical reasons, as these parameters are expected to be used for design of large reaction vessels. Therefore, the accuracy and precision of the reactor design depends on the quality of available kinetic parameters, justifying all efforts made to evaluate the precision of the obtained parameter. However, derivation and use of Equation (5.49) in real problems can be infeasible, given the large number of variables involved, the nature of the mathematical functions that transform the measured variables into the model parameters and the fact that measurements can be correlated to each other and sufficiently large to render the differential procedure useless. In these cases, the combination of parameter estimation procedures and stochastic simulation tools can provide much simpler numerical strategies for computation of parameter uncertainties, as described in detail in Chapter 11.

In order to illustrate how a stochastic procedure can be used to evaluate the precision of the calculated parameters, let us assume that Equations (5.44) and (5.45) should be combined to generate the kinetic parameter k, by assuming that experimental measurements are allowed to vary in the form:

$$P_0 = 4 + (random() - 0.5)\sqrt{12s_{P_0}^2} \text{ atm} \tag{5.52a}$$

$$P_0 = 8 + (random() - 0.5)\sqrt{12s_P^2} \text{ atm} \tag{5.52b}$$

$$t = 8 + (random() - 5)\sqrt{12st^2} \text{ min} \tag{5.52c}$$

where the function **random** of the Excel spreadsheet provides random numbers uniformly distributed in the interval $[0,1]$. The radicand term is used to correct the variance of the simulated data, as the variance of the uniform distribution is equal to $1/12$. The following procedure is then repeated 1000 times: (i) random numbers are generated; (ii) values of P_0, P and t are calculated; (iii) k is calculated and saved. Finally, the average value of k and respective sample variance are calculated as usual and described in Equations (1.9.1) and (1.9.2) of Section 1.9. These numerical tasks can be easily performed in a standard spreadsheet. The obtained results for one realization were $\bar{k} = 0.139\,\text{min}^{-1}$, $s_k^2 = 2.14 \times 10^{-4}\,\text{min}^{-2}$ and $s_k = 0.015\,\text{min}^{-1}$, which were very similar to the results obtained previously. Based on the uniform distribution and discarding the 2.5% (25) smallest values of k and 2.5% (25) largest values of k values of k, to respect the confidence level of 95%, it was possible to obtain:

$$0.114 < k = 0.139 < 0.167\,\text{min}^{-1} \text{ with 95\% confidence}$$

One must observe that the confidence interval of the kinetic parameter calculated with the stochastic procedure was not symmetrical, as predicted by the normal assumption, because of the many nonlinear transformations of the measured variables that were needed to generate the value of k. Based on the uniform distribution and discarding the 0.5% (5) smallest values of k and 0.5% (5) largest values of k to respect the confidence level of 99%, it was possible to obtain:

$$0.110 < k = 0.139 < 0.172\,\text{min}^{-1} \text{ with 99\% confidence}$$

As the uncertainty of the kinetic parameter is located in the second decimal place, it is better to write:

$$0.11 < k = 0.14 < 0.17\,\text{min}^{-1} \text{ with 95\% confidence}$$

$$0.11 < k = 0.14 < 0.17\,\text{min}^{-1} \text{ with 99\% confidence}$$

which are very similar to the ones calculated previously, although the confidence intervals were a bit larger when the normal assumption was used because the normal distribution admits that measurements can be infinitely large, whereas the uniform distribution admits the existence of well-defined boundaries for maximum allowed fluctuations.

5.7   REPARAMETERIZATION OF POWER-LAW RATE EQUATIONS

Existence of high correlations among the parameter estimates of a mathematical model can cause significant numerical problems during estimation of model parameters (as discussed in detail in Chapter 11), since minimization of the objective function may become inefficient (Espie and Macchietto, 1988) and the statistical characterization of the final parameter estimates may lack significance (Watts, 1994). For example, some of these difficulties can be observed during estimation of parameters of power-law rate equations, utilized throughout Chapter 5, because of the usually very high correlations among the obtained parameter estimates. As discussed in the previous sections, power-law rate equations are commonly used in kinetic problems to describe rate expressions and equilibrium conditions. Some examples are the Freundlich equation, used to describe adsorption of gases and liquids onto solid surfaces (Guo et al., 2006), and the well-known nth-order reaction rate models:

$$y = kx^n \tag{5.7.1}$$

where x and y are the independent and the dependent measured variables, and k and n are two power-law model parameters. The correlation between the parameter estimates for k and n are usually very high because they exert similar effects over y, which tends to increase with x when either k or n are increased. In order to minimize the correlation between the two parameter estimates, Equation (5.7.1) can be rewritten in the forms (Schwaab and Pinto, 2008):

$$y = k_{ref} \left(\frac{x}{x_{ref}} \right)^n \tag{5.7.2}$$

$$y = \exp\left[A_{ref} + n \ln\left(\frac{x}{x_{ref}} \right) \right] \tag{5.7.3}$$

where x_{ref} is the reference variable. One must observe that model responses are not affected by reparameterization, since inclusion of the reference variable leads to redefinition of the model parameters, k_{ref} and A_{ref}, as:

$$k_{ref} = kx_{ref}^n \tag{5.7.4}$$

$$A_{ref} = \ln\left(kx_{ref}^n \right) \tag{5.7.5}$$

Particularly, Schwaab and Pinto (2008) showed that correlation between parameters A_{ref} and n might be nullified when:

$$x_{ref} = \exp\left[\frac{\sum_{i=1}^{NE} y_i^2 \ln(x_i)}{\sum_{i=1}^{NE} y_i^2} \right] \tag{5.7.6}$$

It is important to note that Equation (5.7.6) depends only on the available experimental data and can be computed before the estimation of the model parameters, as also discussed previously for the Arrhenius equation in Section 3.3.3. Besides, after determination of parameters A_{ref} and n, the kinetic parameter k can be easily computed in the form:

$$k = \frac{e^{A_{ref}}}{x_{ref}^n}. \tag{5.7.7}$$

The introduction of the reference variable x_{ref} into the power-law rate equation allows for elimination of the parameter correlation and, consequently, for reduction of the computational effort required for estimation of model parameters

(Espie and Macchietto, 1988; Schwaab and Pinto, 2008). Particularly, Schwaab and Pinto (2008) showed that the introduction of the reference variable can simultaneously improve the precision of parameter estimates. For this reason, in the following sections the reparametrized form of the power-law rate equation will be used frequently for quantitative analyses of kinetic data.

When Equation (5.7.1) is combined with the Arrhenius equation in the form

$$y = k_0 \exp\left(-\frac{\Delta E}{RT}\right) x^n \qquad (5.7.8)$$

the reparameterized form of Equation (5.7.3) becomes particularly useful, because Equation (5.7.8) can be written as

$$y = \exp\left[A_{ref} - \frac{\Delta E}{RT}\left(\frac{1}{T} - \frac{1}{T_{ref}}\right) + n \ln\left(\frac{x}{x_{ref}}\right)\right] \qquad (5.7.9)$$

with the simultaneous reparameterization of the Arrhenius equation (see Section 3.3.3) and of the power-law rate function. In this case,

$$A_{ref} \ln(k_0) - \frac{\Delta E}{RT_{ref}} + n \ln(X_{ref}) \qquad (5.7.10)$$

Chapter 6

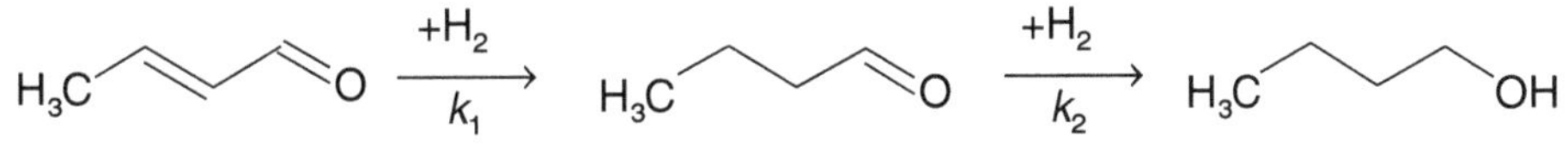

Kinetics of multiple reactions

The multiple reactions involve parallel, series, and mixed reactions. They consist of complex reactions in which the specific rates of each reaction should be determined. They often occur in industrial processes. These reactions can be simple, elementary, irreversible and reversible, or even nonelementary. Some cases are:

Series reaction

| Crotonoaldehyde | Butiraldehyde | Butanol |

Series-paralel reactions

$$C + H_2O \xrightarrow{k_1} CO + H_2$$

$$CO + H_2O \xrightarrow{k_2} CO_2 + H_2$$

Paralel reactions   Fischer Tropsch

$$CO + 3H_2 \xrightarrow{k_1} CH_4 + H_2O$$

$$CO + 2H_2 \xrightarrow{k_2} [C_nH_{2n}]_n + H_2O$$

6.1 SIMPLE REACTIONS IN SERIES

Consider here the simplest irreversible and first-order reactions. If the reaction is consecutive or in series of the type:

$$A \xrightarrow{k_1} R \xrightarrow{k_2} S$$

The corresponding rates for each component in a system at constant volume are:

$$-\frac{dC_A}{dt} = k_1 C_A \tag{6.1}$$

$$\frac{dC_R}{dt} = k_1 C_A - k_2 C_R \tag{6.2}$$

$$\frac{dC_S}{dt} = k_2 C_R \tag{6.3}$$

Considering that in the beginning of the reaction there is only pure reactant, we have:

$$C_A + C_R + C_S = C_{A0} \tag{6.4}$$

Defining new variables:

$$\varphi_A = \frac{C_A}{C_{A0}} \tag{6.5}$$

and analogously φ_R and φ_S, the initial concentration C_{A0} is always used as a reference. Making the time dimensionless ($\theta = k_1 t$) and substituting these new variables into Equations 6.1 and 6.4, we obtain the following solutions:

$$\varphi_A = e^{-\theta} \tag{6.6}$$

$$\varphi_R = \frac{1}{(\kappa - 1)}[e^{-\theta} - e^{-\chi\theta}] \tag{6.7}$$

$$\varphi_S = 1 + \frac{1}{(\chi - 1)}e^{-\chi\theta} - \frac{\chi}{(\chi - 1)}e^{-\theta} \tag{6.8}$$

Considering that

$$\chi = \frac{k_2}{k_1}$$

where χ is a parameter of the specific rate.

By solving Equations 6.6–6.8, we obtain φ_A, φ_R, and φ_S as a function of θ and hence the concentrations of each component as a function of the time, represented by the kinetic curves shown in Figure 6.1. The curves show that the concentration profile of A decreases exponentially, and A totally disappears when $\theta \to \infty$. On the other hand, the concentration of R increases initially and then decreases, since R will be formed over time and transformed into S. Note that the curve of R shows a maximum and depends on the parameter χ, relating the specific rates of the reactions. The time corresponding to this concentration can be determined as follows:

$$\frac{dC_R}{dt} = 0 \quad \text{or} \quad \frac{d\varphi_R}{d\theta} = 0 \tag{6.9}$$

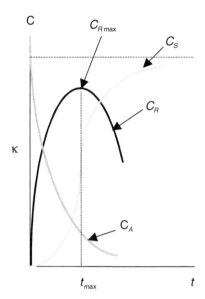
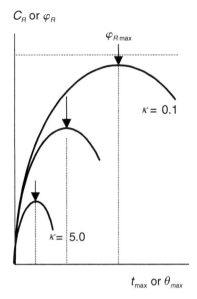

Figure 6.1 Concentration profiles.

Differentiating Equation 6.7, we obtain:

$$\frac{d\varphi_R}{d\theta} = \frac{1}{(\kappa - 1)}[\chi\, e^{-\chi\theta} - e^{-\theta}] = 0 \tag{6.10}$$

Consequently, we can determine the time corresponding to this maximum, i.e.:

$$\theta_{max} = \frac{\ln \chi}{(\chi - 1)} \tag{6.11}$$

Substituting θ_{max} into Equation 6.7, we obtain the maximum concentration of R:

$$\varphi_{R\,max} = \exp(-\chi\theta_{max}) \tag{6.12}$$

Substituting θ_{max} (Equation 6.11) into Equation 6.12, we finally obtain:

$$\varphi_{R\,max} = \chi^{[\chi/(1-\chi)]} \tag{6.13}$$

or

$$\frac{C_{R\,max}}{C_{A0}} = \chi^{[\chi/(1-\chi)]} \tag{6.14}$$

Observe that φ_{Rmax} or the maximum concentration of the intermediate C_{Rmax} changes, when the parameter χ is varied. This means that if the specific rate k_2 of reaction is higher than the specific rate k_1 of formation of $R(\,\cdot\,)$, the maximum concentration of the intermediate product R and the corresponding time t_{max} decrease. Figure 6.1b shows the behavior of the concentration of R as a function of χ.

We can experimentally determine the specific rates of the two reaction steps through the kinetic curves. Upon reaching the maximum concentration of R ($C_{R\,max}$) and the corresponding time t_{max}, the parameter χ is calculated through Equation 6.13 or 6.14. Returning to Equation 6.11, θ_{max} is calculated. By the definition of $\theta_{max} = k_1 t$, the specific rate k_1 is directly determined. Knowing χ, the specific rate k_2 is calculated.

Example

E6.1 Consider the transformation of isopropylbenzene (A) in the presence of hydrochloric acid in isopropyl-sec-butylbenzene (R) and isopropyl di-sec-butylbenzene. Considering the following data obtained in laboratory, determine the specific rates.

t (min)	0.5	1.0	1.5	2.0	2.23	2.5	3.0
C_A mol/L	0.60	0.37	0.22	0.14	0.107	0.08	0.05
C_R mol/L	0.38	0.58	0.68	0.71	0.715	0.71	0.69
C_S mol/L	0.02	0.05	0.10	0.15	0.178	0.21	0.26

Considering that $C_{A0} = 1$ (mol/L), we can observe from Figure E6.1 that the maximum concentration of R is $C_{R\,max} = 0.715$ mol/L and the corresponding time is $t_{max} = 2.23$ min.

Thus,

$$\varphi_{max} = 0.715$$

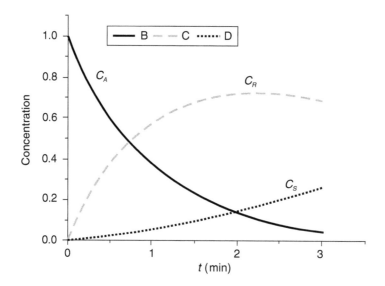

Figure E6.1   Concentration profiles of the reaction in series.

From Equation 6.13, we have:

$$\chi = 0.15$$

Knowing χ, we can calculate through Equation 6.11:

$$\theta_{max} = 2.23$$

Consequently, the reaction rate constants are determined as:

$$k_1 = 1.0 \, min^{-1}$$
$$k_2 = 0.15 \, min^{-1}$$

6.2   SIMPLE PARALLEL REACTIONS

The parallel reactions occur when the initial reactants make part of more than one reaction. The reactant A is transformed into several different products.

When there is only a transformation of reactants into products through two irreversible first-order reactions, we obtain:

$$A \xrightarrow{k_1} R \tag{6.15}$$

$$A \xrightarrow{k_2} S \tag{6.16}$$

whose rates are, respectively:

$$(-r_A) = k_1 C_A + k_2 C_A \tag{6.17}$$

$$r_R = k_1 C_A \tag{6.18}$$

$$r_S = k_2 C_A \tag{6.19}$$

If the reactions occur in gas or liquid phase at constant volume or in a batch reaction, we have:

$$-\frac{dC_A}{dt} = (k_1 + k_2) C_A \tag{6.20}$$

$$\frac{dC_R}{dt} = k_1 C_A \tag{6.21}$$

$$\frac{dC_S}{dt} = k_2 C_A \tag{6.22}$$

Using the same variables defined previously (Equation 6.5) and solving the equation, we obtain the following solutions:

$$C_S = \chi C_R \tag{6.23}$$

and

$$-\ln \varphi_A = -\ln \frac{C_A}{C_{A0}} = (1 + \chi)k_1 t \tag{6.24}$$

where

$$\chi = \frac{k_2}{k_1}$$

From Equations 6.23 and 6.24, we can directly obtain the values of χ and k_1 as shown in Figure 6.2.

Figure 6.2 shows the influence of χ on the concentration of each component. For small values of χ, the specific rate k_1 is higher than k_2, i.e., the transformation of reaction $A \xrightarrow{k_1} R$ is faster than that of $A \xrightarrow{k_2} S$. Consequently, the concentration of R is much higher than that of S, since the reaction slows down the process and thus is the controlling step. By increasing the value of χ, the transformation of reaction $A \xrightarrow{k_2} S$ is faster than that of $A \xrightarrow{k_1} R$. Therefore, the accumulation of S, as well as its concentration, is much higher than the concentration of the product R. Note also that the concentration of A decreases rapidly.

The experimental data of concentrations as a function of the time are used to calculate the specific rates k_1 and k_2. χ can be directly determined through the concentrations of the products R and S (if known) or through the angular coefficient of Figure C_S versus C_R. On the other hand, the specific rate k_1 can be determined by using Equation 6.24 or through the angular coefficient according to Figure 6.2. Knowing k_1 and χ allows for determination of k_2.

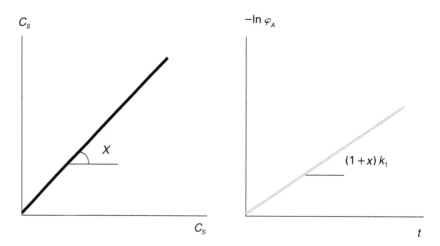

Figure 6.2 Determination of the specific rates of the parallel reactions.

Example

E6.2 Decomposition of A into R and S. Starting from 1 mole of A, the following experimental results were obtained (expresses as molar fraction).

T (min)	1	2	3	4	5	6	7	8	10	12	14	15
φ_A	0.74	0.548	0.406	0.306	0.223	0.162	0.122	0.090	0.067	0.050	0.027	0.015
φ_R	0.172	0.300	0.395	0.465	0.517	0.556	0.585	0.606	0.621	0.633	0.648	0.65
φ_S	0.026	0.150	0.197	0.232	0.258	0.278	0.292	0.303	0.301	0.316	0.324	0.32

Initially, we should construct Figure E6.2 plotting of the concentrations versus time and Figure E6.3 the concentrations φ_R versus φ_S, from which we can obtain the angular coefficient χ.

Then, ploting in Figure E6.4 eq. 6.24 $-\ln\varphi_A$ versus t the angular coefficient is determined

$$(1 + \chi)k_1.$$

The angular coefficient $\chi = 0.575$.
The angular coefficient $(1 + \chi)\, k_1 = 0.273$. Thus,

$$k_1 = 0.173\,\text{min}^{-1}$$

$$k_2 = 0.099\,\text{min}^{-1}$$

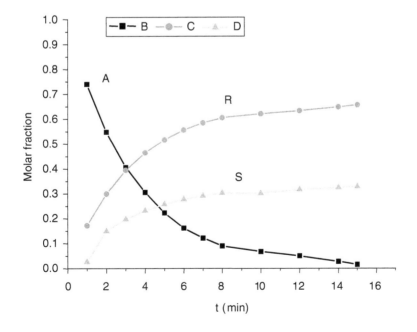

Figure E6.2 Concentration profiles—parallel reaction.

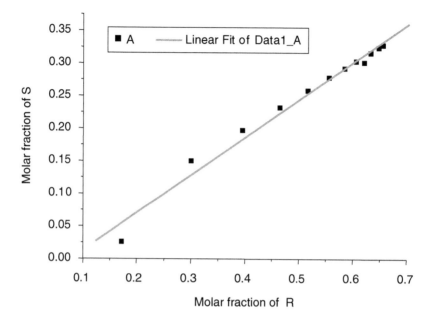

Figure E6.3 Relation between the molar fractions of R and S.

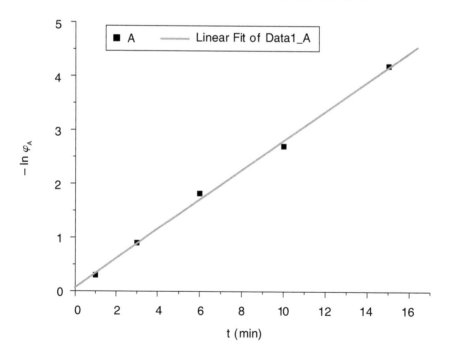

Figure E6.4 Determination of the specific rates.

6.3 CONTINUOUS SYSTEMS

In most cases, a *continuous reactor* is used and the reactions are carried out in gas phase. Under these conditions, volume change may occur due to expansion or contraction of volume. Analogously, we can use the integral or differential methods. For the determination of kinetic parameters, a differential reactor is used.

Consider two gas-phase reactions of the type:

$$A \xrightarrow{k_1} R + S \tag{6.25}$$

$$A \xrightarrow{k_1} 2T \tag{6.26}$$

The rates are, respectively:

$$(-r_A) = k_1 C_A + k_2 C_A \tag{6.27}$$

$$r_R = k_1 C_A$$

$$r_T = k_2 C_A \tag{6.28}$$

Defining the conversion X_{A1} of the first reaction and X_{A2} of the second reaction, we obtain the corresponding molar fluxes:

$$F_R = F_{A0} X_{A1} \tag{6.29}$$

$$F_S = F_{A0} X_{A1} \tag{6.30}$$

$$F_A = F_{A0} - F_{A0} X_{A1} - F_{A0} X_{A2} \tag{6.31}$$

The total molar flow will be:

$$F_A = F_{A0} + (1 - 2X_A) \tag{6.32}$$

Knowing that

$$C_A = \frac{F_A}{v} \tag{6.33}$$

where v is the volumetric flow (L/h) in which the volume variation ε is considered. Therefore:

$$C_A = \frac{F_A}{v} = \frac{F_{A0}(1 - 2X_A)}{v_0(1 + \varepsilon_{A1}(X_{A1} + \varepsilon_{A2} X_{A2}) + \varepsilon_{A2} X_{A2})} \tag{6.34}$$

Thus, the corresponding rates in Equations 6.27 and 6.28 are transformed into:

$$r_R = k_1 \frac{F_{A0}(1 - 2X_A)}{v_0(1 + \varepsilon_{A1} X_{A1} + \varepsilon_{A2} X_{A2})} \tag{6.35}$$

$$r_T = k_2 \frac{F_{A0}(1 - 2X_{A1})}{v_0(1 + \varepsilon_{A1} X_{A1} + \varepsilon_{A2} X_{A2})} \tag{6.36}$$

The rates in a differential reactor in relation to R and T are:

$$\frac{dF_R}{dV} = r_R \tag{6.37}$$

$$\frac{dF_T}{dV} = r_T \tag{6.38}$$

Substituting the rates r_R and r_T in Equations 6.37 and 6.38 and considering that:

$$dF_R = F_{A0}\, dX_{A1} \tag{6.39}$$

$$dF_T = 2F_{A0}\, dX_{A2} \tag{6.40}$$

Two differential equations are obtained.

In the differential reactor, the experimental rates are determined directly.

In the integral reactor, the system is solved by integrating the expressions above and determining the kinetic curve of the conversions as a function of the space time. From Equations 6.35 and 6.40, the following relation is obtained after integration:

$$X_{A1} = \frac{k_2}{k_1} X_{A2} \tag{6.41}$$

Substituting X_{A2} in Equations 6.35 and 6.36, and integrating, according to Equation 5.3:

$$\tau = C_{A0} \int_0^{X_A} \frac{dX_A}{r_j} \tag{5.3}$$

we obtain:

$$\frac{(1+\varepsilon_{A1}) + (1+\varepsilon_{A2})\kappa}{(1+\kappa^2)} \ln\left[1 - (1+\kappa X_{A1}) - \frac{(\varepsilon_{A1}+\varepsilon_{A2})\kappa}{(1+\kappa)} X_{A1}\right] = \tau k_1 \tag{6.42}$$

where:

$$X_{A1} + X_{A2} = X_A$$

$$\kappa = \frac{k_2}{k_1}$$

With the experimental values of the conversions as a function of the space time, the constants k_1 and k_2 are determined, respectively.

Example

E6.3 The *m*-xylene is decomposed into benzene and methane and at the same time into *p*-xylene according to equations: (Adapted from Fogler)

$$m - X \longrightarrow Benzene + CH_4$$

$$m - X \longrightarrow p - X$$

The reaction was conducted in a tubular reactor at 673°C. *m*-X was introduced with 25% of inert, with a concentration of 0.05 mol/L and flux of 2.4 L/min. The reactor has a volume of 1 mL. In the outlet, the measured molar flux of *p*-X was equal to 0.077 mol/L. Calculate the specific reaction rates.

Solution

From the reactions, we can see that $\varepsilon_{A2} = 0$. So the value of ε_{A1} should be calculated, i.e.:

Moles: basis 1 mol

m-X	Benzene	Methane	Inert	Total
0.75	0	0	0.25	I
0	0.75	0.75	0.25	1.75

Therefore:

$$\varepsilon_{A1} = 0.75$$

Calculating:

$$F_{A0} = C_{A0} v_0 = 0.05 \times 2.4 = 0.12 \, \text{mol/min}$$

$$F_B = C_{A0} X_{A2} = 0.077$$

Thus:

$$X_{A2} = 0.64$$

and

$$X_{A1} = 0.35$$

Calculation of τ:

$$\tau = \frac{V}{v_0} = 0.0416 \, \text{min}^{-1} = 2.5 \, \text{s}^{-1}$$

However, from Equation 6.41, we have:

$$\frac{k_1}{k_2} = 0.55$$

Substituting all values into Equation 6.42, we obtain:

$$k_1 = 0.742 \, \text{s}^{-1}$$

$$k_2 = 1.35 \, \text{s}^{-1}$$

6.4 KINETICS OF COMPLEX REACTIONS

The irreversible and reversible complex or multiple reactions have a different behavior and as such cannot be solved by simple integral methods. In most cases, numerical methods are employed. These reactions can occur in series, parallel, or a combination of both. The goal is to determine the rate constants for reactions of any order. Although the order of these reactions is not integer, we can assume that it is entire in the different steps. In such cases, an analytical solution can be obtained. The most complex solutions of generic order will not be studied in this chapter. Consider the three cases as follows.

6.4.1 Decomposition reactions

$$A \xrightarrow{k_1} P_1 \xrightarrow{k_2} P_2 \dots \dots \xrightarrow{k_n} \to P_n$$

Consider P_n as a final polymeric product and k_1, k_2, and k_n are unknown.
The rates are, respectively:

$$-\frac{dC_A}{dt} = k_1 C_A \tag{6.43}$$

$$\frac{dC_{P_1}}{dt} = k_1 C_A - k_2 C_{P_1} \tag{6.44}$$

$$\frac{dC_{P_2}}{dt} = k_2 C_{P_1} - k_3 C_{P_2} \tag{6.45}$$

$$\frac{dC_{P_n}}{dt} = k_n C_{P_n} \tag{6.46}$$

Defining the new variables:

$$\varphi_A = \frac{C_A}{C_{A0}} \tag{6.47}$$

And, analogously, always considering the initial concentration C_{A0} as a reference, the dimensionless time $(\theta = k_1 t)$ should be used in this case. Substituting these new variables and considering the boundary condition that in the beginning of the reaction there is only pure reactant, we obtain the following integrated solutions:

$$\varphi_A = e^{-\theta} \tag{6.48}$$

$$\varphi_{P_1} = \frac{1}{(\kappa_1 - 1)} [e^{-\theta} - e^{-\kappa\theta}] \tag{6.49}$$

and

$$\frac{\varphi_{P_2}}{d\theta} = \frac{\kappa_1}{(\kappa_1 - 1)} [e^{-\kappa\theta} - e^{-\kappa_1\theta}] - \kappa_n \varphi_{P_2} \tag{6.50}$$

where:

$$\kappa_1 = \frac{k_2}{k_1}, \kappa_n = \frac{k_n}{k_1} \tag{6.51}$$

where κ_i is the measurement parameter of the rates of transformation in each step.

Table 6.1   Reaction Rate Constants.

κ_1	$\kappa_2 \kappa_n$	$\varphi_{P_1\,max}$	$\theta_{P_1\,max}$	$\varphi_{P_2\,max}$	$\theta_{P_2\,max}$
0.2	0.8	0.668	2.0	0.162	2.3
0.8	0.2	0.668	2.0	0.65	2.3
2.0	0.2	0.250	0.69	0.77	1.27
0.4	1.5	0.542	1.53	0.16	1.2

The solutions for these systems show that there are maxims for P_1 and P_2 and the corresponding maximum times, θ_1 and θ_2, which can be obtained by differentiating the rates and making them zero. We then obtain:

$$\theta_{P_1\,max} = \frac{\ln \kappa_1}{(\kappa_1 - 1)} \tag{6.52}$$

$$\theta_{P_2\,max} = \frac{\ln \frac{\kappa_n}{\kappa_1}}{(\kappa_n - \kappa_1)} \tag{6.53}$$

Therefore, if the experimental values of the concentrations P_1 and P_2 are known, the corresponding values of $\varphi_{P_1\,max}$ and $\varphi_{P_2\,max}$ are determined and the values of κ_1 and κ_2 are calculated. Consequently, the values of $\theta_{P_1\,max}$ and $\theta_{P_2\,max}$ and finally the constants are determined. Note that the maximum values depend on the relation between the reaction rate constants and therefore on the values of κ_2 and κ_1, according to Table 6.1.

The results show that if $k_2 > k_1$, as in the first case, the maximum concentration of P_2 is lower, indicating that the transformation of P_2 is fast and that the limiting step of the reaction is the transformation of P_1 into P_2. In the second case, both steps are limiting and the transformation of A into P_1 is the only fast step. The time that corresponds to the maximum concentration depends on the reaction rates in the different steps.

6.4.2   Parallel reactions

$$A \rightarrow P_1 \tag{6.54}$$

$$A + B_2 \rightarrow P_2$$

Consider that P_1 and P_2 are the final products and that the constants k_1, k_2 are unknown.

The corresponding rates are:

$$-\frac{dC_A}{dt} = k_1 C_A - k_2 C_A C_{B_2} \tag{6.55}$$

$$-\frac{dC_{B_2}}{dt} = k_1 C_A C_{B_2} \tag{6.56}$$

where:

$$C_A + C_{P_1} + C_{P_2} = C_{A0} \tag{6.57}$$

$$C_{B_2} + C_{P_2} = C_{B20} \tag{6.58}$$

By making dimensionless, dividing and integrating Equations 6.56 and 6.57, we obtain:

$$\varphi_A = (1 - N) + \varphi_{B2} + (R/\kappa_1) \ln(\varphi_{B2}/N)$$ (6.59)

where:

$$\kappa_1 = \frac{k_2}{k_1}$$

To determine κ_1, we can use Equation 6.59 from the product concentrations P_2 substituting when the time $t \to \infty$, and therefore, $\varphi_A = 0$.
Thus:

$$\kappa = R \frac{\ln\left(1 - \frac{\varphi_{P_2}}{N}\right)}{(\varphi_{P_2} - 1)}$$ (6.60)

We can then determine the concentration of B_2, substituting Equation 6.60 into Equation 6.59 and into Equation 6.56. After integration, we have:

$$C_{A0}\theta = -\int_{0}^{\varphi_B} \frac{d\varphi_{B2}}{\kappa \varphi_{B2}[(1 - N) + \varphi_{B2} + (R/\kappa_1) \ln(\varphi_{B2}/N)]}$$ (6.61)

where:

$$R = \frac{1}{C_{A0}}, N = \frac{C_{B2}}{C_{A0}}, \quad C_{B_2} = N\varphi_{P_2}$$

With this relation, we can express the above equation as a function of: φ_{P_1}.
Considering $\theta = k_1 t$ and knowing the concentration of the products P_1 and P_2, we can determine κ_1. The constant k_1 can be determined by Equation 6.61.

Example

E6.4 Consider a reaction of the type in dilute solution. The concentrations of the products were measured at 40°C and are displayed in Table E6.1.

$$A + H_2O \to P_1$$

$$A + B_2 \to P_2$$

The initial concentrations of A and B_2 are 1 mole/L. Determine the reaction rate constants k_1 and k_2.

Table E6.1

t (min)	1.41	2.9	7.6	19.4	43
[P₂] (mol/L)	0.117	0.2	0.35	0.46	0.49

Solution

The data allow us to calculate κ through Equation 6.60.

The concentration of $[P_2]$ for $t \to \infty$ will be in the dimensionless form as follows:

$$\varphi_{P_2} = 0.5$$
$$N = 1$$
$$R = 1$$

Thus, by using Equation 6.60:

$$\kappa = 1.42$$

Integrating Equation 6.61, we have:

$$C_{A0} \cdot k_1 = 2.57$$
$$k_1 = 2.57 \, \text{min}^{-1}$$
$$k_2 = \kappa k_1 = 3.65 \, \text{L} \, (\text{mol}^{-1} \, \text{min}^{-1})$$

Figures E6.5 and E6.6 show the values of φ_{P_2} versus t from Table E6.1 and as function of Equation 6.61, respectively.

Integration of Data1_B from zero:

$$i = 1 > 10$$
$$x = 0 > 0.49$$

Area	Peak at	Width	Height
2.57802	0.49	0.01	117.76805

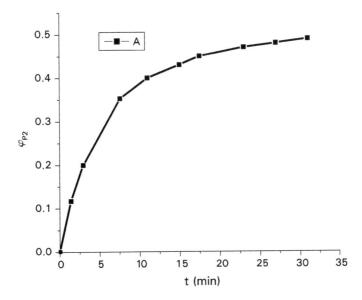

Figure E6.5 φ_{P_1} versus time.

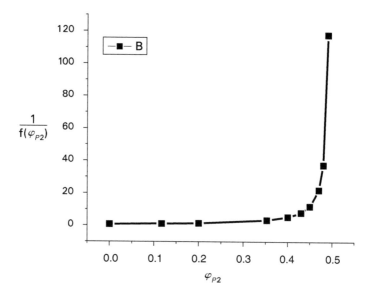

Figure E6.6 According to Equation 6.61.

6.4.3 Series–parallel reactions

$$A + B \rightarrow P_1$$

where k are the rate constants.

$$P_1 + B \rightarrow P_2 \qquad k_1$$

. . . .

$$P_{n-1} + B \rightarrow P_n \qquad k_{n-1}$$

This is the scheme of a polymerization reaction by addition of radicals. Although this system is complex and usually solved by numerical methods, the general solution using the integral method will be shown here. This is the easiest way to identify the kinetic parameters involved and indicate a general solving method for complex reactions of this type, although the numerical solution is more appropriate. We should start from a batch system (constant volume), whose equations for the rates of reactants and products are described as follows:

$$-\frac{dC_A}{dt} = k_0 C_A C_B \tag{6.62}$$

$$\frac{dC_{P_1}}{dt} = k_0 C_A C_B - k_1 C_{P_1} C_B \tag{6.63}$$

. . . .

$$\frac{dC_{P_{n-1}}}{dt} = k_{n-2}C_{P_{n-1}}C_B - k_{n-1}C_{P_{n-1}}C_B \tag{6.64}$$

$$\frac{dC_{P_n}}{dt} = k_{n-1}C_{P_{n-1}}C_B \tag{6.65}$$

Considering that:

$$C_A + C_{P_1} + C_{P_2} + \cdots + C_{P_{n-1}} + C_{P_n} = C_{A0} \tag{6.66}$$

Since for each molecule of P_1 a molecule of B is consumed and for each molecule of P_2 two molecules are consumed and so on, the global balance for B will be:

$$C_B + C_{P_1} + 2C_{P2} + \cdots + (n-1)C_{P_{n-1}} + nC_{P_n} = C_{B0} \tag{6.67}$$

Using the same dimensionless variables defined previously, ie, $\varphi_A = \frac{C_A}{C_{A0}}$, $\varphi_B = \frac{C_B}{C_{B0}}$, etc.

A generic parameter is also defined for the relation between the rate constants, i.e.:

$$\kappa_i = \frac{k_i}{k_0}$$

where $i = 1, 2, \ldots, n-1$.

Equations 6.62–6.66 are transformed and after solving them, we have:

$$\frac{d\varphi_{P_1}}{d\varphi_A} = \kappa_1 \frac{\varphi_{P_1}}{\varphi_A} - 1 \tag{6.68}$$

By using the integration factor, we obtain the solution with the boundary condition that $\varphi_A = \varphi_{P_1} = 0$:

$$\varphi_{P_1} = \frac{1}{(\kappa - 1)}(\varphi_A - \varphi_A^{\kappa_1}) \tag{6.69}$$

By relating φ_{P_1} and φ_{P_2} with φ_A, we analogously obtain:

$$\frac{d\varphi_{P_2}}{d\varphi_A} = \kappa_1 \frac{\varphi_{P_1}}{\varphi_A} - \kappa_2 \frac{\varphi_{P_2}}{\varphi_A} \tag{6.70}$$

Substituting from Equation 6.69, we have:

$$\frac{d\varphi_{P2}}{d\varphi_A} - \kappa_2 \frac{\varphi_{P2}}{\varphi_A} = -\frac{\kappa_1}{1 - \kappa_1}\left[1 - \varphi_A^{(\kappa_1 - 1)}\right] \tag{6.71}$$

Using the integral factors with the boundary conditions $= 0$, $\varphi_A = 1$, we obtain the generic solution:

$$\varphi_{P_2} = \frac{\kappa_1}{(1 - \kappa_1)(1 - \kappa_2)(\kappa_1 - \kappa_2)}\left[(\kappa_1 - \kappa_2)\varphi_A - (1 - \kappa_2)\varphi_A^{\kappa_1} + (1 - \kappa_1)\varphi_A^{\kappa_2}\right] \tag{6.72}$$

Knowing the concentration φ_{Pi} as a function of the concentration of the component φ_A, we can determine the concentration of P_i by integrating Equation 6.71, which is generically given by:

$$\varphi_{P_{i+1}} = -\kappa_i \varphi_A^{\kappa_{i+1}} \int \frac{1}{\varphi_A^{\kappa_{i+1}}} f(\varphi_A) d\varphi_A + C\varphi_A^{\kappa_{i+1}} \tag{6.73}$$

where C is a constant. Therefore, to determine the reaction rate constants, we proceed as in the case for reactions in series, in which the concentration of P_i increases, reaches a maximum and afterward decreases. From the condition of maximum concentration, φ_{Pi+1} is differentiated and equaled to zero.

For P_1, the maximum concentration will be:

$$\varphi_{P\,max} = \kappa_1^{\left(\frac{\kappa_1}{1-\kappa_1}\right)} \tag{6.74}$$

We observe that $\varphi_{Pi\,max}$ depends on the relation between the reaction rate constants κ_1. Analogously, $\varphi_{Pi+1\,max}$ will be determined only as a function of the relations κ_i.

Knowing the experimental values of C_A, C_{Pi} as a function of the time t_{max}, we can separately calculate $\varphi_{P\,max}$ and consequently the rate constants. To determine the reaction time t, we should start from Equation 6.68 and substitute ϑ_B as a function of φ_{Pi}. We finally obtain:

$$\tau = \int_A^1 \frac{1}{f(\varphi_A)} d\varphi_A$$

In this expression, we only have the distribution of the reactant A as a function of the other components.

Chapter 7

Non-elementary reactions

Nonelementary reactions can be identified when the order of reaction rate does not match the stoichiometry of the reaction. An apparent elementary reaction does not show the actual mechanism that may involve several intermediate reactions, which in turn are elementary reactions. The overall reaction can have different intermediate mechanisms. To understand the kinetics of this reaction is necessary to know the intermediate steps, and, consequently, indicate which of these steps are crucial to the overall process.

There are several classical examples:

1. $2NO + 2H_2 \rightarrow N_2 + 2H_2O$
   The kinetics is:

   $$r_{N_2} = k[NO]^2[H_2]$$

   where [...] represents the concentrations of the component.
2. $H_2 + Br_2 \leftrightarrow 2HBr$
   The kinetics is:

   $$r_{HBr} = k[H_2][Br_2]^n[HBr]^m$$

   where n and m are not integer and, therefore, do not match the stoichiometry of the reaction.
3. $N_2 + 3H_2 \leftrightarrow 2NH_3$
   The kinetics is:

   $$r_{NH_3} = k[N_2][H_2]^n[NH_3]^m$$

If the exponents are not an integer and does not match the stoichiometry of the reaction.

What is the explanation? Stable molecules hardly react, only when they form free radicals, carbenes or complex intermediates, ions, or valences. These very reactive species combine easily with molecules or other intermediate species, reacting in successive or parallel steps. These intermediate mechanisms should be known to determine the reaction kinetics. This proves that the overall reaction rate constant is not always true but includes several other constants relative to different intermediate steps of the mechanism.

These constants depend on the energy barrier or activation energy. Each step of the mechanism has a different energy barrier. The higher the energy barrier, the more difficult the particular step occurs. This implies that the energy involved in the process is determinant to the reaction mechanism. The activation energy of the overall reaction is not an average of the energies of each step, but a summation of all the energies involved. For an initial analysis, we present some general hypothesis to deduce a kinetic model of a nonelementary reaction.

Hypothesis

(a)   All steps are assumed to be elementary and irreversible.
(b)   One of the steps in slow and, therefore, determinant.
(c)   The concentration of the intermediate is very low, since your formation and decomposition are very fast (close to 10^{-9} s).
(d)   The resulting rate of the intermediate is very fast and, in this case, we assume a pseudo-equilibrium state.
(e)   The resulting rate of the intermediate species is the sum of the intermediate rates, i.e.:

$$r_j = \sum_{1}^{n} r_{ji} \tag{7.1}$$

where j in the intermediate species and i is the reaction of the intermediate step of the global mechanism.

We can also demonstrate that the resulting rate is composed by several series and parallel intermediate steps, analogous to the resistivity:

$$r_j = \frac{1}{(1/r_1 + 1/r_2 + \cdots + 1/r_i + \cdots)} \tag{7.2}$$

If either step is slow, the resulting rate is equal to the rate of this slow step, i.e.:

$$r_j = r_{RLS} \tag{7.3}$$

A pseudo-equilibrium state is defined when the resulting rate of the intermediate species is zero, i.e.:

$$r_j = 0 \tag{7.4}$$

To compose a reaction mechanism involving intermediate species is necessary to know some general rules:

(a)   Establish the mechanism with all possible intermediate reactions, composed by elementary steps.
(b)   Know the electronic and atomic structure.
(c)   Identify the products and intermediates.
(d)   The intermediates are highly reactive.
(e)   Know the chemistry of the intermediate steps.

Consider the classic case as follows:
The reaction:

$$2NO + 2H_2 \rightarrow N_2 + 2H_2O \tag{a}$$

The observed rate was:

$$r_{N_2} = k_1[NO]^2\,[H_2] \tag{b}$$

where the symbol $[I_j]$ represents the concentration of the component I_j.
Note that the reaction does not match the stoichiometry, and, therefore, we have a non-elementary reaction.
There are two mechanisms:

1. $2NO + H_2 \xrightarrow{\;k_1\;} N_2 + H_2O_2$ slow step

$$\qquad H_2O_2 + H_2 \xrightarrow{\;k_2\;} H_2O \tag{c}$$

2. $2NO \xleftarrow{\;k_3\;} N_2O_2 \tag{d}$

$$\quad N_2O_2 + H_2 \xleftarrow{\;k_5\;} N_2 + H_2O_2 \quad \text{slow step} \tag{e}$$

$$\quad H_2O_2 + H_2 \xleftarrow{\;k_6\;} 2H_2O \tag{f}$$

For the mechanism (1), the slow reaction is the one and, therefore, determinant step (c). Thus, by hypothesis, the resulting rate is equal to the limiting reaction rate:

$$r_{N_2} = k_1[NO]^2[H_2] \tag{g}$$

This rate is in line with the observed rate (b).
For the mechanism (2), the limiting reaction is the second one (e) and, therefore:

$$r_{N_2} = k_5[N_2O_2][H_2]$$
$$\Downarrow \tag{h}$$
$$\text{unknown}$$

An intermediate compound that appears in the reactions (e) and (f) is formed. The resulting rate of this intermediate is unknown, but by the hypothesis of pseudo-equilibrium should be zero. Therefore:

$$r_{N_2O_2} = k_3[NO]^2 - k_4[N_2O_2] - k_5[N_2O_2][H_2] = 0$$

Thus:

$$[N_2O_2] = \frac{k_3}{(k_4 + k_5[H_2])}[NO]^2$$

$$r_{N_2} = \frac{k_3 k_5}{(k_4 + k_5[H_2])}[NO]^2[H_2] \tag{i}$$

Substituting (i) into (h), we have:

$$r_{N_2} = \frac{k_3 k_5}{(k_4 + k_5 [H_2])} [NO]^2 [H_2] \tag{j}$$

The constant k_5 of the slow step is compared with the constant k_4. In this case, we assume that $k_5 < k_4$ and k_5 is neglected. In these conditions, the rate (Equation j) is simplified to:

$$r_{N_2} = \frac{k_3 k_5}{k_4} [NO]^2 [H_2] \tag{k}$$

This rate coincides with the observed rate (b). However, in this case, the reaction rate constant includes the constants of all intermediate steps.

7.1 CLASSICAL KINETIC MODEL

Based on the classical kinetic theory and focusing on the previous example, it is possible to foresee a reaction mechanism for activated molecules.

It consists of three steps. In the first one, a collision of molecules occurs to an energy level that allows the formation an activated molecule activated A^*. In turn, this molecule does not have enough energy to be transformed into product and therefore loses energy and is decomposed to the original molecule. This process is known as the deactivation step of the activated intermediate molecule. Finally, part of these activated molecules acquires enough energy to be transformed into product or be decomposed into one or more products, undergoing a chemical reaction. The molecules surpassed the kinetic energy barrier to react or be transformed into product. Summarizing we have:

1. Activation of the molecules:

$$A + A \xrightarrow{\quad k_1 \quad} A + A^* \tag{7.5}$$

whose rate relative to the activated molecule is represented by:

$$r_{A^*} = k_1 [A]^2$$

2. Deactivation of the activated molecules:

$$A^* + A \xrightarrow{\quad k_2 \quad} A + A \tag{7.6}$$

The corresponding rate will be:

$$r_{A^*} = k_2 [A^*][A]$$

3. Transformation of the activated molecule:

$$A^* \xrightarrow{\quad k_3 \quad} R + S \tag{7.7}$$

With the following formation rate of S:

$$r_S = k_3 [A^*] \tag{7.8}$$

As an example, we have the following overall reaction:

$$(CH_3)_2N_2 \rightarrow C_6H_6 + N_2$$

Showing the following intermediate steps:

$$(CH_3)_2N_2 + (CH_3)_2N_2 \rightarrow (CH_3)_2N_2 + [(CH_3)_2N_2]^*$$
$$(CH_3)_2N_2 + [(CH_3)_2N_2]* \rightarrow 2(CH_3)_2N_2$$
$$[(CH_3)_2N_2]^* \rightarrow [C_3H_6] + N_2$$

To determine the rate of product (S or R) formation, according to Equation 7.8, it is necessary to know the concentration of intermediate species A^*. Thus, the resulting rate of A^* will be equal to the sum of the rates in each intermediate reaction and, by the hypothesis of pseudo-equilibrium state, will be zero:

$$r_{A^*} = k_1 [A]^2 - k_2 [A^*][A] - k_3 [A^*] = 0$$

Therefore:

$$[A^*] = \frac{k_1 [A]^2}{(k_3 + k_2 [A])} \tag{7.9}$$

Substituting this expression into Equation 7.8, we obtain the formation rate of the product S, i.e.:

$$r_S = \frac{k_1 k_3 [A]^2}{(k_3 + k_2 [A])} \tag{7.10}$$

This expression is equal to the equation of the rate (k), shown in the example above.

The rate of formation of S can be simplified for certain conditions of concentrations, as in the following cases:

(a) If the concentration of [A] is low: $k_2 [A] \ll k_3$
     Thus:

$$r_S = k_1[A]^2 \tag{7.11}$$

In this case, the reaction kinetics is of second order.

(b) If the concentration of [A] is high: $k_3 \ll k_2 [A]$.
     Thus:

$$r_S = \frac{k_1 k_3}{k_2} [A] \tag{7.12}$$

In this case, the reaction kinetics is of first order, with an apparent constant which includes the three constants of the intermediate reactions.

Therefore, depending on the experimental conditions, the nonelementary reaction can be represented by a simple kinetics of first or second order. Herewith, we have apparent reaction rate constants.

For a wide concentration range, we should use the expression of the overall rate (Equation 7.10). It is observed that if the concentration of $[A]$ appears in the denominator, it is accompanied by constant k_2, indicating that this species is deactivated (intermediate) and that the molecules did not reach enough energy to be transformed into product, returning to the initial state. On the other hand, if the constant k_3 appears both in the numerator and in the denominator, there was a transformation of the intermediate species into the final product, indicating that part of these species has reached a minimum energy level that which exceeds the energy barrier.

7.2   CHAIN REACTIONS

In the beginning of the last century, Bodenstein observed that when the molecules of H_2 and Cl_2 were subjected to absorption of one photon, they reacted to form HCl. This reaction hardly occurs in the gas phase, since the energy required to transform the reactants would be of the order of 200 KJ/mol. Nernst (1916), at the same time, also explained that this is a chain reaction, in which radicals are initially formed and then react with the molecules forming new radicals and subsequently forming products. Bodenstein showed that with the absorption of one photon, it is possible to form 10^6 molecules of HCl and this would only be possible by lowering the energy barrier. The following steps occur:

(a)   *Initiation*:

$$Cl_2 \xrightarrow{\ h_v\ } 2Cl^* \text{ (radical)}$$

(b)   *Propagation*:

$$Cl^* + H_2 \rightarrow HCl + H^* \qquad \text{(radical)} \quad \Delta H = 25 \text{ kJ/mol}$$
$$H^* + Cl_2 \rightarrow HCl + Cl^* \qquad\qquad\qquad \Delta H = 13 \text{ kJ/mol}$$

.......(cycle of 10^6 times)

(c)   *Termination*:

$$2Cl^* \rightarrow Cl_2$$

If we compare the energies needed to form HCl from the radicals with the total energy required for the formation of HCl from Cl_2 and H_2 molecules, we found that the first is much lower due to the high reactivity of radicals. The radicals are unstable and readily combine with other radical or molecules due to their lower energies.

7.3   THEORY OF THE TRANSITION STATE

To explain the kinetics of the mechanism of chain reactions, we use the theory of the transition state. According to Eyring, Evans, and Polaniy, a transition state with energy E_b is formed. In this case, there is only one degree of freedom corresponding to the energy of vibration, due to the dissociation of the molecule. However, some observations should be made and the following assumptions should be taken into account :

According to the theory of Boltzman, there is a collision between molecules, forming an intermediate state A^* with energy E_b.

If the molecule has energy lower than the energy barrier, it is deactivated and returns to the initial state of the molecule A.

When the molecule exceeds the energy barrier, it does not return to original state.

The activated complexes are in equilibrium with the reactants, satisfying the Boltzman distribution. The movement of molecules causes an increase of energy and hence an increase in temperature of the environment.

To overcome the energy barrier, several other independent products can be formed. The mechanism can be represented as follows:

$$A \xrightleftharpoons{k_2, k_1} A^* \xrightarrow{k_3} R \tag{7.13}$$

Energetically, the following sequence of steps is represented:

(a)   Formation of radical:

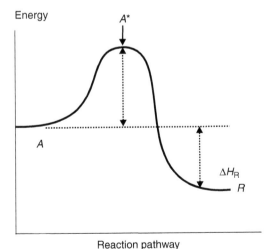

Reaction pathway

(b)   Equilibrium state:

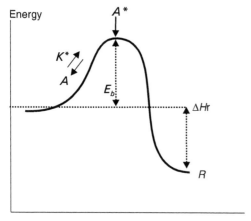

Reaction pathway

(c)   Decomposition of the radical into products:

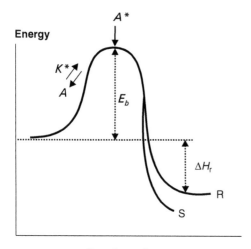

Reaction pathway

In the initial state, there is a formation of radicals due to the collision of molecules with energy $E < E_b$.

In the second stage, there is equilibrium between the molecules and the radicals formed. The resulting rate in the equilibrium is zero. Therefore:

$$r = k_1[A] - k_2[A^*] = 0$$

Thus:

$$[A^*] = \frac{k_1}{k_2}[A] = K^*[A] \tag{7.14}$$

Where the equilibrium constant is:

$$K^* = \frac{k_1}{k_2}$$

Formation of the product:

$$r_R = k_3[A^*] \tag{7.15}$$

Substituting Equation 7.14 into Equation 7.15, we have:

$$r_R = K^* k_3[A]$$

Note that K^* is the thermodynamic equilibrium constant, whereas k_3 is a rate constant. This rate constant can be calculated from theory of collisions and depends on temperature, on the constants of Boltzman k_B and of Planck h_ν, i.e.:

$$k_3 = \frac{k_B T}{h_\nu} \tag{7.16}$$

From the thermodynamics, the equilibrium constant is given by:

$$K^* = K_0^* e^{-\Delta G^0/RT} \tag{7.17}$$

Considering that:

$$\Delta G^0 = \Delta H^0 - T\Delta S^0 \tag{7.18}$$

Substituting this expression into Equation 7.17, we have:

$$K^* = K_0^* e^{-\Delta H^0/RT} e^{\Delta S^0/R} \tag{7.19}$$

In this case, the pre-exponential factor depends on the entropy of the system, i.e.:

$$K_0 = K_0^* e^{\Delta S^0/R} \tag{7.20}$$

We define, therefore, a constant of the transition state: as:

$$k_{TST} = \frac{k_B T}{h_\nu} K_0 e^{-\Delta H^0/RT} \tag{7.21}$$

When compared with the Arrhenius equation :

$$k_{Arr} = k_0 e^{-E^*/RT} \tag{7.22}$$

We can conclude that the activation energy E^* is equal to the energy of formation of the activated intermediate state and, therefore, is lower than the energy barrier. Herewith, the final state can be directly reached.

Thus, the rate of formation of product R or S, expressed as a function of the constant k_{TST} will be:

$$r_R = k_{TST}[A] \tag{7.23}$$

This equation explains the kinetics from the point of view of the concept of transition state.

Example

E7.1 Determine the rate of formation of HBr considering the mechanism of chain reaction, i.e.:

(a)   Initiation:

$$Br_2 \xrightarrow{\ k_1\ } 2Br^*$$

(b)   *Propagation of the chain:*

$$Br^* + H_2 \xleftrightarrow{\ k_2/k_2'\ } HBr + H^* \quad (radical)$$
$$H^* + Br_2 \xrightarrow{\ k_3\ } HBr + Br^*$$
$$(\text{cycle of } 10^6 \text{ times})$$

(c)   *Termination:*

$$2Br^* \xrightarrow{\ k_4\ } Br_2$$

Solution
The rate of formation of HBr will be:

$$r_{HBr} = k_3[H^*][Br_2] + k_2[Br^*][H_2] - k_2'[HBr][H^*] \tag{7.24}$$

If the concentrations of the radicals [H*] and [Br*] are unknown, the rates in relation to the radicals are determined separately:

$$r_{H^*} = k_2[Br^*][H_2] - k_2'[HBr][H^*] - k_3[H^*][Br_2] = 0 \tag{7.25}$$

$$r_{Br^*} = 2k_1[Br_2] - 2k_4[Br^*]^2 + k_2'[HBr][H^*] - k_2[Br^*][H_2] + k_3[H^*][Br_2] = 0 \tag{7.26}$$

Summing up the two expressions, we obtain:

$$[Br^*] = \left(\frac{k_1}{k_4}\right)^{1/2} [Br_2]^{1/2} \tag{7.27}$$

Substituting Equation 7.27 into Equation 7.25 and rearranging, we have:

$$[H^*] = \frac{k_2[H_2] \cdot (k_1/k_4)^{1/2}[Br_2]^{1/2}}{k_2'[HBr] + k_3[Br_2]} \tag{7.28}$$

Substituting Equations 7.28 and 7.27 into Equation 7.24 and solving, we obtain the rate:

$$r_{HBr} = \frac{2k_2k_3(k_1/k_4)^{1/2}[H_2][Br_2]^{3/2}}{k_2'[HBr] + k_3[Br_2]} \tag{7.29}$$

Where

$$k^* = 2k_2k_3(k_1/k_4)^{1/2}$$

Then, dividing by $[Br_2]$, we obtain the usual expression:

$$r_{HBr} = \frac{k^*[H_2][Br_2]^{3/2}}{k_2'[HBr] + k_3[Br_2]} \tag{7.30}$$

Example

E7.2 The formation of HBr is not elementary, as seen in Equation 7.24. Moreover, the rate expressed in Equation 7.30 was confirmed experimentally. However, in order to check this expression, some experimental observations were taken into account. The following experimental observations were made (adapted from Fogler):

1. In the first experiment, it was observed that the rate is independent of the concentration of $[H_2]$. In this experiment, the concentrations of the other components were kept constant.
2. In the second series of experiments, two observations were made:
    For low concentrations of $[HBr]$, the rate is independent of its concentration. When the concentration of $[HBr]$ is high, the rate increases inversely with its concentration.
3. In the third series of experiments, two other observations were made:
    For low concentration of $[Br_2]$ the rate varies proportionally to the concentration of $[Br_2]^{3/2}$.
    If the concentrations]are high, the rate varies with the concentration of $[Br_2]^{1/2}$. Verify the expression of the rate in Equation 7.30.

Solution
From the kinetic model, according to Equation 7.10 for each component separately, we have:

$$r_{HBr} = \frac{k_1k_3[A_j]^2}{(k_3 + k_2[A_j])}$$

In relation to $[H_2]$:

From experiment 1, we can say that the rate is directly proportional to $[H_2]$ or of first order in relation to $[H_2]$.

In relation to Br_2

From experiment 3 for low concentrations, the rate, $r \approx [Br_2]^{3/2}$ and therefore, is in the numerator. For high concentrations, we can say from Equation 7.10 that $k_3 \ll k_2[Br_2]$. Thus, the resulting rate:

$$r \approx \frac{[Br_2]^{3/2}}{[Br_2]} \approx [Br_2]^{1/2} \tag{7.31}$$

From experiments 1 and 3, the rate can be expressed as follows:

$$r_{HBr} = \frac{k_1 k_2 [H_2] [Br_2]^{3/2}}{(k_3 + k_2[Br_2])} \tag{7.32}$$

In relation to [HBr], we have:

$$r_{HBr} = \frac{1}{(k + k' [HBr])} \tag{7.33}$$

From experiment 2, if the concentration is low, then and the rate is independent of its concentration. On the other hand, if the concentration of [HBr] is high, then and, consequently, the rate is inversely proportional to its concentration, i.e.:

$$r_{HBr} \approx \frac{1}{[HBr]} \tag{7.34}$$

Gathering Equations 7.32 and 7.34, the following rate is obtained:

$$r_{HBr} = \frac{k_1 k_2 [H_2] [Br_2]^{3/2}}{(k_3[HBr] + k_2[Br_2])} \tag{7.35}$$

Dividing by $[Br_2]$, we obtain the expression of the rate which confers to the one deduced theoretically, i.e., Equation 7.30:

$$r_{HBr} = \frac{k^*[H_2] [Br_2]^{1/2}}{k_3 + k_2 \frac{[HBr]}{[Br_2]}} \tag{7.36}$$

where k^*, k_2, and k_3 are the values determined experimentally.

7.4 REACTIONS OF THERMAL CRACKING

These reactions are known in industrial processes and occur simply by decomposition and in gas phase. However, it is a nonelementary chain reaction and occurs in different elementary steps. For reactions such as thermal cracking in gas phase, there are several kinetic models, but the most interesting one is the model of Rice–Herzfeld. The main idea is that in the first step of the reaction, the abstraction of the H from the reactant molecule occurs and subsequently there is decomposition with formation of a new radical. The process takes place through the steps of initiation, propagation, and termination.

(a) Initiation:

$$A \xrightarrow{k_1} 2R_1^* \text{ (radical)}$$

(b) Propagation of the chain:

$$R_1^* + A \xrightarrow{k_2} R_1H + R_2^* \text{ (radical)}$$

$$R_2^* \xrightarrow{k_3} P_1 + R_1^*$$

. . . .

(cycles of n times)

(c) Termination:

$$R_1^* + R_2^* \xrightarrow{k_4} P_2$$

The most common radicals are $CH_3^*, C_2H_5^*, H^*$, etc.
The corresponding rates are written:

$$(-r_A) = k_1[A] + k_2[R_1^*][A]$$

Since the initiation rate is low, we can neglect the first term.

$$(-r_A) = k_2[R_1^*][A] \tag{7.37}$$

The corresponding rates to the radicals are, respectively:

$$r_{R_1^*} = 2k_1[A] - k_2[R_1^*][A] + k_3[R_2^*] - k_4[R_1^*][R_2^*] = 0$$
$$r_{R_2^*} = k_2[R_1^*][A] - k_3[R_2^*] - k_4[R_1^*][R_2^*] = 0 \tag{7.38}$$

Summing them up, we have:

$$[R_2^*] = \frac{k_1[A]}{k_2[R_1^*]} \tag{7.39}$$

Substituting Equation 7.36 into Equation 7.35, simplifying and considering that the initial rate is low compared to the rates of propagation and termination, we obtain the following expression:

$$[R_1^*]^2 = \frac{k_1 k_3}{k_2 k_4} \tag{7.40}$$

Substituting Equation 7.40 into Equation 7.37, we obtain:

$$(-r_A) = \sqrt{\frac{k_1 k_2 k_3}{k_4}} [A] \tag{7.41}$$

The rate of conversion of A is of first order in relation to the concentration of $[A]$, but the reaction rate constant encompasses the four specific rates of the mechanism. The apparent constant is determined experimentally, i.e.:

$$k_a = \sqrt{\frac{k_1 k_2 k_3}{k_4}} \tag{7.42}$$

Note that the specific rates depend on the temperature and have their own activation energies t, satisfying the Arrhenius equation, i.e.:

$$k_j = k_0 \exp(-E_j/RT) \tag{7.43}$$

Thus, substituting Equation 7.43 into Equation 7.42 and taking the natural log on both sides of the equation, we can obtain the apparent activation energy:

$$E_a = \frac{E_1 + E_2 + E_3 - E_4}{2} \tag{7.44}$$

The apparent frequency constant will be:

$$k_{a0} = \sqrt{\frac{k_{10} k_{20} k_{30}}{k_{40}}} \tag{7.45}$$

Example

E7.1 Consider the decomposition of acetaldehyde, whose reaction mechanism is known:

$$CH_3CHO \xrightarrow{k_1} CH_3^* + CHO^*$$
$$CH_3^* + CH_3CHO \xrightarrow{k_2} CH_3^* + CO + CH_4$$
$$CHO^* + CH_3CHO \xrightarrow{k_3} CH_3^* + 2\,CO + H_2$$
$$2\,CH_3^* \xrightarrow{k_4} C_2H_6$$

Determine the rate of decomposition of acetaldehyde or formation of ethane.

Solution

The decomposition rate of acetaldehyde (which is now referred to as A) will be:

$$(-r_A) = k_1[A] + k_2[CH_3^*][A] + k_3[CHO^*][A] \tag{7.46}$$

The corresponding rates to the radicals:

$$r_{CH_3^*} = k_1[A] + k_3[CHO^*][A] - k_4[CH_3^*]^2 = 0 \tag{7.47}$$

$$r_{CHO^*} = k_1[A] - k_3[CHO^*][A] = 0$$

Summing up the two expressions, we obtain:

$$[CHO^*] = \frac{k_1}{k_3} \tag{7.48}$$

Substituting into Equation 7.47, we have:

$$[CH_3^*] = \sqrt{\frac{2k_1[A]}{k_4}} \tag{7.49}$$

Substituting them into Equation 7.10 and neglecting the initiation term, we have:

$$(-r_A) = k_2\sqrt{\frac{2k_1}{k_4}}[A]^{3/2} \tag{7.50}$$

Thus, the decomposition rate of acetaldehyde is of order 3/2. The rate constant encompasses the specific rates of the reaction step.

The rate of formation of ethane will be:

$$r_{C_2H_6} = k_4[CH_3^*]^2 \tag{7.51}$$

Substituting Equation 7.13, we obtain:

$$r_{C_2H_6} = 2k_1[A] \tag{7.52}$$

i.e., is of first order in relation to A, where $A = CH_3CHO$.

Chapter 8

Polymerization reactions

Polymerization reactions are of great industrial interest, and their advancement in the last 50 years has been significant. Particularly, Ziegler and Natta created the most important commercial polymerization reactions, starting from olefins and generating oligomeric products with the formation of carbon chains of high molecular weight, using Ti-based catalysts for this purpose. With these catalysts, polymers of varying degrees of polymerization can be produced. The first important outcome is that products having a certain range of hydrocarbons can be obtained; to this end the catalyst plays a crucial role. As a matter of fact, new catalysts that are highly selective to the desired carbon range have already been developed.

In order to comprehend the behavior of most polymerization systems it is necessary to know the kinetics of the polymerization reactions, which involves different steps and are somewhat similar to that seen in nonelementary reactions. The kinetic rate expressions depend on the mechanism of polymerization, which constitutes a fundamental chemical problem. For this reason, the composition and structure of the different involved steps should be known. Once this mechanism is established, the reaction kinetics and the involved reaction constants can be determined. And although the polymerization reactions are nonelementary, they can be identified when the order of the kinetic rate expression does not match the stoichiometry of the reaction itself.

As a whole, polymerization reactions are those that lead to manufacture of macromolecular chains (or polymers). The small molecules that react to form the polymer chains are usually called monomers, and the small segments that constitute the macromolecular chains and that are derived from the monomers are referred to simply as *mers*. Polymers can be found naturally in living tissues (such as proteins, polysaccharides, DNA and RNA) or can be manufactured through many different chemical mechanisms. Particularly, synthetic polymers can be obtained through chain (or addition) polymerization (such as free radical, ionic and Ziegler–Natta) and step polymerization (such as polycondensation). In chain polymerizations, the meric segments are inserted into the polymer backbone according to a one-by-one mechanism, in the form

$$P_i + M \rightarrow P_{i+1} \tag{8.1}$$

where P_i represents a growing chain of size i, and M represents the monomer. On the other hand, in step polymerizations the meric segments are inserted into the polymer backbone in lumps, in the form:

$$P_i + P_j \rightarrow P_{i+j} \qquad (8.2)$$

Usually, chain reactions depend on the existence of an active species, normally provided by a catalyst or an initiating compound. On the other hand, step reactions are normally dependent on the reactivity of organic functions, such as hydroxyl and carboxyl groups in the case of polyesters or amino groups and carboxyl groups in the case of polyamides. Typical chain polymers are polyvinyl compounds, such as polystyrene, polyethylene, polypropylene and poly(vinyl chloride), among others. Among typical step polymers are polyesters (such as PET, poly(ethylene terephtalate)) and polyamides (such as nylon). Additional information about polymer materials and detailed discussion about chain and step polymerizations can be found in standard polymer text books (Odian, 2004).

Importantly, the final macromolecular structure of the produced polymer material can be quite complex. As it may not be possible to control the reactions described by Equations (8.1) and (8.2), the produced chains can be of different sizes, leading to products that present distinct molecular weight distributions and, therefore, distinct properties. Additionally, when more than one monomer takes part in the reaction, the polymerization is usually named copolymerization and the manufactured products can present distinct copolymer compositions and composition distributions, since the composition can change along the polymer chain. Further, different monomers can form blocks of different sizes and lead to formation of chain branches (due to presence of lateral chemical functions). Additionally, the molecular structure of the final product depends not only on the chemical nature of the reacting monomers but also on polymerization mechanism, polymerization process and reactor configuration, defining the end-use properties of the final product. More interestingly, in order to facilitate the manipulation of the polymer material and the operation of the polymerization process, heterogeneous polymerization processes are widely employed for production of polymer particles (including emulsion, dispersion, precipitation and suspension polymerizations) (Dowding and Vincent, 2000). Each process has peculiar characteristics, which allow for production of different resins with distinct properties, enabling different applications of the produced polymer materials. Manufacture of polymers in dispersed media is almost always advantageous, because the viscosity of polymer solutions grows exponentially with the average chain size, making infeasible the mixing and pumping of these solutions under mild temperature conditions. On the other hand, when these materials are dispersed in a heterogeneous medium, the viscosity of the heterogeneous system approaches the viscosity of the continuous system, making manipulation of the mixture much simpler.

For all the reasons discussed in the previous paragraphs, it is impossible to review the entire field of polymerization in the present chapter, so the reader is encouraged to consult specialized textbooks for detailed discussion about polymerization mechanisms, kinetics and processes (Odian, 2004). Still, the most fundamental points of polymerization kinetics are described in the following sections, assuming that reactions are performed in batch reactors and in solution, according to some relatively simple kinetic mechanisms. Although the presentation here is somewhat restrictive, we believe it will be useful, introducing the reader to the wonderful world of polymerization reactions.

8.1   FUNDAMENTAL ASPECTS OF STEP POLYMERIZATIONS

Typical step polymerization reactions involve monomer molecules that present multiple organic functions that can react with each other, such as the hydroxyl and carboxyl groups that can react to produce an ester group and a short byproduct molecule (in this case, water). Multiple chemical functionalities are needed to generate a macromolecule, as chain growth implies that reaction must take place at both sides of the monomer molecule. For instance, assuming that a monomer molecule contains two chemical functions A and B that can react, the following mechanism can be proposed:

$$A - (R)_i - B + A - (R)_j - B \leftrightarrow A - (R)_{i+j} - B + C \qquad (8.3a)$$

$$P_i + P_j \leftrightarrow P_{i+j} + C \qquad (8.3b)$$

where R represents a repetitive mer and C is a reaction byproduct. One must observe that Equation (8.3) represents infinitely many reaction steps, as i and j can vary from 1 to infinity. Therefore, despite the simple representation of Equation (8.3), the polymerization mechanism is indeed very complex! In addition, one should observe that typical organic reactions are slow and reversible, which means both the direct polymerization step and the inverse depolymerization reaction must be considered. This explains why the byproduct molecule must be removed from the reaction vessel, so high monomer conversions can be achieved. This also explains why step polycondensations are usually performed at high temperatures and under low pressures (vacuum conditions), to enhance the reaction rates and the rates of removal of byproducts. These characteristics are the reasons for the peculiar designs of the vessels used in polymerization reactions.

Based on the proposed reaction mechanism, it becomes possible to write the following set of mass balance equations:

$$\frac{dP_i}{dt} = \frac{\sum_{k=1}^{i-1} k_p \frac{P_k}{V} \frac{P_{i-k}}{V} V}{2} - k_p \frac{P_i}{V} \left[\sum_{k=1}^{\infty} \frac{P_k}{V} \right] V - k_d (i - 1) \frac{P_i}{V} \frac{C}{V} V$$

$$+ \left[\sum_{k=i+1}^{\infty} k_d \frac{2}{(k - 1)} (k - 1) \frac{P_k}{V} \frac{C}{V} \right] V$$

(8.4)

where k_p represents the kinetic rate constant of polymerization (esterification) and k_d stands for the kinetic rate constant of depolymerization (hydrolysis). Many important points regarding Equation (8.4) should be observed:

1.   Equation (8.4) describes an infinite set of coupled mass balance equations that must be solved simultaneously, with i varying from 1 to infinity;
2.   The first term on the right-hand side of Equation (8.4) represents the formation of chains of size i as a consequence of polymerization reactions between chains with size smaller than i;

3. The first term on the right-hand side of Equation (8.4) is divided by 2 because of the inherent symmetry of the proposed problem, as reactions between chains of size k and $i - k$ and between chains of size $i - k$ and k are the same and should not be counted twice;
4. The second term on the right-hand side of Equation (8.4) represents the consumption of chains of size i when they react with other chains to form new chains of larger size;
5. The third term on the right-hand side of Equation (8.4) represents the consumption of chains of size i due to depolymerization;
6. The third term on the right-hand side of Equation (8.4) is multiplied by $(i - 1)$ because chains of size i contain $(i - 1)$ chemical groups that were the result of polymerization reactions and therefore subject to depolymerization;
7. The fourth term on the right-hand side of Equation (8.4) represents the formation of chains of size i because of the depolymerization of larger chains;
8. The fourth term on the right-hand side of Equation (8.4) is multiplied by $2/(k - 1)$ because only two of the $(k-1)$ chemical groups of a chain of size $k > i$ can produce a chain of size i after depolymerization;
9. The volume V is explicitly represented in Equation (8.4) to remind the reader that the density of the reaction system can change during the reaction course, leading to volume changes that should not be neglected *a priori*.

For all the reasons presented above, the system of equations represented by Equation (8.4) does not present a closed analytical solution and should be solved almost always with help of advanced numerical tools. In order to develop simple closed solutions that can be used for analytical interpretation of the reaction effects, some additional assumptions must be made. Initially, to avoid the necessary handling of infinitely many mass balance equations, it may be necessary to monitor the evolution of the concentration of reactive chemical groups, which leads to the functional modeling approach. In this case, Equation (8.3) can be represented in the form:

$$A + B \leftrightarrow AB + C \tag{8.5}$$

In this case, only four mass balance equations are needed to describe the course of the reaction, in the form:

$$\frac{dA}{dt} = -k_p \frac{A}{V}\frac{B}{V}V + k_d \frac{AB}{V}\frac{C}{V}V \tag{8.6}$$

$$\frac{dB}{dt} = -k_p \frac{A}{V}\frac{B}{V}V + k_d \frac{AB}{V}\frac{C}{V}V \tag{8.7}$$

$$\frac{dAB}{dt} = k_p \frac{A}{V}\frac{B}{V}V - k_d \frac{AB}{V}\frac{C}{V}V \tag{8.8}$$

$$\frac{dC}{dt} = k_p \frac{A}{V}\frac{B}{V}V - k_d \frac{AB}{V}\frac{C}{V}V - F_C \tag{8.9}$$

where F_C represents the rate of removal of the byproduct C from the reactor vessel. If C is removed efficiently from the reaction vessel, so that $C = 0$, then the functional mass balance equations become:

$$\frac{dA}{dt} = -k_p \frac{A}{V} \frac{B}{V} V \tag{8.10}$$

$$\frac{dB}{dt} = -k_p \frac{A}{V} \frac{B}{V} V \tag{8.11}$$

$$\frac{dAB}{dt} = k_p \frac{A}{V} \frac{B}{V} V \tag{8.12}$$

$$F_C = k_p \frac{A}{V} \frac{B}{V} V \tag{8.13}$$

Additionally, Equations (8.10–12) can be combined in the form:

$$\frac{dA}{dt} = -k_p \frac{A}{V} \frac{B}{V} V \tag{8.14}$$

$$\frac{d(A - B)}{dt} = 0 \Rightarrow A - B = A_0 - B_0 \tag{8.15}$$

$$\frac{d(A + AB)}{dt} = 0 \Rightarrow A + AB = A_0 + AB_0 \tag{8.16}$$

Inserting Equation (8.15) into Equation (8.14), then:

$$\frac{dA}{dt} = -k_p \frac{A}{V} \frac{(A + B_0 - A_0)}{V} V \tag{8.17}$$

Assuming that volume effects can be neglected (which should not be taken for granted), and if reactants are fed in the stoichiometric conditions, so that $A_0 = B_0$, then:

$$\frac{A}{A_0} = \frac{B}{B_0} = \frac{A_0 - AB}{A_0} = \frac{1}{1 + \frac{k_p A_0 t}{V}} \tag{8.18}$$

On the other hand, if reactant B is added in excess with respect to the stoichiometric conditions (and A is assumed to be the limiting reactant), then:

$$\frac{A}{A_0} = \frac{B + A_0 - B_0}{A_0} = \frac{A_0 - AB}{A_0} = \frac{(B_0 - A_0) \exp\left(-\frac{k_p (B_0 - A_0) t}{V}\right)}{B_0 - A_0 \exp\left(-\frac{k_p (B_0 - A_0) t}{V}\right)} \tag{8.19}$$

If one defines the feed excess F in the form:

$$F = \frac{B_0 - A_0}{A_0} \tag{8.20}$$

then Equation (8.19) becomes:

$$\frac{A}{A_0} = \frac{B + A_0 - B_0}{A_0} = \frac{A_0 - AB}{A_0} = \frac{F\, exp\left(-\frac{kpFA0t}{V}\right)}{F + 1 - exp\left(-\frac{kpFA0t}{V}\right)} \tag{8.21}$$

Figure 8.1 illustrates the effect of the feed excess on the course of polymerization for a particular set of conditions. It can be observed that the increase of F leads to acceleration of the reaction rates, as one might already expect when A_0 is kept constant. As it might also be expected, the concentration of reactant A approaches zero as time evolves, while the concentration of B approaches a value that depends on the feed excess.

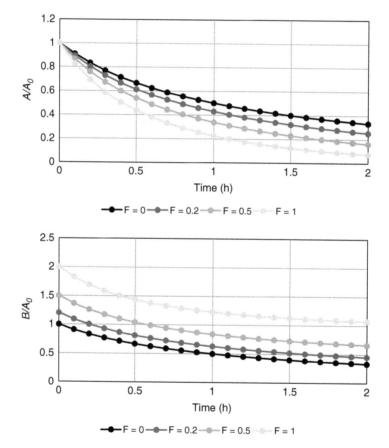

Figure 8.1 Effect of feed excess on the course of polymerization when $A_0 = 1$ mol, $V = 1$ L and $k_p = 1$ L· mol^{-1}·h^{-1}.

8.1.1 The most probable Schulz-Flory distribution

It must be observed that a significant amount of information is lost when the full set of mass balance equations presented in Equation (8.4) is replaced by the functional balances presented in Equations (8.6–9). For instance, Equation (8.4) allows for the calculation of the full molecular weight distribution of the produced polymer material, whereas this is impossible with the functional model. For this reason, it becomes necessary to perform a more involved analysis of the reacting system in order to provide information about the quality of the material that is manufactured when the functional approach is employed. Such an analysis was originally carried out by Wallace Carothers and Paul J. Flory in some of the most fundamental publications of the field (Flory, 1936).

In order to comprehend Flory's approach, it is initially necessary to recognize that the number of molecules in the reacting system decreases continuously when the polymerization reaction takes place. As Equation (8.3) indicates, the number of reacting species present in the reacting mixture is reduced by one unit when a reaction between the chemical groups A and B take place, assuming that the condensate byproduct is properly removed from the system. Therefore, it is possible to write that:

$$N_0 = \sum_{i=1}^{M} N_{i0} = \frac{A_0 + B_0}{f} \tag{8.22}$$

$$f = \frac{\sum_{i=1}^{M} f_i N_{i0}}{\sum_{i=1}^{M} N_{i0}} \tag{8.23}$$

where N_0 is the total number of molecules inside the reacting system at the beginning of the reaction, M is the number of different chemical species that contain the reactive chemical groups, N_{i0} is the total number of molecules of the reactive chemical species i at the beginning of the reaction, f_i is the functionality of the chemical species i (number of reactive functional groups of chemical species i) and f is the average functionality of the system. Therefore, if a single reactive chemical species that contains the reactive groups A and B is added to the mixture, then $M= 1$, $f_1 = 2$ and $f = 2$. Assuming that reactant A is the limiting one, whenever it is consumed in a polymerization step the number of molecules decreases, so that:

$$N_t = N_0 - (A_0 - A) \tag{8.24}$$

where N_t is the number of molecules in the reacting system at time t. If the number of molecules decreases, as the molecular segments remain the same inside the reaction vessel, the average size (or mass) of the molecules must increase. As a matter of fact, it becomes possible to write:

$$\bar{i} = \frac{N_0}{N_t} = \frac{N_0}{N_0 - (A_0 - A)} \tag{8.25}$$

where $\bar{i}$ is the average size of the molecules inside the reactor as a function of the reaction time or of the conversion. Finally,

$$\bar{i} = \frac{\frac{(A_0 + B_0)}{f}}{\frac{(A_0 + B_0)}{f} - A_0 x} = \frac{(2 + F)}{(2 + F) - f x} \tag{8.26}$$

which is a beautiful equation that clearly shows that the average size of the produced polymer material is a function of the feed excess, of the average functionality of the molecular species and of the conversion of the limiting species. In the ideal case, when $F = 0$ and $f = 2$, then

$$\bar{i} = \frac{1}{1 - x} \tag{8.27}$$

which shows that the average molecular size increases with x and approaches infinity as the conversion approaches the limiting value of 1. However, Equation (8.27) also shows that molecular chain size grows very slowly with x in the beginning, reaching the value of 10 (therefore, a mixture of short molecules, or oligomers) when the conversion reaches the value of 90%. This reinforces the previous observation that condensate byproducts must be removed from the reacting mixture to allow the preparation of large molecules, as very large conversion values must be attained to produce long polymer chains.

Example

E8.1 The average functionality of the feed mixture is rarely equal to 2 for many important practical reasons. First, monomers cannot be produced with purity of 100% and are often contaminated with compounds of different functionalities. Besides, as shown in Equation (8.27), conversions must be too high to allow the production of large molecules, which encourages the addition of multifunctional molecular species in the initial reacting mixture to increase the value of f and allow the manufacture of large molecules with lower conversion values. Besides, it may be necessary to compensate for possible variations of F, as preparation of the feed mixture is subject to perturbations due to the finite precision of actual equipment and the monomer feed can be impure. For these reasons, Figures 8.2 and 8.3 show the effects of f and F on the average sizes of produced polymer chains.

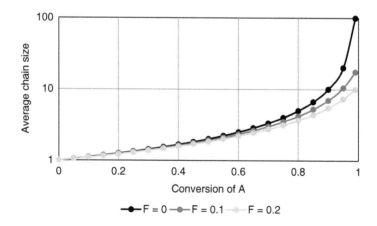

Figure 8.2 Effect of F on average chain sizes ($f = 2$).

Initially, Figure 8.2 shows that the feed excess exerts a highly deleterious effect on produced chain size, leading to dramatic modification of the maximum attainable molecular size and making the manufacture of large polymer chains more difficult. As a matter of fact, when $f = 2$, the maximum attainable chain size predicted with Equation (8.26) becomes equal to

$$\bar{i}_{max} = \frac{(2 + F)}{F}$$

which is equal to 201 when $F = 0.01$. This illustrates the enormous sensitivity of $\bar{i}_{max}$ to variations of F.

On the other hand, Figure 8.3 shows that the increase of the average molecular functionality leads to an increase of the average molecular chain sizes, as one might already expect, since the molecules can grow in more than two directions. This can indeed lead to formation of molecular chains of extremely large sizes at relatively low conversion values, which can cause the formation of gel (polymer material that cannot be dissolved in organic solvents). The formation of gel constitutes a major technological problem in many polymerization systems, as is well discussed in many papers published in the open literature (Nogueira *et al.*, 2005). Particularly, the denominator of Equation (8.26) becomes equal to zero when the conversion attains the gel point, as shown below:

$$x_{gel} = \frac{2 + F}{f}$$

Therefore, the gel point presents a physical significance whenever $f > 2 + F$ and $x_{gel} < 1$.

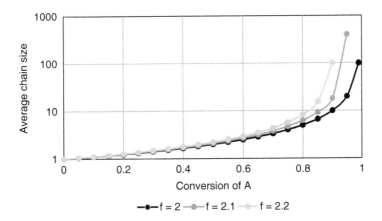

Figure 8.3 Effect of f on average chain sizes ($F = 0$).

From a statistical point of view, if one assumes that all A chemical groups can react with B chemical groups with equal probability, P_A, then the following set of probabilities, P_i, can be assigned to formation of chains of size i in the reacting system:

$$P_i = P_A^{i-1}(1 - P_A) \tag{8.28}$$

as $(i - 1)$ bonds must be formed to produce a chain of size i. As a matter of fact, Equation (8.28) does define a consistent set of probabilities, as:

$$\sum_{i=1}^{\infty} P_i = (1 - P_A) \sum_{i=1}^{\infty} P_A^{i-1} = (1 - P_A) \frac{1}{(1 - P_A)} = 1 \tag{8.29}$$

In order to define how probability P_A can be related to the physical operation variables, it is important to consider that the following mass balance conservation law has to be satisfied:

$$N_0 = \sum_{i=1}^{\infty} i N_i = N_t \sum_{i=1}^{\infty} i P_i = (N_0 - A_0 x) \sum_{i=1}^{\infty} i P_i$$

$$= \left(\frac{A_0 + B_0}{f} - A_0 x \right) \frac{1}{(1 - P_A)} = \left(\frac{(2 + F) A_0}{f} - A_0 x \right) \frac{1}{(1 - P_A)}$$

$$= \left(\frac{A_0 + B_0}{f} - A_0 x \right) \frac{1}{1 - P_A} = \left(\frac{(2 + F) A_0}{f} - A_0 x \right) \frac{1}{1 - P_A} \tag{8.30}$$

Therefore:

$$N_0 = \frac{(2 + F) A_0}{f} = \left(\frac{(2 + F) A_0}{f} - A_0 x \right) \frac{1}{(1 - P_A)} \tag{8.31}$$

and

$$P_A = 1 - \frac{2 + F - fx}{2 + F} = \frac{fx}{2 + F} \tag{8.32}$$

which is simply equal to x when the ideal case is considered, with $f = 2$ and $F = 0$. Based on the previous discussion, the normalized chain size distribution can be obtained by replacing Equation (8.32) into Equation (8.28) in the form:

$$P_i = \left(\frac{fx}{2 + F} \right)^{i-1} \left(\frac{2 + F - fx}{2 + F} \right) \tag{8.33}$$

which is the well-known Schulz–Flory distribution of polymer chain sizes obtained in ideal step polymerizations. Figure 8.4 illustrates the evolution of chain sizes when $F = 0$ and $f = 2$ as a function of conversion. It must be observed that the distribution has an exponential shape and is not symmetrical, thus making it meaningless to represent polymer chain size distributions with the normal distribution.

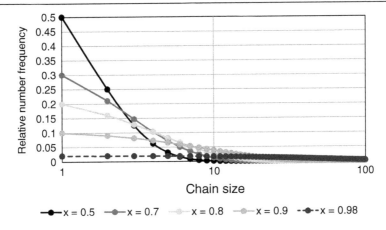

Figure 8.4 Effect of conversion on the chain size distributions ($f = 2, F = 0$).

8.2 FUNDAMENTAL ASPECTS OF CHAIN POLYMERIZATIONS

Typical chain polymerization reactions involve monomer molecules that present unsaturated double bonds (vinyl groups) that can react with active species, including free-radical species (free-radical polymerizations), metal complexes (as in Ziegler–Natta polymerizations) or unstable ionic species (such as carbocations and carbanions in ionic polymerizations). The sequential insertion of monomer molecules into the growing polymer chains characterizes this large set of distinct polymerization mechanisms, in the form:

$$I - (R)_i * + M \leftrightarrow I - (R)_{i+1} * \tag{8.34a}$$

$$P_i + M \leftrightarrow P_{i+1} \tag{8.34b}$$

where I represents a fragment of the original active species, R represents a repetitive mer and * describes the propagation of the chain activity after monomer insertion; otherwise, polymer chains would not be formed. As discussed previously, one must observe that Equation (8.34) represents infinitely many reaction steps, since i can vary from 1 to infinity. Therefore, despite a very simple representation of Equation (8.33), once more the polymerization mechanism is indeed complex! Unlike typical step polymerizations, chain reactions are usually extremely fast and essentially irreversible, making it possible to neglect the inverse depolymerization step in most practical cases. This explains why chain polymerizations are usually performed at milder temperatures than are most step polymerizations.

So, chain polymerizations can involve many reaction steps, including chain initiation, chain propagation, chain termination and chain transfer to other chemical species (Odian, 2004). For this reason, the most important reaction steps are presented in the Sections 8.2.1–5.

8.2.1 Initiation

Initiation consists in the formation of primary active species that can initiate the chain polymerization reaction. Initiators can be provided by manufacturers as active species (for instance, as active Ziegler–Natta catalysts) or they can be produced *in situ* during the reaction (as in free-radical polymerizations). Therefore, the initiation step may be absent in reactions that make use of previously prepared active compounds to initiate the polymer chains. The interested reader should consult standard polymerization textbooks for comprehensive presentation of the very large number of initiators used commercially in polymerization technologies (Odian, 2004).

The most common families of initiators used to initiate polymerization reactions are Ziegler–Natta complexes (frequently titanium complexes containing alkyl groups and chlorine atoms, supported on inorganic materials such as silica and $MgCl_2$) and free-radical initiators (unstable compounds that are subject to homolysis when excited thermally, magnetically, electromagnetically or through ultrasound, among other sources of excitation). The commonest free-radical generators used commercially are peroxides, such as benzoyl peroxide (BPO), and azo compounds, such as azobisisobutyronitrile (AIBN).

Initiation reactions can usually be described in the form:

$$I(+ Cocat) \rightarrow 2R* \qquad (8.35)$$

$$R* + M \rightarrow P_1* \qquad (8.36)$$

where P_1 is a growing chain of size 1. Equation (8.35) indicates that a cocatalyst can be eventually used to provoke the generation of the active species, as in redox initiation mechanisms and Ziegler–Natta reactions (Odian, 2004). The active species then initiates the polymer reaction through addition of a monomer molecule into the polymer chain. When typical thermal initiation is performed, the use of a cocatalyst is not necessary and it becomes possible to write:

$$\frac{dI}{dt} = -k_d I \qquad (8.37)$$

so that

$$I = I_0 e^{-k_d t} \qquad (8.38)$$

where k_d is the kinetic rate constant for the initiator decomposition. Equation (8.38) describes the characteristic exponential decay of free-radical initiators(and of other catalysts that are subject to deactivation or contamination). The half-life time ($t_{1/2}$) of the initiator is the time required for the initiator concentration to reach half the initial concentration, which is equal to:

$$t_{1/2} = \frac{\ln(2)}{k_d} \qquad (8.39)$$

Therefore, the half-life time of an initiating species is also a measure of the stability of the analyzed species, decreasing with the increase of k_d.

The mass balance for the active species $R*$ can be written in the forms:

$$\frac{dR^*}{dt} = 2k_d[I]V - k_1[R^*][M]V \tag{8.40a}$$

$$\frac{dR^*}{dt} = -k_1[R^*][M]V \tag{8.40b}$$

depending on whether the active species is generated *in situ* or fed in the active state. k_1 is the kinetic rate constant for initiation. When radicals are generated *in situ*, the activity is usually so high that active species generate polymer chains almost immediately, allowing the utilization of the quasi-steady state assumption in the form:

$$\frac{dR^*}{dt} \approx 0 \Rightarrow 2k_d[I]V = k_1[R^*][M]V \tag{8.40c}$$

so that the rate of initiation can be assumed to be equal to the rate of generation of active species.

8.2.2 Propagation

This certainly is the most important reaction step of all chain polymerization mechanisms, because it describes the growth of the polymer chains.

$$P_i^* + M \rightarrow P_{i+1}^* \tag{8.41}$$

As infinitely many reaction steps are represented simultaneously in Equation (8.41), the rate equations related to the propagation step can be written in the form:

$$R_{pPi} = k_p[P_{i-1}][M]V - k_p[P_i][M]V \tag{8.42}$$

$$R_{pM} = -k_p \left(\sum_{i=1}^{\infty} [P_i] \right) [M]V = -k_p[P_T][M]V \tag{8.43}$$

as chains of size i are formed through propagation of shorter chains and consumed through reaction with the monomer, while monomer molecules can react with all growing chains of all sizes. k_p is the kinetic rate constant for propagation and represents the total concentration of active growing chains in the reacting mixture. An important assumption in Equations (8.42—43) is that the propagation rate constant does not depend on the size of the growing species, a controversial idea that has been refuted by many studies in the literature (Odian, 2004; Smith *et al.*, 2005; Heuts and Russell, 2006). However, if k_p is assumed to depend on the chain size, the number of kinetic parameters can become too high to be of practical use.

8.2.3 Termination

The termination step describes the loss of activity of the growing polymer chains and leads to formation of inactive (or more commonly dead) polymer chains. Many different reaction steps can lead to inactivation of the growing chains, as shown below:

$$P_i * + P_j * \rightarrow D_i + D_j^=$$
(8.44a)

$$P_i * + P_j * \rightarrow D_{i+j}$$
(8.44b)

$$P_i * + X \rightarrow D_i$$
(8.44c)

$$P_i * \rightarrow D_i$$
(8.44d)

Equation (8.44a) describes the termination by disproportionation and is usually associated with the abstraction of hydrogen atoms from active species, leading to formation of unsaturated dead chains (or macromers, which eventually can be involved in other mechanistic steps and affect the chemical and thermal stability of the final material). This termination mechanism is predominant in polymerizations of vinyl acetate and methyl methacrylate and leads to dead chains that are similar to the growing active chains. Equation (8.44b) describes termination by combination and is predominant in free-radical styrene polymerizations, leading to dead chains that are larger than the original growing chains. Equations (8.44c–d) describe termination steps provoked by smaller molecules (usually a contaminant) and by spontaneous rearrangements of the active center of the growing active chain, respectively, being predominant in ionic and coordinated polymerizations of olefins.

The occurrence of termination steps necessarily causes the appearance of at least one additional family of polymer molecules: the dead polymer chains. For this reason, the kinetic modeling of this reacting system can become quite complex, requiring detailed modeling of infinitely many reacting species of multiple families of polymer chains. For these reasons, closed analytical solutions of the mass balance equations are almost always unavailable. In spite of this, termination rate equations can be written for the species P_i in the form:

$$R_{tPi} = - k_{td}[P_i] \left(\sum_{k=1}^{\infty} [P_k] \right) V - k_{tc}[P_i] \left(\sum_{k=1}^{\infty} [P_k] \right)$$
$$V - k_{tX}[P_i][X]V - k_{ts}[P_i]V$$
(8.45)

where k_{td}, k_{tc}, k_{tX} and k_{ts} stand respectively for the rate constants for termination by disproportionation, termination by combination, termination by X and spontaneous termination. Similarly, for species D_i and the rate equations become:

$$R_{tDi} = k_{td}[P_i]\left(\sum_{k=1}^{\infty}[P_k]\right)V + \frac{k_{tc}}{2}\left(\sum_{k=1}^{i-1}[P_k][P_{i-k}]\right)$$
$$V + k_{tX}[P_i][X]V + k_{ts}[P_i]V \tag{8.46}$$

$$R_{tD=} = k_{td}[P_i]\left(\sum_{k=1}^{\infty}[P_k]\right)V \tag{8.47}$$

In some reaction mechanisms, the termination steps may be absent, leading to so called living polymerization reactions (Odian, 2004). Besides, as discussed previously, the proposed analysis relies on the assumption that termination rate constants do not depend on the size of the growing species, an idea that also constitutes a controversial subject and which also has been refuted by many studies in the literature (Odian, 2004; Smith et al., 2005; Heuts and Russell, 2006). However, if k_{td}, k_{tc}, k_{tX} and k_{ts} are assumed to depend on the chain size, as said previously the number of kinetic parameters can become too high to be of practical use.

8.2.4 CHAIN TRANSFER

Among the many parallel reaction steps that can occur during chain polymerizations, chain transfer to other molecules are certainly the most important ones. The chain transfer step describes the transfer of activity from the growing polymer chain to another molecule, leading to formation of a dead polymer chain and a short active species that reinitiates polymerization. Although this reaction step exerts little effect on the overall rates of polymerization (as the number of active species is not changed), it can significantly affect the quality of the obtained material (as it interrupts the growth of a polymer chain), being of great importance at industrial sites for the manufacture of distinct polymer grades with different properties. Chain transfer reactions can be represented in the general form:

$$P_i * + X \rightarrow D_i + P_1* \tag{8.48a}$$

$$P_i * \rightarrow D_i + R* \tag{8.48b}$$

with rate equations given in the form:

$$R_{trPi} = - k_{trX}[P_i][X]V - k_{trs}[P_i]V \tag{8.49a}$$

$$R_{trDi} = k_{trX}[P_i][X]V + k_{trs}[P_i]V \tag{8.49b}$$

where k_{trX} and k_{trs} are kinetic rate constants for chain transfer to X and spontaneous chain transfer. Typical chain transfer agents (X) for free-radical polymerizations are mercaptans and alcohols, while hydrogen is the most important chain transfer agent in coordinated polymerizations. Besides, ionic and coordinated polymerizations are also subject to spontaneous chain transfer reactions. Monomer and polymer can also be important chain transfer agents in many reacting systems. For instance, the monomer vinyl chloride is a natural chain transfer agent in vinyl chloride polymerizations, while poly(vinyl acetate) is an important chain transfer agent in vinyl acetate polymerizations. Particularly, occurrence of chain transfer to polymer leads to formation of branched macromolecules, with great impact on the final properties of the obtained materials.

8.2.5 QUASI-STEADY STATE BALANCES

When the active growing species is very active, produced *in situ* and subject to termination reactions, accumulation in the reacting mixture is not likely to occur. This is exactly the case of the vast majority of the free-radical reaction mechanisms. In this case, the sum of all reaction rates is expected to be very close to zero, allowing the proposition of quasi-steady state balances in the form:

$$R_{pPi} + R_{tPi} + R_{trPi} = 0 \tag{8.50}$$

If Equations (8.42), (8.45) and (8.49) are added, one can obtain the following quasi-steady state balance equation:

$$k_p[P_{i-1}][M]V - k_p[P_i][M]V - k_{td}[P_i]\left(\sum_{k=1}^{\infty}[P_k]\right)V - k_{tc}[P_i]\left(\sum_{k=1}^{\infty}[P_k]\right)V$$
$$- k_{tX}[P_i][X]V - k_{ts}[P_i]V - k_{trX}[P_i][X]V - k_{trs}[P_i]V = 0 \tag{8.51}$$

which can be written in the form

$$[P_i] = q[P_{i-1}] \tag{8.52}$$

$$q = \frac{k_p[M]}{k_p[M] + (k_{td} + k_{tc})[P_T] + (k_{tX} + k_{trX})[X] + (k_{ts} + k_{trs})} \tag{8.53}$$

q is normally named as the propagation probability, as it is a positive number constrained in the interval [0,1] and represents the frequency ratio of the propagation reaction, when compared to the remaining mechanistic steps. Equation (8.52) leads to a very simple iterative equation, that can be written in the form:

$$[P_i] = q^{i-1}[P_1] \tag{8.54}$$

that indicates that the molecular weight distribution of growing polymer chains also follows the Schulz—Flory distribution! It is amazing that the Schulz—Flory distribution appears in both the analyzed polymerization mechanisms by way of such different reasonings. However, one should not be tempted to conclude that all polymerization mechanisms lead to Schulz–Flory size distribution, although it is true that it appears naturally in many different problems.

Equation (8.54) is inadequate because it depends on [P1] and because Equation (8.53) depends on [P_T]. Therefore, it is necessary to provide closed solutions for these two values. Initially, it is simple to show that:

$$[P_T] = \sum_{i=1}^{\infty} [P_i] = [P_1] \sum_{i=1}^{\infty} q^{i-1} = \frac{[P_1]}{1-q} \qquad (8.55)$$

so that

$$[P_i] = (1-q) \, q^{i-1} [P_T] \qquad (8.56)$$

Finally, if Equation (8.51) is summed for all values of i ranging from 1 to infinity, one gets (and note the fact that when $i=1$ some additional rate equations are needed because of the mechanistic steps described in Equations (8.36) and (8.48)):

$$2k_d[I] = (k_{td} + k_{tc}) \, [P_T]^2 + (k_{tX} + k_{ts}) \, [P_T] \qquad (8.57)$$

which has the general solution

$$[P_T] = \sqrt{\frac{(k_{tX} + k_{ts})^2 + 8 \, (k_{td} + k_{tc}) \, k_d[I]}{4 \, (k_{td} + k_{tc})^2}} \qquad (8.58)$$

Because k_{tX} and k_{ts} are usually equal to zero in most free-radical polymerizations, and since part of the active free-radical generated by the initiator are consumed in lateral reactions with impurities, solvents and reactants, Equation (8.58) is usually presented in the form:

$$[P_T] = \sqrt{\frac{2f_I k_d[I]}{k_{td} + k_{tc}}} \qquad (8.59)$$

where f_I is the initiator efficiency, a positive number in the interval [0,1], usually in the interval [0.6, 0.8], that indicates the fraction of active species that actually initiate growing polymer chains.

Finally, Equation (8.59) can be combined with Equation (8.43) in the form:

$$R_{pM} = -k_p \sqrt{\frac{2f_I k_d[I]}{k_{td} + k_{tc}}} [M]V \qquad (8.60)$$

to provide the rate of polymerization of free-radical polymerizations and other polymerization mechanisms that are subject to bimolecular terminations. In this case, it becomes possible to write the monomer balance equation in the form:

$$\frac{dM}{dt} = -k_p \sqrt{\frac{2f_I k_d[I]}{k_{td} + k_{tc}}} [M]V = -k_p \sqrt{\frac{2f_I k_d[I_0] e^{-k_d t}}{k_{td} + k_{tc}}} M \qquad (8.61)$$

$$\frac{dM}{M} = -Ke^{-\frac{k_d t}{2}}$$
(8.62)

with

$$K = k_p \sqrt{\frac{2 f_I k_d [I_0]}{k_{td} + k_{tc}}}$$
(8.63)

where K is the parametric lump that controls the course of the rate of polymerization. It is important to observe that the rate of polymerization depends on many distinct kinetic parameters and operation conditions, as shown in parameter K. The solution of Equation (8.62) is:

$$\ln\left(\frac{M}{M_0}\right) = \frac{2K}{k_d}\left(e^{-\frac{k_d t}{2}} - 1\right)$$
(8.64a)

$$M = M_0 e^{\left[\frac{2K}{k_d}\left(e^{-\frac{k_d t}{2}} - 1\right)\right]}$$
(8.64b)

It is interesting to observe in Equation (8.64b) that the monomer mass is not depleted even after infinite reaction time. As t goes to infinity, the monomer mass goes to:

$$M = M_0 e^{\left[-\frac{2K}{k_d}\right]}$$
(8.65)

tabulation the importance of developing operation strategies to remove residual monomer from final polymer product.

Based on Equation (8.54) it is possible to conclude that the average chain-length of living chains can be given in the form:

$$\bar{i} = \frac{\sum_{i=1}^{\infty} i[P_i]}{\sum_{i=1}^{\infty}[P_i]} = \frac{\frac{[P_T]}{(1-q)}}{[P_T]} = \frac{1}{(1-q)}$$
(8.66)

As q usually is very close to 1 (typically above 0.99), living chains are very large from the very beginning of the reaction (conversion equal to zero), constituting a major difference with respect to step polymerizations, as average molecular weights increase with conversion in the last case and become large only when conversions are high.

8.3   DIFFUSIVE LIMITATIONS

The viscosity of the reacting mixture can become so high during polymerizations that the diffusive movement of the chemical species, and particularly of the larger molecules, can become limited, leading to significant reduction of reaction rates. As a matter of fact, in most polymerization processes, all elementary reaction steps can be eventually affected by diffusional limitations when the polymer concentration is high. Therefore, the increase of the medium viscosity caused by the increase of the polymer concentration can greatly affect overall reaction rates and molecular structure of the final material.

The gel effect (or Trommsdorff effect) occurs typically when monomer conversion (or, more precisely, polymer concentration) reaches a range between 20 to 40 wt% and the mobility of the longer chains is reduced by the high viscosity, leading to drastic reduction of bimolecular termination rates, resulting in accumulation of growing radicals and the auto-acceleration of the rate of polymerization (Odian, 2004). Particularly, the final properties of the produced polymer materials can be significantly affected by this phenomenon. Moreover, this can also lead to production of resins with broad MWDs and cause the sudden elevation of the reactor temperature and development of thermal runaway conditions. (Hui and Hamielec, 1972; Pollaco *et al.*, 1996; Alexopoulos and Kiparissides, 2007). Additionally, the gel effect can induce the development of relevant nonlinear behavior in the dynamics of free radical polymerizations, especially when the increasing viscosity also causes the reduction of the effective heat transfer coefficient between the reactor walls and the coolant fluid in the reactor jacket. This can lead to a significant decrease of the heat removal capacity of the reactor, resulting in thermal runaway, development of self-sustained oscillations and appearance of multiple steady states (Pinto, 1995; Pinto and Ray, 1995a, 1995b, 1996).

The glass effect occurs when the bimolecular reaction steps that involve small molecules, such as the propagation and chain transfer steps, are also affected by the increasing viscosity of the reaction medium, causing the subsequent reduction of the reaction rates and average molecular weights of the formed chains. As in the previous case, these phenomena lead to enlargement of the MWD and strongly affect the final properties of the produced polymer. The glass effect occurs when the reaction temperature is below the glass transition temperature of the polymer, usually at monomer conversions (or, more precisely, polymer concentrations) above 80 wt%. The consequence of this phenomenon is the "freezing" of the reaction mixture and attainment of a maximum conversion limit.

Modeling of diffusional limitations in polymerization reactions still constitutes a technical challenge and most of the models that have been published in this field are empirical or semi-empirical, depending on model fitting to available experimental data. The proposed models usually have the general form:

$$k = k_0 g(w) \tag{8.67}$$

tabulation k is a generic kinetic rate constant, k_0 is the kinetic rate constant when the polymer concentration is null and is an empirical or semi-empirical function that corrects the kinetic rate constant when the polymer concentration changes. For instance, Freitas Filho *et al.* (1994) showed that the expression:

$$g(w) = (1 - w)^{\gamma} \tag{8.68}$$

where γ is an adjustable parameter that characterizes a certain reaction step and polymerization system, might be used to fit many experimental data sets available in the literature.

8.4 DEPOLYMERIZATION

Although the assumption of irreversible reaction steps in most polymerization studies is reasonable when temperatures are mild and condensate byproducts of step polymerizations are absent, the fact is that this assumption can be bad when temperatures are high and condensate products are present in the reaction medium. For this reason, it is useful to define the ceiling temperature as the temperature when rates of polymerization and depolymerization are equal at typical polymerization conditions, preventing the growth of the polymer chains. In this case:

$$\Delta G = \Delta H - T_c \Delta S = 0 \Rightarrow T_c = \frac{\Delta H}{\Delta S} \tag{8.69}$$

As the polymerization reaction is usually quite exothermic and leads to significant reduction of entropy (as molecules get bonded in the macromolecular chain), T_c values are generally high, above 200 °C. Despite this, some exceptions include the well-known polymerization of α-methyl styrene, whose ceiling temperature is close to 40 °C, posing many technological challenges to performing polymerization under normal industrial conditions (Odian, 2004).

Interest in depolymerization reactions has increased steadily over the years. This interest has been boosted by the increasing worldwide production of plastic goods and the necessity to develop environmentally correct technologies for management and disposal of plastic wastes. Chemical recycling techniques constitute possible solutions for recycling of plastic wastes, converting plastic wastes into raw materials that can be used commercially in other industrial processes, reducing the potentially harmful effects of discarded materials. The most popular route for chemical recycling of plastic wastes is thermal degradation, which can be accomplished through pyrolysis, partial oxidation or gasification. Essentially, when submitted to high temperatures, polymer chains depolymerize (usually though free radical mechanisms) into lighter fractions, like that which is performed in typical petrochemical processes. However, thermal degradation of macromolecules can be significantly affected by the treated material. For instance, thermal treatment of polyolefins produces hydrocarbons in the range of gasoline, whereas poly(methyl methacrylate) decomposes almost completely into the original monomer above 400°C, offering an alternative for feedstock recycling (Braido et al., 2018; Ros et al., 2019; Monteiro et al., 2019). Solvolysis and hydrolysis are also popular chemical routes for depolymerization of step polymers, while enzymatic enhancement of depolymerization mechanisms has evolved rapidly (Monteiro et al., 2019).

Depolymerization reactions usually follow the random scission mechanism:

$$D_i \rightarrow P_k + P_{i-k} \tag{8.70a}$$

$$D_i + C \rightarrow P_k + P_{i-k} \tag{8.70b}$$

or the depropagation scheme:

$$D_i \rightarrow D_{i-1} + M \tag{8.71}$$

although depolymerization mechanisms normally comprise many distinct parallel reactions that take place simultaneously, leading to a broad mixture of products that must be purified for end use, as is well-known in the oil industry.

8.5 CONCLUDING REMARKS

In this chapter, the fundamental aspects of typical polymerization reactions were presented. It was shown that step polymerization reactions can be described with help of the functional modeling approach, revealing that the final properties of the produced material are quite sensitive to modification of the feed composition and functionality of the reacting monomers. Particularly, it was also shown that obtained polymer materials are mixtures of chains of different sizes, following the Schulz–Flory size distribution. It was also shown that chain polymerization reactions involve many different chemical species and reaction steps, although the adoption of the quasi-steady state assumption, valid when the active growing chains are very reactive and subject to termination, indicates that living polymer chains also follow the Schulz–Flory distribution. As the propagation probabilities of chain reactions usually are very close to 1, living chains are very large from the very beginning of the chain polymerization reaction, while average molecular weights increase slowly with conversion in step polymerizations and become large only when conversions are high. Moreover, it was shown that polymerization reactions can be subject to diffusive limitations, which can significantly affect the course of the polymerization and the behavior of industrial reactors. Finally, it was discussed that depolymerization reaction steps can be very important when temperatures are high or condensate products are present, an idea that has been explored in the chemical recycling of plastic wastes.

Example

E8.2 In the polymerization of methyl methacrylate in benzene, in the presence of azobiisobutylnitrile, the following data were generated in laboratory (adapted from Incomplete reference).

$[M]$ (mol/m^3)	$[I_2]$ (mol/m^3)	$(-r_M)$ (mol/m^3) s^{-1}
9.04	0.235	0.193
8.63	0.206	0.170
7.19	0.250	0.165
6.13	0.228	0.129
4.82	0.313	0.122
4.80	0.192	0.0937
4.22	0.235	0.0867
4.19	0.581	0.130
3.26	0.248	0.0715
2.07	0.211	0.0415

It is known that the reaction mechanism occurs in the following steps:

(a) Dissociation of the inhibitor into radicals.
(b) Addition of the monomer M to the radical.
(c) Chain propagation of radicals.
(d) Termination by addition of radicals.

Analyzing the data from the table, is it possible to propose a polymerization rate of the monomer including the different stages of the process?

With the data presented in the table, is it possible to determine the apparent reaction rate constant?

Solution

The problem data allow us to verify some of the reaction rate expressions derived previously, assuming that the initiation rate varies little with time.

Analyzing the effect of initiator concentration on the rate, we see that:

1. Choosing the same value of the concentration of M:

$[M]$ (mol/m^3)	$[I_2]$ (mol/m^3)	$(-r_M)$ (mol/m^3) s^{-1}
4.82	0.313	0.122
4.80	0.192	0.0937

or

$[M]$ (mol/m^3)	$[I_2]$ (mol/m^3)	$(-r_M)$ (mol/m^3) s^{-1}
4.22	0.235	0.0867
4.19	0.581	0.130

From the first data set, we observe that for the same concentrations of $[M]$, when the initiator concentration drops 1.63 times, its square root decreases 1.27 times. This approximately corresponds to a decrease in the rate of 1.3 times. In the second data set, we can see that when the initiator concentration is increased by 2.5 times, its square root increases by 1.57 times. This corresponds to an increase in the rate of 1.49 times. These data show that the rate is directly proportional to $[I_2]^{1/2}$. This in agreement with Equation (8.60), which can be written in the form:

$$(-r_M) = k_P[M]\sqrt{\frac{\gamma k_0[I_2]}{k_t}} \qquad (8.72)$$

2. Effect of the monomer concentration:
    Choosing approximately the same concentration value of $[I_2]$, we observe how the rate varies with the concentration $[M]$.

$[M]$ (mol/m^3)	$[I_2]$ (mol/m^3)	$(-r_M)$ (mol/m^3) s^{-1}
4.22	0.235	0.0867
9.04	0.235	0.193

or

[M] (mol/m³)	[I₂] (mol/m³)	(−r_M) (mol/m³) s⁻¹
7.19	0.250	0.165
3.26	0.248	0.0715

In this example, for both cases, maintaining the same concentration of $[I_2]$, the concentration of the monomer $[M]$ is varied. In the first data set, the concentration $[M]$ increases by 2.1 times while the rate increases by 2.2 times. In the second data set, the concentration of $[M]$ decreases by 2.2 times and the rate also decreases by about 2.3 times. This shows that the rate is directly proportional to the concentration of $[M]$, as described in Equation (8.60).

By making:

$$k^* = k_P \sqrt{\frac{2\gamma k_0 [I_2]}{k_t}} \tag{8.73}$$

Choosing any value from the for example:

[M] (mol/m³)	[I₂] (mol/m³)	(−r_M) (mol/m³) s⁻¹
9.04	0.235	0.193

Substituting the values

$$(-r_M) = k^*[M] \tag{8.72}$$

Then, one can obtain:

$$k* = 0.0213 \, \text{s}^{-1}$$

Substituting the values of k^* and $[I_2]$ into Equations (8.72) or (8.73), one can also obtain:

$$k_a = k_P \sqrt{\frac{k_0}{k_t}} = 0.146 \, (\text{m}^3/\text{mol s})$$

Example

E8.3 In the previous example, the values of the constants were calculated as:

$$k_0 = 10^{-8} \, \text{s}^{-1}$$
$$k_p = 10^{-2} \, \text{L/mol s}^{-1}$$
$$k_t = 5.107 \, \text{L/mol s}^{-1}$$

It is then proposed to calculate the concentration of the monomer after 7.2 min, starting from the following values and assuming that 1% of the initiator has been consumed:

$[M]$ (mol/m^3)	$[I_2]$ (mol/m^3)	$(-r_M)$ (mol/m^3) s^{-1}
9.04	0.235	0.193

Solution

Starting from Equation 8.64a or 8.72:

$$- \ln \frac{M}{M_0} = 2k_P \sqrt{\frac{2\gamma k_0 [I_{20}]}{k_t}} t$$

where

$$k^* = 2k_P \sqrt{\frac{2\gamma k_0 [I_{20}]}{k_t}} = 1.91 \times 10^{-3} \ (m^3/mol\,s)$$

Thus,

$$- \ln \frac{M}{M_0} = 0.822$$

But

$$[M] = 3.97 \ (mol/m^3)$$

And

$$X_M = \frac{M_0 - M}{M_0}$$

Thus, the conversion is equal to 0.56 or 56%.

Chapter 9

Kinetics of liquid-phase reactions

9.1 ENZYMATIC REACTIONS

The enzymatic reactions are of great interest and have great prospects for the future of biotechnology. In general, the enzymatic processes are well known in the alcohol fermentation and biological processes (e.g., physiology). The enzymatic fermentation processes can be promoted by microorganisms, such as bacteria and must, or by enzymes which are produced chemically.

Generally, a fermentation process is represented by the transformation of organic matter, which in the presence of enzymes or bacteria forms noble products of great utility in food, pharmaceutical, and alcohol production industries.

The main types of enzymatic and fermentation reactions that occur are:

Soluble enzyme + insoluble substrate
Insoluble enzyme + soluble substrate
Soluble enzyme + soluble substrate

The reactions with both soluble enzyme and substrate occur in homogeneous liquid phase in biological reactions, mainly used in the pharmaceutical industries. When the enzymes are insoluble, the reactions are heterogeneous and the great advantage is that these enzymes are not lost. In general, enzymes are anchored on solid materials with high surface area and low pore volume. Insoluble substrates and soluble enzymes are commonly used in the detergent industry.

Currently, biological reactions with animal cells are being studied. The kinetics of these particular reactions is equally complex but also belongs to the same reaction category.

The action of enzymes is also explained by the energies involved and by the transition state theory, as described before. Intermediate complexes are formed, which in turn have lower energy barrier, thus allowing its transformation into products.

The kinetic model can be represented schematically following the same approach as in Chapter 7, i.e.:

$$S + E \underset{}{\overset{k_2/k_1}{\longleftrightarrow}} ES^* \overset{k_3}{\longrightarrow} P + E \qquad (9.1)$$

where

> $S = $ substrate (reactant)
> $E = $ enzyme
> $ES^* = $ enzymatic complex.

Note that the presence of water in these reactions is essential. Equation 9.1 is represented by the following sequence:

(a)   Formation of the complex:

$$ES^*$$

(b)   Equilibrium state:

$$S + E \xrightleftharpoons[]{k_2/k_1} ES^* \tag{9.2}$$

(c)   Decomposition of the complex into products:

$$ES^*(+H_2O) \xrightarrow{k_3} P \tag{9.3}$$

$$ES^*(+H_2O) \xrightarrow{k_4} S$$

Figure 9.1 shows the potential energy over the course of the reaction. Without the enzymes, the energy barrier (or activation energy) is high and the complex ES^* formed is unstable and quickly deactivates. This energy barrier is drastically reduced in the presence of enzymes, facilitating the transformation of the complex ES^* into final irreversible products P and S. This behavior is explained by the transition state theory.

9.1.1 Kinetic model

The reaction rate relative to the substrate is obtained by Equation 9.1, i.e.:

$$(-r_S) = k_1[E][S] - k_2[ES^*] \tag{9.4}$$

$$r_P = k_3[H_2O][ES^*] \tag{9.5}$$

where $[\cdot]$ represents the concentration.

In relation to the complex ES^*, we also have:

$$r_{ES^*} = k_1[E][S] - k_2[ES^*] - k_3[ES^*][H_2O] = 0 \tag{9.6}$$

However, not all enzymes are consumed and the complexed enzymes are partly recovered. In this context, the free enzymes are those which participate in the initial reaction. From the balance of enzymes, we have:

$$E_T = E + ES^* \tag{9.7}$$

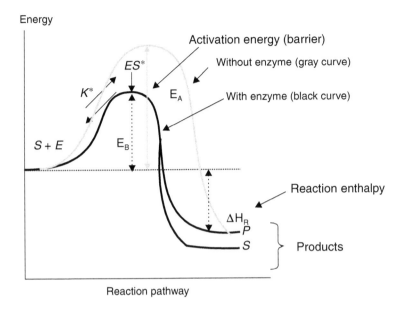

Figure 9.1 Different energy profiles involved in the enzymatic reactions.

From Equations 9.6 and 9.7, the concentrations of the complexed enzymes are determined:

$$[ES^*] = \frac{k_1[E_T][S]}{k_1[S] + k_2 + k_3[H_2O]} \tag{9.8}$$

Substituting Equation 9.8 into Equation 9.5 and considering that $(-r_s) = r_p$, we have:

$$(-r_S) = \frac{k_1 k_3[E_T][S][H_2O]}{k_1[S] + k_2 + k_3[H_2O]} \tag{9.9}$$

However, the amount of water is too large and its concentration practically does not vary compared to the variation of substrate and enzyme concentrations. Therefore, it is assumed that its concentration remains constant. Thus, we can also consider the term $k_3[H_2O] = constant = k_3^*$

Thus:

$$(-r_S) = \frac{k_1 k_3^*[E_T][S]}{k_1[S] + k_2 + k_3^*} \tag{9.10}$$

The Michaelis constant is obtained by dividing the equation by k_1:

$$K_M = \frac{k_2 + k_3^*}{k_1} \qquad (9.11)$$

tabulation the other hand, it is known that the maximum rate is obtained when all enzymes are complexed. Therefore, by making:

$$V_{max} = k_3^*[E_T] \qquad (9.12)$$

Substituting these new parameters in the Equation 9.10, we obtain:

$$(-r_S) = \frac{V_{max}[S]}{[S] + K_M} \qquad (9.13)$$

There are special cases that simplify the expression of the rate.

(a)   If the concentration of S is low, we can simplify the equation which expresses the rate by considering that $S \ll K_M$.
      We then obtain a rate which is directly proportional to the substrate concentration, and a linear variation is observed, i.e.:

$$(-r_S) = \frac{V_{max}[S]}{K_M} \qquad (9.14)$$

(b)   If the substrate concentration is high, we obtain the inverse, and the equation which expresses the rate can be simplified since $K_M \ll S$. In this case, the rate of disappearance of the substrate is equal to the maximum rate. Thus:

$$(-r_S) = V_{max} \qquad (9.15)$$

(c)   It is very common to use the half-life method, assuming that half the maximum rate is reached. This is extremely convenient, since the experiment can be stopped after reaching the concentration corresponding to half of the maximum rate. By adopting this method, we have:

$$(-r_S) = \frac{V_{max}}{2}$$

and substituting into Equation 9.13, we have:

$$K_M = [S]_{1/2} \qquad (9.16)$$

The three cases can be seen in Figure 9.2.

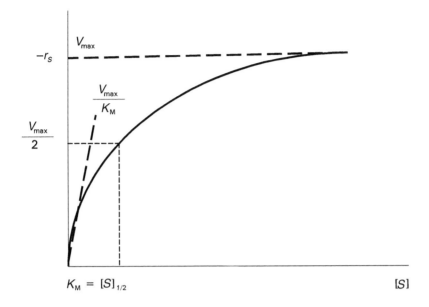

Figure 9.2 Rate versus concentration of substrate.

9.1.2 Determination of the kinetic parameters

For determination of kinetic parameters, we use both the differential and integral methods. In the integral method, the experimental rates are directly determined by using small variations of the concentration. In the case of batch reactors, a strong agitation of the reaction system should be conducted, avoiding problems of diffusion and promoting a homogeneous mixture of the reaction system with no temperature gradient. For low conversions, as we have seen, the rate is directly proportional to the concentration of the substrate.

9.1.2.1 *Differential method*

In the differential method, Equation 9.13 is generally transformed in a more convenient format, i.e.:

$$\frac{1}{(-r_S)} = \frac{K_M}{V_{max}[S]} + \frac{1}{V_{max}} \tag{9.17}$$

When placed on a graph (Figure 9.3), this expression gives a straight line, whose slope can be used to determine the Michaelis constant.

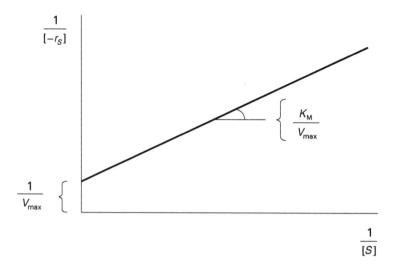

Figure 9.3 Differential method.

9.1.2.2 Integral method

In the integral method, we start from the equation of the rate and the system is chosen. The experiments are usually conducted in batch reactors, and, consequently, we use the equation of the batch reactor (Equation 4.7) as a function of the substrate:

$$t = - \int_{S_o}^{S} \frac{dS}{(-r_S)} \tag{9.18}$$

Or in terms of the conversion, making:

$$X_M = \frac{S_0 - S}{S_0} \tag{9.19}$$

and

$$t = [s_0] \int_{0}^{X_A} \frac{dX_A}{(-r_S)} \tag{9.20}$$

The rate shown in Equation 9.13 is transformed as a function of the conversion and we obtain:

$$(-r_s) = \frac{V_{máx}(1 - X_s)}{(1 - X_s) + \dfrac{K_M}{[S_0]}} \tag{9.21}$$

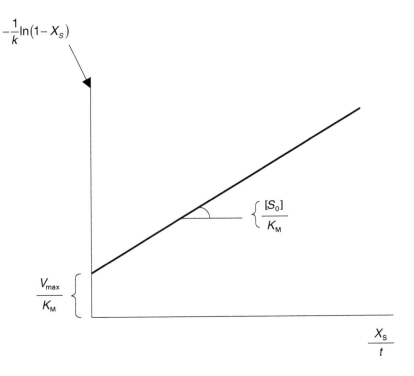

Figure 9.4 Integral method.

Substituting the rate into Equation 9.20 and integrating it between 0 and X_A, we obtain:

$$t = \frac{[S_0]}{V_{max}} X_S - \frac{K_M}{V_{max}} \ln (1 - X_S) \qquad (9.22)$$

Rearranging and placing in Figure 9.4, the constant K_M is determined:

$$-\frac{1}{t} \ln (1 - X_S) = \frac{V_{max}}{K_M} - \frac{[S_0]}{K_M} \frac{X_S}{t} \qquad (9.23)$$

The parameters K_M and $V_{máx}$ are simultaneously determined with the data of X_S versus t.

We can observe from Equation 9.22 that the rate of this reaction involves a zero-order (first term) and first order (second term) kinetics. For longer times, it satisfies the first order kinetics. On the other hand, for shorter times, the zero-order kinetics predominates.

Example

E9.1 The decomposition of H_2O_2 is conducted with an enzyme E and the products H_2O and O_2 are obtained, according to the following mechanism:

$$H_2O_2 + E \xrightarrow{k_1} [H_2O_2E^*]$$

$$[H_2O_2E^*] \xrightarrow{k_2} H_2O_2 + E$$

$$[H_2O_2E^*] + H_2O \xrightarrow{k_3} H_2O + O_2 + E$$

The experimental results that were obtained are displayed as follows:

t (min)	0	10	20	50	100
$[H_2O_2]$ (mol/L)	0.02	0.0177	0.0158	0.0106	0.005

(a) Calculate the Michaelis constant and the maximum rate.
(b) If the concentration of enzyme is triplicated, what is the time necessary to achieve a conversion of 95%?

Solution

With this mechanism, we obtain a rate equal to Equation 9.13, i.e.:

$$(-r_S) = \frac{V_{max}[H_2O_2]}{[H_2O_2] + K_M}$$

Whose solution in a batch reactor is given by Equation 9.23:

$$-\frac{1}{t} \ln (1 - X_S) = \frac{V_{max}}{K_M} - \frac{[H_2O_2] X_S}{K_M \ t}$$

Considering that:

$$X_S = \frac{[H_2O_2]_0 - [H_2O_2]}{[H_2O_2]_0}$$

where X_S is the conversion
Since $[H_2 O_2]_0 = 0.02$, we obtain:

t (min)	0	10	20	50	100
$[H_2O_2]$ (mol/L)	0.02	0.0177	0.0158	0.0106	0.005
X_S	0	0.115	0.210	0.470	0.750
$1/t \times \ln(1 - X_A) \times 10^2$	–	1.193	1.178	1.269	1.386
$\times 10^2$	–	1.125	1.05	0.94	0.75

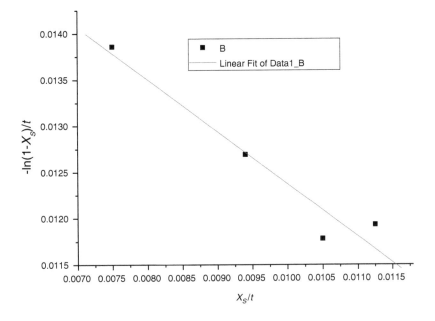

Figure E9.1 Determination of the rate constants.

The angular coefficient is –0.56. Therefore, since $[H_2O_2]_0 = 0.02$, we have:

$K_M = 0.0354$

The linear coefficient is equal to 0.018, therefore:

$V_{max} = 6.37 \times 10^{-4}$ (mol/L) $\times$ min^{-1}

If the enzymes are triplicated, we have:

$V_{máx} = k_3 E_T = 3 \times 6.37 \times 10^{-4} = 1.911 \times 10^{-3}$ (mol/L min)

Thus, the time necessary to reach a conversion (X_S) of 0.95 will be:

$t = 15$ min

9.1.3 Effect of external inhibitors

Enzymes can be competitive and noncompetitive and affect the reaction rate. In certain cases, they work as a blocker, which is advantageous in medicine. On the other hand, it can significantly reduce the capacity of active enzymes.

In the case of competitive enzymes, the substrate competes with the external blocker in the same active enzyme. Therefore, it depends on the adsorption capacity of active enzymes. In the opposite case, both the substrate and the external inhibitors do not compete with the same activated enzyme and the capacity of the reaction will depend on the adsorption strength in the active enzymes.

Note that the kinetic models are different but can be deduced in the same manner. In the competitive model, we have the following system:

$$S + E \overset{k_2/k_1}{\longleftrightarrow} ES^* (+H_2O) \overset{k_3}{\longrightarrow} P + E$$

$$B + E \overset{k'_2/k'_1}{\longleftrightarrow} ES^* \tag{9.24}$$

where

$E =$ free enzymes
ES and $ES^* =$ complexed enzymes
$B =$ blocker.

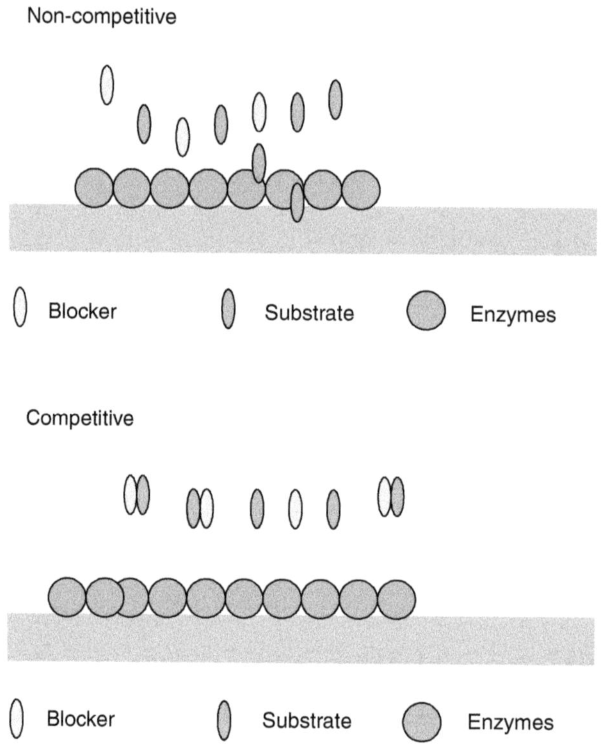

Figure 9.5 Competitive and noncompetitive models.

The rate of formation of the product P can be determined, and the following final expression is obtained:

$$(-r_S) = \frac{k_3 K_1 [S][E_T]}{1 + K_1[S] + K_2[B]} \tag{9.25}$$

where:

$$K_1 = k_1/k_1'$$

$$K_2 = k_2/k_2'$$

Example

E9.2 The data displayed in Figure E9.2 were obtained experimentally. It is known that the reaction mechanism is complex, although the kinetics of the reactions is not known. The following scheme of reactions involved in the transformation of a substrate S into the product P in the presence of an enzyme E is proposed.

$$S + E \xleftrightarrow{k_2/k_1} ES^*$$

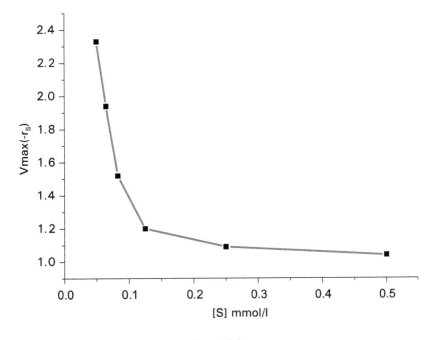

Figure E9.2

$$ES^*(+H_2O)\xrightarrow{k_3} P + E$$

$$ES^* + E \xrightarrow{k_4/k_5} SES^*$$

where S represents the substrate and E represents the enzyme. The experimental results are presented in the following table (adapted from Fogler, 2000):

$V_{max}/(-r_S)$	2.5	1.5	1.0	0.5	0.6	0.6
[S] (moles/L)	0.02	0.04	0.06	0.1	0.3	0.4

Derive the expression of the rate with respect to the substrate S and determine the constants. If possible, make simplifications. Calculate the constants using the data displayed in Figure E9.2.

Solution

$$(-r_S) = k_1[E][S] - k_2[ES^*] \tag{9.26}$$

But:

$$r_{ES^*} = k_1[E][S] - k_2[ES^*] - k_3[ES^*] - k_4[ES^*][S] + k_5[SES^*] = 0 \tag{9.27}$$

$$r_{SES^*} = k_4[ES^*][S] - k_5[SES^*] = 0 \tag{9.28}$$

But:

$$[E_T] = [E] + [ES^*] + [SES^*] = 0 \tag{9.29}$$

From Equation 9.28, we have:

$$[SES^*] = \frac{k_4}{k_5}[ES^*][S] \tag{9.30}$$

Substituting Equations 9.30 and 9.29 into Equations 9.27 and 9.26, we have:

$$(-r_S) = k_1[S][E_T]\frac{k_3 - \dfrac{k_1k_4}{k_5}[S]^2}{k_1[S] + k_2 + k_3}$$

But the reverse constant favors the formation of the complexed enzyme $[ES^*]$ and consequently $k_5 \gg k_4$. The last term of the numerator is neglected and after transforming we obtain:

$$(-r_S) = \frac{k_1[S][E_T]}{k_1[S] + K_M}$$

where:

$$K_M = \frac{(k_2 + k_3)}{k_3'} \text{ constant of Michaelis}$$

in which:

$$k_1[E_T] = V_{max}$$

We then have:

$$(-r_S) = \frac{V_{max}[S]}{k_1[S] + K_M}$$

Transforming this equation we obtain:

$$k_1 + \frac{K_M}{[S]} = \frac{V_{max}}{-r_S} \tag{9.31}$$

From the table we obtain:

$V_{max}/(-r_s)$	2.5	1.5	1.0	0.5	0.6	0.6
$\frac{1}{[S]}$ (moles/L)$^{-1}$	50	25	16	10	3.3	2.5

Placing the values of the table in Figure E9.3, we obtain:

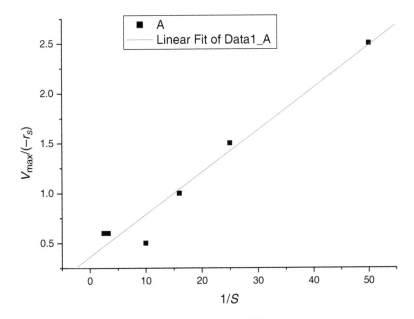

Figure E9.3   Determination of the constants.

From this figure and linear regression we obtain the following results:

$K_M = 0.042$

$k_1 = 0.36$

9.1.4   Kinetics of biological fermentation

In biological fermentation, the substrate is in contact with bacteria and therefore the microorganisms grow. Living cells comprise a series of reactions of large application in the health sector, such as insulin, antibiotics, and food production. In the processes with biomass, there are cells which consume nutrients for energy generation and cell growth.

The most common examples are:

Substrate (S) + Cell (C) → Cells (C) + Product (P)
[Fruits] + Bacteria → Bacteria + alcohol
[Organic matter] + Bacteria → [More bacteria] + [Product + CO_2, H_2O]

To monitor the growth of the cells or bacteria, we can use the graph displayed in Figure 9.6, which shows the variation of cell concentration over time. This graph shows the different phases that appear and disappear in a kinetic process.

1.   In phase I, there is an induction period in which the cells get adapted to the culture medium. The function of this phase is to transfer nutrients (proteins) for the cells, thus forming enzymes and new substrates and causing a multiplication

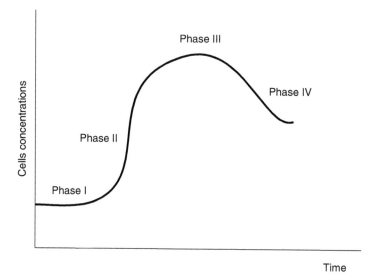

Figure 9.6 Growth of cells.

of new cells. This largely depends on several factors such as the reaction medium, substrate concentration, and pH.

2. In phase II, there is a period of cell growth. This means that there are enough nutrients and cells enabling rapid cell growth.

3. In phase III, there is a stable period and there are not enough nutrients for cell growth. In this stage, equilibrium between the formation of cells and transport of nutrients is established.

4. In phase IV, a period of cells deactivation (death) occurs, since there are no more nutrients. Products and subproducts are formed, and they can alter the properties of the reaction medium, especially the pH, which may cause cell deactivation or death.

The rate of cell growth practically follows the same expression of the rate, according to Michaelis equation. However, we must also take into account the concentration of cells, as follows:

$$r_C = \frac{k[S][C]}{[S] + K_M} \tag{9.32}$$

These constants must be determined. In most cases, empirical rates of cell growth are used, and according to Monod equation, we have the following expression:

$$r_C = \mu[C] \tag{9.33}$$

where μ is an empirical coefficient and the unit is given in s^{-1}, since the concentration is measured in g/dm^3. This coefficient was determined by Monod equation:

$$\mu = \mu_{max} \frac{[S]}{[S] + K_{Monot}} \tag{9.34}$$

where:

$K_{Mon} = $ Monod constant.

Thus:

$$r_C = \mu_{max} \frac{[S][C]}{[S] + K_{Monot}} \tag{9.35}$$

Therefore, this is similar to the Michaelis equation.

There are several other empirical expressions, as the Moser equation:

$$r_C = \mu_{max} \frac{[C]}{(1 + k[S]^{-\lambda})} \tag{9.36}$$

where

k and λ are the empirical constants.

9.1.5 Mass balance

Note that the mass balance in the reactors should take into account the rates of living, generated, and dead cells. Both in the batch reactor and in the CSTR, we have the following expression in relation to cells mass flow:

$$[G_{inlet}] - [G_{outlet}] + [r_{produced}] = [r_{accumulated}] \tag{9.37}$$

However, the resulting rate will be:

$$[r_{generation\ of\ live\ cells}] = [r_{generated}] - [r_{dead}] = (r_g - r_m) \tag{9.38}$$

Therefore, from Equation 9.37, we have:

$$[G]_0 - [G]_s + (r_g - r_m) = [r_a] \tag{9.39}$$

In a liquid-phase continuous system, we have the following expression as a function of the cells concentration:

$$([C]_0 - [C]_s)v_0 + (r_g - r_m) = \frac{dC}{dt} \tag{9.40}$$

Since there is no flow in a batch reactor, we have:

$$(r_g - r_m) = \frac{dC}{dt} \tag{9.41}$$

The same balance should be done in relation to the substrate, but in this case, the rate will be the transformation of the substrate, i.e.:

$$(-r_S)V = \frac{d[S]}{dt} \tag{9.42}$$

With these equations, the rate constants can be determined by using the differential and integral methods (Chapter 5).

Example

E9.3 We need to determine the kinetic parameters of an enzymatic reaction with the substrate S and different concentrations of cells, according to the following table:

Experiment	Substrate [S] (g/L)	Cells [C] (g/L)	Rate r (g/L s)
1	2	0.9	1.4
2	4	0.45	1.25
3	8	0.25	1.10
4	10	0.15	0.80
5	12	0.10	0.80

By using the Monod equation, calculate the constants.

Solution

From Equation 9.35:

$$r_C = \mu_{max} \frac{[S][C]}{[S] + K_{Monot}}$$

(9.35)

Transforming:

$$\frac{K_{Monot}}{\mu_{max}[S]} + \frac{1}{\mu_{max}} = \frac{[C]}{r_C}$$

(E9.3)

$$1 + \frac{K_M}{[S]} = \frac{V_{max}}{-r_S}$$

(9.28)

Substituting the data, we obtain the following table:

Experiment	$1/[S]$	Cells [C] (g/L)	[C]/r
1	0.5	0.9	0.642
2	0.25	0.45	0.360
3	0.125	0.25	0.227
4	0.10	0.15	0.187
5	0.0833	0.10	0.125

In the graphical form, we obtain Figure E9.3.
Solving it, we obtain:

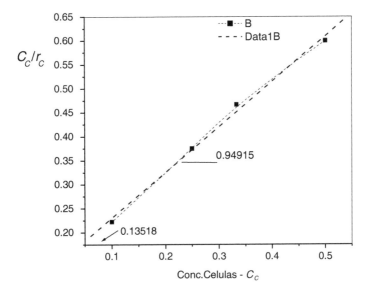

Figure E9.4   Graph constructed from Equation E9.3.

From this figure we obtain:

$$\frac{K_{Monot}}{\mu_{max}} = 0.95$$

$$\frac{1}{\mu_{max}} = 0.135$$

By linear regression:

$$Y = A + B * X$$

Parameter	Value
$A =$	0.135
$B =$	0.94
$R =$	0.99

Therefore,
$$\mu_{max} = 7.4$$
$$K_{Mon} = 7.03$$

9.2 LIQUID-PHASE REACTIONS

Liquid-phase reactions are very important because they occur in different forms. The presence of ions in solutions can significantly affect the dielectric constant of the solvent. These reactions are very fast due to energy transfer efficiency, primarily due to collision between the molecules and the small distance between the molecules in the liquid phase.

There is also the effect of the acidic medium on the solvation due to the formation of complexes and transfer of protons. A classic example is presented as follows:

$$CH_3I + Cl^- \rightarrow CH_3Cl + I^-$$

This reaction is 10^6 times greater in a solution of dimethylacetamide than in acetone. The collision between the molecules is very intense due the approximation of the molecules with ions in solution, influencing other molecules.

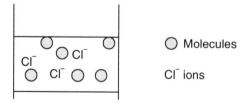

The major problem is how to calculate the kinetics of these reactions. There are several proposals and one of them is gas collision theory, because they are of the same order of magnitude. Collisions occur repeatedly with high frequency, so they constitute multiple collisions. It is estimated that the distances between the molecules in the liquid phase are approximately equal, whereas in the gas phase are quite different. Moreover, besides the attractive forces, there are also repulsive forces, but the important thing is that for the reaction to occur it is necessary to overcome the energy barrier E_b. Differently from the reactions in the gas phase, the collisions in the liquid phase are 10–1000 times higher, but depend on other properties such as the viscosity.

The formation of radicals is also important since they are not in equilibrium and therefore have a great energy capacity and a greater degree of freedom that allows overcoming the energy barrier, but is much lower when compared to the equilibrium state. The collision of the molecules favors the formation of these radicals which are far more numerous in the liquid phase. On the other hand, diffusional problems exist and, in certain cases, may be limiting in the process.

We saw earlier that in the photochemical reactions, an initiation step occurs due to absorption of a photon, favoring the homolytic breaking and forming two free radicals. In the liquid phase, these radicals are neighbors and can recombine themselves or combined with other radicals or molecules.

9.2.1 Liquid solutions

It is a common practice to try to relate the activity coefficients of ions with the composition and dielectric constant under certain conditions of temperature. The most indicated theory is that of Debye–Huckel, for dilute solutions. This theory is not valid when the concentrations are high.

A certain analogy is made with the transition state theory to predict the influence of ionic strength on the reaction rate constant. The model is the same as for the theory of transition state presented previously. An intermediate complex is formed (Figure 9.7), according to the reaction:

$$A_{(l)} + B_{(l)} \overset{k_2/k_1}{\rightleftharpoons} AB^* \overset{k_3}{\longrightarrow} R \qquad (9.43)$$

We will not make the deduction, and details can be found in Hill's book. According to this theory, the rate constant for a bimolecular reaction would be given by:

$$k = k_0 \frac{\gamma_A \gamma_B}{\lambda_{AB^*}} \qquad (9.44)$$

where:

$k_0 =$ constant when it is infinite
γ_A and $\gamma_B =$ activity coefficients of A and B, respectively
$\lambda_{AB^*} =$ activity coefficient of the activated complex.

In general, these activity coefficients are not thermodynamically known and can be determined by analogy with the existing data for the compounds, with the exception

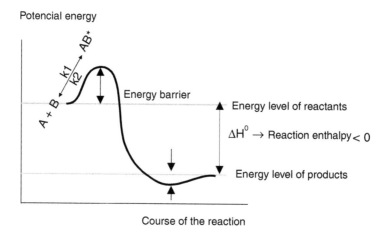

Figure 9.7 Energy barrier.

for the activated complexes. On the other hand, the theory of collisions is used to determine the rate constant, and we therefore obtain [Hill]:

$$\log_{10}^{k} = \log_{10}k_0 + 2z_A.z_B\sqrt{\mu}\left[\frac{q^3\sqrt{8\pi N_0/1000}}{2(k_B T\varepsilon)^{3/2}}\right] \tag{9.45}$$

where:

z_A and z_B = charges of the ions A and B, respectively
q = charge of the electron
ε' = dielectric constant
k_B = Boltzman constant
μ = ionic strength.
N_0 = Number of Avagadro

This ionic strength depends on the concentration of the reactants and the charge of the respective ions. Usually, it can be expressed by the following equation:

$$\mu = \frac{1}{2}\left(\sum_i [C_i]z_i^2\right)$$

where:

$[C_i]$ = molar concentration of the component.

When the solution is diluted, we can calculate the constants above 25°C and the value of the Equation 9.44 is determined. Thus:

$$\log_{10}^{k} = \log_{10}k_0 + 2z_A.z_B.0.509\sqrt{\mu} \tag{9.46}$$

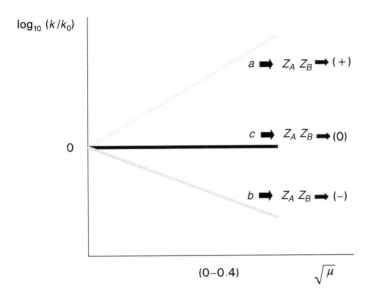

$\log_{10} (k/k_0)$

a ➡ $Z_A\ Z_B$ ➡ (+)

c ➡ $Z_A\ Z_B$ ➡ (0)

0

b ➡ $Z_A\ Z_B$ ➡ (−)

(0–0.4) $\sqrt{\mu}$

Figure 9.8 Constants as a function of the ionic potential (adapted from Hill, 1977).

This is known as the equation of Brönsted–Bjerrum for dilute solutions.

It is observed that the rate constant depends on the ionic strength, which in turn, can vary substantially from one solution to another. Experimental results showed that depending on the ionic charges, the rate constants may vary positively or negatively with the ionic strength, as shown in Figure 9.8:

There are other important variables that can affect the rate constants. One of them is the pressure. When the pressure is low, it does not affect, but when it is high, the rate can increase or decrease, depending on the volume of the system.

Example

E9.4 The reaction (Adapted from Hill):

$$H_2O_2 + 2H^+ + 2I^- \rightarrow 2H_2O + I^2$$

We observed the following rate at 25°C:

$$r = k_1[H_2O_2][I^-] + k_2[A^*] - k_3[H_2O_2][I^-][H^+]$$

And the following experimental data:

Check if these results are consistent with the kinetic model?

Solution

We start from Equation 9.46:

$$\log_{10} k = \log_{10} k_0 + 2z_A.z_B.0.509\sqrt{\mu}$$

Table E9.1 (Hill)

μ (kmol/m³)	k_1 (m³/kmol s)	k_2 (m³/kmol)² s⁻¹
0	–	19
0.0207	0.658	15
0.0525	0.663	12.2
0.0925	0670	11.3
0.1575	0.694	9.7
0.2025	0.705	9.2

Note that

Case (a)—the first term contains only the constant k_1.

Thus, in this case, the equation of Brönsted–Bjerrum contains only a negative ion $z_{I^-} = -1$.

Therefore:

$$z_{H_2O_2} \cdot z_{I^-} = 0$$

In this case, the constant k_1 is independent of ionic strength, i.e., for any value of μ the constant k_1 has the same value. Table E9.1 gives that the values of k_1 are approximately the same.

Case (b)—the second term k_2 contains the ions $[I^-][H^+]$

Thus:

$$z_{[I^-]} z_{[H^+]} = -1$$

thus, the equation becomes:

$$\log_{10} k = \log_{10} k_0 - 2 \times 0.509 \sqrt{\mu}$$

Substituting the values, we have:

When $\mu = 0$, $k_2 = 19$. Thus, $k_0 = 19 k_0$.

Therefore:

$$\log_{10} k = 1.278 - 1.018 \sqrt{\mu}$$

For the other values of the table and Figures E9.5 and E9.6, k_2 is calculated, i.e.:

μ (kmol/m³)	k_2 (experimental)	k_2 (calculated)
0	19	19
0.0207	15	13.5
0.0525	12.2	11.1
0.0925	11.3	9.31
0.1575	9.7	7.49
0.2025	9.2	6.61

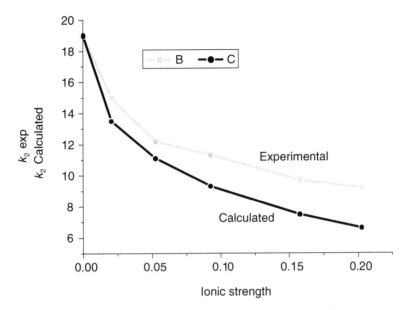

Figure E9.5   Calculated versus experimentally determined constants k_2 as a function of the ionic strength.

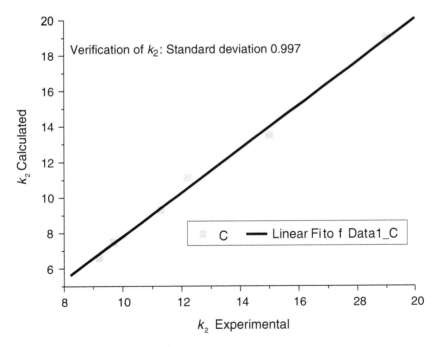

Figure E9.6   Constant k_2: calculated and experimental determined.

9.2.2 Acid—base reactions

These reactions occur by proton transfer. The protonated or nonprotonated species or perhaps the intermediate species react with the other components of the reaction. According to Brönsted, the following reaction scheme occurs:

$$HX + Y \rightarrow HY + X \tag{9.47}$$

Brönsted + Conjugated base   Conjugated acid + Acid

Unlike the previous case, this reaction does not involve electrons, and, therefore, there is no repulsion. The molecules get polarized and have a high degree of solvation in a polar solvent.

The rate or rate constant follows the theory of Bronsted–Lowry and is a function of the ions, in particular of hydronium ion, which is a combination of the proton with water. It also depends on the concentrations of conjugate acids. Its expression is [Hill]:

$$k = k_0 + k_{H^+}[H_3O^+] + k_{OH^-}[OH^-] + \sum k_{HX}[HX_j] + \sum k_{X_j}[X_i] \tag{9.48}$$

Hidronic íon   acids   bases

This constant depends on the temperature and mainly on the pH of the solution. When weak acids and bases are involved, we can simply have,

$$k = k_0 + k_{H^+}[H_3O^+] + k_{OH^-}[OH^-] + k_{HA}[HA] + k_{A^-}[A^-] \tag{9.49}$$

The contribution of the terms depends on the experimental conditions and, depending on the case, may be discarded and simplified. Thus, when an acid is weak,

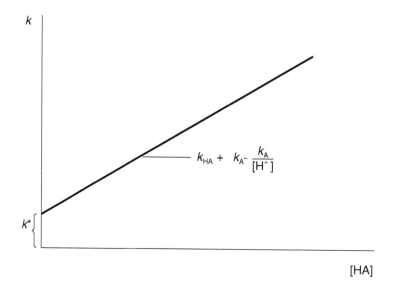

Figure 9.9 Dependence of k* with acid strength.

as for example CH_3COOH (HA^- neutral) and its conjugate base CH_3COO^- (A^-), we have in the equilibrium:

$$[H^+] = K_A \frac{[HA]}{[A^-]} \tag{9.50}$$

In this case, we can simply have the expression above, i.e.:

$$k = k_0 + k_{H^+}[H_3O^+] + k_{OH^-}[OH^-] + k_{HA}[HA] + k_{A^-}[A^-] \tag{9.51}$$

where

$$k^* = k_0 + k_{H^+}[H_3O^+] + k_{OH^-}[OH^-]$$

Thus,

$$k = k^* + [HA] + \left(k_{HA} + k_{A^-}\frac{K_A}{[H^+]}\right) \tag{9.52}$$

Placing eq. 9.52 in a graph (Figure 9.9), we obtain the constants, i.e.:

The constants k^* and the acid constants can be determined separately from the angular coefficient.

9.3 REPARAMETERIZATION OF THE MICHAELIS-MENTEN EQUATION

As seen in Section 3.3.3 and Section 5.7, and discussed in detail in Chapter 11, nonlinear models may rely on parameters that present high correlations among themselves, which can cause significant numerical problems during estimation of model parameters, as the minimization of the objective function may become inefficient (Espie and Macchietto, 1988), and the statistical characterization of the final parameter estimates may lack significance (Watts, 1994). Some of these difficulties can be observed during the estimation of parameters of rate equations that are based on the Michaelis–Menten equation, utilized throughout Chapter 9, because of usually high correlations among obtained parameter estimates. As discussed in the previous sections, rate equations based on the Michaelis–Menten expression are commonly used in enzymatic and fermentation problems to describe rate expressions and equilibrium conditions.

Initially, the Michaelis–Menten equation can be written in different forms, including:

$$R_S = -\frac{K_1[S]}{1 + K_2[S]} \tag{9.53}$$

$$R_S = -\frac{[S]}{K_3 + K_4[S]} \tag{9.54}$$

$$R_S = -\frac{K_5[S]}{K_6 + [S]} \tag{9.55}$$

$$R_S = -\frac{[S]}{A + B\left([S] - [S]_{ref}\right)} \tag{9.56}$$

where R_S is the rate of consumption of the substrate S; $K_1, K_2, K_3, K_4, K_5, K_6, A$ and B are model parameters and $[S]_{ref}$ is a reference numerical value. Additionally:

$$K_3 = \frac{1}{K_1}; K_4 = \frac{K_2}{K_1}; K_5 = \frac{K_1}{K_2}; K_6 = \frac{1}{K_2} \tag{9.57}$$

$$A = \frac{1 + K_2[S]_{ref}}{K_1}; B = \frac{K_2}{K_1} \tag{9.58}$$

Using the approach developed originally by Schwaab and Pinto (2007, 2008) to perform the reparameterization of the Arrhenius equation (see Section 3.3.3) and of the power-law rate model (see Section 5.7), and consequently to remove the undesirable correlation between model parameters and improve parametric precision, it is necessary to understand that the covariance matrix of parameter uncertainties $(\mathbf{V}_\alpha)$ can be calculated in the form (see Chapter 11):

$$\mathbf{V}_\alpha = \begin{bmatrix} \sigma_{11}^2 & \sigma_{12}^2 \\ \sigma_{12}^2 & \sigma_{22}^2 \end{bmatrix} = \begin{bmatrix} \sum\limits_{i=1}^{NE} \left(\frac{\partial y}{\partial \alpha_1}\right)_i^2 & \sum\limits_{i=1}^{NE} \left(\frac{\partial y}{\partial \alpha_1}\right)_i \left(\frac{\partial y}{\partial \alpha_2}\right)_i \\ \sum\limits_{i=1}^{NE} \left(\frac{\partial y}{\partial \alpha_1}\right)_i \left(\frac{\partial y}{\partial \alpha_2}\right)_i & \sum\limits_{i=1}^{NE} \left(\frac{\partial y}{\partial \alpha_2}\right)_i^2 \end{bmatrix}^{-1} \tag{9.59}$$

when the model output y depends on two model parameters, α_1 and α_2 Therefore, in order to remove the parameter correlation and improve the quality of obtained model parameters it is necessary that:

$$\sum_{i=1}^{NE} \left(\frac{\partial y}{\partial \alpha_1}\right)_i \left(\frac{\partial y}{\partial \alpha_2}\right)_i = 0 \tag{9.60}$$

As one can observe, Equation (9.60) cannot be satisfied when the model output is R_S and Equations (9.53–56) are used to describe the rate equations, as derivatives of these equations have always the same sign and the sum of the derivatives cannot become equal to zero. This explains mathematically why the respective model parameters are correlated strongly and the final parameter estimates frequently lack statistical significance. On the other hand, when Equation (9.56) is used as reference:

$$\sum_{i=1}^{NE} \left(\frac{\partial R_S}{\partial A}\right)_i \left(\frac{\partial R_S}{\partial B}\right)_i = \sum_{i=1}^{NE} \frac{[S]_i}{\left[A + B\left([S]_i - [S]_{ref}\right)\right]^2} \frac{\left([S]_i - [S]_{ref}\right)}{\left[A + B\left([S]_i - [S]_{ref}\right)\right]^2}$$

$$\cong \sum_{i=1}^{NE} \frac{R_{Si}^2}{[S]_i} \frac{R_{Si}^2 \left([S]_i - [S]_{ref}\right)}{[S]_i^2} = 0 \tag{9.61}$$

Therefore:

$$[S]_{\text{ref}} \cong \frac{\sum_{i=1}^{NE} \frac{R_{S_i}^4}{[S]_i^2}}{\sum_{i=1}^{NE} \frac{R_{S_i}^4}{[S]_i^3}} \tag{9.62}$$

tabulation provides a suitable and original reparameterization of the Michaelis–Menten equation when rate data are used for purposes of parameter estimation.

Example

E9.5 The data presented in Table 9.53 represent the rates of consumption of a certain substrate in a fermentation process. The data must be used for estimation of the parameters of the Michaelis–Menten equation, using the techniques described in detail in Chapter 11. The obtained results are shown in Table 9.54 and Figures 9.53–5.

Table 9.2 Rates of consumption (R_S) of substrate S as a function of the concentration.

[S] mol/L	R_S mol/L·h
0.1	0.204
0.2	0.298
0.3	0.425
0.4	0.449
0.5	0.460
0.6	0.570
0.7	0.580
0.8	0.591
0.9	0.624
1	0.692

Table 9.3 Parameter estimates for different parametrizations of the Michaelis–Menten equation.

Model Equation	α_1	α_2	σ_{11}^2*	σ_{22}^2*	$\rho_{12} = \frac{\sigma_{12}^2}{\sigma_{11}\sigma_{22}}$
Equation (9.53) ($\alpha_1 = K_1, \text{h}^{-1}; \alpha_2 = K_2, \text{L/mol}$)	2.296	2.509	∞	∞	1
Equation (9.54) ($\alpha_1 = K_3, \text{h}; \alpha_2 = K_2, \text{L·h/mol}$)	0.436	1.092	2.6×10^{-3}	7.3×10^{-3}	-0.92
Equation (9.55) ($\alpha_1 = K_5, \text{mol/L·h}; \alpha_2 = K_6, \text{mol/L}$)	0.915	0.399	6.3×10^{-3}	6.8×10^{-3}	0.98

*Based on the linear approximation of the objective function.

It is important to observe that application of Equation (9.62) leads to $[S]_{\text{ref}} = 0.542$ mol/L, which is the value that leads to essentially null correlation between parameters A and B, as expected, clearly indicating the validity of Equation (9.62). Besides,

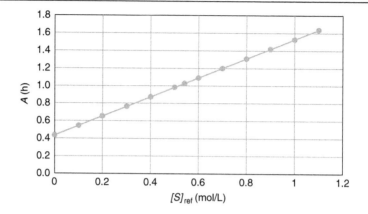

Figure 9.54 Estimated parameter A as a function of $[S]_{ref}$ in Equation (9.56).

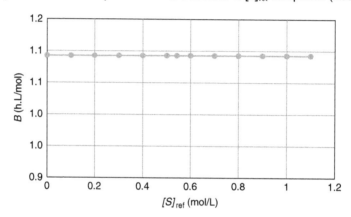

Figure 9.55 Estimated parameter B as a function of $[S]_{ref}$ in Equation (9.56).

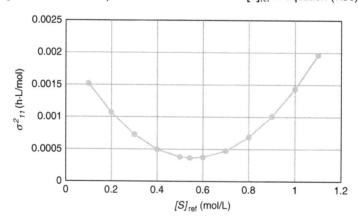

Figure 9.56 Variance of parameter A as a function of $[S]_{ref}$ in Equation (9.56).

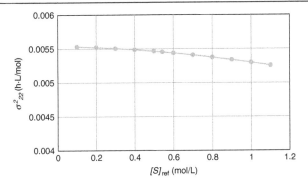

Figure 9.57 Variance of parameter B as a function of $[S]_{ref}$ in Equation (9.56).

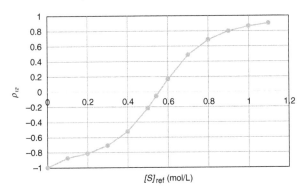

Figure 9.58 Correlation between parameters A and B as a function of $[S]_{ref}$ in Equation (9.56).

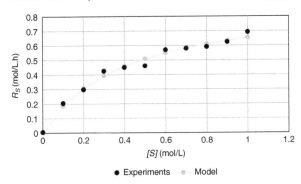

Figure 9.15 Fit of experimental data in Table 9.53 with Equations (9.62).

this parametrization of the Michaelis-Menten equation also leads to the most precise parameter values, as shown in Figure 9.55, clearly indicating that removal of the parameter correlation allows the estimation of more meaningful model parameters. Figure 9.57 shows the quality of the model fit, which is the same in all cases, as the parametrization does not change the relationship between input variables and model responses.

One criticism that can be raised regarding Equation (9.62) is the fact that reaction rates are usually not measured in the lab. As a matter of fact, experimental procedures usually provide concentrations or mass values, so that the dynamic mass balance equations must be considered in the form:

$$\frac{d[S]}{dt} = -\frac{[S]}{A + B\left([S] - [S]_{ref}\right)} \tag{9.63}$$

tabulation leads to the following implicit dynamic equation

$$\left(A - B[S]_{ref}\right) \ln\left(\frac{[S]_t}{[S]_0}\right) + B\left([S]_t - [S]_0\right) = -t \tag{9.64}$$

tabulation must be solved numerically to provide the desired solution. Although Equation (9.64) is implicit, it can be derived with respect to parameter A, leading to:

$$\frac{\partial [S]_t}{\partial A} = -\frac{[S]_t \ln\left(\frac{[S]_t}{[S]_0}\right)}{\left(A + B\left([S]_t - [S]_{ref}\right)\right)} = \frac{d[S]}{dt}\bigg|_t \ln\left(\frac{[S]_t}{[S]_0}\right) \tag{9.65}$$

and with respect to parameter B, leading to:

$$\frac{\partial [S]_t}{\partial B} = -\frac{[S]_t \left\{[S]_{ref} \ln\left(\frac{[S]_t}{[S]_0}\right) - \left([S]_t - [S]_0\right)\right\}}{\left(A + B\left([S]_t - [S]_{ref}\right)\right)} = \frac{d[S]}{dt}\bigg|_t$$
$$\left\{[S]_{ref} \ln\left(\frac{[S]_t}{[S]_0}\right) - \left([S]_t - [S]_0\right)\right\} \tag{9.66}$$

Therefore:

$$\sum_{i=1}^{NE} \frac{\partial [S]_t}{\partial A}\bigg|_{t_i} \frac{\partial [S]_t}{\partial B}\bigg|_{t_i} = \sum_{i=1}^{NE} \frac{d[S]}{dt}\bigg|_{t_i}^2 \ln\left(\frac{[S]_{t_i}}{[S]_0}\right)$$
$$\left\{[S]_{ref} \ln\left(\frac{[S]_{t_i}}{[S]_0}\right) - \left([S]_{t_i} - [S]_0\right)\right\} = 0 \tag{9.67}$$

and

$$[S]_{ref} = \frac{\sum_{i=1}^{NE} \frac{d[S]}{dt}\bigg|_{t_i}^2 \ln\left(\frac{[S]_{t_i}}{[S]_0}\right)\left([S]_{t_i} - [S]_0\right)}{\sum_{i=1}^{NE} \frac{d[S]}{dt}\bigg|_{t_i}^2 \left[\ln\left(\frac{[S]_{t_i}}{[S]_0}\right)\right]^2} \tag{9.68}$$

tabulation suggests the following discretized form for calculation of the reference concentration value:

$$[S]_{ref} = \frac{\sum_{i=1}^{NE} \left(\frac{[S]_{i+1}-[S]_{i-1}}{t_{i+1}-t_{i-1}}\right)^2 \ln\left(\frac{[S]_i}{[S]_0}\right)([S]_i - [S]_0)}{\sum_{i=1}^{NE} \left(\frac{[S]_{i+1}-[S]_{i-1}}{t_{i+1}-t_{i-1}}\right)^2 \left[\ln\left(\frac{[S]_i}{[S]_0}\right)\right]^2}$$

(9.69)

Example

E9.6 In order to illustrate the robustness of the parameterization proposed in Equation (9.69), the enzymatic model described below

$$\frac{d[S]}{dt} = -\frac{[S]}{1+[S]}; [S]_0 = 1\,\text{mol/L}$$

tabulation sampled in the interval [0, 2] h with sampling times of 0.01 h (201 points), 0.1 h (21 points) and 0.5 h (5 points). Then Equation (9.69) was used to calculate the values of $[S]_{ref}$, which were shown to be equal to 0.684787 mol/L, 0.685421 mol/L and 0.691386 mol/L, indicating that the reference parameter is not very sensitive to modification of the sampling frequency. A similar exercise was performed through manipulation of the final reaction time, which was made equal to 1 h (11 points), 10 h (101 points) and 100 h (1001 points), keeping the sampling interval equal to 0.1 h. In this case, the reference values were respectively equal to 0.830011 mol/L, 0.532027 mol/L and 0.532026 mol/L, indicating that the reference value changes with experimental batch time (because it is an average of measured concentrations) but converges fast to an asymptotic value (because it is an average that is weighted by the reaction rates that converge to zero if the reaction time is too long). As a matter of fact, the following empirical and very simple equation can be suggested for $[S]_{ref}$:

$$[S]_{ref} = \frac{[S]_0 + [S]_{NE}}{2}$$

tabulation $[S]_0$ is the initial concentration and $[S]_{NE}$ is the last experimental point of the batch. In the analyzed cases, these values would be equal to 0.634624 mol/L in the first three numerical experiments (constant batch time) and 0.780604 mol/L, 0.5000 mol/L and 0.5000 mol/L in the remaining three numerical experiments (constant sampling time).

Chapter 10

Heterogeneous reaction kinetics

Heterogeneous reactions occur in gas or liquid phase or in both phases in the presence of a solid as a catalyst or as reactant, which depends on the process in use. Usually, the reactions in gas–solid phase or gas–liquid–solid phases depend on the industrial reaction conditions. In catalytic processes, the catalyst promotes the reaction rate in the gas or in the liquid phase.

The ammonia synthesis $N_2 + 3H_2 \rightarrow 2NH_3$ is a classic example of a catalytic reaction in gas phase, as shown in Figure 10.1.

The ammonia synthesis is one of the most important processes in the production of fertilizers. The reaction is exothermic ($\Delta H = -46\,kJ/mol\,NH_3$) and runs at relatively low temperatures ($\approx 400°C$) and high pressures (≈ 60–$100\,atm$). The kinetics is an example of a heterogeneous system occurring on iron-based catalyst.

The limiting step of the kinetics is the chemical reaction occurring at the surface. However, there are diverse physical and chemical phenomena, besides mass transfer and internal diffusion inside pore particles affecting the global reaction rate. There are different steps:

1. Diffusion of reactant molecules through the fluid in the direction of the surface.
2. Diffusion of molecules from the surface through pores.
3. Chemical reaction on surface sites.
4. Diffusion of molecules (not reacted and products) from the sites through pores in the direction of the external surface.
5. Diffusion of molecules (not reacted and products) from the surface in the outflow direction.

However, the main question is what reaction kinetics occurs on the surface sites? What is the expression of the total reaction rate and what parameters or variables are involved?

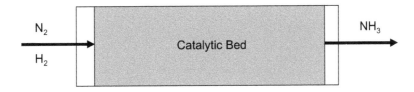

Figure 10.1 Ammonia synthesis.

10.1   EXTERNAL PHENOMENA

Imagine a cross section of the reactor, as shown in Figure 10.2. Reactants and products flow through the reactor containing solid particles but the reaction occurs at the surface.

There are mass transport in the fluid flowing in the direction to the surface and diffusion inside pores, as shown in Figure 10.3, with concentration varying along this path.

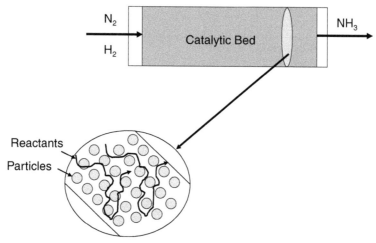

Figure 10.2   Cross section of a reactor.

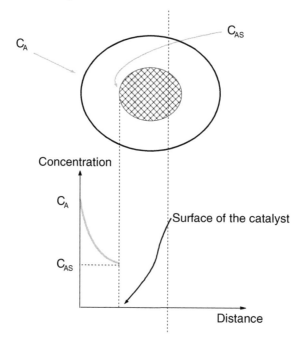

Figure 10.3   Concentration profile through the film.

The film at the surface and its thickness depend on the hydrodynamics in the reactor. Reactants reaching the surface react forming products that must return in the outflow.

There are two possibilities:

1. The mass transfer through the film is facile. In this case, the global reaction rate is determined by the chemical reaction rate at the surface of the catalyst. Therefore, the chemical reaction is the limiting step of the process.
2. The mass transfer through the film is slow and, therefore it is the limiting step. In this case, there is a diffusional barrier in the film surrounding the surface of the catalyst.

This latter case is unwanted for kinetic data acquisition and must be eliminated or avoided. Experiments must run at high velocity or high Reynolds number, diminishing the film surrounding the surface.

To verify experimentally the mass or diffusion phenomena, one measures the conversion by varying the molar flow rate or mass, changing height and diameter of the reactor, as illustrated in Figure 10.4.

In the first case, one fixes the dimensions (L/D) by varying the mass (m) or molar flow rate (F), and measure the conversion. If for successive experiments the conversion

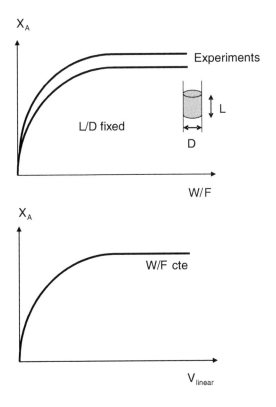

Figure 10.4 Mass transfer effects.

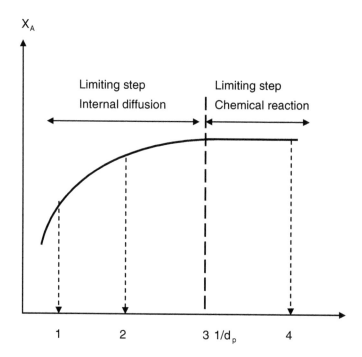

Figure 10.5 Effect of particle sizes on the conversion due to diffusion.

changes, there occur mass transfer effects. However, if conversion did not change, these effects are negligible. In the second case, we change the linear velocity by modifying the diameter, keeping the ratio W/F, and if conversion varies there are mass transfer effects.

Figure 10.5 illustrates the concentration gradients formed in the film surrounding the solid for different cases.

10.2 INTERNAL DIFFUSION PHENOMENA

Catalysts contain active sites which are located inside the pores. Therefore, molecules must first diffuse into the pores. Figure 10.6 shows reactant A diffusing in the pores until reaching the sites before reaction.

The internal molecular diffusion inside pores can be the limiting step and therefore it is unwanted. Basically, there are three forms of diffusions occurring in the pores of the catalyst:

1. *Molecular diffusion:* may occur in big pore diameters (1–10 μm) where the mean free path of the molecules is small compared to the pore diameter.
2. *Knudsen diffusion:* exists in middle pore diameters (10–1.000 Å) where the mass transport happens by collision between the molecules and the wall.
3. *Configuration diffusion:* exists in pores where the diameter of molecules is of the same order of the pore diameter.

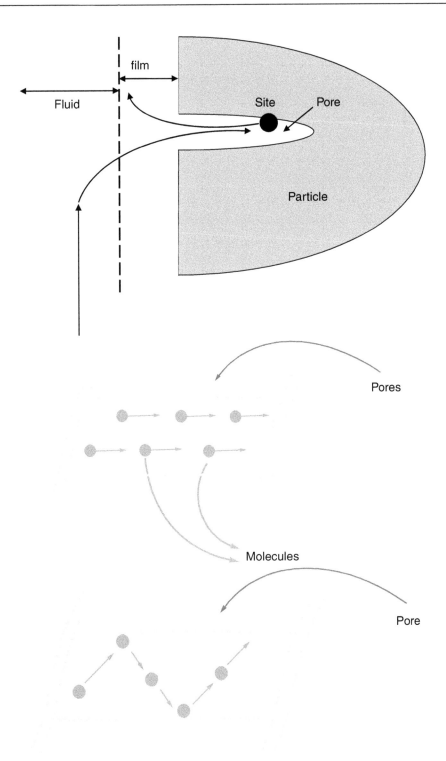

Figure 10.6 Diffusion of molecule A in the pore.

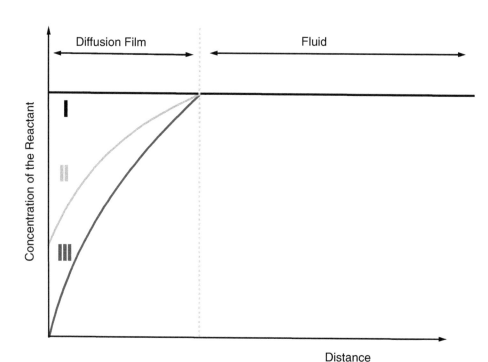

Figure 10.7 Concentration gradient in the neighborhood of the surface.

How can we verify diffusion in pores and when it is the limiting step of the process? There are two experimental procedures:

1.   Measuring the conversion as a function of the particle size—the reaction is performed by varying the particle sizes d_P, keeping constant all other variables (temperature, pressure, flow rate, and total mass of the sample). If diffusion limitations exist, the conversion increases with successive decreasing particles d_P, until attaining a constant value, as shown in Figure 10.5. Any diameter less or equal is free from diffusion resistance and hence, the kinetic regime is the limiting step of the process.
2.   Measuring the activation energy of the reaction from the kinetic constants at different temperatures—graph $\ln(k)$ versus $1/T$, which allows to calculate the activation energy. If the energy of activation is lower at higher temperatures compared to the value at lower temperatures, there are diffusional limitations, which must be eliminated or avoided as shown in Figure 10.8.

Therefore for kinetic measurements, high temperatures are not recommended to avoid diffusional effects. Table 10.1 presents the different possibilities that may occur in a gas phase catalytic reactions.

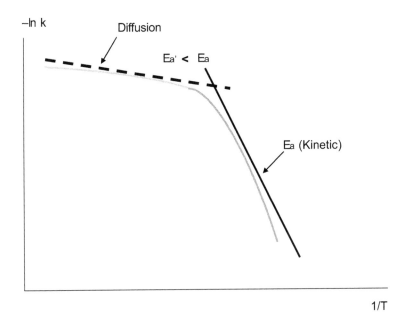

Figure 10.8 Arrhenius ln(k) versus 1/T diagram.

Table 10.1 Comparing Different Reaction Regime Conditions in the Gas Phase.

Limiting Step	Energy of Activation	Influence of Particle Sizes	Influence of Flow
Reaction (pure kinetics)	E	Nil	Nil
Internal diffusion	$E/2$	$1/d_p$	Nil
External diffusion	$E \leq 5$ kcal/moles	$1/d$	$v_{0,6}$

10.3 ADSORPTION–DESORPTION PHENOMENA

Aggregated systems (solids, liquids, and gases) have two energy types:

1. Kinetic energy, E_C, or thermal energy–inducing particles (atoms or ions) with different possibilities of motions such as translation, rotation, and vibration.
2. Interaction energy between particles, E_i, such as van der Waals and electrostatic forces.

The order of magnitude of these energies defines the system.

(a) When $E_C \gg E_i$, the molecules have maximum freedom, since the distance between molecules varies. The volume of the system is a function of the *temperature, pressure, and number of molecules*. It is classified as an *ideal gas system*.

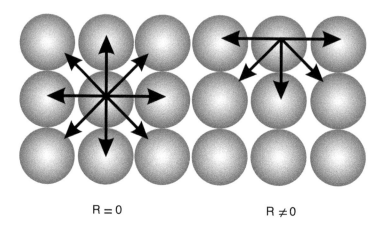

$$R = 0 \qquad\qquad R \neq 0$$

Figure 10.9 Interaction between particles in a condensed system (solid or liquid).

(b) When $E_C \approx E_i$, the molecules or atoms have less degree of freedom. Molecules move but the distance between them varies little. The volume of the system is a function of the *temperature, number of molecules, and ions,* independent *of pressure.* It represents the liquid state.

(c) When $E_C \ll E_i$, the volume depends only on the *number of molecules.* Molecules and atoms have only vibration-free movement, and the kinetic energy is attributed to the vibration of molecules around the same position. It represents the *adsorption of molecules on solids.*

Systems (b) and (c) are called condensed matter state, since the volumes are defined and less sensible to pressure variation. For condensed systems, the molecules on the surface or at the interface of another system present different situations:

1. Particles located inside the system are susceptible to forces in all directions and the resultant force is zero.
2. Particles at the surface have forces deriving only from inner particles.

Figure 10.9 shows the different systems.

Therefore, as shown in Figure 10.9, particles at the surface have an excess of energy, denominated as superficial energy, E_S. This energy is responsible for the surface tension of the liquid and adsorption of fluid over solids.

A catalytic reaction involves physical–chemical phenomena of adsorption and desorption besides chemical reaction. As shown in Figure 10.10, the energy barrier or activation energy of a catalytic reaction is lower than the activation energy of a noncatalytic reaction due to the adsorption and desorption phenomena.

The adsorption in a catalytic reaction is an exothermic phenomenon. However, either the adsorption of molecules at the surface or desorption from the surface occurs under different strengths, decreasing the degree of freedom that facilitates the reaction. The energy of activation of a catalytic reaction is, therefore, lower than the energy

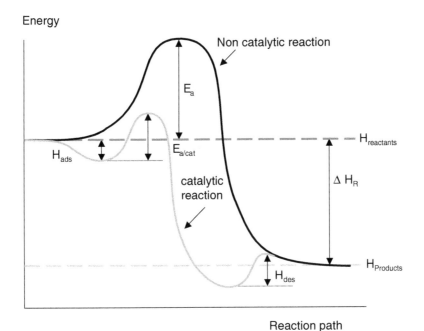

Figure 10.10 Comparison of a catalytic and a noncatalytic reaction. Potential energy changes during the reaction path.

barrier. Thus, it is necessary to determine the adsorption and desorption rates in a catalytic process.

Thermodynamically, we do explain the adsorption phenomenon of a fluid at the surface from the Gibbs free energy. It should be spontaneous and, thus, $\Delta G_{ads} < 0$. However, the final entropy of the system diminishes too, since the disorder is lower when molecules are adsorbed, or $\Delta S < 0$.

$$\Delta G = \Delta H - T\Delta S \qquad (10.1)$$

Adsorption:

$$\Delta H_{ads} = \Delta G_{ads} + T\Delta S_{ads} \qquad (10.2)$$

Since $\Delta G_{ads} < 0$ and $\Delta S < 0$, then the enthalpy change of the system is less than zero, and thus the adsorption is an exothermal phenomenon.

Depending on involved nature of forces, there are two types of adsorptions.

10.3.1 Physical adsorption or physisorption

Its characteristic is low interaction between molecules and solid surfaces. The resulting forces are of the same order of the van der Waals forces, and the enthalpy of adsorption is in the same range of the condensation enthalpy or evaporation gases

(−0.5–5 kcal/mol). In this kind of adsorption, several layers of molecules can be formed and the adsorption force decreases with increasing layers.

The physisorption occurs at low temperatures and is more intense if the temperature is near the condensation. Since the interaction energy with the surface is small and due to the inexistent activation energy of adsorption, the physisorption attains quickly the equilibrium, and thus it is reversible. However, materials having very small pores (zeolites, carbons) exhibit slow physisorption, which indicates that the diffusion in pores is the limiting step of this process.

The physisorption of gases on solids is frequently used for textural analyses of catalysts and solids, such as surface area and pore distribution and sizes.

10.3.2 Chemical adsorption or chemisorption

The chemisorption characteristic is the strong interaction between molecules and surfaces. The chemisorption enthalpy $(-\Delta H_{ads})$ is of the order of 10–100 kcal/mol, thus, of the same order of the reaction involved in chemical bindings.

Different from the physisorption, the chemisorption is irreversible and occurs at higher temperatures than the condensation temperature, and since the interaction is specific between molecules and solids, the adsorbed molecules form a monolayer.

The chemisorption processes need longer times to attain the equilibrium condition, in particular at low temperatures. There are two types of chemisorptions:

- *Activated chemisorption*: The adsorption rate varies with temperature and with self-activation energy. It follows the Arrhenius equation.
- *Nonactivated chemisorption*: This occurs quickly and indicates very low energy of activation or zero-activation energy.

10.3.3 Comparing physical and chemical adsorptions

The adsorption phenomena can be illustrated by the potential energy curve, as shown in Figure 10.11. When a gas approaches the surface of a metal, occurs dissociative adsorption of a diatomic gas X_2 over the metal M.

Curve F shows the physisorption pathway of a gas X_2 toward the metal and curve Q the chemisorption pathway when the gas molecule is first dissociated as $X–X$. One may conclude that when the gas approaches the surface, the physical adsorption is important for the next step, the chemical adsorption. In reality, all molecules are initially physisorbed approaching the surface with less energy. If not, it is necessary to activate the gas molecule, providing high energy for dissociation. However, molecules firstly physisorbed must overcome an energy barrier and then chemisorb. The barrier is located at the intersection of both curves F and Q. When this intersection is located above the axis, according to Figure 10.10, then the activation energy is E_a, and adsorption follows reaching the maximum adsorption state with an adsorption enthalpy ΔH_{ads}, highly exothermic, but when located below the axis, it is not activated. Noteworthy is that the energy barrier of activation is much lower than the enthalpy needed for direct dissociation of the molecules.

Chemisorption of gases over solids is often used for metallic surface area and dispersion of particles on supported catalysts analyses. Table 10.2 presents some properties comparing the physical and chemical adsorptions.

Energy

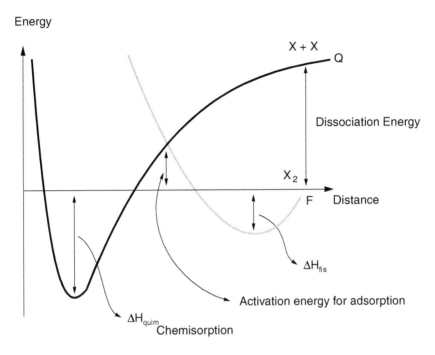

Figure 10.11 Potential energy curve for physisorption (F) and chemisorption (Q).

Table 10.2

Properties	Physisorption	Chemisorption
Solids	All solids	Depends on the gas
Gas	All gases	Depends on the solid
Temperature	Close to the boiling gas	Higher temperatures
Coverage	Multilayers	Monolayer
Reversibility	Reversible	Irreversible
Activation energy	Zero	Not zero
Heat of adsorption	Low (−0.5 to 5 kcal/mol)	High (−10 to −100 kcal/mol)

10.4 ADSORPTION ISOTHERMS

The amount of gas adsorption in a solid is proportional to the mass and depends on the temperature, pressure, solid, and gas nature. Thus, the number of moles of gas adsorption at the surface is:

$$n = f(P, T, \text{gas, solid}) \qquad (10.3)$$

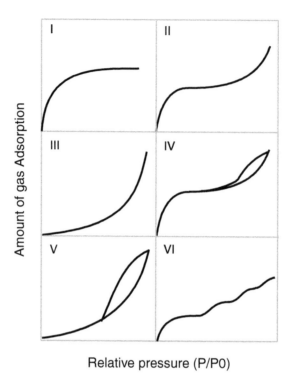

Relative pressure (P/P0)

Figure 10.12 Adsorption isotherms.

For constant temperature, the equation is simplified as:

$$n = f(P)_{T,\text{gas,solid}}$$

which is called "adsorption isotherm," relating the amount of gas adsorbed with the equilibrium pressure at constant temperature.

The experimental results are classified into six different types of adsorption isotherms and are shown in Figure 10.12.

The isotherm of type I indicates chemisorption, where saturation occurs at relatively low pressures, with the formation of a complete monolayer. These isotherms represent microporous material. Isotherms classified as type II and VI represent the physical adsorption. Isotherms II and III indicate infinite adsorption as $P/P_0 \rightarrow 1$, corresponding to physical adsorption on multilayers occurring on nonporous or macroporous materials.

Isotherms IV and V are equivalent to isotherms II and III but with finite adsorption, which indicates pore fill of the macro- or mesoporous materials.

The last isotherm type VI (in steps) occurs on nonporous uniform surfaces and indicates the adsorption layer by layer, where each step corresponds to the maximum adsorption capacity of a monolayer.

10.5 ADSORPTION MODELS

Some models have been proposed interpreting the adsorption–desorption phenomena. The most important models are described by the isotherms introduced by Langmuir, Freundlich, and Temkin.

10.5.1 Langmuir model

The first quantitative theoretical model of gas adsorption in solids was proposed by Langmuir in 1916. Langmuir assumed the following hypotheses in this model:

1. The solid surface has a finite number of adsorption sites.
2. Each site can only adsorb one molecule.
3. All sites have equivalent energies or equal adsorption enthalpies.
4. The adsorption is independent of neighboring adsorbed species or the enthalpy is independent of the surface coverage.
5. The adsorption and desorption rates are equal at equilibrium.
6. At equilibrium and constant temperature and pressure, the number of adsorbed molecules at the surface is n_A. The fraction of sites occupied by A molecules is θ_A:

$$\theta_A = \frac{n_A}{n_m} \tag{10.4}$$

7. At constant temperature, the adsorption rate of A molecule varies with the partial pressure P_A and the number of nonoccupied sites. The desorption rate varies with the number of occupied sites.

10.5.1.1 Adsorption of a single molecule

Assume the adsorption of a single molecule A at the surface. The fraction of sites occupied and unoccupied are:

   $\theta_A =$ fraction of occupied sites of A molecules.
   $(1 - \theta_A) =$ fraction of unoccupied sites or free sites.
where

$$\theta_A = \frac{n_{ads}}{n_m}$$

and
   $n_m =$ number of molecules of a monolayer.
   Schematically,

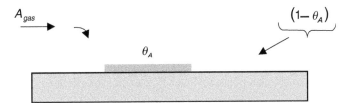

The adsorption rate of A is directly proportional to the partial pressure of A and to the fraction of free surface sites. This proportional factor is called adsorption constant k_a. Thus,

$$r_{ads} = k_a p_A (1 - \theta_A) \tag{10.5}$$

However, simultaneously occurs desorption of adsorbed A on surface sites. The desorption rate of A is directly proportional to the fraction of occupied sites or adsorbed A molecules, and k_d is the desorption constant. Thus,

$$r_{des} = k_d \theta_A \tag{10.6}$$

Of course, the resulting rate is the difference of adsorption and desorption rates; however, it is assumed that they are in equilibrium, which means that molecules are constantly adsorbed and desorbed allowing the adsorption of new molecules and free sites. Thus, at equilibrium, we have:

$$r_{ads} = r_{des}$$

Hence, we can determine the fraction of occupied sites of A molecules, according to the equation:

$$\theta_A = \frac{k_a p_A}{k_d + k_a p_A}$$

This equation can be written as function of the equilibrium adsorption–desorption constant, which is a thermodynamic property. Defining,

$$K_A = \frac{k_a}{k_d}$$

Thus,

$$\theta_A = \frac{K_A p_A}{1 + K_A p_A} \tag{10.7}$$

It is the Langmuir equation of a single adsorption molecule. Note that the fraction of surface sites can be determined experimentally by measuring adsorbed volumes in a system of constant pressure and temperature, according to the ideal gas law, since

$$\theta_A = \frac{V_{ads}}{V_{monolayer}} \tag{10.8}$$

The order of magnitude of the equilibrium constant K_A indicates the affinity of the gas with the solid. For high K_A values, the equilibrium is shifted toward adsorption and vice versa. There are two extreme cases to be considered:

1. If $K_A p_A \approx 0$, the equation becomes $\theta_A = K_A p_A$, i.e., the surface coverage is proportional to pressure A. This situation occurs when $p_A \approx 0$, i.e., on the initial stretch

of the isotherm, or when $K_A \approx 0$, i.e., when the gas affinity with the solid is very small.

2. If $K_A p_A \gg 0$, the equation becomes $\theta_A = 1$, or, the coverage is a complete mono-layer. It occurs when $p_A \gg 0$, i.e., near the vapor pressure of A, or when $K_A \gg 0$, i.e., when high affinity of gas with the surface.

10.5.1.2 Dissociative adsorption

Assuming the adsorption of a molecule A_2 on the site $*$ which occurs according to:

$$A_2(g) + 2^*(s) \rightleftarrows A^*_{(ads)} \tag{10.9}$$

In this case, the adsorption and desorption rates are, respectively:

$$r_{ads} = k_a p_A (1 - \theta_A)^2 \tag{10.10}$$

$$r_{des} = k_d \theta_A^2 \tag{10.11}$$

The dissociation of molecules occurs on two sites simultaneously. At equilibrium, we have:

$$r_{ads} = k_a p_A (1 - \theta_A)^2 = k_d \theta_A^2$$

Since

$$K_A = \frac{k_a}{k_d}$$

Rearranging, we determine the fraction of occupied sites due to dissociation, thus,

$$\theta_A = \frac{\sqrt{K_A p_A}}{1 + \sqrt{K_A p_A}} \tag{10.12}$$

This is the Langmuir equation for dissociative adsorption.

10.5.1.3 Adsorption of n-molecules

Going on, adsorption–desorption of several molecules, reactants or products, we have similar situation, as shown in the following scheme:

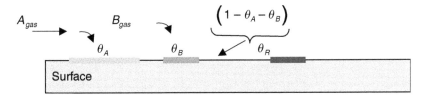

Then, the fraction of free sites is:

$$\theta_v = (1 - \theta_A - \theta_B - \theta_R \dots)$$

where θ_B and θ_R are the fractions of occupied sites by reactant B and product R, respectively.

If the adsorption and desorption rates are similar to the rates presented in Equations 10.5 and 10.6, for each component and considering the equilibrium constant similarly, we get the following system:

$$
\left.\begin{array}{l}
\theta_A = K_A p_A \theta_v \\
\theta_B = K_B p_B \theta_v \\
\theta_R = K_R p_R \theta_v \\
\cdots\cdots\cdots
\end{array}\right\} +
$$

$$
\sum \theta_i = \underbrace{(K_A \cdot p_A + K_B \cdot p_B + K_R \cdot p_R + \cdots)}_{\sum K_i p_i} \underbrace{\theta_v}_{(1-\sum \theta_i)}
$$

Thus,

$$
\sum \theta_i = \frac{\sum K_i p_i}{1 + \sum K_i p_i} \tag{10.13}
$$

and

$$
\theta_v = \frac{1}{1 + \sum K_i p_i} \tag{10.14}
$$

Therefore, we can determine the fraction of occupied sites of each component from Equation 10.14, since

$$
\theta_i = K_i p_i \theta_v
$$

Thus, for example, for component A, we get:

$$
\theta_A = \frac{K_A p_A}{(1 + K_A \cdot p_A + K_B \cdot p_B + K_R \cdot p_R)} \tag{10.15}
$$

If one of the components is dissociated, we substitute the corresponding term by square root. For example, if A is dissociated, then

$$
\theta_A = \frac{\sqrt{K_A p_A}}{(1 + \sqrt{K_A p_A} + K_B \cdot p_B + K_R \cdot p_R)} \tag{10.16}
$$

These expressions of occupied fraction sites by associated or dissociated molecules are important for the determination of the adsorption and the desorption rates, as shown in Equations 10.5 and 10.6. Moreover, they are also important for the determination of the reaction rates occurring simultaneously with adsorption and desorption, as presented in the next sections.

10.5.2   Other chemisorption models

The Langmuir adsorption model fails because:

- Not all sites are active.
- The adsorption enthalpy depends on the surface coverage degree. Adsorbed molecules may interfere on the adsorption of neighboring sites.

The dependence of the surface coverage with the heat of adsorption was accounted on two other models. The Freundlich model presents the coverage surface as:

$$\theta = kP^{1/n} \tag{10.17}$$

where k and n are constants and assume values higher than one. Freundlich's equation was originally introduced as an empirical correlation from experimental data. However, it can be derived theoretically with following considerations.

The heat of adsorption decreases logarithmically with the surface coverage,

$$\Delta H_{ads} = A \ln \theta \tag{10.18}$$

where θ assumes values varying between 0.2 and 0.8.

Temkin's model assumes a decreasing linear relationship of the heat of adsorption with the surface coverage:

$$\Delta H_{ads} = \Delta H_0 (1 - \beta\theta) \tag{10.19}$$

where ΔH_0 is the initial heat of adsorption. The degree of surface coverage is, in this case, given as follows:

$$\theta = \frac{RT}{\beta\Delta H_0} \ln (aP) \tag{10.20}$$

where a is the constant related to the enthalpy of adsorption and β is constant.

Noteworthy is that despite of all limitations, the Langmuir model is used in majority for kinetic models.

10.6   MODEL OF HETEROGENEOUS REACTIONS

Molecules are bonded at the surface (on active sites) by both physical and chemical adsorption processes. The nature and strengths of molecular bindings at the surface sites are of fundamental importance for the occurring reaction. The catalytic activity depends on the adsorption strengths, as shown in Figure 10.13. The activity is low when the adsorption strength is weak, which indicates the presence of van der Waals forces. However, for strong adsorptions bindings (chemical bonds), the activity is also very low. Therefore, intermediate adsorption strengths favor higher activity.

Moreover, products which are strongly bonded may make the desorption-freeing sites for the adsorption of new molecules difficult.

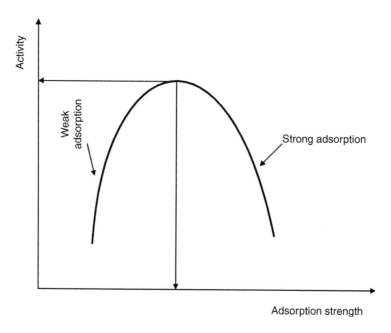

Figure 10.13 Activity vs Adsorption strength.

10.6.1 Langmuir–Hinshelwood–Hougen–Watson-model (LHHW)

This model consists of sequential steps which depend on the molecular or dissociated adsorption forms and the nature of one or more active sites at the surface. One determines the rates for each step and assumes what the rate limiting step is.

Irreversible mono- and bimolecular reactions

For the decomposition reaction of a reactant over one kind of surface active site where both reactant and product are likely adsorbed, the reaction occurs with the adsorbed species.

$$A \xrightarrow{k} R$$

Neglecting external mass transfer and diffusion effects, and assuming the chemical reaction as the limiting step, there are three main steps:

1. *Adsorption of A at the surface*

$$A + {}^* \overset{k_a/k_d}{\longleftrightarrow} A^*$$

where $^* =$ surface active site, $A^* =$ adsorbed molecule at the surface, k_a and $k_d =$ adsorption and desorption constants, respectively.

2. *Surface chemical reaction*
    The decomposition occurs with the adsorbed reactant species transforming into product adsorbed species on free site. Thus,

    $$A^* \xrightarrow{k} R^*$$

    where k is the specific rate constant of a monomolecular irreversible reaction, where the unit is square meters per mole time, and R^* is the product adsorbed species.

3. *Desorption of the product*
    The adsorbed R^* species is desorbed, releasing one active site and producing R gas molecule.

    $$R^* \xrightleftharpoons{k_a/k_d} R^* + {}^*$$

Schematically we obtain:

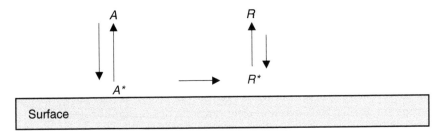

There are three possibilities:

(a) Adsorption of A is the limiting step.
(b) Chemical reaction is the limiting step.
(c) Desorption of R is the limiting step.

Here, we illustrate one possibility:
(b) *Chemical reaction as the limiting step*
Assuming that the surface chemical reaction relative to step (2) is irreversible, first order, and the limiting step. Then, the reaction rate is:

$$r = k\theta_A \tag{10.21}$$

But, the surface fraction of adsorbed A species was calculated assuming adsorption–desorption rates of reactant A and product R in equilibrium, according to Equation 10.15. Then, by substituting we obtain:

$$(-r_A) = \frac{kK_A p_A}{(1 + K_A \cdot p_A + K_R \cdot p_R)} \tag{10.22}$$

For a bimolecular reaction we proceed similarly, with a second component B reacting with A and the formation of a product R, or

$$A + B \xrightarrow{k} R$$

Assuming that both reactants and products are adsorbed and in analogy, we obtain the surface coverage of each adsorbed species.

If the adsorption–desorption rates of species are in equilibrium, we can calculate the corresponding surface fractions from Equation 10.15.

If the bimolecular reaction is irreversible and limiting step, then the reaction rate is proportional to the surface fractions of adsorbed species A and B. Thus,

$$r = k\theta_A\theta_B \tag{10.23}$$

The fractions of adsorbed species A and B were calculated considering all adsorbed species, according to Equation 10.15. Thus,

$$(-r_A) = \frac{kK_A K_B p_A p_B}{(1 + K_A \cdot p_A + K_B \cdot p_B + K_R \cdot p_R)^2} \tag{10.24}$$

This expression is valid only for case (b), when the surface chemical reaction is the limiting step. When the adsorptions or desorptions either of reactant or products are the limiting steps, then the corresponding rates are not in equilibrium.

When the adsorption of A is the limiting step, then:

$$r_{ads} \neq r_{des}(\text{adsorption-desorption of A})$$

$$r_{reação} = 0$$

$$r_{ads} = r_{des}$$

Equations 10.22 and 10.24 are similar and thus generalized in the following form:

$$r = \frac{(\text{Kinetics}) (\text{potencial strength})}{(\text{adsorption} - \text{desorption term})^n}$$

Thus,

1. The numerator contains the kinetic constants and the *potential strength* of a catalytic reaction.
2. The exponent n in the denominator represents the number of active sites, or the number of sites involved in the reaction.
3. The denominator represents the contribution of adsorption–desorption steps in equilibrium.

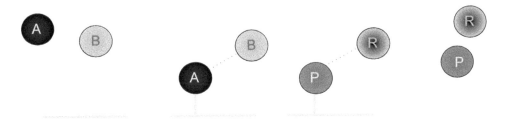

Figure 10.14 Eley-Rideal model.

10.6.2 Eley–Rideal model

This model derives from the previous one. For example, the catalytic reaction in the gas phase is:

$$A(g) + B(g) \rightarrow P(g) + R(g)$$

For this model, we assume that one of the reactants (products also) is adsorbed at the surface site while the other is not, as shown in Figure 10.14.
Thus, B and R do not adsorb and Equation 10.23 becomes:

$$r = k\theta_A p_B \tag{10.25}$$

where

$$K_B = 0$$
$$K_R = 0$$

Thus, similar to Equation 10.24, we obtain:

$$(-r_A) = \frac{kK_A p_A p_B}{(1 + K_A \cdot p_A + K_R \cdot p_R)} \tag{10.26}$$

10.6.3 Effect of the temperature and energies

As already known, both the kinetic constants and equilibrium adsorption–desorption constants depend on the temperature. However, the kinetic constant is also a function of the activation energy, following the Arrhenius equation, while the adsorption–desorption constants depend on the heat of adsorptions or desorptions, respectively, both exothermic. Thus, starting from Equation 10.26, we obtain:

$$k = k_0 \cdot \exp(-E/RT)$$

$$K_i = K_{i0} \cdot \exp(\Delta H_i/RT)$$

where $\Delta H_i < 0$.

For a catalytic reaction, the rate is given by:

$$r_{cat} = k_{cat} \cdot \exp(E_{cat}/RT) \tag{10.27}$$

Equating 10.26, after substitution of the corresponding kinetic and adsorption–desorption constants, we obtain the apparent activation energy:

$$E_{cat} = E - \Delta H_A + \Delta H_R \tag{10.28}$$

As seen, the apparent activation energy involves the kinetic activation energy as well as the adsorption–desorption enthalpies of adsorbed species. The rate of a catalytic reaction is easier than for a noncatalytic reaction since the energetic barrier is lower, increasing the activity of the reaction which explains the catalytic effect.

10.7 DETERMINATION OF THE CONSTANTS

For determining the rate constants, we use similar methodology as presented previously: the integral or differential methods. The differential method is frequently used and easily visualized in the graphic solution after transforming the rate equation. For example, for a monomolecular, irreversible, and first-order reaction, the rate is expressed in Equation 10.22. It was deduced assuming that the reaction rate is the limiting step and both reactants and products adsorbed. Thus,

$$\frac{1}{kK_A} + \frac{1}{k}p_A + \frac{K_R}{kK_A}p_R = \frac{p_A}{(-r_A)} \tag{10.29}$$

It is illustrated graphically in Figure 10.15.

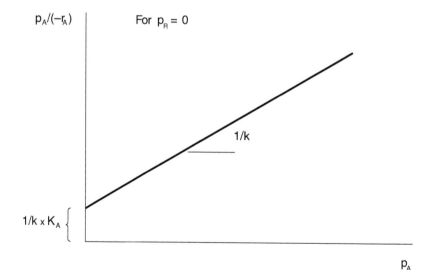

Figure 10.15 Determination of kinetic parameters.

Example

E10.1 Experimental results of a catalytic reaction are presented in the following table. The conversion was kept around 5%, and therefore, it should be considered as a differential reactor. Mass or diffusion effects were eliminated. The reaction is:

$$Cyclohexanol \longrightarrow cyclohexene + water$$

It suggests a reaction rate, where adsorption–desorption and reaction occur, and the reaction is the limiting step. Try to estimate the constants with the following data, assuming that the adsorption–desorption constants of the products are approximately similar [Hill].

Experiments	Rate $r \times 10^5$ (mol/L s)	P_{CH} (atm)	$P_{\overline{CH}}$ (atm)	P_{H2O} (atm)
1	3.3	1	1	1
2	1.05	5	1	1
3	0.565	6	1	1
4	1.826	2	5	1
5	1.49	2	10	1
6	1.36	3	0	5
7	1.08	3	0	10
8	0.862	1	10	10
9	0	0	5	8
10	1.37	3	3	3

Solution

$$A \rightarrow R + S$$

First, verify from this table what components are adsorbed.

1.   For A:
        From the equation 10.22, we consider the adsorption–desorption of component A, keeping the partial pressures of products or other components constant. Thus,

$$r = \frac{kK_A p_A}{(1 + K_A \cdot p_A + K_R \cdot p_R)}$$

From the experiments:

Experiments	Rate $r \times 10^5$ (mol/L s)	P_{CH} (atm)	$P_{\overline{CH}}$ (atm)	P_{H2O} (atm)
1	3.3	1	1	1
2	1.05	5	1	1
3	0.565	6	1	1

- Note that if $P_{CH^-} = P_{H2O} = 1$ then:
- Increasing the pressure five times, the rate decreases three times.
- Increasing (experiments 2 and 3) pressure 1.2 times, the rate falls ≈ 2 times.
- Increasing (experiments 1 and 3) pressure six times, the rate falls ≈ 6 times.

In fact with increasing pressure, the rate is not directly proportional to the partial pressure but inversely proportional, although not equal, which suggests adsorption–desorption of A in the denominator term. Therefore, A is strongly adsorbed.

2. For S (H_2O):

Keeping the partial pressures of A and R constant, from experiments:

Experiments	Rate $r \times 10^5$ (mol/L s)	P_{CH} (atm)	$P_{\overline{CH}}$ (atm)	P_{H2O} (atm)
6	1.36	3	0	5
7	1.08	3	0	10

Increasing the pressure two times, the rates do not change too much (practically 1.3 times). The adsorption is weak, decreasing slightly, which suggests negligible adsorption of S.

3. For R (CH^-):

Keeping the partial pressure of the other components constant, from experiments we observe:

Experiments	Rate $r \times 10^5$ (mol/L s)	P_{CH} (atm)	$P_{\overline{CH}}$ (atm)	P_{H2O} (atm)
1	3.3	1	1	1
8	0.862	1	10	10
4	1.826	2	5	1
5	1.49	2	10	1

- From experiments 1 and 8 → Increasing the pressure 10 times, the rate falls four times.
- From experiments 4 and 5 → Increasing the pressure two times, the rate falls 1.2 times.

It suggests that component R is adsorbed; however, compared with adsorption of A its adsorption is weak, although greater than that of S.

We observe that all components are adsorbed and that the adsorptions of the products are almost of the same order, weakly adsorbed. Therefore, assuming that the reaction is irreversible, first order, and the limiting step, we obtain from Equation 10.22:

$$r = \frac{kK_A p_A}{(1 + K_A \cdot p_A + K_R \cdot p_R)} \tag{10.30}$$

or rearranging, we get:

$$\frac{1}{kK_A} + \frac{1}{k}p_A + \frac{K_R}{kK_A}p_R + \frac{K_S}{kK_A}p_S = \frac{p_A}{r} \tag{10.31}$$

$$\underbrace{\phantom{\frac{1}{kK_A}}}_{a} \quad \underbrace{\phantom{\frac{1}{k}p_A}}_{b} \quad \underbrace{\phantom{\frac{K_R}{kK_A}p_R}}_{c} \quad \underbrace{\phantom{\frac{K_S}{kK_A}p_S}}_{d}$$

$$a + by_1 + cy_2 + dy_3 = X$$

From the experiments:

Experiments	Rate $r \times 10^5$ (mol/L s)	P_{CH} (atm)	$P_{\overline{CH}}$ (atm)	P_{H2O} (atm)
1	3.3	1	1	1
2	1.05	5	1	1
3	0.565	6	1	1

$$P_R = P_S = 1.0$$

Thus,

$$y_2 = y_3 = 1$$

$$a * + by_1 = X$$

where

$$c* = cy_2 + dy_3$$

$$a* = a + c*$$

$$a* = 0$$

$$b = 1.146 \times 10^5$$

$$k = 1/b = 8.72 \times 10^{-6} (\text{L/moles}) \text{ s}^{-1}$$

10.8 NONCATALYTIC HETEROGENEOUS REACTIONS

Noncatalytic heterogeneous reactions are reactions of gases reacting with the solid. These reactions are very important in the regeneration of catalysts, elimination of coke or carbons at the surface blocking pores after the deactivation process. Also, for the combustion of particulates from exhaust gases which are eliminated by trucks, due to the incomplete combustion of diesel, releasing huge amounts of fine carbon particles, and therefore causing environmental problems.

There are several other cases of solid reactions, such as elimination of H_2S using ZnO solids, which transforms the oxide into sulfte that are discarded after completion of reaction. Some typical examples are as follows:

1. Burning coke from catalysts

$$C + O_2 \rightarrow CO_2 \tag{10.32}$$

2. Elimination of sulfur compounds forming H_2S

$$ZnO + H_2S \rightarrow ZnS + H_2O \tag{10.33}$$

The question here is to know the kinetics of these reactions involving diffusion phenomena during the reaction. The solid particle is consumed due to the chemical reaction, but as it proceeds transforms the material into ashes or inert material whereby gas flows and diffuses until reaching the surface for reacting. Thus, there is another limiting step which is the diffusion of gas through the solid. On the other hand, the adsorption–desorption phenomena of gases may occur at the surface and can also be limiting or not. In general, the surface is nonhomogeneous and these phenomena occur on different surface sites.

For simplifying, we disregard the adsorption–desorption phenomena, focusing mainly the diffusion and surface reaction steps. Evidently, time is an important variable, since we want to know how long it takes for total consumption of the solid or transformation of the solid. An example is shown in Figure 10.16.

There are two important steps:

1. Diffusion of gas (O_2) through the inert layer (ash) until the interface.
2. Reaction at the interface with the formation of products and diffusion.

From the mass balance of gas (O_2), we get:

$$\dot{M}_{O_2}(4\pi r^2)\big|_r - \dot{M}_{O_2}(4\pi r^2)\big|_{r+dr} = 0 \tag{10.34}$$

Since no generation or accumulation of gas (O_2) occurs, we obtain the equation in the differential form:

$$\frac{d(\dot{M}_{O_2} r^2)}{dr} = 0 \tag{10.35}$$

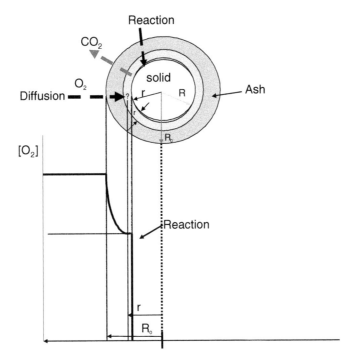

Figure 10.16 Non catalytic reaction model.

However, from Fick's law, we have:

$$\dot{M}_{O_2}\Big|_r = D\frac{dC_{O_2}}{dr} \tag{10.36}$$

Substituting Equation 10.36 into Equation 10.35, we obtain:

$$\frac{d(\dot{M}_{O_2}r^2)}{dr} = 0$$

Rearranging, we get:

$$\dot{M}_{O_2}\Big|_r = -D\frac{dC_{O_2}}{dr} \tag{10.37}$$

Solving the equation for the boundary conditions, we have:

$$\frac{d}{dr}\left(-D\frac{dC_{O_2}}{dr}r^2\right) = 0$$

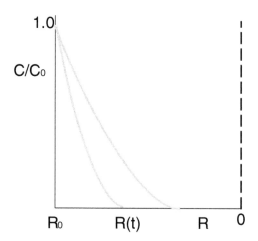

Figure 10.17 Concentration profile of oxygen.

We obtain:

$$\frac{d}{dr}\left(r^2\frac{dC_{O_2}}{dr}\right)=0 \tag{10.38}$$

T
  With boundary conditions:

$$r=R_0 \quad C_{O_2}=C_{[O_2]0}$$

$$r=R_{interface} \quad C_{O_2}=0$$

We obtain:

$$\frac{C_{O_2}}{C_{[O_2]0}}=\phi=\frac{1/R-1/r}{1/R-1/R_0}$$

The concentration profile of O_2 in the solid is shown in Figure 10.17.
3. Reaction at the interface with the formation of product and diffusion.
    (a) Oxygen flows at the interface
    From the derivation of Equation 10.38, we obtain:

$$-D\frac{dC_{O_2}}{dr}r^2\bigg|_{interface}=-D\frac{C_{[O_2]0}}{\left(\frac{1}{R}-\frac{1}{R_0}\right)r^2} \tag{10.39}$$

4. Carbon balance
    [Generation rate] = [Accumulation rate]

$$r'_C\,(4\pi R^2)=-D\frac{d(4/3\pi R^3\rho_C\varepsilon)C_{O_2}}{dt} \tag{10.40}$$

where

r'_C = solid reaction rate per unit area (g/m^2 min)
ρ_C = density of carbon (g/m^3)
ε = fraction of carbon
D = diffusion coefficient of oxygen.

From this expression, we get the consumption of carbon at the interface:

$$\frac{dR}{dt} = \frac{r'_C}{(\rho_C \varepsilon)} \tag{10.41}$$

Carbon/gas (O_2) balance at the interface
We know that for each mole of oxygen reacting 1 mole of CO_2 is released.
[Rate of carbon C disappearing] = [O_2 flow at the interface]

$$r'_C = -\dot{M}_{O_2}\Big|_{R_{\text{interface}}} = D \frac{C_{[O_2]0}}{\left(R - \dfrac{R^2}{R_0} \right)} \tag{10.42}$$

Then, substituting Equation 10.42 into Equation 10.41 for $r=0$, $R=R_0$, we obtain:

$$\frac{dR}{dt} = D \frac{C_{[O_2]0}}{(\rho_C \varepsilon)\left(\dfrac{1}{R} - \dfrac{1}{R_0} \right)} \tag{10.43}$$

With the boundary condition $r = 0$ and $R = R_0$ and integrating, we obtain:

$$t = \frac{(\rho_C \varepsilon)R_0^2}{6DC_{[O_2]0}}\left[1 - 3\left(\frac{R}{R_0} \right)^2 + 2\left(\frac{R}{R_0} \right)^3 \right] \tag{10.44}$$

The time needed for the total consumption of carbon is obtained when $R = 0$. Thus,

$$t = \frac{(\rho_C \varepsilon)R_0^2}{6DC_{[O_2]0}} \tag{10.45}$$

If the conversion is defined as:

$$X = 1 - \frac{(4/3)\pi R^3}{(4/3)\pi R_0^3} = 1 - \left(\frac{R}{R_0} \right)^3 \tag{10.46}$$

Substituting in Equation 10.44, it results as:

$$t = \frac{(\rho_C \varepsilon)R_0^2}{6DC_{[O_2]0}}[1 - 3(1 - X)^{2/3} + 2(1 - X)] \tag{10.47}$$

Example

E10.2 Calculate the time needed for burning carbon particulates from the exhaust gases of trucks using data at 700°C and 850°C as shown in the following table.

T (°C) 700 t (min)	T (°C) 850 t (min)	Conversion X
0	0	0
2	4	0.00054
6	8	0.00238
10	15	0.00572
16	30	0.01086
25	45	0.01821
38	60	0.02839
55	80	0.04242
70	120	0.06215
90	150	0.07533
130	200	0.09211

From 10.47, we calculate:

$$F(X) = \left[1 - 3(1 - X)^{2/3} + 2(1 - X)\right] \tag{10.48}$$

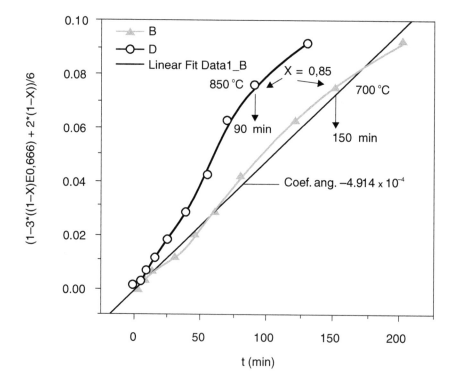

Figure 10.18 Plot of equation 10.47.

Thus,

$$\frac{t \cdot 6DC_{[O_2]0}}{(\rho_C \varepsilon) R_0^2} = F(X) \qquad (10.49)$$

Figure E10.18 shows the results.
From the angular coefficient, we calculate the parameter β.

$$\beta = \frac{DC_{[O_2]0}}{(\rho_C \varepsilon) R_0^2} = 4.914 \times 10^{-4}$$

10.9 REPARAMETERIZATION OF THE LHHW EQUATION

As seen in Section 9.3 and discussed in detail in Chapter 11, nonlinear kinetic models may rely on parameters that present high correlations among themselves, leading to significant numerical problems during the estimation of model parameters and prejudicing the statistical characterization of the final parameter estimates (Espie and Macchietto, 1988; Watts, 1994). Some of these difficulties can also be observed during the estimation of parameters of LHHW rate equations, utilized throughout Chapter 10, because of the usually high correlations among the obtained parameter estimates. As discussed in the previous sections, LHHW rate equations are based on reaction mechanisms that explicitly consider adsorption steps and occurrence of chemical interactions between the active catalyst sites and the reacting chemical species. In its simplest form, reaction rates can be controlled by adsorption phenomena, often described in terms of the Langmuir equation as:

$$C_A^e = \frac{K_1 C_A^b}{1 + K_2 C_A^b} \qquad (10.50)$$

$$C_A^e = \frac{C_A^b}{K_3 + K_4 C_A^b} \qquad (10.51)$$

$$C_A^e = \frac{K_5 C_A^b}{K_6 + C_A^b} \qquad (10.52)$$

$$C_A^e = \frac{C_A^b}{A + B \left(C_A^b - C_{A\text{ref}}^b \right)} \qquad (10.53)$$

where C_A^e is the concentration of reactant A on the catalyst surface; C_A^b is the bulk concentration of reactant A in the reacting system; K_1, K_2, K_3, K_4, K_5, K_6, A and B are model parameters; and $C_{A\text{ref}}^b$ is a reference numerical value. Additionally:

$$K_3 = \frac{1}{K_1}; K_4 = \frac{K_2}{K_1}; K_5 = \frac{K_1}{K_2}; K_6 = \frac{1}{K_2} \qquad (10.54)$$

$$A = \frac{1 + K_2 [S]_{\text{ref}}}{K_1}; B = \frac{K_2}{K_1} \tag{10.55}$$

As a matter of fact, Equations (10.50–5) are quite similar to the ones described and studied previously in Section 9.3, when the Michaelis–Menten equation was used to represent the rates of substrate consumption in enzymatic and fermentation problems. Therefore, very similar numerical approaches can be used for reparameterization of the Langmuir equation and estimation of more precise model parameters.

Example

$E10.3$ Using the approach developed originally by Schwaab and Pinto (2007, 2008) to perform the reparameterization of the Arrhenius equation (see Section 3.3.3) and of the power-law rate model (see Section 5.7), it was shown in Section 9.3 that the Michaelis–Menten equation could be better described in the form:

$$R_S = -\frac{[S]}{A + B \left([S] - [S]_{\text{ref}} \right)}$$

with

$$[S]_{\text{ref}} \approx \frac{\sum_{i=1}^{NE} \frac{R_{S_i}^4}{[S]_i^2}}{\sum_{i=1}^{NE} \frac{R_{S_i}^4}{[S]_i^3}}$$

when experimental rate data are available for the respective substrate concentrations. Therefore, in Equation (10.53), the value of $C_{A\text{ref}}^b$ that allows the calculation of independent parameters A and B with maximum accuracy is:

$$C_{A\text{ref}}^b \cong \frac{\sum_{i=1}^{NE} \frac{\left(C_A^e\right)_i^4}{\left(C_A^b\right)_i^2}}{\sum_{i=1}^{NE} \frac{\left(C_A^e\right)_i^4}{\left(C_A^b\right)_i^3}}$$

when equilibrium concentration data are available for the respective bulk concentrations in adsorption experiments.

Example

$E10.4$ In adsorption experiments, the experimenter usually defines the initial bulk concentration and measures the final attained concentration so that:

$$V_0 C_{A0} = V_0 C_A^b + \frac{M_{\text{ads}} C_A^b}{A + B \left(C_A^b - C_{A\text{ref}}^b \right)}$$

where V_0 is the volume of the solution that contains the adsorbate A at concentration C_{A0} in the beginning of the experiment and M_{ads} is the mass of adsorbent put in contact with the adsorbate solution. Therefore:

$$C_{A0} = \frac{C_A^b[A + B(C_A^b - C_{Aref}^b)] + \left(\frac{M_{ads}}{V_0}\right)C_A^b}{A + B(C_A^b - C_{Aref}^b)}$$ (10.56)

or

$$\frac{M_{ads}}{V_0}\left(\frac{M_{ads}}{V_0}\right) = \frac{(C_{A0} - C_A^b)[A + B(C_A^b - C_{Aref}^b)]}{C_A^b}$$ (10.57)

or

$$B\left(C_A^b\right)^2 + \left(A - B\left(C_{A0} + C_{Aref}^b\right) + \frac{M_{ads}}{V_0}\right)C_A^b - AC_{A0} + BC_{Aref} = 0$$ (10.58)

or

$$\left(C_A^b\right)^2 + \left(A^* - \left(C_{A0} + C_{Aref}^b\right) + B^*\frac{M_{ads}}{V_0}\right)C_A^b - A^*C_{A0} + C_{Aref} = 0$$ (10.59)

where

$$A^* = \frac{A}{B}; B^* = \frac{1}{B}$$ (10.60)

Although the investigation certainly defines V_0, M_{ads} and C_{A0} as inputs (x) and C_A^b as output (y), from a mathematical point of view Equations (10.56–10) are completely equivalent. In spite of that, Equations (10.56) and (10.58–10) are significantly more complex because they do not allow the independent evaluation of the adsorption parameters, unless previous knowledge of the parameter values is available. However, when Equation (10.57) is considered a reference for estimation of model parameters of the form:

$$y_k = \frac{x_{1k}(x_{2k} - x_{3k})[A + B(x_{3k} - C_{Aref}^b)]}{x_{3k}}$$ (10.61)

where $x_{1k} = V_{0k}$; $x_{2k} = C_{A0k}$; $x_{3k} = C_{Ak}^b$; and $y_k = M_{adsk}$. In this case,

$$\left.\frac{\partial y}{\partial A}\right|_k = \frac{x_{1k}(x_{2k} - x_{3k})}{x_{3k}}$$ (10.62)

$$\left.\frac{\partial y}{\partial B}\right|_k = \frac{x_{1k}(x_{2k} - x_{3k})(x_{3k} - C_{Aref}^b)}{x_{3k}}$$ (10.63)

so that, in order for

$$\sum_{k=1}^{NE} \frac{\partial y}{\partial A}\Big|_k \frac{\partial y}{\partial B}\Big|_k = \sum_{k=1}^{NE} \frac{x_{1k}^2 \, (x_{2k} - x_{3k})^2 \, \left(x_{3k} - C_{Aref}^b\right)}{x_{3k}^2} = 0 \qquad (10.64)$$

$$C_{Aref}^b = \frac{\sum_{k=1}^{NE} \frac{x_{1k}^2 (x_{2k} - x_{3k})^2}{x_{3k}}}{\sum_{k=1}^{NE} \frac{x_{1k}^2 (x_{2k} - x_{3k})^2}{x_{3k}^2}} = 0 \qquad (10.65)$$

Example

E10.5 An adsorption experiment was performed, using $V_0 = 100$ mL, $C_{A0} = 10$ g/L and different masses of adsorbent, as shown in Table 10.50.

Table 10.50 Equilibrium bulk adsorbate concentrations for different masses of adsorbent.

M_{ads} (g)	0	0.25	0.50	0.75	1.00	1.25	1.50	1.75	2.00	2.25
C_A^b (g/L)	10.36	8.02	5.97	4.05	2.68	1.85	1.30	1.00	0.80	0.68

According to Equation (10.65), $C_{Aref}^b = 0.9314$ g/L can allow the estimation of independent model parameters when $y = M_{ads}$. For this reason, Equations (10.56–60) were used for estimation of the model parameters and the parameter correlations are shown in Figure 10.50. The obtained results confirmed the expectations that $y = M_{ads}$ would provide independent parameters and showed that nearly independent parameters could be obtained with the same value of $C_{Aref}^b = 0.9314$ g/L when $y = C_{A0}$. On the

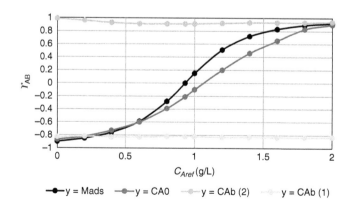

Figure 10.50 Parameter correlations for different formulations of the parameter estimation problem using the Langmuir equation.

other hand, the parameters obtained with $y = C_A^b$ were strongly correlated to each other when either Equation (10.58) or Equation (10.57) were used. It must be emphasized that, despite the modification of the analyzed output model responses and objective functions, parameter values and model performances were very similar in all cases, as illustrated in Figure 10.51. However, use of $y = M_{ads}$ led to much better statistical behavior of obtained parameter estimates, as shown in Figure 10.50

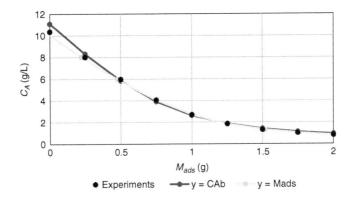

Figure 10.51 Model performances for different formulations of the parameter estimation problem using the Langmuir equation.

In general, LHHW kinetic rate expressions have the general form:

$$R_k = \frac{\sum_{i=1}^{n} K_i^N \prod_{j=1}^{NC} C_j^{p_{ij}}}{1 + \sum_{i=1}^{m} K_i^D \prod_{j=1}^{NC} C_j^{q_{ij}}} \tag{10.66}$$

where k represents one particular reactant, and m are the numbers of terms in the numerator and denominator of Equation (10.66), respectively, K_i^N nd K_i^D are the kinetic rate constants associated with the ith terms of the numerator and denominator of Equation (10.66), respectively,;C is the number of reactants.;nd p_{ij} and q_{ij} are the kinetic rate orders associated with component j at the ith terms of the numerator and denominator of Equation (10.66), respectively. Based on Equation (10.66), it is not possible to derive a general solution for the reparameterization of a general LHHW reaction rate equation. In spite of that, reparameterization can be performed iteratively with the help of appropriate numerical procedures, as discussed by Schwaab *et al.* (2008). In this case, a suitable objective function must be proposed (for instance, the summation of the absolute values of all parameter correlations) and then the reparameterizing values (for instance, the reference concentration used in Equations (10.56–60)) must be manipulated in order to reduce the value of the proposed objective function iteratively, until attainment of the minimum possible correlation values. This is exactly the procedure illustrated in Figure 10.50, which shows parameter correlation values as a function of reference concentration.

Part III

Parameter estimation and experimental design

Chapter 11

Determination of kinetic parameters through parameter estimation

As we had the opportunity to observe in the previous sections, in order for reaction rate expressions to be used for kinetic analyses and design of chemical reactors (as discussed in detail in the following chapters), it is necessary to determine first the kinetic rate constants with the help of available experimental data. Formally, this numerical procedure that transforms available experimental data into useful model parameters that can be used to perform simulations and equipment design is named **parameter estimation**.

Parameter estimation procedures are important in many different chemical engineering fields, and particularly in the field of chemical kinetics, for development of mathematical models (such as reaction rate expressions), since design, optimization and advanced control of chemical processes (and chemical reactors in particular) usually rely on mathematical models, which in turn depend on parameter values obtained with the help of available experimental data. For this reason, the parameter estimation problem is discussed in this section more formally.

Before initiating the presentation of mathematical details, it is important to recognize that chemists and engineers frequently search for correlations among the variables involved in a problem in order to develop and implement model-based process simulation and process design strategies. As a matter of fact, accurate models can be very useful for simulating experimental responses under the desired experimental conditions and to design process equipment of distinct sizes without incurring the expensive costs of experimentation. For these reasons, mathematical models find ample use in most engineering areas, especially to control, design and optimize process operations.

Models can be developed based on theoretical grounds (for instance, detailed reaction mechanisms and mass balance equations) or on empirical premises (for instance, power-law rate expressions), and they are useful if they can accurately represent the actual experimental behavior of the analyzed system. Consequently, theoretical models are not necessarily better than empirical models, as theoretical assumptions can be incorrect or incomplete to describe the considered problem. Thus, in the present section both theoretical and empirical reaction rate expressions are equally valued. Moreover, mathematical models are usually classified, in terms of time dependence, as dynamic (when the model can be used to follow changes during the simulation time) or stationary (when model responses do not change with time). Analyses of batch reaction problems usually make use of dynamic reaction rate models, while design of chemical reactors normally employ stationary versions of the reaction rate models. Most times, numerical procedures are also needed to solve the model equations and calculate the output model responses, so that the mathematical model can be defined less strictly as the set of equations and numerical tools that are needed to solve the mathematical equations and describe the analyzed experimental problem.

Almost always the mathematical model depends on variables that cannot be measured experimentally and must be determined with the help of experimental data, like the kinetic rate constants and respective activation energies. These variables are generically called model parameters and must be estimated before there can be any meaningful use of the mathematical model. For this reason, the fundamental aspects of the parameter estimation problem are discussed in the next sections.

11.1 DEFINITION OF THE PARAMETER ESTIMATION PROBLEM

During the proposition of mathematical models, two distinct entities must be defined: the model structure (the mathematical relationship that correlates output model responses, y, and model inputs, x) and the parameters of the model. For example, linear and exponential models are presented in Equations (11.1) and (11.2):

$$y = \alpha \cdot x + \beta \tag{11.1}$$

$$y = \alpha \cdot \exp(\beta \cdot x) \tag{11.2}$$

In reaction rate problems, x and y involve pressures, concentrations and temperatures at distinct times, the parameters α and β usually are reaction rate constant and activation energy, respectively. As one can see in Equations (11.1–2), the model structure can only be used when the parameter values are also known.

A very simple linear model can be obtained when a reaction mechanism comprises two parallel reactions of the form:

$$\begin{cases} A \xrightarrow{k_1} B \\ A \xrightarrow{k_2} C \end{cases} \tag{11.3}$$

where A, B and C are chemical species and k_1 and k_2 are kinetic rate constants. If reaction conditions are kept constant, the relative amounts of reactants B and C are correlated to each other in the form:

$$C_C = \frac{k_2}{k_1} C_B \tag{11.4}$$

where C_C and C_B are the molar concentrations of the reaction products C and B, and the parameter:

$$\alpha = \frac{k_2}{k_1} \tag{11.5}$$

is a measure of the selectivity. Whereas C_C and C_B can be measured experimentally, α must be estimated through proper manipulation of C_C and C_B as the slope of Equation (11.4), as illustrated graphically in Figure 11.1.

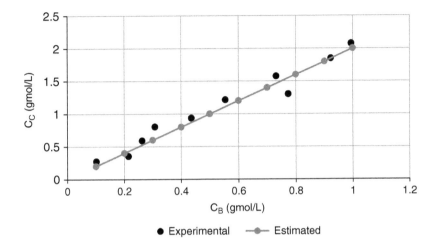

Figure 11.1   Measured concentrations and estimation of parameter α in Equations (11.4–5). α is the slope of the experimental data. Experimental data were generated by assuming that $C_C = 2C_B$ and adding a random noise of amplitude ± 0.1.

As illustrated in Figure 11.1, the estimation of the model parameters involves the determination of the parameter values that lead to the smallest difference (or highest similarity, or likelihood) between the values predicted by the model and the values observed experimentally. Therefore, the parameter estimation procedure should manipulate the parameter values in order to minimize the difference between the model responses and the experimental data, leading to a definition of an optimization problem. However, in order to minimize the difference between model responses and observed data, it is necessary first to define how this difference can be calculated. In the area of parameter estimation, this difference (or distance) is normally named as the objective function.

One must observe that measurement uncertainties, as defined in Section 1.9, must also be considered during the parameter estimation procedure, because unavoidable experimental fluctuations can affect the experimental data and therefore the obtained parameter estimates. Besides, due to unavoidable experimental uncertainties, parameter estimates will also be uncertain to some extent, leading to uncertain model predictions. In order to show this, Figure 11.2 reproduces the information content of Figure 11.1, assuming the occurrence of much larger experimental fluctuations. As one can observe, proper identification of the slope in this case can be much more difficult.

Based on the above, the parameter estimation procedure can usually be divided into three main steps: (i) definition of the objective function that evaluates the distance between model responses and experimental observations; (ii) determination of the parameter values that minimize the objective function (usually performed with the help

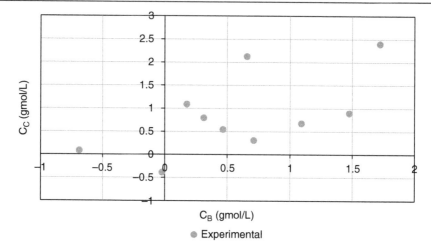

Figure 11.2 Measured concentrations and estimation of parameter α in Equations (11.4–5). α is the slope of the experimental data. Experimental data were generated by assuming that $C_C = 2C_B$ and adding a random noise of amplitude ± 1.

of a suitable numerical procedure); and (iii) statistical interpretation of the obtained results. These three steps are discussed in the following sections.

11.2 THE OBJECTIVE FUNCTION

The definition of the objective function is important and must provide an unambiguous measure of the distance between the values calculated with the model, y^*, and the values obtained experimentally, y^e. From a mathematical standpoint, the distance between y^* and y^e, $d(y^*, y^e)$, must satisfy the following axioms (Kreyszig, 1978):

(i) $d(y^*, y^e) \in \mathcal{R}$ and $d(y^*, y^e) \geq 0$ (11.6)

so that distance is a real and positive number;

(ii) $d(y^*, y^e) = 0$ iff $y^* = y^e$ (11.7)

so that distance equals zero if and only if elements y^* and y^e are equal;

(iii) $d(y^*, y^e) = d(y^e, y^*)$ (11.8)

so that distance is commutative; and

(iv) $d(y^*, y^e) \leq d(y^*, z) + d(z, y^e)$ (11.9)

so that the triangle inequality is satisfied, indicating that the shortest trajectory between two points is the one that connects the points directly.

   It must be observed that in real problems y^* and y^e contain many values, so that $d(y^*, y^e)$ must be computed with vectors containing NE data points. It must be observed that any distance function that satisfies the axioms (i–iv) of Equations (11.6–9) is suitable for proposition of the parameter estimation problem, so that it is not possible to say that one objective function is better than another if both satisfy the axioms. However, the objective function should also contain statistical meaning, which can be regarded as a fifth technical requirement for function $d(y^*, y^e)$.

In order to introduce statistical meaning into the objective function it is convenient to formulate some hypotheses. The first hypothesis is the perfect model assumption, which considers that the proposed mathematical model explains the relationships among the variables accurately. Based on this assumption, observed experimental data and values calculated with the model are not equal because of unavoidable experimental fluctuations, represented in the form:

$$y^e = y^* + \varepsilon \tag{11.10}$$

where ε represents the unknown experimental fluctuations. In fact, ε also contains modeling errors and variability associated with the model predictions, which are neglected at this point because model inaccuracies cannot be defined *a priori* without meaningful comparison with available experimental data, and because there is no reason to believe that the investigator is interested in a model that is known to be inadequate or incorrect. Therefore, although a perfect model does not exist, the perfect model assumption seems to be appropriate for purposes of model building.

The second hypothesis is the perfect experiment assumption, which considers that the collected experimental data represent accurately and precisely the real behavior of the analyzed system. Therefore, the perfect experimental hypothesis assumes that experimental fluctuations observed during the measurement stage are due to the unavoidable uncertainties associated with the measurement process only, as discussed in Section 1.9. Also, this assumption considers that the probability of finding the measured experimental values is maximum and that no other alternative set of experiments would represent the analyzed system better.

Combining the perfect model and the perfect experiment assumptions, it becomes possible to write:

$$E\{y^e\} = E\{y^* + \varepsilon\} = E\{y^*\} + E\{\varepsilon\} = E\{y^*\} = y^* \tag{11.11}$$

where $E\{\blacksquare\}$ represents the average of $\blacksquare$, so that the variance of the experimental data can be written as:

$$s_{y^e}^2 = \frac{\sum_{i=1}^{NE}(y_i^e - y_i^*)^2}{NE - NP} \tag{11.12}$$

where NP is the number of estimated parameter values, in accordance with Equations (1.9.1–2) of Section 1.9. The denominator of Equation (11.12) is related to the loss of degrees of freedom when one uses an additional parameter to adjust available data with a mathematical model (Himmelblau, 1970; Bard, 1974; Schwaab and Pinto, 2007). In Equation (11.12), $s_{y^e}^2$ is a measure of the experimental precision of available experimental data, if the two underlying assumptions are appropriate. Therefore, it becomes natural to propose:

$$F_{obj} = d\left(y^e, y^*\right) = s_{y^e}^2 = \frac{\sum_{i=1}^{NE}(y_i^e - y_i^*)^2}{NE - NP} \tag{11.13}$$

or simply

$$F_{obj} = \sum_{i=1}^{NE}(y_i^e - y_i^*)^2 \tag{11.14}$$

Equations (11.13–14) satisfy the four previously proposed axioms of the distance (objective function) and contains significant statistical meaning, as minimization of F_{obj} is equivalent to assuming experimental fluctuations should not be higher than the maximum admissible value, given the proposed hypotheses. Equation (11.14) is known generically as the least-squares function, and estimation of parameter values through minimization of Equation (11.14) is called as the least-squares estimation problem.

Example

E11.1 The widespread use of the least-squares procedure is associated with the fact that it allows the development of analytical solutions when the model is linear in respect to its parameters. For example, let us assume that:

$$y^*(x, \alpha) = \sum_{j=1}^{NP} \alpha_j f_j(x) \tag{11.15}$$

where $f_j(x)$ are functions of the input variables x (commonly named variable effects) and α_j are the model parameters. In this case:

$$F_{obj} = \sum_{i=1}^{NE}\left(y_i^e - \sum_{j=1}^{NP}\alpha_j f_j(x_i^e)\right)^2 \tag{11.16}$$

In order to minimize the objective function:

$$\frac{\partial F_{obj}}{\partial \alpha_k} = \sum_{i=1}^{NE} 2\left(y_i^e - \sum_{j=1}^{NP}\alpha_j f_j(x_i^e)\right)(-f_k(x_i^e)) = 0, k = 1, \ldots, NP \tag{11.17}$$

so that the matrix representation of the solution can be written in the form:

$$\alpha = M^{-1}Y_f \tag{11.18}$$

where:

$$
M = \begin{bmatrix} \sum_{i=1}^{NP} f_1(x_i^e) f_1(x_i^e) & \cdots & \sum_{i=1}^{NP} f_1(x_i^e) f_{NP}(x_i^e) \\ \vdots & \ddots & \vdots \\ \sum_{i=1}^{NP} f_{NP}(x_i^e) f_1(x_i^e) & \cdots & \sum_{i=1}^{NP} f_{NP}(x_i^e) f_{NP}(x_i^e) \end{bmatrix} \tag{11.19}
$$

$$
Y_f = \begin{bmatrix} \sum_{i=1}^{NP} y_i^e f_1(x_i^e) \\ \vdots \\ \sum_{i=1}^{NP} y_i^e f_{NP}(x_i^e) \end{bmatrix} \tag{11.20}
$$

In the case of Equation (11.4), which has the form of Equation (11.15), and Figures 11.1 and 11.2, the solution becomes:

$$
\alpha = \left[\alpha = \frac{k_2}{k_1}\right] = M^{-1} Y_f = \left[\sum_{i=1}^{NE} C_{Bi}^2\right]^{-1} \left[\sum_{i=1}^{NE} C_{Ci} C_{Bi}\right] \tag{11.21}
$$

Using real numbers, in Figure 11.13:

C_{Bi} (g/mol)	C_{ci} (g/mol)	C_{Bi}^2 (g/mol)	$C_{ci} C_{Bi}$ (g/mol)
0.148	0.155	0.02191	0.02291
0.204	0.445	0.04181	0.09109
0.229	0.564	0.05250	0.12929
0.452	0.833	0.20458	0.37654
0.455	0.999	0.20705	0.45473
0.644	1.227	0.41435	0.78969
0.607	1.495	0.36888	0.90806
0.823	1.656	0.67664	1.36246
0.847	1.718	0.71738	1.45522
1.085	2.094	1.17757	2.27256

Then:

$$
\left[\alpha = \frac{k_2}{k_1}\right] = \left[\sum_{i=1}^{NE} C_{Bi}^2\right]^{-1} \left[\sum_{i=1}^{NE} C_{Ci} C_{Bi}\right] = 2.025 \tag{11.22}
$$

The reader can observe that, although the estimated value of was close to the true value of 2, the obtained parameter estimate was perturbed by the experimental errors.

When the least-squares procedure is considered, there is an implicit assumption that experimental measurements are independent (that is, the uncertainties that affect a certain measurement do not simultaneously affect a second measurement) and constant throughout the experimental region; otherwise, the use of Equations (1.9.1–2) of Section 1.9 and of Equations (11.12–14) would not be possible (Schwaab and Pinto, 2007). However, actual measurement systems can provide data values that correlate and be subject to errors that change with the experimental condition. In other words, the uncertainties of temperature and pressure measurements can be correlated if, for example, the measurements are affected by a common heat source placed at the sampling process lines, for instance. Furthermore, high temperature values may be more prone to measurement fluctuations than low temperature values due to heat dissipation to the environment. Therefore, there are many incentives for the formulation and proposition of more involving objective functions, ones that can take these important statistical effects into account (Larentis et al., 2003; Alberton et al., 2009; Da Ros et al., 2017).

Although the development of distinct objective functions based on rigorous statistical principles falls beyond the scope of the present text, it is important to point out that the **maximum likelihood method** can be used to formulate objective functions that present different statistical properties for data measurements subject to uncertainties (Schwaab and Pinto, 2007). Among the infinitely many objective functions that can be formulated and adjusted to the characteristics of the particular measurements being analyzed, those based on Gaussian behavior (or normal behavior) are the most common. Skipping some formal aspects of the derivation, if the hypothesis of normal behavior is added to the previous perfect model and perfect experiment assumptions, then the following objective function can be proposed:

$$F_{\text{obj}}(\alpha) = \left(\mathbf{y}^e - \mathbf{y}^*(\alpha)\right)^{\text{T}} \mathbf{V}_y^{-1} \left(\mathbf{y}^e - \mathbf{y}^*(\alpha)\right) \tag{11.23}$$

where α is the vector of parameters (with dimension NP), $\mathbf{y}^e$ is the vector of experimental measurements (with dimension $NY \times NE$, where NE is the number of experiments and NY is the number of responses at each individual experiment), $\mathbf{y}^*(\alpha)$ is the vector of model responses (with dimension $NY \times NE$), and $\mathbf{V}_y$ is the covariance matrix of measurement fluctuations, which has the general form:

$$\mathbf{V}_y = \begin{bmatrix} s_{1,1}^2 & \cdots & s_{1,NY}^2 \\ \vdots & \ddots & \vdots \\ s_{NY,1}^2 & \cdots & s_{NY,NY}^2 \end{bmatrix} \tag{11.24}$$

where $s_{i,j}^2$ can be calculated with replicates in the form:

$$s_{i,j}^2 = \frac{\sum_{k=1}^{NR}\sum_{l=1}^{NR}\left(y_{i,k} - \bar{y}_i\right)\left(y_{j,l} - \bar{y}_j\right)}{NR - 1} \tag{11.25}$$

where NR is the number of replicates, $s_{i,j}^2$ is the sample covariance between variables y_i and y_j, $y_{i,k}$ is the kth replicate of variable y_i, and $\bar{y}_i$ is the sample average of variable y_i, defined in Equation (1.9.2) of Section 1.9. $s_{i,j}^2$ is a measure of the dependence between variables y_i and y_j: if $s_{i,j}^2$ is close to zero, variables y_i and y_j are independent and do not fluctuate together; if $s_{i,j}^2$ is positive, variables y_i and y_j tend to fluctuate along the same direction; if $s_{i,j}^2$ is negative, variables y_i and y_j tend to fluctuate in opposite directions.

Example

E11.2 Larentis *et al.* (2003) studied the combined reforming of carbon dioxide and partial oxidation of methane over Pt/γ-Al$_2$O$_3$ catalyst and performed replicated experiments in order to calculate the covariance matrix of measurement fluctuations associated with the determination of concentrations of the many chemical species involved in the chemical reaction system being analyzed. The experimental data collected at 800 °C at the output stream of a lab-scale reactor are presented below, with the respective sample averages, variances and standard deviations.

T (°C)	CH$_4$ (gmol/m^3)	O$_2$ (gmol/m^3)	CO (gmol/m^3)	CO$_2$ (gmol/m^3)	H$_2$ (gmol/m^3)	H$_2$O (gmol/m^3)	N$_2$ (gmol/m^3)
800	0.629	0	3.008	3.868	0.194	0.161	3.496
800	0.815	0	2.726	3.805	0.301	0.318	3.401
800	0.608	0	3.163	4.006	0.093	0	3.476
800	0.52	0	3.128	4.095	0.091	0.126	3.397
800	0.377	0	3.104	4.23	0.07	0.089	3.486
800	0.437	0	3.049	4.238	0.077	0	3.555

Species	Average $\bar{y}_i$ (gmol/m^3)	Variance $s_{i,i}^2$ (gmol/m^3)^2	Standard deviation $s_{i,i}$ (gmol/m^3)
CH$_4$	0.564333	0.024439	0.156330
O$_2$	0.000000	0.000000	0.000000
CO	3.029667	0.025206	0.158764
CO$_2$	4.040333	0.032859	0.181270
H$_2$	0.137667	0.008457	0.091960
H$_2$O	0.125667	0.014124	0.128802
N$_2$	3.468500	0.003654	0.060447

Finally, the covariance matrix of measurement fluctuations associated with the determination of concentrations is equal to:

$$\begin{bmatrix}
0.0204 & 0.0000 & -0.0154 & -0.0225 & 0.0109 & 0.0123 & -0.0043 \\
0.0000 & 0.0000 & 0.0000 & 0.0000 & 0.0000 & 0.0000 & 0.0000 \\
-0.0154 & 0.0000 & 0.0210 & 0.0157 & -0.0123 & -0.0135 & 0.0028 \\
-0.0225 & 0.0000 & 0.0157 & 0.0274 & -0.0123 & -0.0131 & 0.0046 \\
0.0109 & 0.0000 & -0.0123 & -0.0123 & 0.0070 & 0.0082 & -0.0022 \\
0.0123 & 0.0000 & -0.0135 & -0.0131 & 0.0082 & 0.0128 & -0.0041 \\
-0.0043 & 0.0000 & 0.0028 & 0.0046 & -0.0022 & -0.0041 & 0.0030
\end{bmatrix}$$

Therefore, one can see that the calculation of V_y in Equation (11.23–25) is simple when replicated experiments are available. As one can also see above, matrix V_y is not necessarily diagonal (that is, measurement fluctuations are not necessarily independent and variables tend to fluctuate together, in response to similar sources of perturbations, such as the unavoidable fluctuation of catalyst activity) and can contain significantly different covariance values. For this reason, the hypothesis of independent measurements and constant variances should always be evaluated carefully, especially if convincing experimental replicates are not available.

Despite the results shown in Example 11.2, if it is assumed (or shown experimentally, as advised) that measurement fluctuations are independent and not correlated to each other, matrix V_y becomes diagonal and Equation (11.23) can be written in the form:

$$F_{obj}(\alpha) = \sum_{i=1}^{NE} \sum_{j=1}^{NY} \frac{\left(y_{i,j}^e - y_{i,j}^*(\alpha)\right)^2}{s_{i,j}^2} \tag{11.26}$$

which has the form of a weighted least-squares function whose weights are the experimental variances of the obtained experimental responses. According to Equation (11.26), the importance of the experimental response for computation of the objective function diminishes with the increase of the experimental variance, a fact that makes a lot of sense. According to Equation (11.26), the most important experimental points are the ones measured more precisely. Therefore, proper characterization of experimental variances is extremely important when one is interested in explaining available experimental data with help of a kinetic model. More interestingly, if a single experimental response is considered $(NY = 1)$ and all experimental variances are equal (when all experiments are measured with the same precision), then Equation (11.26) can be reduced to Equations (11.13–14), indicating that the least-squares procedure is equivalent to the maximum likelihood procedure when measurement uncertainties follow the normal behavior, sample variances are equal, measurement fluctuations are independent, and perfect model and perfect experiment assumptions are good. This clearly shows that the least-squares procedure should be used only after careful thought about the statistical nature of available measurements! And, Equations (11.23–26) explain why proper characterization of experimental uncertainties is so important for quantitative analyses of actual kinetic data.

Example

E11.3 In order to illustrate the importance of proper characterization of experimental errors for quantitative analyses, a simple exercise is proposed below. Let us first consider that a chemical reaction can be described in the form:

$$A \xrightarrow{k_1} B \tag{11.27}$$

Assuming that the reaction is carried out in solution and in batch mode, then the following set of mass balance equations can be written:

$$\frac{dC_A}{dt} = -k_1 C_A \Rightarrow C_A = C_{A0}e^{-k_1 t} \tag{11.28a}$$

$$\frac{dC_B}{dt} = k_1 C_A = k_1 C_{A0}e^{-k_1 t} \Rightarrow C_B = C_{A0}\left(1 - e^{-k_1 t}\right) \tag{11.28b}$$

where C_{A0} is the initial concentration of reactant A, and it is assumed that the initial concentration of product B is equal to zero. It is also considered here that the following set of experimental data are available:

t (h)	C_A^e (mol/L)	C_B^e (mol/L)	s_A (mol/L)	s_B (mol/L)
0	1	0	0.01	0.01
1	0.53	0.55	1	1
2	0.08	0.85	0.1	0.1

The reader should observe that the sum of C_A and C_B is not C_{A0}, due to the unavoidable measurement fluctuations. In addition, the reader should observe that the standard errors of the distinct measurements are not equal, due to the variation of measurement performances in the experimental grid, and that the second experimental point is measured with higher precision than the first. Based on the available data, the following objective function can be proposed:

$$F_{obj}(k_1) = \frac{\left(C_{A1}^e - C_{A0}e^{-k_1 t_1}\right)^2}{s_{A1}^2} + \frac{\left(C_{A2}^e - C_{A0}e^{-k_1 t_2}\right)^2}{s_{A2}^2}$$
$$+ \frac{\left(C_{B1}^e - C_{A0}\left(1 - e^{-k_1 t_1}\right)\right)^2}{s_{B1}^2} + \frac{\left(C_{B2}^e - C_{A0}\left(1 - e^{-k_1 t_2}\right)\right)^2}{s_{B2}^2} \quad (11.29)$$

Then, minimization of Equation (11.29) (which must be performed with help of numerical methods, as discussed in the following sections) leads to $k_1 = 1.07$ h^{-1}. Interestingly, if the experimental data are changed as:

t (h)	C_A^e (mol/L)	C_B^e (mol/L)	s_A (mol/L)	s_A (mol/L)
0	1	0	0.01	0.01
1	0.53	0.55	0.1	0.1
2	0.08	0.85	1	1

then minimization of Equation (11.29) leads to $k_1 = 0.72$ h^{-1}. One may wonder why the parameter value changed if the concentration values remained the same. The answer is provided by Figure 11.3. Although concentration values remained the same, measurement precisions were different in both analyzed cases and the kinetic model attempted to remain closer to the most precise experimental values, which makes a lot of sense! Therefore, one should not neglect the proper characterization of experimental measurements during quantitative kinetic analyses, as obtained results can be quite sensitive to variations in measurement precision.

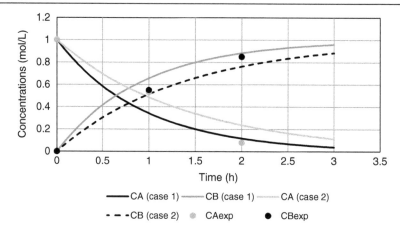

Figure 11.3 Sensitivity of parameter estimates and model responses to modification of measurement precision.

11.3   ERROR PROPAGATION AND PARAMETERIZATION OF THE ESTIMATION PROBLEM

Based on the discussion presented in the previous sections, one might be tempted to write Equation (11.28) in the form of straight lines, as:

$$z_A = \ln\left(\frac{C_A}{C_{A0}}\right) = -k_1 t \tag{11.30a}$$

$$z_B = \ln\left(\frac{C_B}{C_{A0}}\right) = -k_1 t \tag{11.30b}$$

in order to allow the derivation of an analytical solution for the problem proposed in Example 11.3. As a matter of fact, similar strategies were suggested by many investigators in the past, when computer resources were not so well developed as they are today. In the case analyzed in Example 11.3, this "analytical" approach would probably lead to formulation of the following objective function and set of transformed experimental data:

t (h)	z_A^e	z_B^e
0	0	0
1	−0.63488	−0.79851
2	−2.52573	−1.89712

$$F_{obj}(k_1) = \left(z_{A1}^e - (-k_1 t_1)\right)^2 + \left(z_{A2}^e - (-k_1 t_2)\right)^2$$
$$+ \left(z_{B1}^e - (-k_1 t_1)\right)^2 + \left(z_{B2}^e - (-k_1 t_2)\right)^2 \tag{11.31}$$

whose minimization can be performed analytically in the form

$$k_1 = \frac{z_{A1}^e t_1 + z_{A2}^e t_2 + z_{B1}^e t_1 + z_{B2}^e t_2}{2t_1^2 + 2t_2^2}$$

(11.32)

and leads to $k_1 = 1.03 \text{ h}^{-1}$, which is different from both the values obtained in Example 11.5 and is not consistent with the measurement precision of observed experimental responses.

As a matter of fact, for many reasons no one should feel compelled to transform experimental data nowadays. First of all, computer resources and numerical techniques have seen tremendous developments recently, so that the use of analytical solutions that are based on gross approximations of the experimental scenario are no longer necessary in many mathematical problems and therefore should be avoided. Besides, it must be clear that the statistical nature of the measurement system can be completely modified when experimental data are transformed. To illustrate this, let us assume that a transformation of the measured data is proposed in the form:

$$z = f(y)$$

(11.33)

where y is the measured experimental value (C_A, for example) and z is the transformed measured response ($\ln\left(\frac{C_A}{C_{A0}}\right)$, for example). f represents a mathematical function that maps y into z ($z = \ln(y)$, for example). Assuming y is subject to small fluctuations of order dy (and this is an important assumption, as the likely occurrence of large measurement deviations may render the proposed analysis even more complex), then it becomes possible to write:

$$dz = \frac{df}{dy} dy$$

(11.34)

tabulation can be used to interpret the transformation of statistical responses in the form:

$$s_z^2 \cong E\{dz^2\} = E\left\{ \left(\frac{df}{dy}\right)^2 dy^2 \right\} \cong \left(\frac{df}{dy}\right)^2 s_y^2$$

(11.35)

In other words, the function f transforms simultaneously the experimental data and the respective statistical characteristics of the measurement system, leading to a completely different estimation problem and to a distinct set of parameter estimates. This explains why Equation (11.32) leads to a different kinetic rate constant value, when compared to the results presented in Example 11.3. In the analyzed case:

$$dz_A = \frac{1}{C_A} dC_A \Rightarrow s_{z_A}^2 = \left(\frac{1}{C_A}\right)^2 s_{C_A}^2$$

(11.36a)

$$dz_B = \frac{1}{C_{A0} - C_B} dC_B \Rightarrow s_{z_B}^2 = \left(\frac{1}{C_{A0} - C_B}\right)^2 s_{C_B}^2$$

(11.36b)

so that the uncertainty of the transformed variable z_A increases when the measured value of C_A diminishes, while the opposite behavior is observed for the transformed variable z_B. Therefore, large absolute z_A and z_B values are not informative for purposes of parameter estimation and should be properly weighed during numerical calculations. Unless this artificial transformation effect is taken into consideration (which is very rare in the field), as a rule of thumb the reader should always avoid the artificial manipulation of experimental data and instead make use of available numerical procedures for quantitative analyses whenever necessary.

Example

E11.4 A popular procedure employed in many temperature-dependent kinetic problems is the transformation of the Arrhenius equation in the form:

$$z = \ln\left(k\right) = \ln\left(k_0\right) - \frac{E}{RT} = A - Bx \tag{11.37}$$

tabulation z and x are the transformed experimental variables and A and B are the estimated model parameters. However, the real experimental variables are k and T, which can be obtained with different error contents. Using simple error propagation analysis:

$$dz = \frac{1}{k}dk \Rightarrow s_z^2 = \left(\frac{1}{k}\right)^2 s_k^2 \tag{11.38a}$$

$$dx = -\frac{1}{T^2}dT \Rightarrow s_x^2 = \left(\frac{1}{T}\right)^4 s_T^2 \tag{11.38b}$$

tabulation clearly show that the uncertainties of both z and x change considerably in the experimental range, even when the precision of both k and T measurements can be considered constant in the analyzed experimental grid. Therefore, the use of Equation (11.37) and of estimation procedures based on Equation (11.37) should be avoided whenever possible.

11.4 NUMERICAL MINIMIZATION OF THE OBJECTIVE FUNCTION

According to the definition of the objective function, $F_{obj}(\alpha)$ must be minimized through manipulation of parameter values (α). Thus, it becomes necessary to choose a suitable optimization method to determine the unknown value of α that leads $F_{obj}(\alpha)$ to its minimum value. However, this can constitute a complex numerical task: the number of experimental data values ($NE \times NY$) and model parameters (NP) can be large; the proposed mathematical model can comprise a large set of nonlinear equations; the model parameters can be highly correlated (as discussed afterwards); and the objective function can present multiple local minima. For all these reasons, different numerical approaches have been proposed for minimization of the objective function and esti-mation of model parameters, as discussed in many scientific articles and textbooks (Himmelblau, 1970; Bard, 1974; Schwaab and Pinto, 2007).

It is not intended to review this very wide area of knowledge in the present text. Nevertheless, it is important to understand that the numerical methods employed most often for estimation of kinetic parameters are based on derivative-based procedures such as the Newton and the Gauss–Newton methods. According to these methods, min-imization must be performed along a direction in the parameter space that combines a gradient vector (the vector of first derivatives with respect to the model parameters, which indicates the direction of fastest variation of $F_{obj}(\alpha)$) and the Hessian matrix (the matrix of second derivatives in respect to model parameters, which indicates the curvature of $F_{obj}(\alpha)$ and is used to evaluate the point of minimum along the descend-ing direction). On the other hand, direct search methods perform the minimization of the objective function based only on sampling of $F_{obj}(\alpha)$, without calculation of derivatives. Both gradient and direct search methods may be classified as local deter-ministic search procedures, since the search starts from an initial parameter guess and then evolves to a minimum, following an iterative trajectory that is determined by the initial guess. For this reason, it must be emphasized that proper selection of a good initial guess can be of fundamental importance to achieve convergence, which can be a serious drawback of this class of numerical methods, since good initial guesses are often unavailable in real kinetic problems. Additionally, these methods are not able to discriminate between local and global minima, making it necessary to perform the minimization several times, using different initial guesses in order to verify if the calculated minimum point represents a global character.

The Newton method is based on the quadratic approximation of the objective function and can be presented in the form:

$$\alpha^{k+1} = \alpha^k - \mathbf{H}_{\alpha^k}^{-1} \nabla \mathbf{f}_{\alpha^k} \tag{11.39}$$

where

$$\nabla \mathbf{f}_{\alpha^k} = \left[\left. \frac{\partial F_{obj}}{\partial \alpha_1} \right|_{\alpha_1 = \alpha_1^k} \cdots \left. \frac{\partial F_{obj}}{\partial \alpha_N} \right|_{\alpha_N = \alpha_N^k} \right]^{\mathrm{T}} \tag{11.40}$$

$$\mathbf{H}_{\alpha^k} = \begin{bmatrix} \frac{\partial^2 F_{obj}}{\partial \alpha_1^2} & \cdots & \frac{\partial^2 F_{obj}}{\partial \alpha_1 \partial \alpha_N} \\ \vdots & \ddots & \vdots \\ \frac{\partial^2 F_{obj}}{\partial \alpha_N \partial \alpha_1} & \cdots & \frac{\partial^2 F_{obj}}{\partial \alpha_N^2} \end{bmatrix} \tag{11.41}$$

are respectively the gradient vector and the Hessian matrix of the objective function. At the point of minimum of $F_{obj}(\alpha)$, the gradient vector is equal to zero, so that the size of the gradient vector can be used as a convergence criterion to evaluate the rate of convergence and the quality of the approximation. Given an initial guess for the optimization variables, α^0, Equation (11.39) must be employed recursively until the convergence criterion is attained.

The main advantage of the Newton method is the fast rate of convergence when the initial guess is sufficiently close to the unknown desired solution, something that is unfortunately rare. Additionally, the method requires the calculation of first- and second-order derivatives of $F_{obj}(\alpha)$ with respect to the NP searched parameters and the inversion of the Hessian matrix at each numerical iteration. Even when the computation of the derivatives is performed numerically, through small perturbations of the optimization variables around the analyzed guess, the calculation of derivatives can be computationally expensive. This explains why many numerical procedures have been developed to circumvent the calculation of derivatives and inversion of the Hessian matrix. A popular alternative numerical approach consists in sampling the objective function around the initial guess and selecting the next guess as the sample that provides the lowest value for the objective function. Additional information regarding derivative-based and direct search methods can be found in standard parameter estimation textbooks (Himmelblau, 1970; Bard, 1974; Schwaab and Pinto, 2007).

An additional group of optimization procedures that can be used for estimation of kinetic parameters is formed by stochastic optimization algorithms. This group includes many distinct techniques, including Monte Carlo procedure, genetic algorithm, simulated annealing, differential evolution and particle swarm optimization (PSO), among many others (Kirkpatrick *et al.*, 1983; Goldberg, 1989; Kennedy and Eberhart, 1995; Storn and Price, 1997; Schwaab *et al.*, 2008; Schwaab and Pinto, 2007). These algorithms are characterized by the random character of the search and the large number of evaluations of the objective function, which assure a high probability of finding the global point of minimum. Plus, these algorithms do not require the definition of initial guesses for parameter values and do not use derivatives, making the numerical performance more predictable and less dependent on the model structure and experience of the analyst (Schwaab and Pinto, 2007). Despite that, it is important to emphasize that stochastic optimization algorithms do not ensure that the global minimum will indeed be attained and that these techniques can lead to very low rates of convergence. This explains why hybrid optimization algorithms have been proposed often (Schwaab *et al.*, 2008). Hybrid algorithms associate the global search characteristics of the stochastic algorithms (usually in the first stages of the iterative numerical procedure) with the fast convergence features of the local search algorithms (usually in the last stages of the numerical search, when good guesses are expected to be available for parameter values). Nevertheless, the main advantage of stochastic procedures is the fact that these techniques allow more rigorous statistical analyses of parameter estimates, which can be very important in kinetic problems for reasons that are discussed in the following sections.

According to typical stochastic procedures, the optimization process starts with the random generation of Npt candidates in the parameter space. Each candidate is then used for computation of $F_{obj}(\alpha)$ and afterwards the obtained results are ranked with respect to the obtained objective function values (the best candidates are the ones that lead to the smallest values of the objective function). Finally, the best set of candidates is used to update the full list of candidates and the search region. The criteria used for updating the list of candidates and search region constitute the main characteristics of each considered stochastic procedure.

Example

E11.5 A very simple and useful stochastic procedure is the Monte Carlo algorithm. According to this technique, candidates must be initially generated within the search region with the help of a random number generator. For example, if an Excel spreadsheet is used to implement the procedure, the expression has the general form:

$$\alpha_i^k = \alpha_{min}^k + \left(\alpha_{max}^k - \alpha_{min}^k\right) random\,(),i=1,\ldots,Ncand \tag{11.42}$$

tabulation α_{min}^k is the minimum admissible parameter value at iteration k, α_{max}^k is the maximum admissible parameter value at iteration k, α_i^k is the ith parameter candidate at iteration k, $random\,()$ is a function that generates random numbers uniformly in the interval [0, 1] and $Ncand$ is the total number of candidates generated by the numerical scheme. According to Equation (11.42), the parameters that characterize the numerical procedure are α_{min}^0, α_{max}^0 and $Ncand$, which define the search region and the refinement of the numerical search. Then, the objective function must be computed for each parameter candidate, resulting in an ordered list of $Ncand$ parameter values and respective objective function values. If the best candidate (α_{opt}^k, the one that leads to the smallest objective function value) is used as reference, then the search region can be updated in the form:

$$\alpha_{min}^{k+1} = \alpha_{opt}^k - \frac{\left(\alpha_{max}^k - \alpha_{min}^k\right)}{2}r \tag{11.43a}$$

$$\alpha_{max}^{k+1} = \alpha_{opt}^k + \frac{\left(\alpha_{max}^k - \alpha_{min}^k\right)}{2}r \tag{11.43b}$$

tabulation r is an additional numerical parameter $(0 < r < 1)$ that characterizes the contraction of the search region and controls the speed of convergence. Typically, r is close to 0.9, as larger r values lead to slow convergence, while smaller r values lead to inefficient exploration of the search region. After calculation of the new boundaries of the search region, the procedure must be repeated. The iterative procedure must be interrupted when the search region is sufficiently small or when the objective function values stop changing significantly, according to an additional numerical parameter that specifies the tolerance of the iterative scheme.

11.5 STATISTICAL CHARACTERIZATION OF MODEL ADEQUACY

After finding the minimum of the objective function, it becomes necessary to evaluate the statistical significance of the obtained results. This evaluation must consider model ability to explain available experimental data, quality of estimated model parameters and uncertainty of model prediction.

As discussed in the previous sections, an important assumption considered the model to be perfect. The quality of this hypothesis should be verified *a posteriori*, at the end of the parameter estimation phase, by comparing the experimental data with the model predictions. In short, the model cannot be discarded if differences between model predictions and available data are comparable to the unavoidable fluctuations of the experimental measurements. Once more, the reader should observe that the appropriate characterization of experimental fluctuations is of fundamental importance for final evaluation of obtained results. For example, if the least-squares objective function of Equation (11.14) is used to perform the estimation of the model parameters, then:

$$\hat{s}_y^2 = \frac{F_{obj}}{NE - NP} \tag{11.44}$$

tabulation $\hat{s}_y^2$ is a measure of prediction variance, which can be compared with the experimental variance, s_y^2, with the help of the standard statistical F-test, as illustrated in Equation (11.14) (Himmelblau, 1970; Bard, 1974; Schwaab and Pinto, 2007):

$$F_{NR-1,NE*NY-NP}^{(\frac{\varphi}{2})} < \frac{s_y^2}{\hat{s}_y^2} = F < F_{NR-1,NE*NY-NP}^{(1-\frac{\varphi}{2})} \tag{11.45}$$

where $1 - \varphi$ represents the confidence level (for 99% confidence, $\varphi = 0.01$) and $NE*NY - NP$ and $NR - 1$ represent the degrees of freedom of prediction variance and experimental variance, respectively. The limits of the cumulative F distribution in Equation (11.45) can be calculated with the help of standard functions provided by the vast majority of commonly used spreadsheets. For example, in the Excel spreadsheet the reader will find the functions DIST.F and INV.F, which can be used for computation of the limits of Equation (11.45).

If the ratio between variances should fall within the acceptable interval, meaning the variances are not considered statistically different within the specified level of confidence, one may conclude that the perfect model assumption cannot be ruled out and that the model is good. On the other hand, if the ratio between the variances should fall outside the acceptable interval, meaning both variances can be considered statistically different within the specified level of confidence, one may conclude that two cases are possible.

When prediction variance is higher than experimental variance, it is possible to conclude that the model is not able to explain the available data within the experimental precision, so that additional efforts should be made to improve model performance (for instance, changing the kinetic rate equations). However, it can also indicate that experimental errors have been underestimated (especially when a small number of replicates is considered) and that efforts should be made to characterize the experimental fluctuations more accurately. When model prediction variance is lower than experimental variance, it is possible to conclude that model performance is unexpectedly good, suggesting the model is possibly overparameterized and that efforts should be made to improve model performance (for instance, reducing the number of parameters in the kinetic rate equations). However, this result can also indicate superestimation of experimental errors (especially when a small number of replicates is considered) and that efforts should be made to characterize the experimental fluctuations more accurately.

If the maximum likelihood function of Equations (11.23–26) is used to perform the estimation of the model parameters, then the objective function can be interpreted as the χ^2 function in the form:

$$\chi_{NE*NY-NP}^{2(\frac{\varphi}{2})} < F_{obj} < \chi_{NE*NY-NP}^{2(1-\frac{\varphi}{2})} \tag{11.46}$$

The limits of the cumulative χ^2 distribution in Equation (11.46) can also be calculated with help of standard functions provided by the vast majority of commonly used spreadsheets. For example, in the Excel spreadsheet the reader will find the functions DIST.CHI and INV.CHI, which can be used for computation of the limits of Equation (11.46). If the final objective function value falls within the acceptable interval, meaning the differences between model responses and experimental data are acceptable statistically within the specified level of confidence, then one may conclude that the perfect model assumption cannot be ruled out and that the model is good. On the other hand, if the final objective function value falls outside the acceptable interval, then differences between model responses and experimental data are not acceptable statistically within the specified level of confidence, so two cases are possible.

When the final objective function value is higher than the acceptable limit, it is possible to conclude that the model is not able to explain the available data within the experimental precision, so that additional efforts should be made to improve model performance (for instance, changing the kinetic rate equations). However, as in the previous case this outcome can also indicate that experimental errors have been underestimated (especially when a small number of replicates is considered) and that efforts should be made to characterize the experimental fluctuations more accurately. When the final objective function value is lower than the experimental variance, it is possible to conclude that the model performance is unexpectedly good, suggesting that the model is possibly overparameterized and that efforts should be made to improve model performance (for instance, reducing the number of parameters in the kinetic rate equations). However, once more this can also indicate superestimation of experimental errors (especially when a small number of replicates is considered) and that efforts should be made to characterize the experimental fluctuations more accurately.

It is also common to report the correlation coefficient between model responses and respective experimental outputs in the form:

$$\rho = \frac{\sum_{i=1}^{NE}\sum_{j=1}^{NY}(y_{i,j}^e - \bar{y}_j^e)(y_{i,j}^* - \bar{y}_j^*)}{\sqrt{\left[\sum_{i=1}^{NE}\sum_{j=1}^{NY}(y_{i,j}^e - \bar{y}_j^e)^2\right]\left[\sum_{i=1}^{NE}\sum_{j=1}^{NY}(y_{i,j}^* - \bar{y}_j^*)^2\right]}} \tag{11.47}$$

The correlation coefficient ρ is a normalized form of the covariance (as defined in Equation (11.47)), varying inside the interval $[-1, 1]$. When experimental and calculated data are similar, ρ is close to 1 (in other words, experimental and calculated data are correlated to each other). For this reason, when ρ is higher than 0.9, it is normally assumed that the model is adequate; when ρ is lower than 0.7, it is normally assumed that the model is poor. However, it must be observed that the correlation coefficient does not take into account experimental fluctuations or the number of model parameters, meaning that it must be employed with care. For example, high correlation coefficients can indicate the occurrence of superparameterized models (and the model should thus be regarded as poor). Additionally, low correlation coefficients can indicate the occurrence of large measurement fluctuations (and the model thus should not be blamed for the lack of fit). For these reasons, the correlation coefficient should not be employed as a universal criterion for specification of model adequacy and should preferentially be combined with the other statistical analyses to provide a more conclusive evaluation of model adequacy.

11.6 THE CONFIDENCE REGION OF PARAMETER ESTIMATES

Generally speaking, a good kinetic model should be able to explain the available experimental data within the precision of the measurements and simultaneously rely on parameters that are statistically significant (in other words, are statistically different from zero) and independent from each other (in other words, parameters whose values can be determined independently from the values of the remaining parameters). As shown in Section 5.6, parameters can be subject to uncertainties because of the unavoidable occurrence of experimental fluctuations; however, in some cases parameter uncertainties can be so large that the model parameters can be confused with the null value. In other cases, estimated parameter values can be correlated strongly, meaning that modification of one parameter of the model can be compensated by the manipulation of the remaining model parameters, making the proper understanding of variable effects uncertain. Furthermore, the existence of high correlation between some of the model parameters can introduce severe numerical difficulties into the parameter estimation problem, usually resulting from an inappropriate mathematical representation of the model, poor experimental design or the intrinsic nonlinear nature of the system (Schwaab *et al.*, 2007; Schwaab and Pinto, 2008).

In order to evaluate the points raised in the previous paragraph, it is necessary to determine first the confidence region of the parameter estimates; in other words, the region of the parameter space where parameter values can be regarded as statistically acceptable. The parameter estimates are significant when the confidence region (within the specified confidence level) does not contain the null value.

Parameter uncertainties and parameter correlations can be calculated with the help of the covariance matrix of the parameters estimates, V_α. The meaning of V_α is similar to the meaning of the covariance matrix of measured model responses, as defined in Equation (11.24), replacing the experimental measurement fluctuations by the parameter uncertainties (caused by the experimental uncertainties). Omitting the technical details, the covariance matrix of parameter uncertainties can be shown to present the approximate form (Himmelblau, 1970; Bard, 1974; Schwaab and Pinto, 2007):

$$V_\alpha = 2H_\alpha^{-1} \qquad (11.48)$$

tabulation H_α is the Hessian matrix defined in Equation (11.41). Therefore, Equation (11.48) establishes a clear connection between the characteristics of the estimation problem (objective function, model, experimental errors) and the precision of the model parameters. Even more interestingly, when model sensitivity to variation of model parameters increases (derivatives become larger), parameter uncertainties decrease (as V_α is proportional to the inverse of H_α). Therefore, Equation (11.48) reinforces the importance of having a proper definition for the objective function and indicates that the functional representation of the model affects the quality of the obtained model parameters, as proposed by distinct reparameterization techniques. It must be emphasized that the main diagonal of the covariance matrix V_α contains the variances of the parameter estimates, while the off-diagonal elements of V_α characterize the covariances between pairs of parameters, which measure the degrees of dependence among the parameter estimates. In an ideal estimation problem, the obtained covariance matrix of parameter estimates, V_α, should be diagonal, as model parameters are expected to be independent of each other. However, Equation (11.48) shows that this ideal result is rarely obtained. As a matter of fact, the correlation ρ_{ij} between parameters i and j can be defined as:

$$\rho_{ij} = \frac{v_{ij}}{\sqrt{v_{ii}v_{jj}}} \tag{11.49}$$

tabulation v_{ij} is the element of the ith row and jth column of matrix V_α. The closer the absolute value of ρ_{ij} is to 1, the worse the quality of the obtained model parameters is. It can be proposed heuristically that absolute values of ρ_{ij} lie below 0.7.

Calculation of H_α and V_α at the end of the estimation process also allows the determination of confidence intervals for the parameter estimates with the help of the Student's t-distribution, as shown in the form:

$$\alpha_i \pm t_{NY \cdot NE - NP}^{1-\frac{\varphi}{2}} (v_{ii})^{\frac{1}{2}} \tag{11.50}$$

where $t_{NY \cdot NE - NP}^{1-\frac{\varphi}{2}}$ is the t value of the Student's t-distribution, defined for $NY \cdot NE - NP$ degrees of freedom and cumulative probability of $1 - \frac{\varphi}{2}$. As reported before, $1 - \varphi$ represents the confidence level (for confidence of 99%, φ is equal to 0.01). The limits of the Student's t-distribution in Equation (11.50) can be calculated with help of standard functions provided by the vast majority of commonly used spreadsheets. For example, in the Excel spreadsheet the reader will find the functions DIST.T and INV.T, which can be used for computation of the limits of Equation (11.50).

Example

E11.6 In Example 11.3, the objective function was defined in the form:

$$F_{obj}(k_1) = \frac{\left(C_{A1}^e - C_{A0}e^{-k_1t_1}\right)^2}{s_{A1}^2} + \frac{\left(C_{A2}^e - C_{A0}e^{-k_1t_2}\right)^2}{s_{A2}^2}$$

$$+ \frac{\left(C_{B1}^e - C_{A0}\left(1 - e^{-k_1t_1}\right)\right)^2}{s_{B1}^2} + \frac{\left(C_{B2}^e - C_{A0}\left(1 - e^{-k_1t_2}\right)\right)^2}{s_{B2}^2} \tag{11.51}$$

so that

$$\frac{\partial F_{obj}}{\partial k_1} = \frac{2\left(C_{A1}^e - C_{A0}e^{-k_1t_1}\right)\left(C_{A0}t_1e^{-k_1t_1}\right)}{s_{A1}^2} + \frac{2\left(C_{A2}^e - C_{A0}e^{-k_1t_2}\right)\left(C_{A0}t_2e^{-k_1t_2}\right)}{s_{A2}^2}$$

$$+ \frac{2\left(C_{B1}^e - C_{A0}\left(1 - e^{-k_1t_1}\right)\right)\left(-C_{A0}t_1e^{-k_1t_1}\right)}{s_{B1}^2}$$

$$+ \frac{2\left(C_{B2}^e - C_{A0}\left(1 - e^{-k_1t_2}\right)\right)\left(-C_{A0}t_2e^{-k_1t_2}\right)}{s_{B2}^2} \tag{11.52}$$

an

$$\frac{\partial^2 F_{obj}}{\partial k_1^2} = \frac{2\left(C_{A0}t_1 e^{-k_1 t_1}\right)^2}{s_{A1}^2} + \frac{2\left(C_{A1}^e - C_{A0}e^{-k_1 t_1}\right)\left(-C_{A0}t_1^2 e^{-k_1 t_1}\right)}{s_{A1}^2} + \frac{2\left(C_{A0}t_2 e^{-k_1 t_2}\right)^2}{s_{A2}^2}$$

$$+ \frac{2\left(C_{A2}^e - C_{A0}e^{-k_1 t_2}\right)\left(-C_{A0}t_2^2 e^{-k_1 t_2}\right)}{s_{A2}^2} + \frac{2\left(C_{A0}t_1 e^{-k_1 t_2}\right)^2}{s_{B1}^2}$$

$$+ \frac{2\left(C_{B1}^e - C_{A0}\left(1 - e^{-k_1 t_1}\right)\right)\left(C_{A0}t_1^2 e^{-k_1 t_1}\right)}{s_{B1}^2} + \frac{2\left(C_{A0}t_2 e^{-k_1 t_2}\right)^2}{s_{B2}^2}$$

$$+ \frac{2\left(C_{B2}^e - C_{A0}\left(1 - e^{-k_1 t_2}\right)\right)\left(C_{A0}t_2^2 e^{-k_1 t_2}\right)}{s_{B2}^2} \tag{11.53}$$

$$\mathbf{V}_\alpha = \left[s_{(k_1)}^2\right] = 2\left[\frac{\partial^2 F_o bj}{\partial k_1^2}\right]^{-1} \tag{11.54}$$

Equations (11.52–53) show explicitly that the uncertainties of the model parameters depend on the available data points, on the mathematical structures of the proposed objective function and kinetic model and also on the experimental precision. As one can see, the calculations associated with the determination of parameter uncertainties can be quite complex even in very simple kinetic problems; therefore, investigators involved with kinetic modeling and determination of kinetic parameters should be encouraged to develop fundamental programming skills and to rely on the support of appropriate computer software.

In order to illustrate how the formulation of the estimation problem can affect the calculation of parameter uncertainties and distort the statistical significance of the obtained estimates, Equation (11.31) is also used to calculate $\mathbf{V}_\alpha$. In this case:

$$\frac{\partial F_{obj}}{\partial k_1} = 2t_1\left(z_{A1}^e - (-k_1 t_1)\right) + 2t_2\left(z_{A2}^e - (-k_1 t_2)\right)$$

$$+ 2t_1\left(z_{B1}^e - (-k_1 t_1)\right) + 2t_2\left(z_{B2}^e - (-k_1 t_2)\right) \tag{11.55}$$

$$\frac{\partial^2 F_{obj}}{\partial k_1^2} = 4t_1^2 + 4t_2^2 \tag{11.56}$$

The differences between Equations (11.52) and (11.55) are enormous!! This shows once more that the estimation problem should not be manipulated only to allow the derivation of analytical expressions, without sound comprehension of the consequences for interpretation of the statistical significance of the obtained parameters.

Finally, for the sake of completeness, in case 1 of Example 11.3, $0.548\ \text{h}^{-1} < k_1 = 1.072\ \text{h}^{-1} < 1.596\ \text{h}^{-1}$; in case 2 of Example 11.3, $0.126\ \text{h}^{-1} < k_1 = 0.715\ \text{h}^{-1} < 1.304\ \text{h}^{-1}$; and in the case of Equation (11.31), $0.022\ \text{h}^{-1} < k_1 = 1.028\ \text{h}^{-1} < 2.034\ \text{h}^{-1}$. To perform the calculations, it was necessary to define the confidence level (assumed to be equal to

95%, so that $\varphi = 0.05$) and the degrees of freedom ($NE * NY - NP = 2 * 2 - 1 = 3$). Thus, in this case the t value becomes equal to 3.18, as calculated with the cumulative Student's t-distribution in the Excel spreadsheet. The reader must observe that the parameter estimate is significant in all cases within the established confidence level of 95%, as the confidence intervals do not include the null value.

Although the computation of confidence limits for model parameters can be important for statistical interpretation of the actual physical meaning of the obtained parameter estimates, confidence limits calculated with Equations (11.48–50) are rough approximations of the real confidence limits of the parameters. This is because derivation of Equation (11.48) assumes that measurement fluctuations are small and because Equation (11.50) neglects the correlations among the many possible model parameters (Himmelblau, 1970; Bard, 1974; Schwaab and Pinto, 2007). For these reasons, parameter uncertainties can be characterized better when confidence regions, rather than confidence intervals, can be computed. The confidence region of parameter estimates can be defined as the set of all parameter values that can provide model predictions comparable to the available experimental data, within the specified statistical level of confidence.

Assuming the objective function can be approximated around the point of minimum by a quadratic function, as described in Equations (11.56–57), where represents the parameter estimates:

$$F_{\mathrm{obj}}(\boldsymbol{\alpha}) \cong F_{\mathrm{obj}}(\hat{\boldsymbol{\alpha}}) + (\boldsymbol{\alpha} - \hat{\boldsymbol{\alpha}})^{\mathrm{T}} \frac{\partial F_{\mathrm{obj}}}{\partial \boldsymbol{\alpha}} + \frac{1}{2}(\boldsymbol{\alpha} - \hat{\boldsymbol{\alpha}})^{T} \mathbf{H}_{\hat{\alpha}}(\boldsymbol{\alpha} - \hat{\boldsymbol{\alpha}}) \tag{11.57}$$

$$F_{\mathrm{obj}}(\boldsymbol{\alpha}) - F_{\mathrm{obj}}(\hat{\boldsymbol{\alpha}}) \cong (\boldsymbol{\alpha} - \hat{\boldsymbol{\alpha}})^{\mathrm{T}} \mathbf{V}_{\alpha}^{-1}(\boldsymbol{\alpha} - \hat{\boldsymbol{\alpha}}) \tag{11.58}$$

The right-hand side of Equation (11.57) represents an ellipsoid in the parameter space, justifying the usual assumption that the confidence region of parameter estimates has the shape of an ellipsoid (which can only be confirmed when Equations (11.56–57) provide good local approximations of the objective function, which is usually not the case!). If one assumes that parameter values follow the Gaussian distribution (which must not be true either!), the right-hand side of Equation (11.57) defines the chi-square variable computed with NP degrees of freedom (Himmelblau, 1970; Bard, 1974; Schwaab and Pinto, 2007). Consequently, the confidence region of the parameter estimates can be calculated in the form:

$$(\boldsymbol{\alpha} - \hat{\boldsymbol{\alpha}})^{\mathrm{T}} \mathbf{V}_{\alpha}^{-1}(\boldsymbol{\alpha} - \hat{\boldsymbol{\alpha}}) \leq X_{NP}^{2(1-\phi)} \tag{11.59}$$

After minimization of $F_{\mathrm{obj}}(\boldsymbol{\alpha})$ and calculation of $\hat{\boldsymbol{\alpha}}$, $\mathbf{V}_{\alpha}^{-1}$ and $X_{NP}^{2(1-\phi)}$, parameter values that satisfy Equation (11.58) can be obtained, reported and plotted. However, as the covariance matrix of experimental fluctuations $\mathbf{V}_y$ is usually characterized poorly (or is even unknown) and the covariance matrix of parameter estimates represents only a rough approximation of the true matrix $\mathbf{V}_{\alpha}$, Equation (11.58) is usually presented in alternative forms, such as (Himmelblau, 1970; Bard, 1974; Schwaab and Pinto, 2007):

$$(\alpha - \hat{\alpha})^T V_\alpha^{-1}(\alpha - \hat{\alpha}) \le \frac{NP}{NE * NY - NP} F_{obj}(\hat{\alpha}) F_{NP,NE*NY-NP}^{(1-\phi)} \tag{11.60}$$

$$F_{obj}(\alpha) \le F_{obj}(\hat{\alpha})\left(1 + \frac{NP}{NE * NY - NP} F_{NP,NE*NY-NP}^{(1-\phi)}\right) \tag{11.61}$$

The use of Equations (11.58–60) is facilitated tremendously by stochastic procedures, such as the Monte Carlo algorithm presented in Example 11.4, as many samples (often thousands of samples) of $F_{obj}(\alpha)$ and respective values of α become available after convergence. This can be regarded as a major advantage of most stochastic algorithms employed for estimation of kinetic parameters. Additionally, it is important to emphasize that Equation (11.60) does not make use of the quadratic approximation of Equation (11.56), which allows the computation of more realistic nonelliptical confidence regions of parameter estimates than the ones computed with Equations (11.58–59). The reader must observe that the continual development of computer resources is promoting the fast and simultaneous development of numerical techniques used for characterization of kinetic parameters, so that this particular area of knowledge now constitutes an effervescent field of investigation.

Example

E11.7 The estimation problem proposed in Example 11.3 was solved once more with help of a Monte Carlo procedure. The parameter k_1 was searched in the interval $[0, 2]$ and 1000 candidates were used at each iteration to perform the numerical search. The adopted absolute numerical tolerance was set to 0.001 both for the size of the search region and the modification of the objective function. Finally, the contraction rate was set to 0.9. Based on the data reported in Example 11.3, $NE * NY - NP = 3$ and $NP = 1$. Assuming a confidence level of 95% ($\varphi = 0.05$), then according to the cumulative F-distribution available in the Excel spreadsheet:

$$F_{1,3}^{(0.95)} = 10.128$$

$$F_{obj}(\alpha) \le F_{obj}(\hat{\alpha})\left(1 + \frac{1}{3}F_{1,3}^{(0.95)}\right) = 4,376 F_{obj}(\hat{\alpha}) \tag{11.62}$$

At the point of minimum for case 1, $k_1 = 1.072 \, h^{-1}$ and $F_{obj}(k_1) = 0.2928$. Therefore, all k_1 values that lead to $F_{obj}(k_1) \le 1.3028$ are statistically acceptable in case 1. At the point of minimum for case 2, $k_1 = 0.716 \, h^{-1}$ and $F_{obj}(k_1) = 0.3534$. Therefore, all k_1 values that lead to $F_{obj}(k_1) \le 1.5465$ are statistically acceptable in case 2. Figure 11.4 shows acceptable parameter values and respective objective function values calculated with the help of the Monte Carlo procedure in both cases. The reader must observe that the confidence limits are not symmetrical, as imposed by the quadratic approximation.

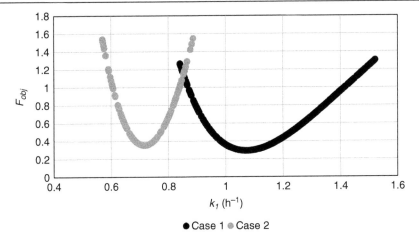

Figure 11.4 Confidence region of parameter values and respective objective function values calculated with help of the Monte Carlo procedure in both cases of Example 11.3.

11.7 UNCERTAINTIES OF MODEL PREDICTIONS

After evaluation of model adequacy and quality of model parameters, it becomes necessary to characterize the intrinsic uncertainties of the predictions provided by the model. Given the uncertainties of the parameter estimates and the use of the parameters for calculation of model responses, it seems clear that model responses are also uncertain to some extent. In particular, if one assumes that parameter uncertainties are small (and Figure 6 shows that uncertainties can be large), it is possible to write:

$$\hat{\mathbf{V}}_y = \mathbf{B}\mathbf{V}_\alpha \mathbf{B}^T + \mathbf{V}_y \tag{11.63}$$

where $\hat{\mathbf{V}}_y$ is the covariance matrix of model responses and

$$\mathbf{B} = \left[\begin{pmatrix} \frac{\partial \hat{y}_1}{\partial \alpha_1} & \cdots & \frac{\partial \hat{y}_1}{\partial \alpha_{NP}} \\ \vdots & \ddots & \vdots \\ \frac{\partial \hat{y}_{NY}}{\partial \alpha_1} & \cdots & \frac{\partial \hat{y}_{NY}}{\partial \alpha_{NP}} \end{pmatrix} \right] \tag{11.64}$$

so that matrix $\mathbf{B}$ contains model sensitivities to changes of parameter estimates (Himmelblau, 1970; Bard, 1974; Schwaab and Pinto, 2007). Based on the covariance matrix of model responses, the confidence intervals of model predictions can be defined as:

$$y_i^* \pm t_{NE*NY-NP}^{1-\frac{\varphi}{2}} (v_{ii})^{\frac{1}{2}} \tag{11.65}$$

tabulation v_{ii} is the ith diagonal element of matrix $\hat{\mathbf{V}}_y$, $t_{N-NP}^{1-\frac{\varphi}{2}}$ is the Student's t variable defined for $NE * NY - NP$ degrees of freedom of the associated estimation problem and cumulative confidence level of $1 - \varphi/2$.

As described previously for the parameter estimates, Equations (11.62–63) assume that measurement fluctuations are small, while Equation (11.64) neglects correlations among the many possible model responses (Himmelblau, 1970; Bard, 1974; Schwaab and Pinto, 2007). For these reasons, the uncertainties of model responses can be characterized better when confidence regions (rather than confidence intervals) of parameter estimates are considered. In this case, if stochastic parameter estimation procedures are used, execution of this task becomes relatively simple, as all parameter candidates that satisfy Equation (11.60), for instance, can be used to calculate the respective model responses, which are statistically meaningful model outputs that belong to the confidence region of model predictions.

Example

E11.8 The estimation problem proposed in Example 11.3 and solved in Example 11.7 with help of a Monte Carlo procedure led to the confidence regions of parameter estimates shown in Figure 11.4. The parameter values shown in Figure 11.4 were then used for computation of the model outputs, leading to the model prediction boundaries presented in Figure 11.5, which constitute the confidence regions of model responses for cases 1 and 2.

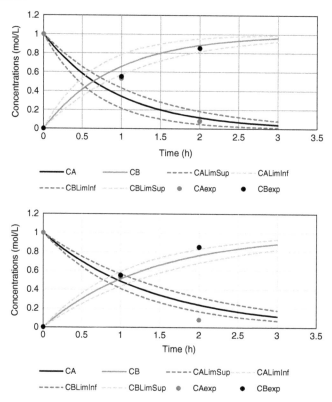

Figure 11.5 Confidence region of model responses calculated with help of the Monte Carlo procedure in both cases of Example 11.3: (a) Case 1 and (b) Case 2.

11.8 PARAMETERIZATION OF THE ARRHENIUS EQUATION

As shown in Section 3.3.3, the Arrhenius equation can be written in different forms, including the one shown below:

$$k = \exp\left[A - \frac{E}{R}\left(\frac{1}{T} - \frac{1}{T_{ref}}\right)\right]$$

(11.66)

tabulation

$$A = \ln\left(k_{ref}\right) = \ln\left(k_0\right) - \frac{E}{RT_{ref}}$$

(11.67)

and E is the activation energy. T_{ref} is an arbitrary reference value that can be used to normalize the Arrhenius equation.

Using the numerical tools presented in the previous sections, Schwaab and Pinto (2007b) and Schwaab et al. (2008) investigated the estimation of A and E in different temperature-dependent kinetic problems. Particularly, A and E were estimated for the problem presented in Example 11.3, assuming that concentrations of species A were available at different reaction temperatures and using the usual least-squares objective function of Equation (11.14). The authors were able to show both analytically and numerically that there always is an optimum value of T_{ref} that allows the estimation of model parameters with minimum correlation and maximum precision and that this T_{ref} value is placed inside the interval $[T_{min}, T_{max}]$, where T_{min} and T_{max} are the minimum and maximum experimental temperatures used to generate the experimental data employed in the estimation procedure.

In order to illustrate the points raised in the previous paragraph, Figure 11.6 shows the effect of T_{ref} on the parameter correlation between A and E in Equation (3.3) for a certain set of experimental data, clearly showing that the parameters can be estimated independently if T_{ref} is selected appropriately. In other words, Figure 11.6 clearly shows

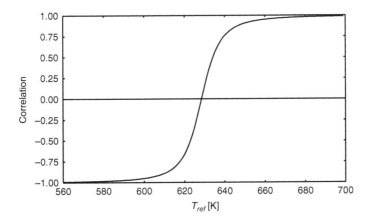

Figure 11.6 Correlation between A and E in Equation (3.3) as a function of T_{ref}, after estimation of model parameters in a temperature-dependent kinetic problem described by a first-order reaction rate equation as shown in Example 11.3 (Schwaab and Pinto, 2007b).

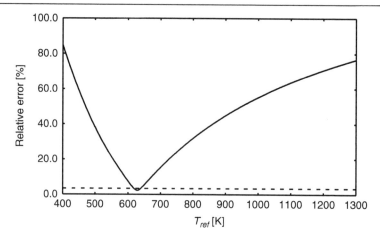

Figure 11.7 Relative uncertainties of parameters k_{Tref} (line) and E (dashed line) in Equation (3.3) as functions of T_{ref}, after estimation of model parameters in a temperature-dependent kinetic problem described by a first-order reaction rate equation as shown in Example 11.3 (Schwaab and Pinto, 2007b).

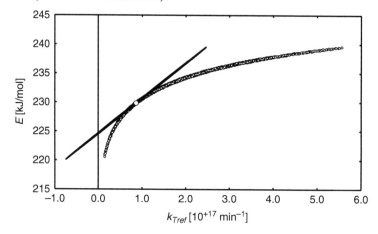

Figure 11.8 Elliptical approximation and likelihood confidence regions of parameters k_{Tref} and E calculated with the traditional Arrhenius equation, after estimation of model parameters in a temperature-dependent kinetic problem described by a first-order reaction rate equation as shown in Example 11.3 (Schwaab and Pinto, 2007b).

that proper manipulation of the parametric form of the model can allow the estimation of independent model parameters, which is certainly amazing and encourages the development of different parameterized versions of kinetic rate models. Additionally, Figure 11.7 shows that the point of minimum correlation between A and E is also the point where the error contents of both parameters are minimum, showing that manipulation of the parametric form of the model can also allow the estimation of more precise model parameters, which is certainly important for practical purposes, as parameters are expected to be as precise as possible. Furthermore, Figures 11.8 and 11.9 illustrate the effect of T_{ref} on the confidence region of the parameter estimates, as

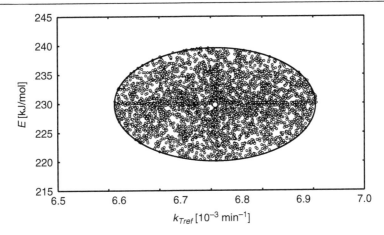

Figure 11.9 Elliptical approximation and likelihood confidence regions of parameters k_{Tref} and E calculated with Equation (3.3) and reference temperature of 628.45 K, after estimation of model parameters A and E in a temperature-dependent kinetic problem described by a first-order reaction rate equation as shown in Example 11.3 (Schwaab and Pinto, 2007b).

calculated with the help of the elliptical approximation and the parameter candidates provided by a stochastic parameter estimation procedure. Particularly, Figure 11.8 shows that the elliptical approximations can indeed provide poor approximations of the real confidence regions of the parameter estimates, as discussed before. However, Figure 11.9 shows once more that proper manipulation of the parametric form of the model can allow the estimation of smaller and well-behaved confidence regions of model parameters, reinforcing the possible benefits of investigating the use of different parameterized versions of kinetic rate models.

11.9 CONCLUDING REMARKS

In Chapter 11, the parameter estimation problem was presented and used to estimate kinetic rate constants in kinetic problems. As discussed in previous sections, the parameter estimation procedure comprises three main stages: (i) definition of the objective function; (ii) minimization of the objective function; and (iii) statistical analyses of the obtained results. Mathematical tools were then developed and used to perform the tasks of the different estimation steps. Particularly, the maximum likelihood method was presented as a powerful technique to estimate kinetic rate parameters, combining experimental data, model equations and statistical characteristics of the measuring system. As a matter of fact, it was shown that the precise determination of experimental fluctuations is of fundamental importance for model building, evaluation of model adequacy, characterization of parameter quality and determination of confidence regions for both model parameters and model responses. Finally, it was shown that proper manipulation of the parametric form of the model can allow the estimation of independent and more precise model parameters, encouraging the development of different parameterized versions of kinetic rate models.

Chapter 12

Experimental design

In Chapter 11, the parameter estimation problem was presented and discussed. Particularly, it was assumed in Chapter 11 that both the experimental kinetic data and the kinetic model were available. Nevertheless, the mathematical structure of the model and the effects of the independent variables on the dependent variables are usually unknown during the earlier stages of an experimental kinetic study. At these initial stages, very frequently it is not even known which variables in a set can indeed affect the analyzed process responses. In these cases, the use of statistical techniques for proposal of experimental designs can be quite helpful (Box *et al.*, 1978; Montgomery and Runger, 2002; Schwaab and Pinto, 2011).

Statistical experimental design constitutes families of experimental design procedures that attempt to maximize some sort of metrics that depend on previous knowledge about the analyzed experimental system and the main result the analyst seeks. As a consequence, the number of distinct design procedures is enormous, so it is not possible to review this entire area in a single chapter of one book. Nevertheless, in the following sections some fundamental aspects of statistical experimental design are discussed, so that the interested reader will be able to characterize the most important features of the particularly analyzed kinetic problem and feel encouraged to search for additional information in the open literature.

In order to organize the presentation of the following sections, four fundamental experimental design problems are considered:

1. Problems where the analyst does not know the effects caused by experimental design variables (x) on response variables (y), so that the investigator wants to characterize the influence of input variables on response variables with maximum precision with help of NE designed experiments $(x_1 \, x_2 \ldots, \, x_{NE})$;

2. Problems where the analyst does know the reference kinetic model that must be used to describe a particular kinetic problem and wants to design a batch of NE experiments $(x_1 \, x_2 \, \ldots, \, x_{NE})$ that can allow the estimation of model parameters (α) with maximum precision;

3. Problems where the analyst has already performed NE experiments $(x_1, x_2, \ldots, x_{NE})$, handles a set of NM model candidates to describe a particular kinetic problem $(y_1^M(x, \alpha_1^M), y_2^M(x, \alpha_2^M), \ldots, y_{NM}^M(x, \alpha_{NM}^M))$ and wants to define the best model candidate through sequential experimentation and selection of one additional experiment (x_{NE+1});

4. Problems where the analyst has already performed NE experiments $(x_1 \ x_2 \ ..., \ x_{NE})$, has selected a reference model to describe a particular kinetic problem $y^M(x, \alpha^M)$ and wants to define the most precise set of model parameters through sequential experimentation and selection of one additional experiment (x_{NE+1})

Almost certainly, the reader will be able to fit the analyzed problem into one of these four large families, as described in the following sections.

12.1 FACTORIAL EXPERIMENTAL DESIGN

When the analyst does not know the effects caused by experimental design variables (x) on response variables (y), the main objective of the investigation probably is the characterization of influence of input variables on response variables with maximum precision, with the help of NE designed experiments $(x_1 \ x_2 \ ..., \ x_{NE})$. In this case, the main objective of the experimental study is the determination of correlations among input variables (in kinetic problems, usually temperature, pressure, catalyst compositions, contact time, spatial velocity, time and feed compositions) and process responses (in kinetic problems, usually reaction rates, selectivities, reaction yields and product compositions). Therefore, x is a vector of NX inputs and y is a vector of NY responses. The experimental design procedure must then provide a set of NE experimental conditions $(x_1 \ x_2 \ ..., \ x_{NE})$ for determination of the influential effects of experimental variables on outputs. Thus, the main product of the experimental design procedure is a spreadsheet X that contains a set of experimental conditions of the form:

$$X = \begin{bmatrix} x_{1,1} & \cdots & x_{1,NE} \\ \vdots & \ddots & \vdots \\ x_{NX,1} & \cdots & x_{NX,NE} \end{bmatrix}_{NX \times NE} \tag{12.1}$$

In classical factorial experimental designs, experimental conditions (experimental condition for variable i at experiment j) must be uniformly distributed over the experimental region. Therefore, the analyst must provide the range of feasible experimental conditions for each variable $x_{(i,min)}$ and $x_{(i,max)}$, $i = 1, \ldots, NX$, and the experiments are selected as all possible combinations among the selected variable levels (experimental conditions) of the independent manipulated variables. Consequently, the investigator must also provide the number of variable levels NL_i for each input variable.

Example

E12.1 In a certain kinetic problem, the investigator wants to determine how reaction yield (Y) responds to modification of reaction temperature (T), reaction pressure (P) and contact time (θ). Therefore, in this case:

The order is wrong. The vectors x and y should be presented before this sentence, which refers to NL, xmin and xmax.

$$x_i = \begin{bmatrix} T \\ P \\ \theta \end{bmatrix}_i, y_i = [Y]_i$$

$$NL = \begin{bmatrix} 3 \\ 2 \\ 2 \end{bmatrix}, x_{min} = \begin{bmatrix} 70 \\ 1 \\ 30 \end{bmatrix}, x_{max} = \begin{bmatrix} 110 \\ 3 \\ 60 \end{bmatrix}$$

tabulation T is given in °C, P in atm and in min. In order to obtain a uniform distribution over the whole range of experimental conditions, the following conditions must be satisfied:

$$NL = \begin{bmatrix} 70,90,110 \\ 1,3, \\ 30,60 \end{bmatrix}$$

and the experimental design becomes

$$X = \begin{bmatrix} 70 & 70 & 70 & 70 & 90 & 90 & 90 & 90 & 110 & 110 & 110 & 110 \\ 1 & 1 & 3 & 3 & 1 & 1 & 3 & 3 & 1 & 1 & 3 & 3 \\ 30 & 60 & 30 & 60 & 30 & 60 & 30 & 60 & 30 & 60 & 30 & 60 \end{bmatrix}$$

with $NE = 3 \times 2 \times 2 = 12$ experiments.

Although factorial designs are quite simple and seem somehow natural, the fact is they rely on a solid theoretical structure, and they can be very helpful. In order to provide the desired correlations among the analyzed variables, empirical linear correlation models are normally used to perform the initial correlation analyses, as described in Example 11.1 of Chapter 11, in the form:

$$y^M(x,\alpha) = \sum_{j=1}^{NP} \alpha_j f_j(x) \tag{12.2}$$

with

$$\alpha = \mathbf{M}^{-1}\mathbf{Y}_f \tag{12.3}$$

where

$$\mathbf{M} = \begin{bmatrix} \sum_{i=1}^{NP} f_1(x_i^e)f_1(x_i^e) & \cdots & \sum_{i=1}^{NP} f_1(x_i^e)f_{NP}(x_i^e) \\ \vdots & \ddots & \vdots \\ \sum_{i=1}^{NP} f_{NP}(x_i^e)f_1(x_i^e) & \cdots & \sum_{i=1}^{NP} f_{NP}(x_i^e)f_{NP}(x_i^e) \end{bmatrix} \tag{12.4}$$

$$\mathbf{Y_f} = \begin{bmatrix} \sum_{i=1}^{NP} y_i^e f_1(x_i^e) \\ \vdots \\ \sum_{i=1}^{NP} y_i^e f_{NP}(x_i^e) \end{bmatrix} \tag{12.5}$$

$$\mathbf{V}_\alpha = 2\mathbf{M}^{-1} \tag{12.6}$$

In Equation (12.2), α_j is an empirical model parameter that measures the relative importance of the experimental effect $f_j(x)$. Usually, Equation (12.2) is presented in the particular form:

$$y^M(x, \alpha) = a_0 + \sum_{i=1}^{NX} a_i x_i + \sum_{i=1}^{NX} \sum_{j>i} b_{ij} x_i x_j \tag{12.7}$$

where parameter a_0 (bias) is associated with the effect $f(x) = 1$; parameters a_i, $i = 1, \ldots, NX$ (linear main effects) are associated with the effects $f_i(x) = x_i$; and parameters $b_{i,j}$, $i = 1, \ldots, NX$, $j > i$ (synergetic, interaction or nonlinear effect) are associated with the effects $f_i(x) = x_i x_j$.

The extremely important property of Equations (12.3–7) is the fact that the design matrix $\mathbf{M}$ (and therefore the covariance matrix of model parameters $\mathbf{V}_\alpha$) is diagonal in factorial designs, allowing the independent determination of the model parameters (and respective variable effects). For this reason, factorial experimental designs in the form of Equations (12.3–7) are called *orthogonal* designs. However, in order to obtain orthogonal designs, it is necessary first to reparameterize the model equations in the form:

$$y^M(z, \alpha) = a_0 + \sum_{i=1}^{NX} a_i z_i + \sum_{i=1}^{NX} \sum_{j>i} b_{ij} z_i z_j \tag{12.8}$$

with

$$z_i = \frac{x_i - \left(\frac{x_{max} + x_{min}}{2}\right)}{\left(\frac{x_{max} - x_{min}}{2}\right)} \tag{12.9}$$

Reparameterization of the model can be fundamental for quantitative analyses for many practical reasons: (i) it allows for calculation of independent model parameters; (ii) it removes physical units of model parameters and variables, allowing direct comparison of obtained parameter values; and (iii) it allows objective ordering of input variables in terms of relative influence on process responses, as shown in the following examples.

Example

E12.2 In Example 12.1, it is possible to write:

$$z_i = \begin{bmatrix} \frac{T-90}{20} \\ \frac{P-2}{1} \\ \frac{\theta-45}{15} \end{bmatrix}_i$$

so that the experimental design becomes:

$$Z = \begin{bmatrix} -1 & -1 & -1 & -1 & 0 & 0 & 0 & 0 & +1 & +1 & +1 & +1 \\ -1 & -1 & +1 & +1 & -1 & -1 & +1 & +1 & -1 & -1 & +1 & +1 \\ -1 & +1 & -1 & +1 & -1 & +1 & -1 & +1 & -1 & +1 & -1 & +1 \end{bmatrix}$$

with

$$y^M = a_0 + a_1 z_1 + a_2 z_2 + a_3 z_3 + b_{13} z_1 z_2 + b_{13} z_1 z_3 + b_{23} z_2 z_2$$

and

$$M = \begin{bmatrix}
\sum_{i=1}^{NE} 1 & \sum_{i=1}^{NE} z_{1,i} & \sum_{i=1}^{NE} z_{2,i} & \sum_{i=1}^{NE} z_{3,i} & \sum_{i=1}^{NE} z_{1,i} z_{2,i} & \sum_{i=1}^{NE} z_{1,i} z_{3,i} & \sum_{i=1}^{NE} z_{2,i} z_{3,i} \\
\sum_{i=1}^{NE} z_{1,i} & \sum_{i=1}^{NE} z_{1,i}^2 & \sum_{i=1}^{NE} z_{1,i} z_{2,i} & \sum_{i=1}^{NE} z_{1,i} z_{3,i} & \sum_{i=1}^{NE} z_{1,i}^2 z_{2,i} & \sum_{i=1}^{NE} z_{1,i}^2 z_{3,i} & \sum_{i=1}^{NE} z_{1,i} z_{2,i} z_{3,i} \\
\sum_{i=1}^{NE} z_{2,i} & \sum_{i=1}^{NE} z_{1,i} z_{2,i} & \sum_{i=1}^{NE} z_{2,i}^2 & \sum_{i=1}^{NE} z_{2,i} z_{3,i} & \sum_{i=1}^{NE} z_{2,i}^2 z_{1,i} & \sum_{i=1}^{NE} z_{1,i} z_{2,i} z_{3,i} & \sum_{i=1}^{NE} z_{2,i}^2 z_{3,i} \\
\sum_{i=1}^{NE} z_{3,i} & \sum_{i=1}^{NE} z_{1,i} z_{3,i} & \sum_{i=1}^{NE} z_{2,i} z_{3,i} & \sum_{i=1}^{NE} z_{3,i}^2 & \sum_{i=1}^{NE} z_{1,i} z_{2,i} z_{3,i} & \sum_{i=1}^{NE} z_{3,i}^2 z_{1,i} & \sum_{i=1}^{NE} z_{3,i}^2 z_{2,i} \\
\sum_{i=1}^{NE} z_{1,i} z_{2,i} & \sum_{i=1}^{NE} z_{1,i}^2 z_{2,i} & \sum_{i=1}^{NE} z_{2,i}^2 z_{1,i} & \sum_{i=1}^{NE} z_{1,i} z_{2,i} z_{3,i} & \sum_{i=1}^{NE} z_{1,i}^2 z_{2,i}^2 & \sum_{i=1}^{NE} z_{1,i}^2 z_{2,i} z_{3,i} & \sum_{i=1}^{NE} z_{2,i}^2 z_{1,i} z_{3,i} \\
\sum_{i=1}^{NE} z_{1,i} z_{3,i} & \sum_{i=1}^{NE} z_{1,i}^2 z_{3,i} & \sum_{i=1}^{NE} z_{1,i} z_{2,i} z_{3,i} & \sum_{i=1}^{NE} z_{3,i}^2 z_{1,i} & \sum_{i=1}^{NE} z_{1,i}^2 z_{2,i} z_{3,i} & \sum_{i=1}^{NE} z_{1,i}^2 z_{3,i}^2 & \sum_{i=1}^{NE} z_{3,i}^2 z_{1,i} z_{2,i} \\
\sum_{i=1}^{NE} z_{2,i} z_{3,i} & \sum_{i=1}^{NE} z_{1,i} z_{2,i} z_{3,i} & \sum_{i=1}^{NE} z_{2,i}^2 z_{3,i} & \sum_{i=1}^{NE} z_{3,i}^2 z_{2,i} & \sum_{i=1}^{NE} z_{2,i}^2 z_{1,i} z_{3,i} & \sum_{i=1}^{NE} z_{3,i}^2 z_{1,i} z_{2,i} & \sum_{i=1}^{NE} z_{2,i}^2 z_{3,i}^2
\end{bmatrix}$$

$$M = \begin{bmatrix}
13 & 0 & 0 & 0 & 0 & 0 & 0 \\
0 & 8 & 0 & 0 & 0 & 0 & 0 \\
0 & 0 & 13 & 0 & 0 & 0 & 0 \\
0 & 0 & 0 & 13 & 0 & 0 & 0 \\
0 & 0 & 0 & 0 & 8 & 0 & 0 \\
0 & 0 & 0 & 0 & 0 & 8 & 0 \\
0 & 0 & 0 & 0 & 0 & 0 & 13
\end{bmatrix}$$

One important problem associated with standard factorial experimental design is the resulting large number of experiments (NE), equal to:

$$NE = \left(\prod_{i=1}^{NX} NL_i \right) NR \tag{12.10}$$

tabulation NR is the number of replicates at each experimental point, which must be equal throughout the experimental grid to assure the orthogonal properties of the final experimental design. This can lead to explosive growth of NE, as NX and NL_i increase. For instance, in Example 12.1 NE becomes equal to 81 if $NL_1 = NL_2 = NL_3 = NR = 3$. For this reason, Schwaab and Pinto (2011) recommended the use of $NL = 2$, when the investigator is interested in determining the coefficients of Equations (12.8–9), or $NL = 3$, when the investigator is also interested in the determination of quadratic coefficients in the form:

$$y^M (z, \alpha) = a_0 + \sum_{i=1}^{NX} a_i z_i + \sum_{i=1}^{NX} \sum_{j>i} b_{ij} z_i z_j + \sum_{i=1}^{NX} c_i z_i^2 \tag{12.11}$$

The use of quadratic coefficients can be important in problems that are associated with the optimization of experimental responses and in which the occurrence of points of maximum or minimum must be found during the investigation. Particularly, Schwaab and Pinto (2011) showed that the precision of the model parameters of Equations (12.8–9) increases very slowly with NL, so that the use of $NL > 2$ is almost always difficult to explain.

In order to reduce the number of experimental points in the design, it is important to observe that factorial designs contain fractions that are also orthogonal and that also allow the computation of independent model parameters. For instance, when $NL = 2$, factorial designs can be divided sequentially into two parts of equal size that remain orthogonal, giving birth to **fractional factorial designs**, represented in the form 2^{NX-Nd}, where Nd is the number of sequential divisions imposed on the original plan. The repartition of the original plan is possible while $NE = 2^{NX-Nd} > NX$ (Schwaab and Pinto, 20011). Obviously, as the number of experimental points is reduced, the number of model effects that can be observed with the empirical model must also be reduced, to assure the existence of appropriate degrees of freedom for the estimation problem and also because some experimental effects become confounded, as observed in Example 12.3.

Example

E12.3 In Example 12.1, when $NL = 2$ for all variables, the experimental design becomes:

$$NL = \begin{bmatrix} 2 \\ 2 \\ 2 \end{bmatrix}, x_{min} = \begin{bmatrix} 70 \\ 1 \\ 30 \end{bmatrix}, x_{max} = \begin{bmatrix} 110 \\ 3 \\ 60 \end{bmatrix}$$

where T is given in °C, P in atm and θ in min. In order to obtain a uniform distribution over the whole range of experimental conditions, the following conditions must be satisfied:

$$NL = \begin{bmatrix} 70,110 \\ 1,3 \\ 30,60 \end{bmatrix}$$

and the experimental design becomes

$$X = \begin{bmatrix} 70 & 70 & 70 & 70 & 110 & 110 & 110 & 110 \\ 1 & 1 & 3 & 3 & 1 & 1 & 3 & 3 \\ 30 & 60 & 30 & 60 & 30 & 60 & 30 & 60 \end{bmatrix}$$

$$Z^1 = \begin{bmatrix} -1 & -1 & -1 & -1 & +1 & +1 & +1 & +1 \\ -1 & -1 & +1 & +1 & -1 & -1 & +1 & +1 \\ -1 & +1 & -1 & +1 & -1 & +1 & -1 & +1 \end{bmatrix}$$

tabulation $NE = 2 \times 2 \times 2 = 8$ experiments (or 24, if each experiment is replicated three times for computation of experimental variances). One must observe that the original full factorial design can be split into two halves that contain four experiments ($NE = 4 > 3 = NX$) of the form:

$$Z^1 = \begin{bmatrix} -1 & -1 & +1 & +1 \\ -1 & +1 & -1 & +1 \\ -1 & +1 & +1 & -1 \end{bmatrix}$$

$$Z^2 = \begin{bmatrix} -1 & -1 & +1 & +1 \\ -1 & +1 & -1 & +1 \\ +1 & -1 & -1 & +1 \end{bmatrix}$$

In the case of the full factorial design, the following empirical model can be used for quantitative analyses:

$$y^M = a_0 + a_1z_1 + a_2z_2 + a_3z_3 + b_{13}z_1z_2 + b_{13}z_1z_3 + b_{23}z_2z_2$$

tabulation the empirical model used as reference must be much smaller when the fractional experimental design is proposed:

$$y^M = a_0 + a_1z_1 + a_2z_2 + a_3z_3$$

Therefore, repartition of the experimental design imposes the necessary simplification of the quantitative analyses. Besides, it is easy to show that in the fractional factorial plan Z^1, $z_1 = -z_2z_3$, $z_2 = -z_1z_3$ and $z_3 = -z_1z_2$. This means that synergetic effects are confounded with the linear main effects in the simpler Z^1 experimental design, indicating that the experimenter will not be able to calculate the synergetic effects and that, if they do exist, they will be interpreted as the effect of another different variable. Similarly, it is easy to show that in the fractional factorial plan Z^2, $z_1 = z_2 z_3$, $z_2 = z_1 z_3$ and $z_3 = z_1 z_2$.

As illustrated in Example 12.3, one important issue to consider when proposing the use of fractional factorial experimental design is the proper characterization of confounding and understanding which variable effects can indeed be calculated with the empirical model of Equations (12.8–9). In order to do this, most textbooks, spreadsheets and statistical applications provide tables and intrinsic functionalities that can be used to characterize the confounding of variable effects.

Example

E12.4 A simple technique that can be used to characterize confounding is computation of correlation matrixes. In this case, each analyzed effect must be associated with a virtual variable, as shown below for plans Z and Z^1:

z_1	z_2	z_3	$z_1 z_2$	$z_1 z_3$	$z_2 z_3$
−1	−1	−1	+1	+1	+1
−1	−1	+1	+1	−1	−1
−1	+1	−1	−1	+1	−1
−1	+1	+1	−1	−1	+1
+1	−1	−1	−1	−1	+1
+1	−1	+1	−1	+1	−1
+1	+1	−1	+1	−1	−1
+1	+1	+1	+1	+1	+1

z_1	z_2	z_3	$z_1 z_2$	$z_1 z_3$	$z_2 z_3$
−1	−1	−1	+1	+1	+1
−1	+1	+1	−1	−1	+1
+1	−1	+1	−1	+1	−1
+1	+1	−1	+1	−1	−1

that lead to the following correlation matrixes:

	z_1	z_2	z_3	$z_1 z_2$	$z_1 z_3$	$z_2 z_3$
z_1	1	0	0	0	0	0
z_2	0	1	0	0	0	0
z_3	0	0	1	0	0	0
$z_1 z_2$	0	0	0	1	0	0
$z_1 z_3$	0	0	0	0	1	0
$z_2 z_3$	0	0	0	0	0	1

	z_1	z_2	z_3	z_1z_2	z_1z_3	z_2z_3
z_1	1	0	0	0	0	−1
z_2	0	1	0	0	−1	0
z_3	0	0	1	−1	0	0
z_1z_2	0	0	−1	1	0	0
z_1z_3	0	−1	0	0	1	0
z_2z_3	−1	0	0	0	0	1

that clearly indicate the occurrence of confounding in the fractional factorial plan.

An additional point to be emphasized is the fact that experiments must be repli-cated to provide experimental precision, as described in Sections 1.8 and 1.9. Although replication of all experiments constitutes a sound experimental strategy, this can lead to a significant increase in experimental effort. In order to reduce the number of replicates and simultaneously keep the orthogonality of the experimental plan, most experimen-tal designs include the center of the experimental region as an additional experiment that must be performed and replicated. In this case, the experimental errors are nor-mally assumed to be the same throughout the experimental grid and equal to the value provided by the replicated central condition, which frequently constitutes a strong simplifying assumption about the behavior of experimental variances and should be considered carefully by the analyst.

Example

E12.5 In Example 12.3, the inclusion of the central point leads to the following experimental design

$$X = \begin{bmatrix} 70 & 70 & 70 & 70 & 110 & 110 & 110 & 110 & 90 & 90 & 90 \\ 1 & 1 & 3 & 3 & 1 & 1 & 3 & 3 & 2 & 2 & 2 \\ 30 & 60 & 30 & 60 & 30 & 60 & 30 & 60 & 45 & 45 & 45 \end{bmatrix}$$

$$Z = \begin{bmatrix} -1 & -1 & -1 & -1 & +1 & +1 & +1 & +1 & 0 & 0 & 0 \\ -1 & -1 & +1 & +1 & -1 & -1 & +1 & +1 & 0 & 0 & 0 \\ -1 & +1 & -1 & +1 & -1 & +1 & -1 & +1 & 0 & 0 & 0 \end{bmatrix}$$

with $NE = 2 \times 2 \times 2 + 3 = 11$ experiments.

Before finishing this section, it is important to observe that, whenever possible, the experiments in experimental design should be performed in random order to avoid confounding with uncontrolled external effects and the undesired capturing of external effects by estimated model parameters.

Example

E12.6 In Example 12.5, let us assume the experiments are performed in the exact order in which they are presented. Let us also assume that contamination of the feed lines takes place after the fifth experiment and goes unnoticed by the investigator. Then, as reaction yields decrease after the fifth experiment, quantitative analysis concludes that reaction yields tend to decrease with temperature, as the last six experiments are performed at higher temperature levels. This is certainly incorrect and does not represent the real effect of temperature on yields but rather the effect of confounding between the reaction temperature and the undesired and unmeasured contamination effect.

Example

E12.7 One might argue about the necessity to implement factorial design plans instead of the usual one-variable-a-time plan (or cross design). For instance, in Example 12.1, let us assume that a cross design approach is used. Then, the following design can be assumed:

$$
Z = \begin{bmatrix}
-1 & +1 & 0 & 0 & 0 & 0 & 0 & 0 & 0 \\
0 & 0 & -1 & +1 & 0 & 0 & 0 & 0 & 0 \\
0 & 0 & 0 & 0 & -1 & +1 & 0 & 0 & 0
\end{bmatrix}
$$

with $NE = 9$ experiments. In this case:

z_1	z_2	z_3	$z_1 z_2$	$z_1 z_3$	$z_2 z_3$
−1	0	0	0	0	0
+1	0	0	0	0	0
0	−1	0	0	0	0
0	+1	0	0	0	0
0	0	−1	0	0	0
0	0	+1	0	0	0
0	0	0	0	0	0
0	0	0	0	0	0
0	0	0	0	0	0

that leads to the following correlation matrixes:

	z_1	z_2	z_3	$z_1 z_2$	$z_1 z_3$	$z_2 z_3$
z_1	1	0	0	−	−	−
z_2	0	1	0	−	−	−
z_3	0	0	1	−	−	−
$z_1 z_2$	−	−	−	−	−	−
$z_1 z_3$	−	−	−	−	−	−
$z_2 z_3$	−	−	−	−	−	−

that clearly indicates that synergetic effects cannot be computed and that the proposed design is unable to capture the existence of nonlinear interaction effects among the many variables of the system. As a matter of fact, cross designs can only be useful during the very early stages of experimental analyses, for initial screening of variable effects, or when determination of individual quadratic coefficients is recommended for further modeling purposes. As factorial designs explore the experimental region much more efficiently and allow the determination of synergetic effects, the use of factorial design approaches is normally much more effective for quantitative analyses than are cross designs. Despite that, it is typical to combine cross design and factorial design when determination of individual quadratic coefficients is sought.

The interested reader is strongly encouraged to consult the literature for illustrative examples of applications of factorial experimental design in kinetic problems (Nele *et al.*, 1999; Larentis *et al.*, 2001; Bhering *et al.*, 2002; Larentis *et al.*, 2003). Example 12.8 illustrates the use of factorial design for interpretation of kinetic problems.

Example

E12.8 Matos *et al.* (2001) studied the effects of reaction operation conditions on slurry propylene polymerizations performed with a standard Ziegler–Natta catalyst. In order to carry out the experimental study, the authors proposed a 2^{7-4} fractional factorial experimental design (7 variables at 2 experimental levels, with 4 sequential repartitions of the original full factorial design) with replicates at the central point, as described in Table 12.1.

Before using the data for quantitative evaluations, it is important to emphasize some relevant practical points related to Table 12.1:

1.   Design variables were selected *a priori*, based on previous knowledge of the authors, and included P (propylene partial pressure, as propylene is the main reactant and P reflects variations of the reactant concentration), PH_2 (hydrogen partial pressure, as hydrogen is known to affect the catalyst activity), T (reaction temperature), $[Ti]$ (the titanium concentration in the slurry), A/T_1 (the ratio between aluminum and titanium molar concentrations in the mixture, as provided by a first aluminum compound that activates the catalyst), t (reaction time, as it is not possible to take samples from the reactor during the reaction course) and A/T_2 (the ratio between aluminum and titanium molar concentrations in the mixture, as provided by a first aluminum compound that activates the catalyst);

2.   Experimental conditions were defined in accordance with the reparameterized plan presented in Table 12.2;

3.   Experiment H0 was an industrial reference, included in the plan to allow the direct comparison with the plant performance and industrial validation of the obtained results;

4.   All experiments were replicated at least twice, although experiment H6 had to be interrupted because of excessive production of polymer material, flooding the reactor and indicating that it should not be repeated;

Table 12.1 Experimental design for investigation of slurry propylene polymerizations performed with Ziegler–Natta catalysts.

Code	Order	P (kgf/cm²)	PH₂ (kgf/cm²)	T (°C)	[Ti] (mg/L)	A/T₁	t (h)	A/T₂	Pol (g)
H0	1	7	1	65	200	0.9	2	2	299
H0	15	7	1	65	200	0.9	2	2	317
H0	28	7	1	65	250	0.9	2	2	353
H0	37	7	1	65	250	0.9	2	2	372
H1	6	6	0	50	200	0.9	1	1.4	54
H1	9	6	0	50	200	0.9	1	1.4	59
H2	31	6	1	70	400	1.13	1	1.4	307
H2	59	6	1	70	400	1.13	1	1.4	340
H3	8	6	1	70	200	0.9	3	2	457
H3	10	6	1	70	200	0.9	3	2	495
H4	32	6	0	50	400	1.13	3	2	318
H4	56	6	0	50	400	1.13	3	2	338
H5	29	8	0	70	200	1.13	1	2	214
H5	58	8	0	70	200	1.13	1	2	211
H6	13	8	0	70	400	0.9	2.5	1.4	860
H7	57	8	1	50	200	1.13	3	1.4	145
H7	60	8	1	50	200	1.13	3	1.4	156
H8	11	8	1	50	400	0.9	1	2	172
H8	13	8	1	50	400	0.9	1	2	195
H9	34	7	0	60	300	1.01	2	1.7	358
H9	35	7	0	60	300	1.01	2	1.7	355
H9	55	7	0	60	300	1.01	2	1.7	363
H9	36	7	0	60	300	1.01	2	1.7	371
H10	38	7	1	60	300	1.01	2	1.7	452
H10	39	7	1	60	300	1.01	2	1.7	426

Table 12.2 Experimental design for investigation of slurry propylene polymerizations performed with Ziegler–Natta catalysts in the reparameterized form.

Code	$P = z_1$	$PH_2 = z_2$	$T = z_3$	$[Ti] = z_4$	$A/T_1 = z_5$	$t = z_6$	$A/T_2 = z_7$
H0	0	+1	+0.5	−1	−1	0	+1
H1	−1	−1	−1	−1	−1	−1	−1
H2	−1	+1	+1	+1	+1	−1	−1
H3	−1	+1	+1	−1	−1	+1	+1
H4	−1	−1	−1	+1	+1	+1	+1
H5	+1	−1	+1	−1	+1	−1	+1
H6	+1	−1	+1	+1	−1	+1	−1
H7	+1	+1	−1	−1	+1	+1	−1
H8	+1	+1	−1	+1	−1	−1	+1
H9	0	−1	0	0	0	0	0
H10	0	+1	0	0	0	0	0

5. The reference experiment H0 was used as a blank (which most frequently is the role of the central point) and was repeated several times during the execution of the plan to ensure that the catalyst remain active;

6. Experiments H9 and H10 were used as central points, because hydrogen partial pressure behaved as a discrete variable for experimental reasons (absence or presence of hydrogen) and could not be manipulated to provide the mean pressure value of 0.5 kgf/cm^2;
7. Experiment H9 was replicated four times in order to better characterize the experimental variances;
8. As recommended, experiments were performed in random order, to prevent the inclusion of external effects on the quantitative analyses;
9. The apparent high number of experiments (60) was due to the execution of other simultaneous experimental studies in the same experimental unit;
10. The main process response was the polymer yield ($y = Pol$).

Based on Table 12.1, the variance of the measured polymer yields was calculated in the form:

$$s_{Pol}^2 = \frac{\sum_{i=1}^{10} (NR_i - 1)\, s_i^2}{\sum_{i=1}^{10} (NR_i - 1)} = \frac{\sum_{i=1}^{10} (NR_i - 1) \sum_{k=1}^{NR_i} \frac{(y_{ki} - \bar{y}_i)^2}{(NR_i - 1)}}{\sum_{i=1}^{10} (NR_i - 1)} = 371 g^2$$

tabulation NR_i represents the number of replicates of experimental condition i, and y_{ki} represents the kth measurement available for experimental condition i. Therefore, assuming measurement fluctuations follow the normal distribution and that the degree of confidence is equal to 95% (see Sections 1.8 and 1.9):

$$\varepsilon = \pm 2s = \pm 39g$$

As the original factorial plan was partitioned four times, the experimental designs presented in Tables 12.1 and 12.2 do not allow the independent calculation of synergetic effects (total of 21), as they are confounded with the main linear effects (total of 7). According to the model presented in Equation (12.11):

$$y^M = a_0 + a_1 z_1 + a_2 z_2 + a_3 z_3 + a_4 z_4 + a_5 z_5 + a_6 z_6 + a_7 z_7$$
$$y^M = (342 \pm 21) + (35 \pm 27) z_1 + (-32 \pm 21) z_2 + (151 \pm 26) z_3$$
$$+ (110 \pm 25) z_4 + (-76 \pm 25) z_5 + (137 \pm 28) z_6 + (-34 \pm 25) z_7$$

Therefore, all analyzed variable effects are significant (as the confidence intervals of model parameters do not include the 0), indicating that the prior knowledge of the authors was important for proper selection of the input variables. Besides, the observed effects indicate that P (z_1), T (z_3), $[Ti]$ (z_4) and t (z_6) exert positive effects on the polymer yields, as it might already be expected, as yields normally increase when reactant concentrations, reaction temperatures, catalyst concentrations and reaction times increase. On the other hand, the observed effects indicate that PH_2 (z_2), A/T_1 (z_5) and A/T_2 (z_7) exert negative effects on the process response, so that the operation levels of these variables should not be kept above levels required to achieve desired process performances. For instance, PH_2 controls the average molecular weight of the product and must increase when products of smaller molecular weights are desired, imposing the increase of the catalyst concentration if polymer yields should remain constant.

On the other hand, as both A/T_1 and A/T_2 are needed to activate the catalyst system, some sort of optimization of the catalyst system formulation must be performed to assure maximum possible catalyst activity. Finally, based on the process responses, reaction temperature can be regarded as the most influential process variable in the analyzed experimental region, as the respective model parameter was found to be the largest one, followed by reaction time and catalyst concentration. As said before, direct comparison of obtained parameters can only be carried out because the model was calculated in the reparameterized form, after removal of the physical units of the input variables.

12.2 OPTIMAL EXPERIMENTAL DESIGN FOR PARAMETER ESTIMATION

A different family of experimental design comprises techniques that are usually named as optimal experimental designs. In this case, the analyst normally knows the reference kinetic model that must be used to describe a particular kinetic problem (perhaps owing to his or her previous experience in the field) and wants to design a batch of NE experiments $(x_1\ x_2 \ldots, x_{NE})$ that can allow the estimation of model parameters (α) of the reference model with maximum precision. In a certain sense, factorial design can also be regarded as optimal experimental design for Equations (12.8–9), although these equations lack physical meaning. Besides, optimal experimental designs are more general than factorial designs, because a smaller number of assumptions are imposed on the design problem.

From a mathematical point of view, optimal designs search for NE experiments $(x_1\ x_2 \ldots, x_{NE})$ that allow the maximization of a certain performance index defined in the experimental region. Therefore, the experimental design becomes equivalent to an optimization problem, where the manipulated variables are the experimental conditions (Box et al., 1978; Edgar and Himmelblau, 1988; Montgomery and Runger, 2002; Schwaab and Pinto, 2011). The performance index is usually related to the quality of the obtained model parameters, as described by the covariance matrix of model parameters, in accordance with Equation (11.48) of Chapter 11 in the form:

$$V_\alpha = 2H_\alpha^{-1} \tag{12.12}$$

which can be reduced to the form:

$$V_\alpha = \left[\sum_{k=1}^{NE} B_k^T B_k \right]_{NP \times NP}^{-1} \tag{12.13}$$

$$B_k = \left[\left. \frac{\partial y_i^M}{\partial \alpha_j} \right|_k \right]_{NY \times NP} \tag{12.14}$$

tabulation the standard least-squares objective function is used for estimation of model parameters and derivatives of higher orders are neglected (Box *et al.*, 1978; Montgomery and Runger, 2002; Schwaab and Pinto, 2011).

Performance indexes (J) are generally formulated in the form:

$$J_1 = \min_{x_1, \ldots, x_{NE}} \det\left(\mathbf{V}_\alpha\right) \tag{12.15}$$

$$J_2 = \min_{x_1, \ldots, x_{NE}} \operatorname{trace}\left(\mathbf{V}_\alpha\right) \tag{12.16}$$

tabulation propose the minimization of the parameter variances (or maximization of the parameter precision). Optimization of Equations (12.15–16) can be computationally expensive, especially when the number of designed experiments (NE) is large and experimental regions are not regular. However, the ability to handle arbitrary NE values and irregular experimental regions constitutes a major advantage of optimal experimental designs when compared to factorial experimental plans. Further, in order to facilitate a design problem, most problems assume that the designed experiments (x_1 x_2 ..., x_{NE}) must be selected from a pool of candidate experiments (x_1 x_2 ..., $x_{NG>NE}$) selected *a priori* by the investigator, according to previous experience and limitations of the experimental setup. This way, optimization must be performed in a discrete set of candidate design plans (N_{plan}), whose number is equal to:

$$N_{plan} = \binom{NG}{NE} \tag{12.17}$$

so that N_{plan} is the combination of NG candidate experiments in sets of NE experiments. The combinatorial nature of the optimization procedure can constitute a numerical difficulty that must be tackled when NG and NE values are large, and this can be regarded as a major disadvantage of the proposed procedure, especially because factorial designs can be proposed easily. Another disadvantage is the necessity to provide the reference model and to compute model derivatives, which can be difficult in more involved kinetic analyses.

Example

E12.9 In Example 11.3 of Chapter 11, the following set of kinetic equations are used to describe the analyzed reaction system:

$$\frac{dC_A}{dt} = -k_1 C_A \Rightarrow C_A = C_{A0} e^{-k_1 t}$$

$$\frac{dC_B}{dt} = k_1 C_A = k_1 C_{A0} e^{-k_1 t} \Rightarrow C_B = C_{A0} \left(1 - e^{-k_1 t}\right)$$

so that

$$y^M = \begin{bmatrix} C_A \\ C_B \end{bmatrix}, x = \begin{bmatrix} t \\ C_{A0} \end{bmatrix}, \alpha = [k_1]$$

Therefore,

$$B = \begin{bmatrix} \frac{\partial C_A}{\partial k_1} \\ \frac{\partial C_B}{\partial k_1} \end{bmatrix} = \begin{bmatrix} -tC_{A0}e^{-k_1 t} \\ tC_{A0}e^{-k_1 t} \end{bmatrix}$$

$$B^T B = 2t^2 C_{A0}^2 e^{-2k_1 t}$$

$$V_\alpha = 2 \left[\sum_{k=1}^{NE} t_k^2 C_{A0k}^2 e^{-2k_1 t_k} \right]^{-1}$$

As the proposed problem involves a single kinetic parameter, performance indexes presented in Equations (12.15–16) are equal. In this case, minimization of the proposed performance index is equivalent to maximization of:

$$J = \max \left(\sum_{k=1}^{NE} t_k^2 C_{A0k}^2 e^{-2k_1 t_k} \right)$$

which leads to:

$$\frac{\partial J}{\partial C_{A0k}} = 2 \sum_{k=1}^{NE} t_k^2 C_{A0k} e^{-2k_1 t_k} = 0$$

$$\frac{\partial J}{\partial t_k} = 2 \sum_{k=1}^{NE} t_k C_{A0k}^2 e^{-2k_1 t_k} (1 - k_1 t_k) = 0$$

The first equation cannot be satisfied, because all variables are positive, indicating that the point of optimality is a boundary of the experimental region. In this case, maximization of J clearly imposes the use of the highest possible values of C_{A0}. This makes sense, as the increase of C_{A0} leads to higher rates of reaction and more precise determination of the kinetic rate constant. On the other hand, assuming that $NE = 1$, it becomes clear that the optimum sampling time is:

$$t_1 = \frac{1}{k_1}$$

tabulation indicates that determination of the optimum sampling time requires some previous knowledge of the kinetic rate constant. When nonlinear reaction models are considered, this is almost always true and explains why optimal experimental designs are not used very frequently for experimental screening and initiation of experimental studies, as it is usually very difficult to obtain good guesses of unknown kinetic rate constant values. For this reason, the use of optimal design plans is more common in kinetic analyses when empirical linear models in the form of Equation (12.2) are considered.

12.3 SEQUENTIAL EXPERIMENTAL DESIGNS

As described in the previous section, it can be difficult to propose optimal experimental designs for real nonlinear kinetic models when some sort of previous knowledge of the kinetic rate constant values is not available. However, in many problems the analyst has already performed NE experiments $(x_1\ x_2\ ...,\ x_{NE})$ and has not yet been able to select the best kinetic model or determine the kinetic parameters with sufficient precision. In these cases, some models and respective kinetic parameters are indeed available, although perhaps discrimination of the best model has not been possible, and the quality of the model parameters must be improved. Therefore, the investigator still handles a set of NM model candidates to describe a particular kinetic problem $(y_1^M(x,\alpha_1^M),\ y_2^M(x,\alpha_2^M),\ ...,\ y_{NM}^M(x,\alpha_{NM}^M))$ and wants to define the best model candidate through sequential experimentation and selection of at least one additional experiment (x_{NE+1}). Otherwise, even if a kinetic model has been already selected, the investigator may want to improve the quality of the model parameters with the help of at least one additional experiment (x_{NE+1}). These problems lead to formulation of sequential experimental design procedures.

According to the sequential experimental design procedure, the available experimental information (due to availability of NE previously performed experiments or to previous experience of the analyst) can be initially used to provide estimates for the parameters of the many candidate models under consideration. Then, in the second step at least one additional experiment (x_{NE+1}) must be designed through maximization of a certain specified performance index, as discussed in the previous section; however, as guesses for model parameters are assumed to be available, numerical computations are indeed possible, and optimization can be performed in a meaningful manner. Then, in the third step the selected experiments are performed in the lab and the set of experiments is enlarged. Finally, in the fourth step the researcher updates the model parameters for the models under consideration and returns to the initial step if the obtained results are not conclusive and recommend a new round of experimentation.

In most problems it is usually much more difficult to perform the experiments than to design the experimental conditions. Therefore, the number of designed experiments depends mainly on the time required to carry out the experiments. If experiments can be performed fast (for instance in a couple of hours of a couple of days), then the investigator will find few reasons to select and perform more than one experiment per iteration. However, if the experiments are too long (as in agriculture, for instance) or a high throughput experimental device is used to perform many experiments simultaneously, the analyst may find it adequate to design many experiments simultaneously.

12.3.1 Sequential experimental design for model discrimination

Many times, and especially during the first stages of the model development process, available experimental data can be explained by different kinetic models. This is quite common in the field of chemical kinetics, as different kinetic mechanisms that lead to different kinetic models can explain available experimental data with similar statistical significance (Knözinger *et al.*, 1973; Schwaab *et al.*, 2006; Froment *et al.*, 2011). Therefore, assuming the investigator considers a set of *NM* reasonable candidate kinetic models to explain the data, Figure 12.1 illustrates how the proposed sequential experimental design procedure should be implemented (Gonzáles-Velasco *et al.*, 1991; Schwaab *et al.*, 2006; Schwaab *et al.*, 2008; Froment *et al.*, 2011; Schwaab and Pinto, 2011; Alberton *et al.*, 2013).

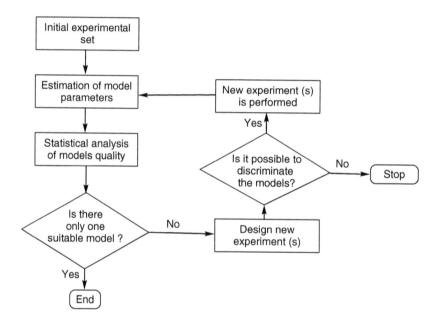

Figure 12.1 Sequential experimental design procedure for model discrimination.

Different performance indexes can be used for design of additional experiments. For example, a performance index can be very simply formulated to represent the average discriminating capacity of the experimental plan in the form (Froment *et al.*, 2011; Schwaab and Pinto, 2011):

$$J = \sum_{k=NE+1}^{NE+NS} \sum_{i=1}^{NM} \sum_{j>i}^{NM} \left(y_i^M (x_k, \alpha_i) - y_j^M (x_k, \alpha_j) \right)^2 \tag{12.18}$$

where maximization of Equation (12.18) assumes that the best experimental conditions for discriminating between candidate models are those that lead to maximum differences among respective model predictions. Although the basic reasoning behind Equation (12.18) is correct, this equation lacks statistical significance. For this reason, many publications have proposed different performance indexes for model discrimination problems, based on more rigorous statistical arguments (Schwaab *et al.*, 2006; Schwaab *et al.*, 2008; Alberton *et al.*, 2013). For example, the discriminating capacity between rival models i and j can be written in the form:

$$J_{i,j}(x) = (P_i P_j)^z [y_i^M(x) - y_j^M(x)]^T V_{i,j}^{-1} [y_i^M(x) - y_j^M(x)] \qquad (12.19)$$

tabulation P_i and P_j are the probabilities for the analyzed models to be the correct models, calculated with help of a chi-square distribution, z is a parameter that defines the relative importance of the rival model probabilities for the design, and $V_{i,j}$ is a matrix that depends on the covariance matrix of model predictions of the form (Schwaab *et al.*, 2006; Schwaab *et al.*, 2008):

$$V_{i,j} = 2V_y(x) + V_i(x) + V_j(x) \qquad (12.20)$$

where V_y is the covariance matrix of experimental fluctuations and V_i is the covariance matrix of model predictions calculated for model i (as described in Chapter 11).

Example

E12.10 Let us assume that two models are being considered for representation of a certain kinetic experiment: a first-order model, of the form:

$$y_1^M = C_A^{(1)} = C_{A0} e^{-k_1^{(1)} t}$$

and a second-order model in the form:

$$y_2^M = C_A^{(2)} = \frac{C_{A0}}{1 + k_1^{(2)} t}$$

According to Equation (12.18) and assuming that a single additional experiment must be designed at each iteration:

$$J = \left(y_i^M(x_k, \alpha_i) - y_j^M(x_k, \alpha_j) \right)^2 = \left(C_{A0} e^{-k_1^{(1)} t} - \frac{C_{A0}}{1 + k_1^{(2)} t} \right)^2$$

As shown in Example 12.9, the performance index increases continuously with C_{A0}, so the initial concentration must be as high as possible. On the other hand, for determination of the best sampling time for model discrimination, the following design equation can be proposed:

$$\frac{\partial J}{\partial t_{NE+1}} = 2 \left(C_{A0NE+1} e^{-k_1^{(1)} t_{NE+1}} - \frac{C_{A0NE+1}}{1 + k_1^{(2)} t_{NE+1}} \right)$$

$$\left(-k_1^{(1)} C_{A0NE+1} e^{-k_1^{(1)} t_{NE+1}} + \frac{k_1^{(2)} C_{A0NE+1}}{\left(1 + k_1^{(2)} t_{NE+1} \right)^2} \right) = 0$$

Assuming the maximum allowed value for $C_{A0} = 1$ mol/L, and that the previous experiments indicate $k_1^{(1)} = k_1^{(2)} = 1$ h^{-1}, it becomes possible to build Figures 12.2 and 12.3.

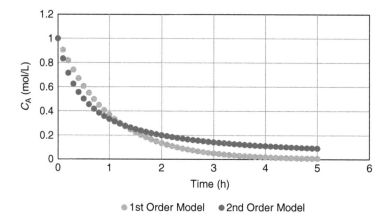

Figure 12.2 Expected model predictions.

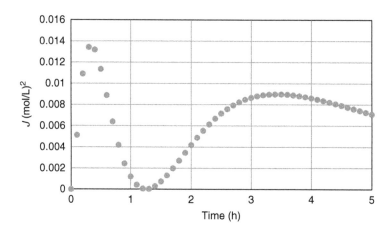

Figure 12.3 Performance index as a function of the sampling time.

Figure 12.2 shows that model responses are indeed very similar, while Figure 12.3 shows that the best experimental condition for discrimination between the two model responses combines the high initial concentration of reactant A with sampling time of $t = 0.35$ h.

12.3.2   Sequential experimental design for parameter estimation

Even when the kinetic model that best describes the available kinetic data has been determined, there may still be incentives for improvement of the parameter estimates. In this case, the investigator normally attempts to reduce the correlations among the model parameters or to enhance the parameter precision (diminish the parameter uncertainties). Figure 12.4 illustrates the sequential design procedure that can be used in this phase of the experimental study to increase the precision of the parameter estimates (Froment *et al.*, 2011; Schwaab and Pinto, 2011).

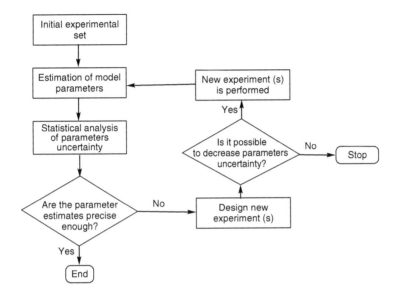

Figure 12.4 Sequential experimental design procedure for parameter estimation.

Different performance indexes have been developed for the sequential design of experiments for estimation of precise parameters, but they are all based on some sort of metrics of Equation (2.13–16), as described in Section 12.2 (Pinto *et al.*, 1990; Pinto *et al.*, 1991; Alberton *et al.*, 2011). The difference now is that previous information is assumed to be available for the estimable model parameters. For this reason, Equation (12.13) is usually written in the iterative form (Froment *et al.*, 2011; Schwaab and Pinto, 2011):

$$V_{\alpha NE+NS}^{-1} = V_{\alpha NE}^{-1} + \sum_{k=NE+1}^{NE+NS} B_k^T B_k \tag{12.21}$$

tabulation *NS* is the number of selected experiments in the new design iteration. It is important to observe that model discrimination and precise parameter estimation can be combined and performed simultaneously with the help of information measures, as described by Alberton *et al.* (2011).

Example

E12.11 In Example 12.9, let us assume that information about the kinetic rate constant is available in the form:

$$\alpha = [k_1] = [1.00], V_{\alpha NE} = [\sigma_1^2] = [0.01]$$

Assuming that a single experiment must be designed to improve the precision of parameter k_1, according to Equation (12.21):

$$V_{\alpha NE+1}^{-1} = V_{\alpha NE}^{-1} + B_{NE+1}^T B_{NE+1}$$

$$J = V_{\alpha NE+1}^{-1} = 100 + t_{NE+1}^2 C_{A0NE+1}^2 e^{-2k_1 t_{NE+1}}$$

Minimization of parameter uncertainty is equivalent to maximization of J in the previous equation, which can be observed graphically in Figure 12.5.

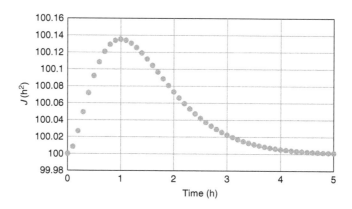

Figure 12.5 Performance index as a function of the sampling time.

As one can see in Figure 12.5, the best experimental condition for estimation of k_1 is $t_{NE+1} = 1$ h. However, according to Figure 12.5, the expected improvement of the parameter accuracy is not significant at these conditions. This is because Equation (12.13) is based on the standard least-squares function. If the maximum likelihood procedure is applied, as discussed in Section 11.2 of Chapter 11, then Equation (12.13) becomes:

$$V_\alpha = \left[\sum_{k=1}^{NE} B_k^T V_y^{-1} B_k \right]_{NP \times NP}^{-1} \qquad (12.22)$$

and Equation (12.21) becomes:

$$V_{\alpha NE+NS}^{-1} = V_{\alpha NE}^{-1} + \sum_{k=NE+1}^{NE+NS} B_k^T V_y^{-1} B_k \qquad (12.23)$$

tabulation V_y is the covariance matrix of the experimental measurements. In this case,

$$J = V_{\alpha NE+1}^{-1} = 100 + \frac{t_{NE+1}^2 C_{A0NE+1}^2 e^{-2k_1 t_{NE+1}}}{\sigma_{CA}^2}$$

and it becomes clear that improvement of parameter precision requires a reduction in the experimental errors (with possible modification of measuring devices and procedures). When $\sigma_{CA}^2 = 0.01$ (mol/L)^2, for instance, Figure 12.6 can be drawn, showing that the parameter variance can be reduced from 1×10^{-2} h^2 to 0.88×10^{-2} h^2 with the additional experimental point.

Figure 12.6. Performance index as a function of the sampling time when $\sigma_{CA}^2 = 0.01$ (mol/L)^2.

12.4   CONCLUDING REMARKS

In Chapter 12, statistical design procedures were presented and discussed. The main features of four families of statistical designs were discussed and illustrated through examples. The first family of statistical designs comprised factorial designs, used most often for initial screening of experimental problems and based on empirical mathematical structures. The advantageous aspects of orthogonal designs were presented and illustrated through simple examples. Then the concept of optimal experimental design was presented and used to introduce the families of sequential experimental designs for model discrimination (in which the investigator handles multiple candidate kinetic models) and for parameter estimation (in which a model has been selected but the investigator attempts to improve the quality of the model parameters). Finally, it was shown that the use of the maximum likelihood approach allows for explicit consideration of experimental measurement errors in experimental planning.

Chapter 13

Kinetic exercises

13.1 SOLUTION OF KINETIC EXERCISES

Example

ER.1 The following reaction was performed in a batch reactor:

$$2N_2O_5 \rightarrow 2N_2O_4 + O_2$$

Calculate the activation energy using the half-life time data for different temperatures.

T (°C)	300	200	150	100	50
$t_{1/2}$ (s)	3.9×10^{-3}	3.9×10^{-3}	8.8×10^{-2}	4.6	780

Solution
As seen, this is a nonelementary reaction and can be a first- or second-order reaction.
Assuming a first-order reaction rate, we calculate the specific rate constants for different temperatures, using Equation 5.18. Results are shown below:

$$k = \frac{0.693}{t_{1/2}}$$

Thus, Table 13.1 presents the values of k_1 varying with temperature:

Table 13.1

T (°C)	T (K)	$t_{1/2}$	k_1	Ln k_1
300	573	0.0039	177.7	5.18
200	473	0.0039	177.7	5.18
150	423	0.088	7.87	2.06
100	373	4.60	0.150	−1.89
50	323	780.0	0.00088	−7.02

Figure 13.1 shows the Arrhenius plot which does not follow a straight line. However, between 50°C and 200°C, the experimental data fit very well in the Arrhenius plot (Figure 13.2). From this figure, we calculate the energy of activation equal to 23,400 cal/mol. In conclusion, experiments were carried out under kinetic conditions in this temperature range. However, for higher temperatures, there are diffusion effects.

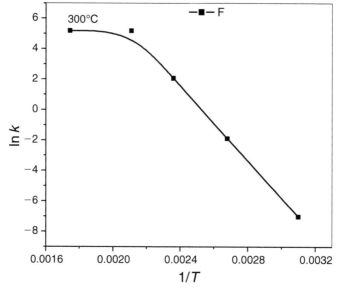

Figure 13.1 Arrhenius plot.

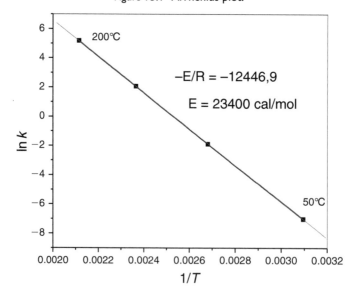

Figure 13.2 Activation energy.

Example

ER.2 Demonstrate that the reaction rate of the product for an autocatalytic reaction:

$$A + B \xrightarrow{k_1} B + B$$

can be expressed as:

$$r_B = k_1 \left[(C_0 - C_B) - \frac{1}{K} C_B^2 \right]$$

where:

$$C_0 = C_{A0} + C_{B0}$$

Show that the maximum concentration of B is equal to:

$$C_{B\,max} = \frac{C_{B0}}{2} \left(1 + \frac{C_{B0}}{C_{A0}} \right)$$

Solution

$$r_B = k_1 C_A + k_1 C_B - k_1' C_B^2 - k_1 C_B$$

Then,

$$r_B = k_1 (C_A + C_B - C_B) - k_1' C_B^2$$

But,

$$C_0 = C_A + C_B$$

Thus,

$$r_B = k_1 \left[(C_0 - C_B) - \frac{1}{K} C_B^2 \right]$$

The maximum concentration of B is given by the following equation:

$$C_{B\,max} = \frac{C_{B0}}{2} \left(1 + \frac{C_{B0}}{C_{A0}} \right)$$

Example

ER.3 The following reversible reaction is carried out in a tubular reactor at 25°C:

The equilibrium constant at 25°C is equal to 1.25. The conversion is 90% of the equilibrium conversion. Pure reactant is fed at 0.36 L/h and with a concentration of 0.2 mol/L. The reactor volume is 0.5 L. Assuming a first-order reaction rate for both direct and reverse reactions, calculate the specific kinetic rate constants.

Solution

As seen this is a reaction type like:

$$2A \Leftrightarrow R$$

The resultant reaction rate is, therefore:

$$r = kC_A - k'C_R$$

This can be rearranged as follows:

$$r = kC_{A0}\left[(1 - X_A) - \frac{1}{2K}X_A\right]$$

Since:

$$C_R = \frac{C_{A0}}{2}X_A$$

However, in equilibrium, the resultant rate is zero and thus:

$$K = \frac{k}{k'} = \frac{X_{Ae}}{2(1 - X_{Ae})} = 1.25$$

or

$$X_{Ae} = 0.714$$

Expressing the resultant rate as a function of the conversion, we obtain for $R = 0$ the following Equation 3.21:

$$r = kC_{A0}\left(1 - \frac{X_A}{X_{Ae}}\right)$$

Substituting this rate expression in the PFR equation (Equation 4.17) and integrating we obtain for $R = 0$ a similar equation as Equation 5.24, namely:

$$\tau(t) = C_{A0}\int_0^{X_A} \frac{dx_A}{r}$$

$$-\ln\left(1 - \frac{X_A}{X_{Ae}}\right) = \frac{k}{(X_{Ae})}\tau(t) \tag{5.24}$$

Since the conversion is 90% of the equilibrium conversion, we get:
where:

$$\frac{X_A}{X_{Ae}} = \frac{0.642}{0.714} = 0.89$$

Then:

$$\frac{k\tau}{X_{Ae}} = 2.207$$

Since $\tau = V/v_0 = 83.3$ min.
   Thus,

$$k = 1.89 \times 10^{-2} \text{ min}^{-1}$$

$$k' = 1.51 \times 10^{-2} \text{ min}^{-1}$$

Example

ER.4 The reaction was performed in a batch reactor and in the gas phase. 10% N_2 was introduced in the reactor at 2 atm and 450°C. After 50 min, the pressure was 3.3 atm. The reaction is irreversible and of first order. Calculate the specific reaction rate constant. If the same reaction would be performed in a piston system with variable volume, how volume changes keeping pressure constant at 2 atm and considering the same conversion as before? Calculate the initial concentration?

$$H_2C-CH_2 \longrightarrow CH_4 + CO$$
$$\backslash_O\!/$$

Solution
The reaction is:

$$A \longrightarrow R + S$$

First, the reaction occurs at constant volume, so we calculate the partial pressures. The initial partial pressures of reactant and N_2 (inert) are:

$$p_{N_2} = 0.1 \times 2 = 0.2 \text{ atm}$$
$$p_{A0} = 1.8 \text{ atm}$$

The partial pressure of A is therefore:

$$p_A = p_{A0} - \frac{a}{\Delta v}(P - P_0)$$

After 50 min, the total pressure of the system is 3.3 atm and the initial pressure 2 atm. Since, $\Delta v = 2 - 1 = 1$ and $a = 1$, then:

$$p_A = 1.8 - 1(3.3 - 2) = 0.5$$

Thus, the conversion is:

$$X_A = \frac{p_{A0} - p_A}{p_{A0}} = 0.72$$

The rate is a first-order reaction and for a constant volume it is equal to:

$$(-r_A) = kC_A = kC_{A0}(1 - X_A)$$

Substitute the rate equation in the batch reactor equation and integrate. Then,

$$t = C_{A0} \int\limits_0^{X_A} \frac{dx_A}{kC_{A0}(1 - X_A)}$$

$$-\ln(1 - X_A) = kt$$

Substituting the conversion and time, we calculate the specific rate constant:

$$k = 0.0256 \text{ min}^{-1}$$

In the second part, we assume a closed piston system and constant pressure; however, volume changes due to the gas expansion. Thus,

$$V = V_0(1 + \varepsilon_A X_A)$$

The expansion factor is:

$$A \longrightarrow R + S$$

	A	R	S	Inert	Total
Initial	0.9	0	0	0.1	1.0
Final	0	0.9	0.9	0.1	1.8

Thus,

$$\varepsilon_A = \frac{V_{X_A=1} - V_{X_A=0}}{V_{X_A=0}} = 0.8$$

Substituting these values, we obtain:

$$V = 0.1(1 + 0.8 \times 0.722) = 0.157 \text{ L}$$

The initial concentration is:

$$C_{A0} = \frac{y_{A0}P_0}{RT} = \frac{0.9 \times 2}{0.082(273+450)} = 3.03 \times 10^{-2} \text{ mol/L}$$

Example

ER.5 The reaction $A \longrightarrow 2R + \frac{1}{2}S$ is carried out in a PFR reactor in the gas phase and under isothermal conditions. Reactant A is fed with 30% inert gas at 10 atm and 800 K. The reaction is irreversible and second order. Data are presented in the table 13.2 below.

Table 13.2

V_e (min^{-1})	0.02	0.0095	0.0062
F_S (mol/min)	0.5	0.8	2.0

The total flow rate is 7.2 mol/min. The energy of activation is 30 kcal/mol. Calculate the specific rate constant and comment results. Diffusion effects were eliminated. What variables must be calculated?

70% A
30% I
10 atm, 527°C
F_S = 7.2 mol/min

Solution
If the rate is a second-order reaction and occurs in the gas phase, we get the following rate expression as function of conversion:

$$(-r_A) = kC_A^2$$

$$(-r_A) = kC_{A0}^2 \frac{(1-X_A)^2}{(1+\varepsilon_A X_A)^2}$$

Substitution in the PFR equation:

$$\tau C_{A0} = \int_0^{X_A} \frac{(1+\varepsilon_A X_A)^2 \, dX_A}{k(1-X_A)^2} dX_A$$

The solution of this integral is:

$$(1+\varepsilon_A)^2 \frac{X_A}{(1-X_A)} + \varepsilon_A^2 X_A + 2\varepsilon_A(1+\varepsilon_A)\ln(1-X_A) = \tau k C_{A0}$$

Calculation of C_{A0}:

$$C_{A0} = \frac{y_{A0}P_0}{RT} = \frac{0.3 \times 10}{0.082 \times (273 + 527)} = 4.57 \times 10^{-2} \, \text{mol/L}$$

Calculate conversion:

$$X_A = \frac{F_{A0} - F_A}{F_{A0}} = \frac{F_R}{2F_{A0}} = \frac{2F_S}{F_{A0}}$$

The initial flow rate of A:

$$F_{A0} = y_{A0}F_0 = 0.7 \times 7.2 = 5 \, \text{mol/L}$$

Calculation of conversions from F_S data is shown in the following table 13.3:

Table 13.3

V_e (min^{-1})	0.02	0.0095	0.0062
F_S (mol/min)	0.5	0.8	2.0
X_A	0.2	0.32	0.80
k (L/mol min)	5.8×10^{-2}	5.9×10^{-2}	6.0×10^{-1}

Calculate ε:

$$A \longrightarrow 2R + \frac{1}{2}S$$

	A	R	S	Inert	Total
Initial	0.7	0	0	0.3	1.0
Final	0	1.4	0.35	0.3	2.05

$$\varepsilon_A = \frac{V_{X_A=1} - V_{X_A=0}}{V_{X_A=0}} = \frac{2.05 - 1}{1} = 1.05 = 1.0$$

The equation becomes:

$$\frac{4X_A}{(1 - X_A)} + X_A + 4\ln(1 - X_A) = \tau k C_{A0}$$

Substituting these values, we calculate k, as shown in the table 13.3.

Observe that the first two columns at 800 K indicate constant k values equal to 5.85×10^{-2} (L/mol min). Since diffusion effects were discarded, probably the temperature was not correct. The third column shows a value which is 10 times greater than the previous ones, and therefore we calculate the correct temperature, since the activation energy is given and constant.

$$k = k_0 e^{-(E/RT)}$$

However, first we must calculate the constant k_0. The specific constant at 800 K is 5.85×10^{-2} (L/mol min). If the activation energy is $E = 30{,}000$ cal/mol, then,

$$k_0 = 8.25 \times 10^6$$

Hence, with $k = 5.85 \times 10^{-2}$, we obtain:

$$k = k_0 e^{-(E/RT)} = 5.85 \cdot 10^{-2} = 8.25 \cdot 10^6 e^{-(30000/2T)}$$

Thus, $T = 912$ K or 639°C.

Example

ER.6 The following reaction mechanism was proposed to explain the formation of polyurethane and polyesters.
Initiation:

$$RSH \xrightarrow{k_1} RS^\bullet + H^\bullet$$

Transfer:

$$H^\bullet + RSH \xrightarrow{k_2} H_2 + RS^\bullet$$

Propagation:

$$RS^\bullet + CH_2 = CHR' \xrightarrow{k_3} RSCH_2 - CH^*R'$$
$$RSCH_2 - CH^*R' + RSH \xrightarrow{k_4} RSCH_2 CH_2 R'$$

Termination:

$$2RS^\bullet \xrightarrow{k_5} RSSR$$

Determine the rate of disappearance of thiol (RSH). The stoichiometry of the global reaction is:

$$RSH + CH_2 = CHR' \xrightarrow{k} RSCH_2 - CH^*R'$$

Check the experimental results and determine the specific kinetic constant, assuming an irreversible reaction. Compare it with the global reaction rate.

Solution:

Naming: $A = RSH$, $R_1 = RS$; $MR_1 = RSCH_2–C^*HR'$, $M = CH_2 = CHR'$, and $P = RSSR$.

Then, the mechanism can be written simply by the following equations:

$$A \xrightarrow{k_1} R_1^{\bullet} + H^{\bullet}$$
$$H^{\bullet} + A \xrightarrow{k_2} H_2 + R_1^{\bullet}$$
$$R_1^{\bullet} + M \xrightarrow{k_3} MR_1^{\bullet}$$
$$MR_1^{\bullet} + A \xrightarrow{k_4} R_1^{\bullet} + R_2$$
$$2R_1^{\bullet} \xrightarrow{k_5} P$$

The rate of disappearance of A is:

$$-r_A = k_1[A] + k_2[A][H^{\bullet}] + k_4[MR_1^{\bullet}][A] \tag{13.1}$$

$$r_{H^{\bullet}} = k_1[A] - k_2[A][H^{\bullet}] = 0 \tag{13.2}$$

$$[H^{\bullet}] = \frac{k_1}{k_2} \tag{13.3}$$

$$r_{MR_1^{\bullet}} = k_3[R_1^{\bullet}][M] - k_4[MR_1^{\bullet}][A] = 0 \tag{13.4}$$

$$r_{R_1^{\bullet}} = k_1[A] + \underbrace{k_2[A][H^{\bullet}]}_{k_1[A]} - k_3[R_1^{\bullet}][M] + k_4[MR_1^{\bullet}][A] - k_5[R_1^{\bullet}]^2 \tag{13.5}$$

Thus,

$$2k_1[A] = k_5[R_1^{\bullet}]^2$$

or:

$$[R_1^{\bullet}] = \sqrt{\frac{2k_1[A]}{k_5}} \tag{13.6}$$

From Equation 13.4 we get:

$$k_3[R_1^{\bullet}][M] = k_4[MR_1^{\bullet}][A]$$

$$[MR_1^{\bullet}] = \frac{k_3[R_1^{\bullet}][M]}{k_4[A]} \tag{13.7}$$

Substituting Equation 13.6 in Equation 13.7, we obtain:

$$[MR_1^{\bullet}] = \frac{k_3[R_1^{\bullet}][M]}{k_4[A]} \sqrt{\frac{2k_1[A]}{k_5}} \tag{13.8}$$

Substituting Equation 13.8 in Equation 13.1, we obtain:

$$-r_A = k_1[A] + k_2[A][H^\bullet] + k_4[MR_1^\bullet][A]$$

$$-r_A = 2k_1[A] + k_3[M]\sqrt{\frac{2k_1[A]}{k_5}}$$

Neglect the first term of this equation, which represents the initial rate, when compared to step 3. Therefore, the rate expression of disappearance of A is as follows:

$$-r_A = k_3[M]\sqrt{\frac{2k_1[A]}{k_5}} \tag{13.9}$$

The concentration of the monomer M in the rate equation indicates that it is a first-order reaction in relation to the monomer.

Note that the concentration of reactant $[A]$ varies with time.

Table below presents data of reactant $[A]$ varying with time.

t (s)	0	350	400	500	640	750	830
[A]	0.874	0.223	0.510	0.478	0.432	0.382	0.343

For the global reaction, the rate of disappearance of $[A]$ is second order:

$$RSH + CH_2{=}CHR' \xrightarrow{k} RSCH_2 - C^\bullet HR'$$
$$A + M \longrightarrow MR$$

Thus, the rate equation is:

$$-r_A = k[A][M] = k[A]^2 \tag{13.10}$$

Assuming equal initial concentrations of A and M: $[A_0] = [M_0]$

Therefore, comparing these equations, we conclude that the rate expressions are different:

$$-r_A = k[A]^{1/2}[M] \tag{13.11}$$

Verifying if the rate of the global reaction 13.10 is correct, assuming a second-order and irreversible reaction.

The expression as function of conversion is:

$$(-r_A) = kC_A^2 = kC_{A0}^2(1 - X_A)^2$$

Substituting in the batch reactor equation, we get:

$$\frac{X_A}{(1 - X_A)} = kC_{A0}t \tag{13.12}$$

Verifying Equation 13.11, assuming proportionality we get:

$$-r_A = k[A]^{1/2}[M]$$

$$(-r_A) = kC_A^{1/2}C_M = kC_{A0}^{3/2}(1 - X_A)^{3/2} \qquad (13.13)$$

Substituting this equation in the batch reactor equation and further integration results in:

$$2 \times \left[\frac{1}{\sqrt{(1 - X_A)}} - 1 \right] = kC_{A0} \cdot t \qquad (13.14)$$

We calculate the specific constants from the following table 13.4 (last two columns).

Table 13.4

t (s)	[A]	X_A	k (Equation 13.12)	k (Equation 13.14)
0	0.8740	0		
350	0.5100	0.416476	0.002333	0.00189
400	0.4780	0.453089	0.00237	0.00188
500	0.4320	0.505721	0.002341	0.00180
640	0.3820	0.562929	0.002303	0.00171
750	0.3430	0.607551	0.002362	0.00170
830	0.3240	0.629291	0.00234	0.00165
		Mean value	0.002341	0.00177

Note that the specific kinetic constant for the global reaction is 2.34×10^{-3} L/mol min, while for the real model it is likely different, or 1.77×10^{-3} L/mol min.

Example

ER.7 The following irreversible reaction $A \longrightarrow 3R$ was studied in the PFR reactor. Reactant A is fed with an inert gas (40%) at 10 atm and 600 K and a flow rate of 1.0 L/min. Product R was measured in the exit gas for different space velocities, as presented in the following table. The rate is a second-order reaction. Calculate the specific rate constants.

v_e (min^{-1})	5×10^{-2}	7.57×10^{-3}	3.52×10^{-3}	1.55×10^{-3}
F_R (mol/s)	0.05	0.15	0.20	0.25

Solution

Since the rate is a second-order reaction and occurs with variable volume, we have:

$$(-r_A) = kC_{A0}^2 \frac{(1 - X_A)^2}{(1 + \varepsilon_A X_A)^2}$$

Substituting in the PFR equation, we obtain:

$$\tau C_{A0} = \int_0^{X_A} \frac{(1 + \varepsilon_A X_A)^2}{k(1 - X_A)^2} dX_A$$

The solution is as follows:

$$(1 + \varepsilon_A)^2 \frac{X_A}{(1 - X_A)} + \varepsilon_A^2 X_A + 2\varepsilon_A(1 + \varepsilon_A) \ln(1 - X_A) = \tau k C_{A0}$$

Calculate ε_A:

$$A \rightarrow 3R$$

	A	R	I	Total
Initial	0.6	0	0.4	1.0
Final	0	1.8	0.4	2.4

Then, $\varepsilon_A = 1.4$
Calculate initial conditions:

- Concentration:

$$C_{A0} = \frac{y_{A0} P_0}{RT} = \frac{0.6 \times 10}{0.082(600)} = 0.121 \text{ mol/L}$$

- Molar flux:

$$F_{A0} = v_0 C_{A0} = 1.0 \times 0.121 = 0.121 \text{ mol/L}$$

- Conversion:

$$X_A = \frac{F_{A0} - F_A}{F_{A0}} = \frac{F_R}{3F_{A0}}$$

$$\frac{5.75 X_A}{(1 - X_A)} + 1.96 X_A + 6.72 \ln(1 - X_A) = \tau k C_{A0}$$

From these data (Table 13.5), we calculate the specific rate constant k:

Table 13.5

V_e	F_R	X_A	k
0.05	0.05	0.137	0.0795
0.00757	0.15	0.413	0.0798
0.00352	0.20	0.550	0.0801
0.00155	0.25	0.688	0.079
		k (mean)=	0.0798

Example

ER.8 The thermal decomposition of isocyanite was performed in a differential reactor and the following data were obtained:

r_0 (mol/L min)	4.9×10^{-4}	6×10^{-5}	1.1×10^{-4}	2.4×10^{-3}	2.2×10^{-2}	1.18×10^{-1}	1.82
C_{A0} (mol/L)	0.2	0.06	0.020	0.05	0.08	0.1	0.06
$T(K)$	700	700	750	800	850	900	950

Determine the reaction order, activation energy, and the specific rate constants for each temperature.

Solution

The rate can be expressed for a general order: $r = k_a C_A^n$

These values correspond to the initial rates and concentrations. Transforming in logarithmic form, we obtain:

$$\ln(r_0) = \ln k_a + n \ln C_{A0}$$

where:

$$k = k_0 e^{-(E/RT)}$$

For determining the reaction order, we use the first two columns in the following table 13.6 and at the same temperature. Thus,

$$\ln(4.9 \times 10^{-4}) = -7.62 \quad \ln(6 \times 10^{-5}) = -9.72$$
$$\ln(0.2) = -1.60 \quad \ln(0.06) = -2.81$$
$$-7.62 = \ln k - 1.6n$$
$$-9.72 = \ln k - 2.81n$$

Thus, $n = 1.73$

Table 13.6 Specific rate constant.

r_0	C_{A0}	T	k	$1/T$
0.00049	0.2	700	0.007942	0.001429
0.00006	0.06	700	0.007802	0.001429
0.00011	0.02	750	0.095652	0.001333
0.0024	0.05	800	0.427807	0.00125
0.022	0.08	850	1.746032	0.001176
0.118	0.1	900	6.344086	0.001111
1.82	0.06	950	236.671	0.001053

Thus, n is not an integral order.

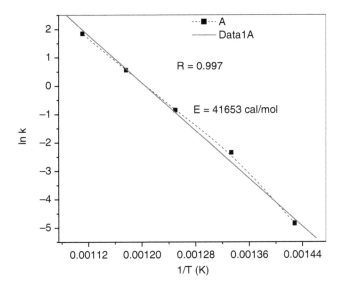

Figure 13.3 Arrhenius plot.

The reaction rate is thus:

$$r_0 = kC_A^{1.73}$$

or:

$$k = \frac{r_0}{C_{A0}^{1.73}}$$

From Table 13.2, we calculate the specific rate constants and plotting these values in Figure 13.3, we calculate the activation energy.

Example

ER.9 The reversible reaction was studied in the PFR reactor:

$$A + B \leftrightarrows R$$

The specific rate constants are unknown; however, the equilibrium constant is 10 L/mol. The exit conversion is 70% of the equilibrium conversion. The feed concentrations of A and B are each one equal to 2 mol/L, and the feed rate is 0.2 L/min. The reactor volume is 0.1 L. Calculate the direct and reverse specific velocities.

Solution

The resultant reaction rate is as follows:

$$r = kC_A C_B - k'C_R$$

However, for equimolar concentrations, we obtain:

$$r = kC_A^2 - k'C_R$$

In terms of conversion it becomes:

$$r = kC_{A0}^2 \left[(1 - X_A)^2 - \frac{1}{KC_{A0}} X_A \right] \tag{13.15}$$

At equilibrium, $r = 0$ and substituting $K = 10$ and $C_{A0} = 2$, we obtain:

$$KC_{A0} = K_c = 20$$

$$0 = \left[(1 - X_{Ae})^2 - \frac{1}{20} X_{Ae} \right]$$

$$20X_{Ae}^2 - 41X_{Ae} + 20 = 0$$

Finally, we get:

$$X_{Ae} = 0.80 \quad \text{and} \quad X'_{Ae} = 1.25$$

Substituting:

$$KC_{A0} = \frac{X_{Ae}}{(1 - X_{Ae})^2} = K_c \tag{13.16}$$

From the Equation 13.5, we get:

$$r = kC_{A0}^2 \left[(1 - X_A)^2 - \frac{(1 - X_{Ae})^2}{X_{Ae}} X_A \right]$$

or:

$$r = kC_{A0}^2 X_{Ae} \left[\left(1 - \frac{X_A}{X_{Ae}} \right) \left(\kappa - \frac{X_A}{X_{Ae}} \right) \right] \tag{13.17}$$

where:

$$\kappa = \frac{X'_A}{X_A}$$

It is the reaction rate as function of the conversion. Substituting Equation 13.17 in the PFR equation, we obtain:

$$\tau = C_{A0} \int_{0}^{X_A} \frac{dX_A}{r} \tag{13.18}$$

After integration, we obtain the solution:

$$\ln \left[\frac{\left(\kappa - \dfrac{X_A}{X_{Ae}} \right)}{\kappa \left(1 - \dfrac{X_A}{X_{Ae}} \right)} \right] = k^* \tau \tag{13.19}$$

where:

$$k^* = (\kappa - 1)kC_{A0} \frac{K_c - 1}{K_c} X_{Ae} \tag{13.20}$$

Since the final conversion is 70% of the equilibrium conversion, we have:

$$\kappa = \frac{X'_{Ae}}{X_{Ae}} = 1.56 \quad \text{and} \quad \frac{X_A}{X_{Ae}} = 0.7$$

Thus,

$$\ln \left[\frac{\left(\kappa - \dfrac{X_A}{X_{Ae}} \right)}{\kappa \left(1 - \dfrac{X_A}{X_{Ae}} \right)} \right] = 0.608 = k^* \tau$$

However,

$$\tau = \frac{V}{v_0} = 2$$

$$0.608 = k^* \tau$$

Then,

$$k^* = 0.304$$

From Equation 13.20, we get:

$$k^* = (\kappa - 1)kC_{A0} \frac{K_c - 1}{K_c} X_{Ae}$$

Substituting the values, we have:

$$k = 3.57\,\text{L/mol min}$$

$$k' = 0.178\,\text{min}^{-1}$$

Example

ER.10 The biological reaction follows the Monod equation and kinetic constants are unknown. From the data in the following table 13.7, calculate the constants.

Table 13.7 Experimental data.

Experiment	$C_{substrate}$ (g/L)	C_{cells} (g/L)	Rate (g/g·s)
1	2	0.9	1.5
2	3	0.7	1.5
3	4	0.6	1.6
4	10	0.4	1.8

Solution
The cell growth rate is expressed by the Equation 9.33:

$$r_C = \mu[C]$$

where:

$$\mu = \mu_{max}\frac{[S]}{[S] + K_{Monod}}$$

Thus,

$$r_C = \mu_{max}\frac{[S][C]}{[S] + K_{Monod}}$$

or:

$$\frac{[S] + K_{Monod}}{\mu_{max}} = \frac{[S][C]}{r_C}$$

This expression is conveniently transformed into:

$$\frac{K_{Monod}}{\mu_{max}[S]} + \frac{1}{\mu_{max}} = \frac{[C]}{r_C} \qquad (13.22)$$

These data are plotted in Figure E13.4.
Therefore,

$$\frac{K_{Monod}}{\mu_{max}} = 0.949$$

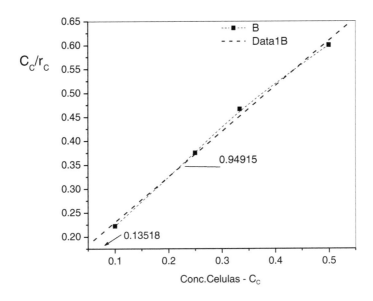

Figure 13.4 Determining rate constants.

$$\frac{1}{\mu_{max}} = 0.13518$$

Thus,

$$\mu_{max} = 7.4$$

$$K_{Mon} = 7.02$$

Example

ER.11 During fermentation, oxygen was consumed at 28°C. The following data were obtained at 23°C. The rate equation follows the basic fermentation equation. Calculate the maximum rate at 28°C. Using the half-time equation calculate the constant K_M.

P_{O_2} (mmHg)	0	0.5	1.0	1.5	2.5	3.5	5.0
r_{O_2} ($\mu L_{O_2}/h\, mg_{bact}$)	0	23.5	33	37.5	42	43	43

Solution
Figure 13.5 displays the rate as function of the partial pressure.

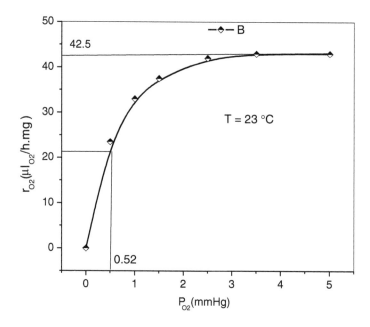

Figure 13.5 Rate versus pressure.

From Equation 9.17, we obtain the following expression, considering that 1 mol of oxygen is consumed by 1 mol of substrate:

$$\frac{1}{(-r_{o2})} = \frac{K_M}{V_{max}[P_{o2}]} + \frac{1}{V_{max}} \tag{13.23}$$

Plotting in Figure 13.6, we get:

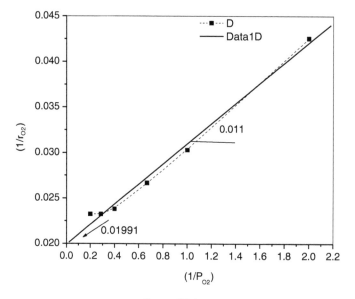

Figure 13.6

From this Figure 13.6 we determine:

$$\frac{1}{V_{max}} = 0.01991$$

And

$$V_{max} = 50.22$$

Since,

$$\frac{V}{T} = \frac{V'}{T'}$$

Then:

$$V'_{max} = 50.22\frac{301}{296} = 51.7,$$

Finally we calculate:

$$\frac{K_M}{V'_{max}} = 0.01107$$

and

$$K_M = 0.565$$

For half-life time, we get from the figure: $-r_{0_2} = 21.3$ and $P_{0_2} = 0.52$.
According to Equation 9.16, we obtain:

$$K_M = [S]_{1/2} = 0.52$$

This value is approximately the same as obtained from the graphic solution.

Example

ER.12 Decomposition of H_2O_2 occurred in the presence of an enzyme E and the corresponding concentrations are presented in the following table. Using the enzymatic kinetic model:

$$H_2O_2 + E \xrightarrow{k_1} [H_2O_2E^*]$$

$$[H_2O_2E^*] \xrightarrow{k_2} H_2O_2 + E$$

$$[H_2O_2E^*] + H_2O \xrightarrow{k_3} H_2O + O_2 + E$$

t (min)	0	10	20	50	100
$[H_2O_2]$ (mol/L)	0.02	0.0177	0.0158	0.0106	0.005

(a) Calculate the Michaelis constant and the maximum rate.
(b) Tripling the enzyme concentration, what time is needed to reach 95% conversion?

Solution

From the classic reaction mechanism, we demonstrated that the reaction rate is expressed by Equation 9.13.

$$(-r_S) = \frac{V_{max}[H_2O_2]}{[H_2O_2] + K_M}$$

The solution in a batch reactor is given by Equation 9.23:

$$-\frac{1}{t}\ln(1-X_S) = \frac{V_{max}}{K_M} - \frac{[H_2O_2]}{K_M}\frac{X_S}{t}$$

where:

$$X_S = \frac{[H_2O_2]_0 - [H_2O_2]}{[H_2O_2]_0}$$

where:

X_S is the conversion of $[H_2O_2]$

Since $[H_2O_2]_0 = 0.02$, we calculate the values in the below table 13.8 and plot Figure 13.7.

Table 13.8

t (min)	0	10	20	50	100
$[H_2O_2]$ (mol/L)	0.02	0.0177	0.0158	0.0106	0.005
X_S	0	0.115	0.210	0.470	0.750
$1/t \ln(1 - X_A) \times 10^2$	–	1.193	1.178	1.269	1.386
$X_S/t \times 10^2$	–	1.125	1.05	0.94	0.75

The angular coefficient is -0.56. Then, if $[H_2O_2]_0 = 0.02$:

$$K_M = 0.00354$$

The linear coefficient is 0.01802 and

$$V_{max} = 6.37 \times 10^{-4} \text{ mol/L min}$$

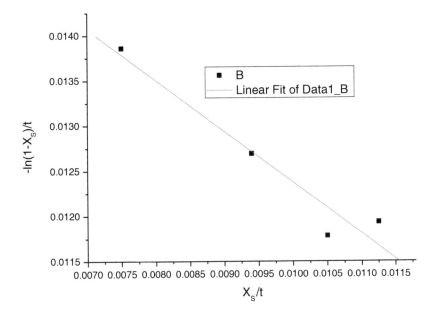

Figure 13.7 Plotting eq. 13.24, from data shown in the table above.

Tripling enzymes we obtain:

$$V_{\max} = k_3 E_T = 3 \times 6.37 \times 10^{-4} = 1.911 \times 10^{-3} \text{ mol/L min}$$

The time needed for a conversion of $X_S = 0.95$ is according eq.9.22:

$$t = 189 \text{ min}$$

Example

ER.13 The following reactions are represented by second-order reaction rates:

$$A \xrightarrow{k_1} 2R$$
$$A \xrightarrow{k_2} S$$

The reaction was performed in a PFR reactor at 500 K. Calculate the specific rate constants for a conversion of 83% of the reactant A. The molar flow rate is 0.181 mol/min and the volume of the reactor 1.0 L. Reactant A is fed with 50% inert gas and the total pressure is 10 atm. The selectivity was 70%. Neglect volume variation.

Solution

The reaction rates are represented as:

$$r_R = k_1 C_A^2$$

$$r_S = k_2 C_A^2$$

Relative to the reactant, it follows:

$$(-r_{A1}) = \frac{r_R}{2}$$

$$(-r_{A2}) = r_S$$

$$(-r_A) = \frac{r_R}{2} + r_S = \left(\frac{k_1}{2} + k_2\right) C_A^2 = k^* C_A^2$$

$$k^* = \left(\frac{k_1}{2} + k_2\right) \tag{8.25}$$

Substituting these rates in the PFR equation, we obtain the following solution (Equation 6.8):

$$\frac{X_A}{(1 - X_A)} = k^* C_{A0} \tau \tag{13.26}$$

The total feed molar flow rate is: $F_0 = 0.0181$ mol/min
Then:

$$F_{A0} = 0.5 F_0 = 0.090 \text{ mol/min}$$

Initial concentration:

$$C_{A0} = \frac{y_{A0} P}{RT} = 9.06 \times 10^{-2} \text{ mol/L}$$

Volumetric flow rate:

$$v_0 = \frac{F_{A0}}{C_{A0}} = 0.993 \text{ L/min}$$

Space time:

$$\tau = \frac{V}{v_0}$$

$$\tau = 1.0$$

For 0.53 conversion, we obtain from eq. 13.26:

$$k^* C_{A0} \tau = 1.127$$

Neglecting volume variation, we calculate the space time as 1.0 min.
  Thus:

$$k^* = 12.4 \, \text{L/mol min}$$

As defined, the selectivity is:

$$S = \frac{r_R}{-r_A} = \frac{k_1}{k^*} = 0.70$$

Thus:

$$k_1 = 0.7k^* = 8.71 \, \text{L/mol min}$$

and from eq. 13.25

$$k_2 = 8.04 \, \text{L/mol min}$$

Example

ER.14 The following reactions occur during the formation of ethyl bromide:

$$C_2H_5Br \xrightarrow{k_1} C_2H_5^* + Br^*$$
$$Br^* + C_2H_5Br \xrightarrow{k_2} HBr + C_2H_4Br^*$$
$$C_2H_4Br^* \xrightarrow{k_3} C_2H_4 + Br^*$$
$$Br^* + C_2H_4Br^* \xrightarrow{k_4} C_2H_4Br_2$$

The activation energies for each step are $E_1 = 20\,\text{kcal/mol}$, $E_2 = 15\,\text{kcal/mol}$, $E_3 = 20\,\text{kcal/mol}$, and $E_4 = 10\,\text{kcal/mol}$, respectively. Determine the decomposition rate of C_2H_5Br and the apparent activation energy.

Solution
Rename

$$M = C_2H_5Br, \ R_1^\bullet = C_2H_5^*, \ R_2^\bullet = Br^*, \ R_3^\bullet = C_2H_4Br^*$$

$$M \xrightarrow{k_1} R_1 + R_2$$
$$R_1 + M \xrightarrow{k_2} HBr + R_3$$
$$R_3 \xrightarrow{k_3} C_2H_4 + R_1$$
$$R_1 + R_3 \xrightarrow{k_4} P_2$$

Thus, the reaction rates are as follows:

$$(-r_M) = k_1[M] + k_2[R_1][M] \approx k_2[R_1][M] \tag{13.27}$$

$$r_{\bar{R}_1} = k_1[M] - k_2[R_1][M] + k_3[R_3] - k_4[R_1][R_3] = 0 \tag{13.28}$$

$$r_{R_3} = -k_3[R_3] + k_2[R_1][M] - k_4[[R_1][R_3] = 0 \tag{13.29}$$

Add Equations 13.28 and 13.29, we obtain:

$$R_3 = \frac{k_1[M]}{2k_4[R_1]} \tag{13.30}$$

Substitute Equation 13.30 in Equation 13.28:

$$[R_1]^2 = \frac{k_1 k_3}{2k_2 k_4} \tag{13.31}$$

Substitute Equation 13.31 in Equation 13.27:

$$(-r_M) = k_2[R_1][M] = \sqrt{\frac{k_1 k_2 k_3}{2k_4}}[C_2H_5Br]$$

Thus, the decomposition rate is a first-order reaction.
Considering that:

$$k_a = k_0 e^{-(Ea/RT)}$$

or

$$k_i = k_0 e^{-(Ei/RT)}$$

Using logarithm and equaling, we obtain:

$$E_a = \frac{E_1 + E_2 + E_3 - E_4}{2} = 27.5\,\text{kcal/mol}$$

$$E_a = 27.5\,\text{kcal/mol}$$

Example

ER.15 The hydrodenitrogenation reaction was carried out in a batch reactor at 100°C, containing 20 g/L CoMo/Al$_2$O$_3$ catalyst. Samples were collected and conversion is calculated with time as shown in the table below:

t (min)	0	5	10	20	30	40	50
X$_A$	0	0.30	0.50	0.73	0.85	0.93	0.97

Determine the reaction order and calculate the specific rate constant.

If the observed reaction rate at 100°C is four times the rate at 80°C and the catalyst concentration is doubled, calculate the reaction rate at 90°C. Calculate the conversion after 10 min.

Solution
Assuming a first-order reaction rate:

$$(-r_A) = kC_A = kC_{A0}(1 - X_A)$$

Then:

$$t = -\frac{1}{k}\ln(1 - X_A)$$

Thus, from the date in table 13.9 we calculate:

Table 13.9

t (min)	X_A	k
0	0	–
5	0.3	0.0713
10	0.5	0.069
20	0.73	0.065
30	0.85	0.063
40	0.93	0.066
50	0.97	0.070
	Mean value	0.067

The mean specific constant value k is:

$$k = 0.0676 \, \text{min}^{-1}$$

However:

$$\frac{r_{100}}{r_{80}} = 4$$

Then:

$$\frac{k_{100}}{k_{80}} = 4$$

Thus:

$$k_{80} = \frac{0.0067}{4} = 0.0167 \, \text{min}^{-1}$$

From these values, we calculate the activation energy:

$$E = R\frac{\ln(k_{100}/k_{80})}{(1/T_{80}) - (1/T_{100})} = 1.80 \times 10^4 \, \text{cal/mol}$$

Calculate k_0:

$$k_0 = ke^{(E/RT)} = 2.83 \times 10^9$$

Therefore:

$$k = 2.83 \times 10^9 e^{-(1.8 \times 10^4/RT)}$$

The specific rate constant at 90°C (363 K) is:

$$k_{90} = 0.0341 \, \text{min}^{-1}$$

How the catalyst concentration affects the rate?

$$\frac{(-r_A)_1}{(-r_A)_2} = \frac{k_1 C^1_{\text{cat}}}{k_2 C^2_{\text{cat}}} = \frac{k_{100}}{k_{90} \times 2} = \frac{0.067}{2 \times 0.034} = 1$$

In fact, the rate was not affected.
The conversion after 10 min at 90°C is:

$$(1 - X_A) = e^{-kt} = e^{-0.034 \times 10} = 0.711$$

$$X_A = 0.288$$

Example

ER.16 The reaction $A + B \rightarrow 2R$ was studied in a differential reactor containing 40 g catalyst. A mixture of A and B is introduced in the reactor and the following data (Table 13.10) were obtained.

Table 13.10

Feed Molar Flow Rate (F_0) (mol/h)	Feed Partial Pressures (atm)		Exit Molar Rate, R
	P_{AO}	P_{BO}	
1.7	0.5	0.5	0.05
1.2	0.5	0.5	0.07
0.6	0.5	0.5	0.16
0.3	0.4	0.6	0.16
0.75	0.6	0.6	0.10
2.75	0.6	0.4	0.06

Assuming the general kinetic rate equation $(-r_a) = kP_A^\alpha P_B^\beta$ determine the parameters α, β, and k.

Solution

$$A + B \rightarrow 2R$$

$$\alpha = \frac{F_{A0} - F}{F_{A0}} = \frac{F_R - F_R}{2F_{A0}}$$

$$X_A = \frac{F_R}{2F_{A0}}$$

P_{A0}	P_{B0}	X_A
0.5	0.5	0.05
0.5	0.5	0.07
0.5	0.5	0.16
0.4	0.6	0.2
0.6	0.6	0.1
0.6	0.4	0.05

From the conversions, we calculate the partial pressures P_A and P_B:

$$P_A = P_{A0}(1 - X_A)$$

$$P_B = P_{B0}(M - X_A), \text{ where } M = P_{B0}/P_{A0}$$

M	X_A	P_A	P_B
1	0.05	0.475	0.475
1	0.07	0.465	0.465
1	0.16	0.420	0.420
1.5	0.2	0.320	0.520
1	0.1	0.540	0.540
0.66	0.05	0.570	0. 370

From the relation:

$$(-r_A) = (-r_B) = (1/2)r_R$$

It follows:

$$(-r_A) = F_{A0}X_A/W = F_R/2W = Y_R F_0/2W, \text{ where } W = 0.040 \text{ kg}$$

F_0	Y_R	$(-r_A)$
1.7	0.05	1.062
1.2	0.07	1.050
0.6	0.16	1.200
0.3	0.16	0.600
0.75	0.10	0.937
2.75	0.06	2.062

Using the least square method and the equation:

$$(-r_A) = k P_A^{\alpha'} P_B^{\beta'}$$

$$\ln(-r_A) = \ln k + a' \ln P_A + b' \ln P_B$$

From Equation 5.38 for N experiments and for j tests, we obtain:

$$Y_j = A_0 + A_1 X_{1j} + A_2 X_{2j}$$

With simultaneous equations, it follows:

$$\sum_{I=1}^{N} Y_j = N A_0 + A_1 \sum_{I=1}^{N} X_{1j} + A_2 \sum_{I=1}^{N} X_{2j}$$

$$\sum_{I=1}^{N} X_{1j} Y_J = A_0 \sum_{I=1}^{N} X_{1j} + A_1 \sum_{I=1}^{N} X_{1j}^2 + A_2 \sum_{I=1}^{N} X_{2j} X_{1j}$$

$$\sum_{I=1}^{N} X_{2j} Y_J = A_0 \sum_{I=1}^{N} X_{2j} + A_1 \sum_{I=1}^{N} X_{2j} X_{1j} + A_2 \sum_{I=1}^{N} X_{2j}^2$$

$(-r_A)$	P_A	P_B	Y_j	X_{1j}	X_{1j2}	X_{2j}	X_{2j2}	$X_{1j} * X_{2j}$
1.0625	0.475	0.475	0.6006	−0.7444	0.5541	−0.7444	0.5541	0.5541
1.05	0.465	0.465	0.0488	−0.7657	0.5863	−0.7657	0.5863	0.5863
1.2	0.420	0.420	0.1823	−0.8675	0.7525	−0.8675	0.7275	0.7275
0.6	0.320	0.520	−0.5108	−1.1394	1.2982	−0.6539	0.4276	0.7450
0.9375	0.540	0.540	−0.0645	−0.6162	0.3797	−0.6162	0.3797	0.3797
2.0625	0.570	0.370	−0.7239	−0.5621	0.3160	−0.9942	0.9884	0.5580
Σ			0.4403	−4.6953	3.8868	4.6419	3.6886	3.5764

$X_{1j} * Y_j$	$X_{2j} Y_j$
−0.0451	−0.0451
−0.0374	−0.0374
−0.1581	−0.1581
−0.5820	−0.3340
−0.0397	−0.0397
−0.4069	−0.7197
−0.0258	−0.5866

Substituting these values, we obtain following equations:

$$0.4403 = 6 A_0 - 4.6953 A_1 - 4.6419 A_2$$

$$-0.0258 = -4.6953 + 3.8868 A_1 + 3.5764 A_2$$

$$-0.5866 = -4.6419 A_0 + 3.5764 A_1 + 3.6886 A_2$$

Solving these equations, it follows:

$A_0 = -0.7423$

$A_1 = 0.9639$

$A_2 = -2.035$

Then:

$K = 0.476$ mol atm/kg h

$a' \approx 1.0$

$b' \approx -2.0$

Example

ER.17 Ethanol dehydrogenation was carried out in the integral reactor at 275°C and following experimental data (Table 13.11) were obtained:

X		0.118	0.196	0.262	0.339	0.446	0.454	0.524	0.59	0.60
P (atm)		1	1	1	1	1	1	1	1	1
W/F_{A0} (kg$_{cat}$ h/kmol)	0.2	0.4	0.6	0.88	1.53	1.6	2.66	4.22	4.54

X		0.14	0.2	0.25	0.28	0.35	0.14	0.19	0.23	0.27	0.32	0.11	0.16
P (atm)		3	3	3	3	3	4	4	4	4	4	7	7
W/F_{A0} (kg$_{cat}$ h/kmol)	0.2	0.4	0.6	0.88	1.6	0.2	0.4	0.6	0.88	1.6	0.2	0.4

X		0.194	0.214	0.254	0.1	0.148	0.175	0.188	0.229
P (atm)		7	7	7	10	10	10	10	10
W/F_{A0} (kg$_{cat}$ h/kmol)	0.6	0.88	1.6	0.2	0.4	0.6	0.88	1.6

The equilibrium constant is $K = 0.589$. The feed is an azeotropic mixture of ethanol and water containing 13.5% (molar) of water. Estimate the adsorption–desorption and reaction parameters, using the conversion as variable regression. Which model fits best the experimental data?

Solution
We may suggest different models as listed below.
Model 1: reversible reaction and adsorbed product

$$r = k\theta_A P_B - k'\theta_R$$

$$(-r_A) = k \left[\frac{kK_A p_A p_B}{(1 + K_A \cdot p_A + K_R \cdot p_R)} - \frac{1}{K} \frac{kK_R p_R}{(1 + K_A \cdot p_A + K_R \cdot p_R)} \right]$$

Assuming a differential reactor and rearranging:

$$\left[\frac{K_R}{kK_A} + \frac{K_R}{kK_A} - \frac{K}{k} \right] X_A + \frac{K}{k} \left(1 + \frac{1}{K_A} \right) = \frac{K}{r} (1 - X_A) P_B$$

Model 2: irreversible reaction and adsorbed product

$$r = k\theta_A P_R$$

Assuming a differential reactor and rearranging:

$$r = \frac{kK_A p_A p_R}{(1 + K_A \cdot p_A + K_R \cdot p_R)}$$

Model 3: irreversible reaction and nonadsorbed product

$$r = k\theta_A P_B$$

Assuming a differential reactor and rearranging:

$$\frac{1}{kK_A} + \frac{1}{k} p_{A0}(1 - X_A) = \frac{p_{A0} p_B}{r} (1 - X_A)$$

From experimental data and the corresponding models, we obtain X_A after linearization.

Model 1: Varying pressures for constant flow rates

P	W	0.2	y
	P_b	X	
1	0.135	0.118	0.0140
3	0.405	0.14	0.0410
4	0.54	0.14	0.0547
7	0.945	0.112	0.098
10	1.35	0.1	0.143

P	W	0.4	y
	P_b	X	
1	0.135	0.196	0.0255
3	0.405	0.2	0.0763
4	0.54	0.196	0.1022
7	0.945	0.163	0.1863
10	1.35	0.148	0.2709

Varying flow rates for constant pressures.

P	W	0.6	y
	P_b	X	
I	0.135	0.262	0.0352
3	0.405	0.25	0.1073
4	0.54	0.235	0.1459
7	0.945	0.194	0.2691
10	1.35	0.175	0.3935

P	W	0.8	y
	P_b	X	
I	0.135	0.399	0.0462
3	0.405	0.286	0.1498
4	0.54	0.271	0.2040
7	0.945	0.214	0.3849
10	1.35	0.188	0.5681

P	W	1.6	y
	P_b	X	
I	0.135	0.454	0.0694
3	0.405	0.352	0.2473
4	0.54	0.32	0.3460
7	0.945	0.254	0.6643
10	1.35	0.229	0.980

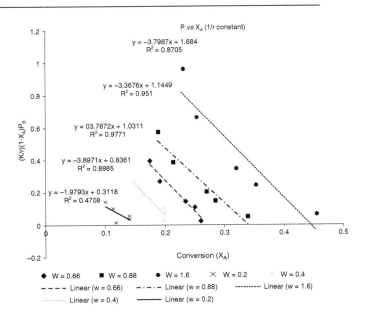

Figure E13.1 Model 1.

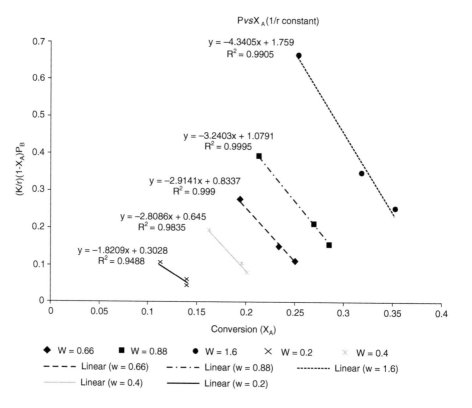

Figure E13.2 Model 2.

w	P_a	P_b	P		y
			x	$(1-x)$	
0.2	0.865	0.135	0.118	0.882	0.0205
0.4	0.865	0.135	0.196	0.804	0.0375
0.6	0.865	0.135	0.262	0.738	0.0517
0.88	0.865	0.135	0.339	0.661	0.0679
1.6	0.865	0.135	0.454	0.546	0.1020

w	P_a	P_b	P	3	y
			x	$(1-x)$	
0.2	2.595	0.405	0.14	0.86	0.180
0.4	2.595	0.405	0.2	0.8	0.336
0.6	2.595	0.405	0.25	0.75	0.472
0.88	2.595	0.405	0.286	0.714	0.660
1.6	2.595	0.405	0.352	0.648	1.089

w	P_a	P_b	$\dfrac{P}{x}$	$\dfrac{4}{(1-x)}$	y
0.2	3.46	0.54	0.14	0.86	0.321
0.4	3.46	0.54	0.196	0.804	0.600
0.6	3.46	0.54	0.235	0.765	0.857
0.88	3.46	0.54	0.271	0.729	1.198
1.6	3.46	0.54	0.32	0.68	2.032

w	P_a	P_b	$\dfrac{P}{x}$	$\dfrac{7}{(1-x)}$	y
0.2	6.055	0.945	0.112	0.888	1.016
0.4	6.055	0.945	0.163	0.837	1.915
0.6	6.055	0.945	0.194	0.806	2.767
0.88	6.055	0.945	0.214	0.786	3.957
1.6	6.055	0.945	0.254	0.746	6.829

w	P_a	P_b	$\dfrac{P}{x}$	$\dfrac{10}{(1-x)}$	y
0.2	8.65	1.35	0.1	0.9	2.101
0.4	8.65	1.35	0.148	0.852	3.979
0.6	8.65	1.35	0.175	0.825	5.780
0.88	8.65	1.35	0.188	0.812	8.344
1.6	8.65	1.35	0.229	0.771	14.405

From Figures E13.1 and E13.2, we obtain the linear and angular coefficients, Model 2.

With these coefficients, we calculated the corresponding constants. Note that the angular coefficients are similar for both Models 1 and 2.

Example

ER.18 The following reaction mechanism denotes different bondings of enzyme E with the substrate S. Two different complexes were formed: a binary reactive complex ES and a nonreactive ternary complex ESS.

$$S + E \xrightleftharpoons{k_2/k_1} ES$$

$$ES + S \xrightarrow{k_3} K_2$$

$$ES \xrightleftharpoons{k_4/k_5} E + P$$

(a)   Describe the reaction rate of the product $P(r_p)$ based on this mechanism?
(b)   Show how inhibition occurs in the product rate equation P when compared to the reaction rate based on the Michaelis-Menten mechanism for a single site.
(c)   Compare the maximum rates for both models and for what S value it occurs $(V_{max} = k_r C_{[E0]})$. Plot r_p as function of $[S]$.

$$S_1 + E \xrightleftharpoons{k_2/k_1} ES_1$$

$$S_2 + E \xrightleftharpoons{k_2/k_1} ES_2$$

$$ES \xrightarrow{k_3} E + P_1$$

$$ES \xrightarrow{k_4} E + P_2$$

Solution

(a)   Rate equations for ES and ESS

$$r_{ES} = k_1[E][S] - k_1'[ES][S] + k_2'[ESS] - k_t[ES] \tag{13.32}$$

$$r_{ESS} = k_2[ES][S] + k_2'[ESS] \tag{13.33}$$

Assuming steady state for $[ES]$ and $[ESS]$, we have:

$$dr_{ES}/dt = 0$$
$$dr_{ES}/dt = 0$$

The total enzyme concentration is:

$$[E_0] = [E] + [ES] + [ESS] \tag{13.34}$$

From Equation (13.32):

$$[ES^*] = \frac{k_1'[ES] + k_2[ES][S] - k_2'[ESS] + k_t[S]}{k_1[S]}$$

From Equation (13.33):

$$k_2'[ESS] = k_2[ES][S]$$

From Equation (13.34):

$$[ESS] = \frac{k_2}{k_2'}[ES][S]$$

Substitution of [ESS] results:

$$[ES] = \frac{[E_0]}{\dfrac{k'_1 + k_t}{k_1[S]} + 1 + \dfrac{[S]}{K_2}}$$

where: $k_m = \dfrac{k'_1 + k_t}{k_1}$

since $k_p = k_r[ES]$ we have:

$$k_p = \frac{k_t[E_0][S]}{\dfrac{k_m}{[S]} + [S] + \dfrac{[S]^2}{K_2}} \tag{13.35}$$

where: $k_r[E_0] = V_{max}$

$$k_p = \frac{V_{max}[S]}{k_m + [S] + \dfrac{[S]^2}{K_2}} \tag{13.36}$$

(b) The nonreactive complex formation promotes inhibition because the amount of enzymes decreases with increasing product formation. Comparing Equation (13.30) with the Michaelis-Menten equation, it shows that the term $[S]^2/K_2$ is responsible for the decreasing rate formation of the product. The inhibition depends on k_m, $[S]$, and K_2. (K_2 decreases with more [ESS] formation).

(c) Since:

$$r_p = \frac{V_{max}[S]}{k_m + [S] + \dfrac{[S]^2}{K_2}} \tag{13.37}$$

The rate is maxima when: $\dfrac{dr_p}{d[S]} = 0$

Thus:

$$r_{pmax} = \frac{k_t[E_0]\sqrt{k_m K_2}}{k_m + \sqrt{k_m K_2} + \dfrac{k_m K_2}{K_2}} \tag{13.38}$$

when $V_{max} = k_r[E_0]$

Therefore:

$$r_{pmax} = \frac{V_{max}\sqrt{k_m K_2}}{k_m + \sqrt{k_m K_2} + \dfrac{k_m K_2}{K_2}}$$

V_{max} is smaller when compared to Michaelis-Menten mechanism.
For the Michaelis-Menten it is: $V_{máx}k_r[E_0]$

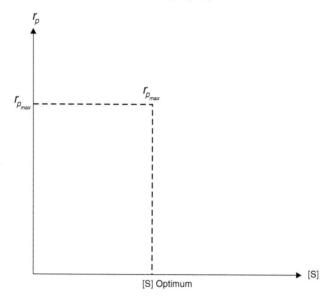

Example

ER.19 Determine the polymerization grade using the rate equation based on free radicals.

The mean polymerization grade X_N is determined by the following expression:

$$X_N = \frac{-r_M}{\sum r_j}$$

Plot the concentration of the monomer with time or space time for the batch, CSTR and PFR systems, separately.

Using 3 CSTR in series of equal volumes, calculate the final volume for a final conversion of 90%.

Plot the concentration of the monomer with space time in the PFR and 90% conversion.

Varying the parameters (k_0, k_p) discuss the results:

Data:

$M_0 = 3 \, mol/dm^3$

$I_{20} = 0.01 \, mol/dm^3$

$k_0 = 10^{-3} \, s^{-1}$

$k_p = 10 \, dm^3/mol \, s$

$k_t = 5 \times 10^7 \, dm^3/mol \, s$

- *Rate of monomer disappearance $(-r_M)$*

  The rate of disappearance of the monomer occurs in the initiation and propagation steps.

  From Equation 8.38, we get:

$$(-r_M) = k_p[M] \sqrt{\frac{2\gamma k_0[I_2]}{k_t}}$$

- *Calculation of the polymerization grade X_N*

  From the definition:

$$X_N = \frac{-r_M}{\sum r_j} = \frac{k_p M \cdot R^*}{2\gamma k_0[I_2] + k_t (R^*)^2}$$

where: $[M] = M$.

When the concentration of the initiator does not vary with time ($[I_2] = [I_2]_0$), then, Equation 8.36 becomes:

$$R^* = \sqrt{\frac{2\gamma k_0[I_2]}{k_t}} = 4.47 \times 10^{-7}\ \text{mol/dm}^3$$

Since $M = M_0(1 - X_M) = 3(1 - X_M)$, the expression of X_N results:

$$X_N = \frac{k_p M \cdot R^*}{2\gamma k_0[I_2] + k_t(R^*)^2}$$

$$X_N = 1.34 \times (1 - X_M)$$

(b) *Concentration of the monomer with time:*

The rate equation of consumption of the monomer or formation of polymer is given as:

$$-r_M = -r_p = k_p M R^*$$

For batch reactor (Equation 4.7):

$$t = M_0 \int_0^{X_M} \frac{dX_M}{(-r_M)}$$

Changing the integration interval:

$$X_M = \frac{M_0 - M}{M_0} \Rightarrow dX_M = -\frac{dM}{M_0}$$

Then:

$$t = \int_{M_0}^{M} \frac{dX_M}{(-r_M)}$$

Substituting $(-r_M)$:

$$t = -\frac{1}{k_p R^*} \ln(M - M_0)$$

Thus, the concentration of the monomer with time in a batch reactor is:

$$M = M_0 \times \exp(-t k_p R^*) \qquad (13.39)$$

- For PFR:

$$\tau = \int_{M_0}^{M} \frac{dX_M}{(-r_M)}$$

Changing the integration interval and substituting expression of $(-r_M)$, after integration, we obtain the concentration of M as function of space time τ

$$M = M_0 \times \exp(-\tau k_p R^*) \qquad (13.40)$$

- For CSTR:

$$\tau = \frac{dX_M}{(-r_M)}$$

Similarly:

$$X_M = \frac{M_0 - M}{M_0}$$

Substituting $(-r_M)$ results:

$$\tau = \frac{M_0 - M}{k_p M R^*}$$

Rearranging, the concentration of the monomer with the space time is:

$$M = \frac{M_0}{\tau k_p R^* + 1} \qquad (13.41)$$

Figures E13.1 and E13.2 display the concentration profiles of batch, PFR, and CSTR systems, respectively.

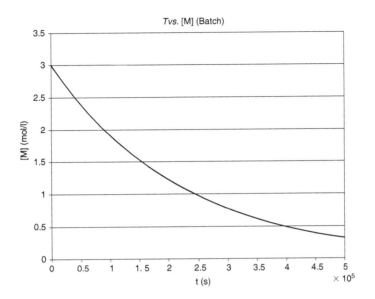

Figure E19.1 Batch.

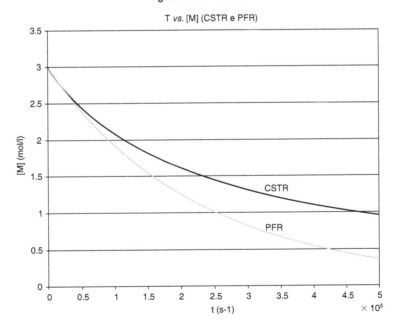

Figure E19.2 PFR and CSTR.

For comparison, we plotted the profiles in the same figure.

As seen, the concentration of the monomer decays more significantly on the PFR reactor.

(c) *Calculation of volumes for a final conversion of 90% on 3 CSTRs.*
Since:

$$\tau_i = \frac{X_{Ai} - X_{Ai-1}}{-r_m}$$

Equating τ_i:

$$\tau_i = \frac{X_{Ai} - X_{Ai-1}}{k_p[M]\sqrt{\dfrac{2\gamma k_0[I_2]}{k_t}}}$$

For each reactor:

$$V_i = v_0 \times \frac{X_{Ai} - X_{Ai-1}}{k_p[1 - X_{Ai}]\sqrt{\dfrac{2\gamma k_0[I_2]}{k_t}}}$$

Equating the volumes $V_1 = V_2$ and $V_1 = V_3$, we obtain:

$$\frac{X_{M1}}{(1 - X_{M1})} = \frac{X_{M2} - X_{M1}}{(1 - X_{M2})}$$

The final conversion is 90% ($X_{M3} = 0.9$). Solving these systems, we obtain X_{M1} and X_{M2}:

$$X_{M1} = 0.535$$
$$X_{M2} = 0.784$$
$$X_{M3} = 0.9$$

Figure E19.3 displays the volume of the reactor (since, $V_1 = V_2 = V_3$) as function of v_0.

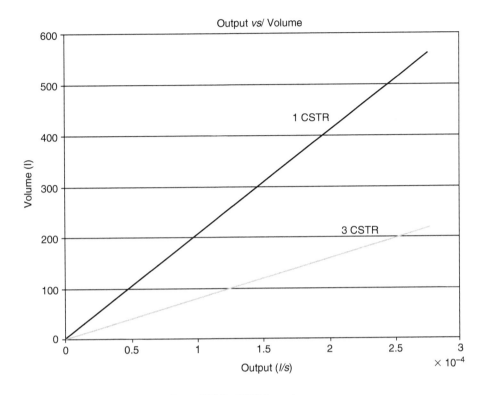

Figure E19.3   CSTR in series v_0.

From this figure, we conclude that the volume in one CSTR reactor (blue line) is greater than in three reactors in series.

(d) *Varying parameters k_0 and k_p.*
   Using the equation of a PFR reactor:

$$\tau = \int_{M_0}^{M} \frac{dX_M}{(-r_M)}$$

Substituting $(-r_M)$ and after integration results:

$$M = M_0 \times \exp(-\tau k_p R^*)$$

where:

$$R^* = \sqrt{\frac{2\gamma k_0 [I_2]}{k_t}}$$

Varying k_0 and k_p, we obtain:

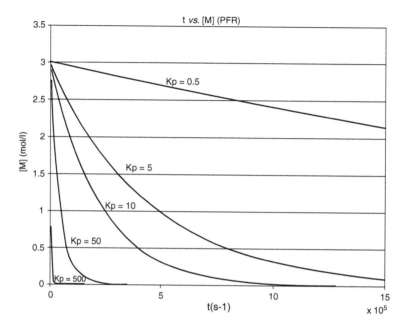

Figure E19.4(a).

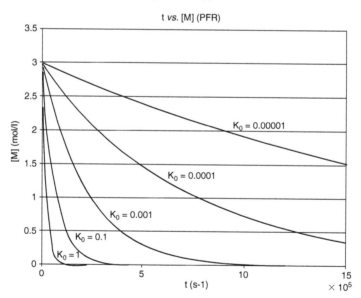

Figure E19.4(b) Influence of parameters K_0 and K_p.

We see that the influence of K_p is more significant than of K_0. Therefore, the parameter controlling the polymerization kinetics is the propagation constant. The concentration is strongly affected for higher K_p values.

Example

ER.20 The polymerization reaction of high carbon chains occurs in the liquid phase varying with space time or very high mean residence times.

(a) In this case, the propagation step is greater than the initiation step:

$$\frac{r_p}{r_i} > 1$$

Or (LCA)

Calculate the concentration profiles of the monomer M in PFR and batch reactor separately.

Solution

Substitution of Equations 8.38 (or 8.39) and 8.30 in the relationship:

$$\frac{r_p}{r_i} > 1$$

gives:

$$\frac{r_p}{r_i} > \frac{M}{\sqrt{I_2}} \sqrt{\frac{k_p^2}{2k_0 k_t}}$$

PFR and batch

$$\frac{dM}{dt} = r_M \tag{8.42}$$

Substituting Equation 8.38 in Equation 4.7:

$$\tau = t = \int_{M_0}^{M} \frac{dX_M}{(-r_M)}$$

Batch and PFR

Equation 8.43

$$\ln \frac{M}{M_0} = k_p \sqrt{\frac{8\gamma[I_{20}]}{k_t k_0}} [e^{-\left(\frac{k_0 t}{2}\right)} - 1]$$

or

$$M = M_0 \times \exp\left(-\tau k_p \sqrt{\frac{2\gamma[I_{20}]}{k_t}} [e^{-\left(\frac{k_0 t}{2}\right)} - 1]\right)$$

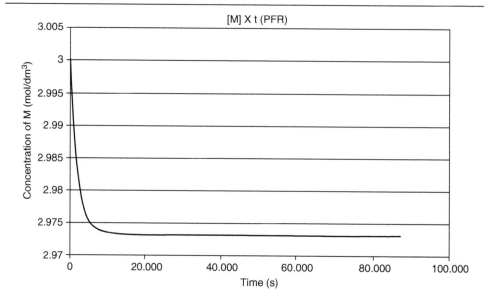

PFR or batch reactors

X_M	0.0089
Polymerization grade	1.0089

X_M—Conversion of the monomer

$$\frac{(-r_A)_1}{(-r_A)_2} = \frac{k_1 C_{cat}^1}{k_2 C_{cat}^2} = \frac{k_{100}}{k_{90} \times 2} = \frac{0.067}{2 \times 0.034} = 1$$

$$(1 - X_A) = e^{-kt} = e^{-0.034 \times 10} = 0.711$$

$$X_A = 0.288$$

13.2 PROPOSED EXERCISES

Example

EP.1 Calculate the equilibrium concentration of the reforming reaction and following feed composition (molar fraction): 0.15 N_2; 0.60 H_2O; 0.25 C_2H_4. The temperature is constant at 527 K at pressure of 264.2 atm. Assume principal reaction.

$$H_2O + C_2H_4 \longleftrightarrow C_2H_5OH$$

Data:

Component	ΔG^0_{298K} (kcal)	ΔH^0_{298K} (kcal)
H_2O	−54.63	−57.79
C_2H_4	16.28	12.49
C_2H_5OH	−40.3	−56.24

Example

EP.2 The reaction occurs in a batch reactor and determines the specific rate constant from the following experimental data.

t (s)	0	27	45	62	79	93	105	120	∞
$\Phi(cm^{-1})$	–	0.352	0.646	0.813	0.969	1.12	1.26	1.40	3.5

Assume a second-order reaction under isothermal conditions at $25°C$. The initial concentration is 3.6×10^{-3} mol/m³.

Example

EP.3 The decomposition of DME (dimethylether) occurs at $504°C$, according to the reaction:

$$(CH_3)_2O \rightarrow CH_4 + H_2 + CO$$

Reactant is fed with 30% inert (N_2). The kinetics is unknown and therefore following experiments were performed in a batch reactor at constant temperature.

t (s)	390	777	1195	3155	∞
P (mmHg)	408	488	562	799	931

Verify if the reaction rate is first or second order?

If the experiment would occur in a continuous system at constant pressure (initial pressure), calculate the space time for a conversion reached before at the end reaction.

If the volume is 5 L, calculate the flow rate needed for the continuous reactor for the same conversion.

Example

EP.4 The specific rate constants at 273 K is 0.001 min^{-1} and at 373 K equal to 0.05 min^{-1}.

(a)   Calculate the activation energy.
(b)   Calculate the collision factor.
(c)   Calculate the rate and compare with item b.
(d)   Calculate the specific rate constant at 700 K. Comment results.

Example

EP.5 It is known that for an increase of 10°C of temperature the rate is doubled. Assuming constant activation energy and constant k_0, show the following relationship.

$$T = \left[\left(\frac{10E}{R \ln 2}\right)^{1/2}\right]$$

Show how the activation energy changes with enthalpy, according to the equation.

$$E = C - \alpha\left(-\Delta H^0\right)$$

where C and α are constants. What is the meaning?

Example

EP.6 Interpret the following graphs and explain the mass transfer effects. Show how the activation energy varies with temperature.

Example

EP.7 The reaction:

$$CH_3COOC_6H_5 + NA^+OC_6H_5 \rightarrow CH_3COO^-Na + 2C_6H_5OH$$

is performed in a batch reactor and in the liquid phase. Feed concentrations are equimolar and equal to 30 mol/m³. During heating up, reaction may occur and when counting begins, the concentration of reactant was 26.9 mol/m³ and of the product phenol was 7.42 mol/m³. Assume irreversible reaction and following rate expression:

$$r = k\frac{C_A C_B}{C_R}$$

where C_A = concentration of ester $CH_3COOC_6H_5$, C_B = concentration of phenolate $Na^+OC_6H_5$, and C_R = concentration of phenol C_6H_5OH.

Following experimental data were obtained by collecting samples with different times:

t (kseg)	0.72	2.16	4.32	8.64	20.6	27.2
C_A (mol/m³)	22	18.2	15.3	12.15	8.5	7.3

Show that the rate of the equation is correct and determine the kinetic specific constants. The temperature is 30°C and the total pressure is 1 atm.

Example

EP.8 The enzymatic reaction is competitive and the reaction mechanism is as follows:

$$E + S \xleftrightarrow{k_1} ES$$
$$E + I \xleftrightarrow{k_2} EI$$
$$ES \xrightarrow{k_3} E + P$$

where E = Enzyme, I = Initiator, and ES, EI = Complexes.

Show that the rate of product P formation is given according to the equation:

$$r_P = \frac{V_{max}[S]}{[S] + K_M\left(1 + \frac{[I]}{K_I k_2}\right)}$$

where K_M = Michaelis constant and K_I = Inhibitor (k_4/k_3) constant.
Do all simplifications needed and justify.

Example

EP.9 The reaction $(CH_3)_3COOC(CH_3)_3 \xrightarrow{k} C_2H_6 + 2CH_3COCH_3$ was performed in a batch reactor, measuring total pressures with time at $50°C$.

t (min)	0	2.5	5	10	15	20
P (mmHg)	7.5	10.5	12.5	15.8	17.0	19.4

Write the rate equation as function of total pressure for n-reaction order.
Show how P varies with time and determine the reaction order.
Choose an approximate integral order and calculate the constant by the integral method.

Example

EP.10 The parallel reactions are irreversible and the $A \xrightarrow{k_1} 2R$ and $A \xrightarrow{k_2} S$ product rates are of second order with respect to reactants. Reaction was performed in a PFR reactor. Feed composition was 50% for A and 50% for Inert. The molecular weights are 40 and 20, respectively. The exit flow rate of product R is 6 mol/h. The total flow rate is 1000 mL/min at 10 atm and constant temperature $400°C$. Assume ideal gases. The reactor volume is 2 L and the selectivity relative to R is 85%. Calculate constants k_1 and k_2.

Example

EP.13 The consecutive reaction is $A \xrightarrow{k_1} R \xrightarrow{k_2} S$ performed in a batch reactor. The reaction rates are first order and irreversible. The ratio at $50°C$ is known as:
At $70°C$, this ratio is 1.5. Calculate the maximum concentration of R, when the time is 30 min at $50°C$. Calculate $\frac{k_2}{k_1} = 3$ the activation energy and the selectivity of R.

Example

EP.12 The hydrodenitrogenation is performed in a PFR reactor at $100°C$ and $60°$atm. The following results were obtained in a batch reactor:

τ (min)	0	5	10	20	30	40	50
X_A	0	0.3	0.5	0.73	0.85	0.93	0.97

Calculate the volume for a conversion of 80% and feed rate of 10 L/min.

Example

EP.13 The propane oxidation to acroleine was done in tubular reactor according to the reaction:

$$CH_3CH=CH_2+O_2 \xrightarrow{k_1} CH_2=CHCHO+H_2O$$

The fixed bed reactor contains 0.5 g of a catalyst and operates at 623 K and constant pressure. The following data were measured:

F_R (min)	0.21	0.48	0.09	0.39	0.6	0.14	1.44
p_P (atm)	0.1	0.2	0.05	0.3	0.4	0.05	0.5
P_0 (atm)	0.1	0.2	0.05	0.01	0.02	0.4	0.5

Determine the reaction orders assuming an irreversible reaction.

Example

EP.14 The reaction is irreversible and the specific rate constant is equal to:

$$C_2H_2Cl_4 \xrightarrow{k} C_2H_2Cl_3+HCl$$

$$k=10^{12}e^{-21940/T}(s^{-1})$$

During the reaction, other subproducts were formed. Cl_2 may poison the catalyst when their concentration is higher than 150 ppm. The Cl_2/HCl ratios were observed with different temperatures, as shown in the following table.

T (°C)	408	440	455
Cl_2/HCl	1.7×10^{-4}	3.2×10^{-4}	4.0×10^{-4}

For the design of a tubular reactor of 0.15 m³ and temperature of 450°C and at 1 atm the molar flow rate is 41.7 mol/kseg. Find out for what conditions catalyst poisoning can be avoided?

Example

EP.15 The dimerization of phenyl isocyanate in the liquid phase is performed in a tubular reactor, according to the following reaction:
The reaction rate is second order and zero order for the catalyst concentration. The reverse reaction is first order. Data are presented below at 25°C:

$$2C_6H_5NCO \longrightarrow C_6H_5N \quad \text{(ring structure with two } C=O\text{)} \quad N-C_6H_5$$

Equilibrium constant: 0.178.
Specific rate constant $k = 1.15 \times 10^{-3}$ L/mol s.
Activation energy: $E = 1.12$ kcal/mol and $E' = 11.6$ kcal/mol.
Pure reactant is fed at 1 mol/L and 0°C. Calculate specific rate constants at 200°C for 80% conversion and exit flow of 0.6 mol/min.

Example

EP.16 The reaction is performed in a batch reactor at 25°C:

Samples were collected with time and measured by conductivity, according to the table:

t (s)	0	27	62	93	120	10000
$\Omega (cm^{-1})$	–	0.352	0.813	1.12	1.40	3.5

Equal concentrations were introduced in the reactor. The reaction is second order. Determine the specific rate constant?

Example

EP.17 The reaction ethanol to diethyl-ether was studied in a differential reactor and data are shown in following table. Using potential law and LH equations determine the reaction order with respect of each component and if possible calculate the kinetic and adsorption-desorption constants.

Experiments	P_A (atm)	P_{DE} (atm)	$P_{água}$ (atm)	Rate (mol/g_{cat} min)
1	1.0	0.0	0	1.34×10^{-3}
2	0.9	0.1	0	1.32×10^{-3}
3	0.7	0.3	0	1.35×10^{-3}
4	0.4	0.6	0	1.31×10^{-3}
5	1.0	0.2	0.8	0.35×10^{-3}
6	0.5	0.2	0.3	0.85×10^{-3}

LH—Langmint–Hinshelwood

Example

EP.18 The catalytic reaction is irreversible. With data presented in the table, determine the reaction mechanism and propose a LH rate expression. The conversion is lower than 5%.

Experiments	P_A (atm)	P_R (atm)	Rate (mol/g_{cat} min)
1	1.0	1	3.3×10^{-5}
2	5	1	1.5×10^{-5}
3	10	1	0.56×10^{-5}
4	1	10	0.86×10^{-5}
5	2	5	1.82×10^{-5}
6	2	10	1.49×10^{-5}

Part IV

Reactors

Chapter 14

Ideal reactors

INTRODUCTION

In the first part of this book, we have studied the kinetics of homogeneous and heterogeneous chemical reactions as well as the influence of parameters on the reaction rates not taking into account diffusion and mass effects. These phenomena are caused by heat and mass transfer limitations. Therefore, the parameters have been determined under kinetic control regime in the absence of transport phenomena effects. When these phenomena take place, the observed rate is lower than the intrinsic kinetic rate. The effects must be determined separately.

Unlike heat or mass exchangers, for reactors we should take into account a generation term in the mass and heat balance due to the chemical reaction, therefore the chemical reaction rate.

The conventional chemical reactors comprise batch and tank or tubular reactors. The batch reactors are used in small-scale processes while tubular reactors have been preferentially applied in large-scale production. The advantages and drawbacks of batch and continuous reactors are shown in the table below:

Batch	Continuous
Advantages	*Advantages*
Small scale	Large scale
Homogeneous phase–liquid or gas	Homogeneous and heterogeneous phases
Any pressure and temperature	Any pressure and temperature
Low cost	High cost
Requires labor	Requires skilled labor
Simple equipment	Appropriate control equipment
Drawbacks	*Drawbacks*
Huge losses (time and material)	Unforeseen problems

The batch reactor is advantageous for small-scale processes, such as drug production and fine chemicals. However, the batch reactor should be constructed in such a way to avoid preferential paths, with very efficient stirring system so that the mixture is homogeneous. Samples may be collected intermittently or at the end of the reaction allowing to obtain the composition of the intermediate or final products. For kinetic studies, sampling is carried out at different times to monitor reaction system as function of time.

The drawbacks depend on the system scale. The reactor has to be loaded, unloaded, and cleaned, sometimes resulting in longer time than the reaction itself. The batch reactor requires less labor, however, requires special care to avoid contamination. Batch reactors are utilized in pharmaceutical industry, fine chemicals, natural products, and processes little known.

The continuous reactors are mostly used in large-scale industries and in numerous industrial processes. They may be used to operate homogeneous and heterogeneous processes for several months or even years producing large amount of products. The operating conditions are wide but require a continuous control and consequently more labor.

Reactors face serious problems of flow including heat and mass transfer limitations. The flow regime affects the system, especially heterogeneous ones. The most important variables such as flow, temperature, and concentration should be monitored continuously requiring high precision instruments.

The main disadvantage is the occurrence of clogging or overloading demanding immediate interruption of the reactor, especially in systems with potentially explosive reactions.

In a continuous tank-type reactor, the flow should not follow preferential paths. In the continuous tubular reactor, the flow can be in extreme cases: laminar (not desired) or turbulent (desired), but without dead volume. The type of flow may cause radial and longitudinal diffusion effects causing radial or axial temperature and concentration gradients and consequently affecting the chemical reaction.

Dimensioning a reactor can be easy when the flow is ideal. To study the flow, one invokes the so-called "population balance."

Let us consider an irreversible first-order reaction in isothermal conditions, whose solution is:

$$- \ln (1 - X_A) = kt$$

where X_A is the conversion of reactant A, k is the rate constant, and t is reaction time. Time t is the contact time or residence time of molecules in the reactor. In batch reactors, the time measured is equal to the average contact time of the molecules. In a continuous system, the residence time of the molecules may be or not the same for all molecules since it depends on their distribution in the reactor, which in turn depends on the fluid flow. Therefore, it is impossible to determine the rate constant without the "true" reaction time.

We can consider the following cases:

(a)   All fluid elements or molecules have the same residence time in the reactor or the same average contact time, characterizing a perfect mixture—ideal batch or tank.

(b)   The fluid going through the reactor has uniform composition at any cross section of the tube characterizing a "plug" flow—ideal tubular reactor.

(c)   The fluid flow or distribution of molecules is not uniform in the cross section; molecules flow with different contact times along the reactor, characterizing the nonideal reactors.

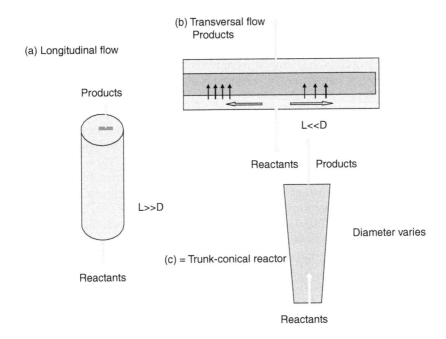

(a) Longitudinal flow

Products

L>>D

Reactants

(b) Transversal flow
Products

L<<D

Reactants   Products

(c) = Trunk-conical reactor

Diameter varies

Reactants

Figure 14.1 Shapes of tubular reactors.

The flow also depends on the reactor shape. The cylindrical tubular reactor is the most used but there are other shapes as conical and cylindrical trunk with transversal flow, which change the velocity profile affecting the flow as shown in Figure 14.1.

The choice of the type of reactor depends on the flow and other factors:

(a)   Residence time.
(b)   Pressure drop in the reactor.
(c)   Type of load in the reactor (catalyst or filling).
(d)   Internal or external heat exchange, which depends whether the reaction is exothermic or endothermic and consequently if the operation is isothermal or adiabatic.

The flow in the reactor and also the superficial velocities in a cross section of the reactor vary with the condition and shape of the reactor. The velocity profile for the laminar flow in a cylindrical tube can be calculated. It presents a parabolic velocity distribution, which is a variable in the cross section, causing different contact times of the molecules along the reactor; this situation characterizes a nonideal reactor. In turbulent flow, when the Reynolds number is high, the velocity distribution varies in the cross section: high velocities near the walls and more uniform velocities in the center of the tube. These velocity profiles can be calculated by Van Karman theory. The larger the Reynolds number the higher the velocity and more uniform the velocity profile in the reactor, approaching the ideal flow and favoring a uniform contact, therefore, allowing an ideal mean residence time of the molecules. The heterogeneous reactors, particularly catalytic reactors, have highly variable velocity distribution, and although it is possible to predict the behavior of the fluid in these systems, it is very difficult to calculate the voids, which create channeling. The flow depends on the type and placement of the catalyst particles inside the reactor. On the other hand, reactors containing catalysts allow higher contact between molecules due to a large contact area promoted by catalyst particles, which enhances chemical reactions between molecules. The higher the Reynolds number and therefore the higher the fluid velocity, the more similar the flow will be to a turbulent flow and to an ideal reactor.

Besides the flow, one should consider the mass and heat transfer limitations. In reactors without bed, one may calculate the heat and mass exchange and determine the conditions for an adiabatic or isothermal operation, since the temperature profile in the reactor is known. For uniform velocities, the heat transfer depends on the heat capacities; if they are constant, the temperature profile is uniform. Otherwise, there are considerable deviations and consequently large temperature variations. In catalytic reactors, there is also the influence of conductive heat of the particles. The temperature affects substantially the rate constant and consequently the reaction rate. At the same time, mass transfer limitation may be present due to convection and diffusion inside the pores of the particles, which depend on the fluid flow and the diffusive properties of molecules. Mass transfer limitation affects significantly the rate constant and consequently the reaction rate causing different residence times of the molecules.

These considerations also apply to continuous-tank reactors, batch or semibatch. However, the manner of contact between molecules also depends on the geometry of the reactors. One should avoid dead volume and for this purpose strong stirring can be used. The higher the stirring the better the contact between molecules and the lower the chance of dead volume. The contact is instantaneous and the concentrations in the tank or in the batch should be uniform, and if possible equal to the reactor outlet. One reaches the ideal condition when the mixture is perfectly uniform. Figure 14.2 illustrates the different cases.

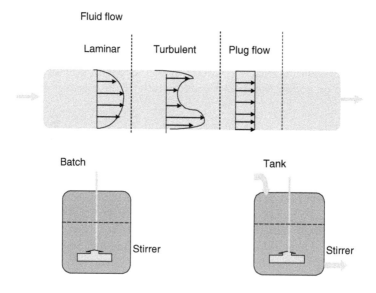

Figure 14.2 Reactors—types of flow.

14.1 TYPES OF REACTORS

The conditions for classifying ideal reactors are as follows:

Tubular reactor: The contact time is the same for all molecules or fluid elements along the reactor when the velocity is uniform in the cross section of the tube, satisfying the plug flow. All molecules have the same velocity. Therefore, the concentration is uniform in a cross section of the tube and varies only along the reactor. In the isothermal case, the temperature remains constant in the longitudinal and radial directions. In the nonisothermal case, the temperature varies along the reactor. This reactor will be denominated ideal PFR (plug flow reactor).

Tank reactor: The molecules should have the same mean residence time in the tank. Therefore, the concentration inside the tank should be equal to the concentration at the reactor outlet, implying in a uniform and perfect mixture. To reach a perfect mixture, dead volume must be avoided so that the mean residence time is uniform. A reactor in these conditions will be an ideal CSTR (continuous stirred-tank reactor).

Batch reactor: The mixture must be perfect leading to a homogeneous concentration throughout the reactor volume. The reactor should be well stirred and dead volume cannot be present. The temperature is also uniform.

14.2 DEFINITIONS AND CONCEPTS OF RESIDENCE TIME

As seen previously, the concentration varies with the reaction time or contact time in the batch reactor.

In continuous reactors, the concentration varies with space time, which is a variable equivalent to time but measured as a function of reactor volume and inlet flow of the fluid (or inlet velocity of the fluid). The flow is measured experimentally by using a rotameter or mass flow meters (MFM) through conductor signals. They are concrete measurements. We can define the new variables as follows:

Space velocity: ratio between the feed volumetric flow and reactor volume. It represents the feed rate per unit volume to achieve a given concentration or conversion of the reactant in the reactor outlet. One can interpret the space velocity as the velocity with which a fluid element or molecules pass through the reactor volume to achieve a certain conversion. If F_{A0} (mol/h) is the molar flow, v_0 (m^3/h) is the volumetric flow at the reactor inlet, and V (m^3) the reactor volume, then:

$$v_e = s = \frac{v_0}{V} (h^{-1}) \tag{14.1}$$

or

$$v_e = s = \frac{F_{A0}}{C_{A0} \times V} (h^{-1}) \tag{14.2}$$

where C_{A0} and F_{A0} are, respectively, the initial concentration and initial molar flow of the reactant A (i.e., at the reactor inlet).

Space time: the inverse of space velocity, i.e., the ratio between the reactor volume and the feed volumetric flow. It represents the time in which a fluid element or molecules pass through the reactor volume to achieve a final concentration or conversion of reactant. Therefore, we have:

$$\tau = \frac{V}{v_0}(\text{h})$$

(14.3)

As an example, $v_e = 5\,\text{h}^{-1}$ means a feed rate of $5\,V\,\text{h}^{-1}$ to process the reaction in order to achieve 80% conversion. On the other hand, $\tau = 5\,\text{min}$ represents the time required to process a feed rate in the volume V and achieve 80% conversion, or each 5 min one processes a load in the reactor volume to achieve 80% conversion.

Residence time: the residence time t takes into account the time in which each fluid element or molecule passes through the reactor and it depends on the molecules velocity inside the reactor; therefore, it depends on the flow in the reactor. Residence time is equal to space time if the velocity is uniform in a cross section of the reactor, as in ideal tubular reactors. This situation is not valid to tank reactors, since the velocity distribution is not uniform. In most nonideal reactors, the residence time is not the same for all molecules, leading to variations in radial concentrations along the reactor; and therefore, the concentration in the tank and at the reactor outlet is not uniform. That means we need to define initially the residence time and calculate the residence time distribution for each system.

Tracers: one can determine experimentally the local and mean residence time in systems where no reaction occurs. One chooses a nonreactive fluid and adds a tracer, measuring its concentration at the outlet of the reactor. In general, a dye is used but other alternatives are possible such as conductivity, radioactive materials, etc. since they can be measured. The nonreactive fluid flows through the reactor and the tracer is introduced as step or pulse. To simplify, we will use a tank reactor with volume V and a nonreactive liquid with inlet volumetric flow v_0. One introduces a dye with concentration C_0 at the reactor inlet and measures its concentration at the reactor outlet from the instant $t = 0$. By the balance, we have:

Step tracer experiment

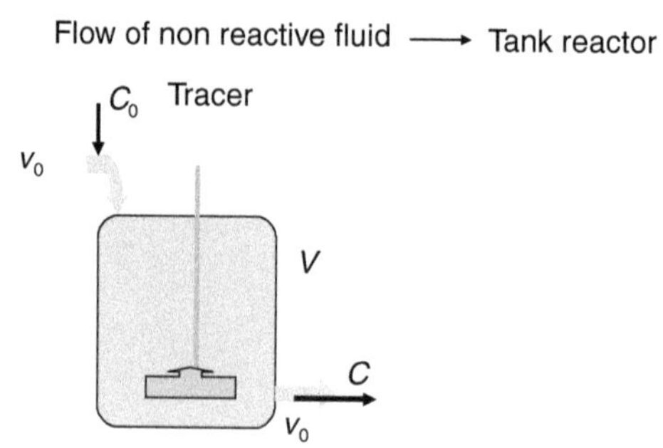

Figure 14.3 Mass balance of nonreactive fluid in the tank reactor.

{molar flow of tracer entering system, mol/h} – {molar flow of tracer leaving system, mol/h} = {molar flow of tracer accumulating, mol/h}

$$v_o C_o - v_o C = V(dC/dt)$$

Integrating for $t = 0$, $C = C_0$, and for $t \neq 0$, $C = C_{any}$ and considering that τ is space time:

$$\tau = \frac{V}{v_0}(h)$$

We obtain:

$$\frac{C}{C_0} = 1 - \exp(-t/\tau) \tag{14.4}$$

The tracer concentration at the reactor outlet varies exponentially with time, indicating that its distribution varies inside the reactor. The molecules have different residence times and leave the reactor with a concentration in exponential form. This is the concentration distribution in an ideal tank reactor.

Generally, one relates the concentration with a residence time distribution function, assuming that a fraction of molecules has a residence time in between t and $t + dt$. At time t, the tracer concentration at the reactor outlet is C. Therefore, one measures a fraction of molecules that remains in the reactor in a time shorter than t and another fraction that remains in the reactor in a time longer than t. The first fraction is represented by the cumulative distribution function $F(t)$ and the second fraction is represented by the difference $(1 - F(t))$ which remains for a time longer than t. This last fraction contains no more C_0 at the reactor outlet. A balance at the reactor outlet leads to:

$$v_0[1 - F(t)] \times C_0 + v_0 F(t)C_0 = v_0 C$$

along with Equation 14.4:

$$F(t) = \frac{C}{C_0} = 1 - \exp(-t/\tau) \tag{14.5}$$

Therefore, the cumulative residence time distribution function is determined by measuring the concentration versus time at the reactor outlet. The function is represented graphically as follows:

We have the mean residence time if residence time is equal to space time $(t = \tau)$ or if Area 1 = Area 2 in Figure 14.4. One can notice that a fraction of molecules in the first area has a residence time lower than t and another fraction in Area 2 has longer time. Similar fractions have a residence time corresponding to the mean value, which we designate $\bar{t}$ (mean residence time).

In the ideal PFR, the residence time is the same for all molecules, assuming constant and uniform velocity. The outlet concentration of the tracer is equal to the inlet concentration. Therefore, $F(t) = 1$ or $t = \tau$. The mean residence time is equal to the space time.

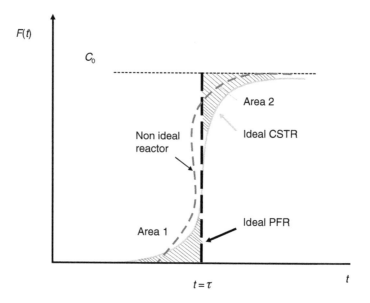

Figure 14.4 Curve of the residence time distribution function.

Any other form of residence time distribution in between ideal CSTR and ideal PFR behaves as a nonideal reactor.

To determine the function $F(t)$ from experimental data, we use the property G, as shown in Chapter 1. If G is any property (conductivity, ionization, wavelength, etc.) proportional to the concentration, in which G_1 is the magnitude at the inlet and G_2 at the outlet, then the cumulative residence time distribution function, which remained in the reactor at an instant shorter than t, will be:

$$F(t) = \frac{G(t) - G_1}{(G_2 - G_1)} \tag{14.6}$$

Example

E14.1 A nonreactive tracer is introduced into an inert fluid that flows through a reactor. The inlet concentration of the tracer is $2\,\text{g/m}^3$ and its concentration C at the reactor outlet is given according to the following table:

t (min)	0.1	0.2	1	2	5	10	20	30
C (g/m³)	1.96	1.93	1.642	1.344	0.736	0.286	0.034	0.004

The reactor volume is 1 m³ and the feed volumetric flow is 0.2 m³/min. Determine the cumulative residence time distribution function (F) and the mean residence time.

Solution

Space time:

$$\tau = \frac{V}{v_0}(h) = \frac{1}{0.2} = 5 \text{ min}$$

Calculation the function F:

$$F(t) = \frac{G(t) - G_1}{(G_2 - G_1)} = \frac{2 - C(t)}{2}$$

t (min)	0.1	0.2	1	2	5	10	20	30
C (g/m³)	1.96	1.93	1.642	1.344	0.736	0.286	0.034	0.004
F(t)	0.02	0.035	0.179	0.328	0.632	0.866	0.983	0.088

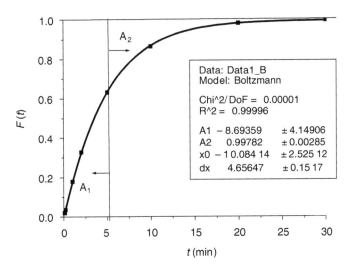

Figure E14.1 Residence time distribution.

The mean residence time can be determined by the graph assuming equal areas A_1 and A_2 according to the dashed line corresponding to $\bar{t} = 5$ min. The cumulative distribution function as a function of time is shown in the above table. Longer times indicate a fraction of molecules that leaves the reactor at a time longer than the mean residence time.

Pulse tracer experiment

Another method to determine the residence time consists in introducing the tracer as a pulse at a short time. The response of tracer concentration at the reactor outlet can be calculated assuming that a fraction of molecules ΔF left the reactor at an interval of time Δt. A fraction of molecules leaves the reactor with concentration C_0 and another fraction leaves without C_0 at the outlet. Therefore, at the interval of time Δt we have the following balance:

$$v_0 \, \Delta F(t) \times C_0 = v_0 C$$

At the limit $\Delta t \to \infty$, we have:

$$\lim_{\Delta t \to 0} \frac{\Delta F(t)}{\Delta t} = \frac{\mathrm{d}F}{\mathrm{d}t} = \frac{C}{C_0} \tag{14.7}$$

The variation of the cumulative residence time distribution function is represented by a Gauss curve, indicating the concentration variation $C(t)$ as a function of time according to Figure 14.5:

Integrating:

$$\int \mathrm{d}F = \int (C/C_0)\mathrm{d}t$$

But from the curve, we conclude that:

$$\int \left(\frac{C}{C_0}\right) \mathrm{d}t = \int \left(\frac{C}{C_0}\right) \tau \, \mathrm{d}\theta = 1 \tag{14.8}$$

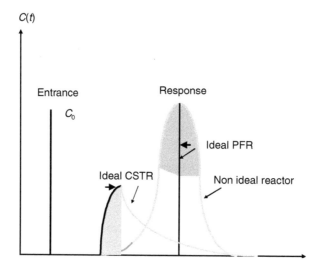

Figure 14.5 Residence time distribution curve.

where:

$$\theta = \frac{t}{\tau}$$

One defines $E(t)$, residence time distribution function as:

$$\frac{dF}{dt} = E(t)$$

or

$$\frac{dF(\theta)}{d\theta} = \tau \times E(t) = \frac{C}{C_0} \qquad (14.9)$$

Therefore:

$$\int dF = \int t \times E(t) dt$$

Then, the mean residence time will be:

$$\bar{t} = \int t \times E(t) \, dt \qquad (14.10)$$

In ideal cases: in the ideal PFR, the response is instantaneous under pulse. On the other hand, in the ideal CSTR, the response will be a noninstantaneous distribution of molecules, as shown in Figure 14.5.

Example

E14.2 Determine the mean residence time in a CSTR reactor using a tracer that provides the following data:

t (min)	0	5	10	15	20	25	30	35
C (mol/m³)	0	84.9	141.5	141.5	113.3	56.6	28.3	0

The tracer concentration has been measured at the reactor outlet. The reactor volume is $2\,m^3$ and the volumetric flow at the outlet is $7.2\,m^3/h$.

Solution

We plot the graph of concentration at the reactor outlet from data in the table. After integrating, we determine the area and calculate τC_0 with Equation 14.7:

$$\tau C_0 = 2.830$$

Then, using Equation 14.9, we have:

$$E(t) = \frac{C(t)}{\tau C_0}$$
(14.11)

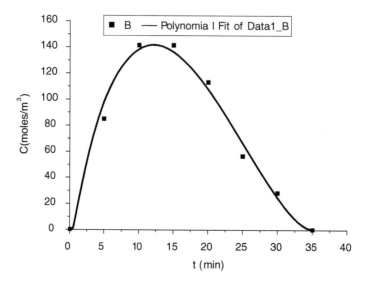

Figure E14.2 Concentration versus time.

t (min)	C (mol/m³)	E (t)	t · E (t)
0	0	0	0
5	84.9	0.03	0.15
10	141.5	0.05	0.5
15	141.5	0.04	0.75
20	113.2	0.02	0.8
25	56.6	0.01	0.5
30	28.3	0	0.3
35	0		0

The mean residence time is determined by Equation 14.10, i.e.:

$$\bar{t} = \int t \cdot E(t)dt$$

Figure E14.3 shows the curve $t, E(t)$ versus t.
By integrating it according to Equation 14.10, we obtain the mean residence time:

$$\bar{t} = 15 \, \text{min}$$

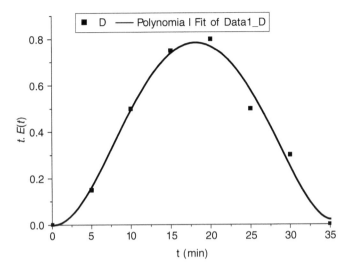

Figure E14.3 Residence time distribution.

Calculating the space time:

$$\tau = \frac{V}{v_0} = \frac{2}{7.2/60} = 16.6 \text{ min}$$

Thus,

$$\tau = 16.6 \text{ min}$$

14.3   IDEAL REACTORS

The kinetics of reactions has been studied for different reaction systems in liquid or gas phase, simple and multiple reactions, taking into account volume variation for different cases, and an important conclusion is that the understanding of kinetics is fundamental to the design of reactors.

To design reactors, we have to calculate molar and energy balances considering that reactions can also take place under nonisothermal conditions. These balances contain always the generation term due to the chemical reaction, which is represented by the reaction rate.

The conventional ideal reactors are batch, continuous, and semibatch. The conditions established for ideal reactors were shown in the previous section, and recapping, tanks should have perfect mixture and tubular reactors should have plug flow.

The molar balance in an open system is shown in the following scheme and Figure 14.6 considering any reaction initially at constant temperature:

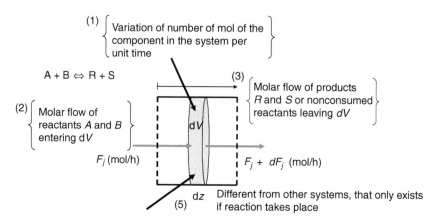

Figure 14.6 Molar balance in an open system.

Molar flow of component j entering system, per unit volume V	−	Molar flow of component j leaving system, per unit volume V	+	Generation or consumption rate of component j due to chemical reaction, per unit volume V	=	Accumulation rate of component j, per unit volume V
[1]		[2]		[3]		[4]

Considering F_j as the molar flow of component j, G_j as the generation or consumption rate, and n_j the number of moles of component j, we have:

$$F_{j0} - F_j + G_j = \frac{dn_j}{dt} \qquad (14.12)$$

The balance can be carried out to any component, reactant or product, and presents the unit mol/time.

The generation or consumption rate is given per unit volume in this system, being represented by the reaction rate within each volume element ΔV. Then,

$$F_{j0} - F_j + \int r_j \, dV = \frac{dn_j}{dt} \qquad (14.13)$$

This is the general equation of a molar balance for any component j of a chemical reaction.

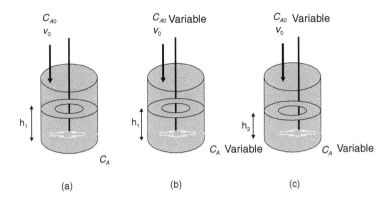

Figure 14.7 Accumulation term and reaction term.

We must distinguish clearly the reaction term from the accumulation term. Let us consider the following schemes in Figure 14.7:

Reactions usually occur in steady state in the continuous tubular and tank reactors, but if there is a disturbance on the system, this disturbance will appear in the accumulation term (4), since the reaction does not depend on the operation. The reaction manifests through the reaction rate, in the generation term of the molar balance (3). In case (a) of Figure 14.7, one feeds the reactant A with concentration C_{A0} and volumetric flow v_0 at the reactor entrance. If the reactor is ideal (perfect mixture), the concentration of A in the tank (C_A) is uniform and equal to the concentration at the reactor outlet. While in steady state, there will be no variation in the tank height (h_1) or in the outlet concentration. In case (b), the initial concentration in the tank varied because of a feed with different concentration from C_{A0}, causing a change in the concentration at the tank outlet; therefore, this disturbance will manifest in the accumulation term. In case (c), there was variation in the volumetric flow caused by failure in the pump or leak in the pipe, which leads to a variation in height of the liquid and consequently variation in the accumulation term. All these disturbances cause variations in the outlet concentrations resulting in a nonsteady state regime.

We will consider systems operating at steady state and therefore the last term will be null. Then:

$$F_{j0} - F_j + \int r_j \, dV = 0 \tag{14.14}$$

14.3.1 Batch reactor

Basically, there is no flow in the batch reactor and we need to determine the total reaction time to calculate the reactor volume that processes a particular reaction and achieves a desired final conversion. We also need to know the reaction rate through the intrinsic kinetics or the opposite: determining the intrinsic kinetics or reaction rate from the reactor volume. In the batch reactor, one disregards the terms (1) and (2), and therefore Equation 14.13 becomes:

$$\int r_j \, dV = \frac{dn_j}{dt}$$

In a closed system, the volume is constant, and thus:

$$r_j V = \frac{dn_j}{dt}$$

Therefore, integrating between C_{j0} and C_j, we calculate the total reaction time:

$$t = \int_{C_{j0}}^{C_j} \frac{dC_j}{r_j}$$

Using the conversion as a variable, we have for liquid or gas systems without contraction or expansion of volume:

$$C_A = C_{A0}(1 - X_A)$$

$$t = C_{A0} \int_0^{X_A} \frac{dX_A}{(-r_A)} \qquad (14.15)$$

This is the time required to reach a final conversion and it can be calculated by determining the reaction rate (r_j).

In a closed gas system, one assumes a piston that keeps the pressure constant but allows variation in the gas volume (expansion or contraction). In this case, Equation 14.15 becomes broader, i.e.:

$$t = n_{A0} \int_0^{X_A} \frac{dX_A}{V(-r_A)} \qquad (14.16)$$

where:

$$V = V_0(1 + \varepsilon_A X_A)$$

And ε_A is the factor of expansion or contraction defined previously.

The reaction time depends on the kinetics and therefore on the reaction type. If the reaction is simple, complex, irreversible, reversible, or multiple, one may integrate the equation using analytical or numerical methods.

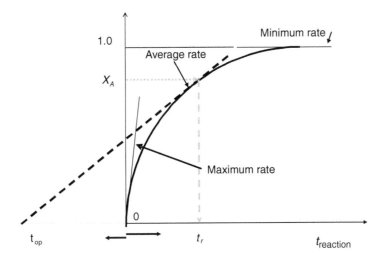

Figure 14.8 Average rate and time in a batch reactor.

To design the batch reactor, one should consider, besides the reaction time t_r, the operation time t_0, which takes into account stops, unloading, cleaning, and reloading of the reactor. The total time will be:

$$t = t_r + t_0$$

Considering the daily capacity of the reactor as G (i.e., the daily mass of desired product, kg/day), the total number of daily batches as N and the average density of the mixture as ρ, then the batch reactor volume will be:

$$V_{reactor} = \frac{\overline{G}}{\rho \cdot N} \tag{14.17}$$

Usually one assumes a safety factor to calculate the reactor volume. This factor is arbitrary and often 100% higher than the calculated volume. However, this consideration is unnecessary since one can predict the reaction time from the reaction kinetics. The reaction rate is high in the beginning but decreases with reaction time. Thus, the initial reaction rate is maximum and then decreases inducing the reaction time to be very long to convert the remaining reactants. If we take into account this time and also the operation time, the process yield may become not viable. Therefore, it is desirable to obtain an average rate, i.e., a value in between the initial (maximum) and the final one. One should choose a rate whose tangent line to the curve in Figure 14.8 has origin in the point corresponding to the operation time (indicated to the left on the abscissa axis) and touches the curve at the point corresponding to the reaction time. The conversion is not complete, but high, remaining little reactant to be converted. The operation time is estimated. One chooses the final reaction time after reaching the conversion X_A corresponding to the tangent to the curve. The remaining reactants can be recycled. The average rate will be:

$$\text{Average production rate: } \frac{G}{t_r + t_0} \text{ (kg/h)} \tag{14.18}$$

Example

E14.3 A reaction $A + B \rightarrow 2R$ takes place in liquid phase in a batch reactor. One introduces the reactants with initial concentrations of 1.5 and 3.0 kmol/cm³, but the reactor already contains the product R whose initial concentration is 2.25 kmol/cm³. Calculate the reactor volume aiming a production of 20 ton/day. The final conversion of reactant A was 80%. The operation time including opening, cleaning, and loading of reactor was 46 min and the density of the final mixture was 880 kg/m³. The reaction rate is known and given by the following expression:

$$r = \frac{k_1 C_A C_B}{k_2 C_R + k_3 C_A}$$

where:
$k_1 = 0.3 \text{ kseg}^{-1}$
$k_2 = 0.625 \text{ cm}^3/\text{kmol}$
$k_3 = 0.36 \text{ cm}^3/\text{kmol}.$

Solution

Rearranging the rate expression as a function of conversion (reactant A):

$$C_A = C_{A0}(1 - X_A)$$
$$C_B = C_{A0}(M - X_A)$$
$$C_R = C_{A0}(R + 2X_A)$$

where C_{A0}, C_{B0}, and C_{R0} are the initial concentrations of reactants A, B, and product R.

$$r_A = \frac{k_1 C_{A0}(1 - X_A)(M - X_A)}{(N + k'X_A)} \tag{14.19}$$

where:
$M = C_{B0}/C_{A0}$
$R = C_{R0}/C_{A0}$
$N = k_2 R + k_3$
$k' = 2k_2 - k_3.$

Substituting the numerical values, we obtain:
$M = 2$
$R = 1.5$
$N = 1.3$
$k' = 0.89.$

Substituting the rate expression 14.19 into Equation 14.15, we obtain:

$$t = C_{A0} \int_0^{0.8} \frac{(N + k'X_A)dX_A}{k_1 C_{A0}(1 - X_A)(M - X_A)} \tag{14.20}$$

Integrating:

$$t = 1.80\,h$$

Total time includes operation time:

$$t_t = 2.58\,h$$

The desired production of R is 20 ton/day. If conversion of A is 80%, it is necessary to have 25,000 kg of reactant/day. In this case, the mass per batch will be 2687 kg, i.e., 1343 kg of A and 1343 kg of B. Using the density of the mixture, we calculate the reactor volume:

$$V = 3.0\,m^3$$

Example

E14.4 For the production of propionic acid, one dissolves a sodium salt into a solution of hydrochloric acid according to the reversible and second-order reaction:

$$C_2H_5COONa + HCl \rightarrow C_2H_5COOH + NaCl$$

In the lab, a student collected samples at different times, neutralizing them with a 0.515N NaOH solution. The initial concentrations at 50°C were the same. The data are shown in the table below:

t (min)	0	10	20	30	50	∞
NaOH (mL)	52.5	32.1	23.5	18.9	14.4	10.5

Calculate the reactor volume to obtain 453 kg/day of product. The loading and unloading time is 30 min. The desired final conversion is 75%. The density of the mixture is 0.21 kg/L.

Solution

The reaction is reversible and of second order, therefore, we have:

$$t = C_{A0} \int_0^{X_A} \frac{dX_A}{(-r_A)} \tag{6.2}$$

where:

$$r = k\left[C_A^2 - \frac{1}{K}C_R^2\right]$$

or

$$r = kC_{A0}^2\left[(1 - X_A)^2 - \frac{1}{K}X_A^2\right] \tag{14.21}$$

Since the initial concentrations of reactants are the same, $C_{A0} = C_{B0}$, we substitute the Equation 14.21 into Equation 6.2 and integrate:

$$k^* = \frac{1}{t}\ln\left[\left(\frac{1}{\kappa}\right)\frac{\left(\kappa - \dfrac{X_A}{X_{Ae}}\right)}{\left(1 - \dfrac{X_A}{X_{Ae}}\right)}\right]$$

where X_{Ae} and X'_{Ae} are, respectively, the real and imaginary equilibrium conversions, given by the following equations:

$$K = \frac{k}{k'} = \frac{X_{Ae}^2}{(1 - X_{Ae})^2} = 15.8$$

$$\kappa = k\frac{2K}{(K-1)X_{Ae}} - 1 = 1.672$$

$$k^* = k_1 - k_2$$

Calculating the conversion:

Number of moles $= V/1000N$ (normality) $= V/1000 = 0.515$

The values are shown below:

t (min)	V (mL)	n_A	X_A	A $1 - (X_A/X_{Ae})$	B $x - (X_A/X_{Ae})$	Ln[$x^{-1} \cdot A/B$]	k^*
0	52.5	0.027	0	1	1.672	–	–
10	32.1	0.0165	0.388	0.514	1.186	0.3224	0.0322
20	23.5	0.0121	0.551	0.31	0.982	0.6393	0.0319
30	18.9	0.00973	0.639	0.20	0.872	0.958	0.0319
50	14.4	0.00741	0.725	0.0926	0.764	1.597	0.0319
∞	10.5	0.00540	0.800	0	–	–	–
			0.75	0.0625	0.7345		

$$C_{A0} = 0.0270\,\text{mol}$$

The mean residence time is 5 s.

(a)   Calculate the conversion and space time.
(b)   If the reaction is carried out in a CSTR under the same conditions, what is the conversion? Compare both cases.

Since it is a first-order reaction, the rate expression will be:

$$(-r_A) = kC_{A0}\frac{(1-X_A)}{(1 + \varepsilon_A X_A)} \tag{14.48}$$

The mean residence time will be:

$$\bar{t} = C_{A0}\int_0^{X_A} \frac{dX_A}{(1 + \varepsilon_A X_A)(-r_A)} \tag{14.47}$$

Where the rate is given by Equation 14.48. Substituting and integrating, we obtain:

$$\bar{t} = -\frac{1}{k}\ln(1-X_A)$$

The rate constant is calculated at 300°C (573 K):

$$k = 0.408\ \text{s}^{-1}$$

Therefore, if $\bar{t} = 5$ s, then:

$$X_A = 0.86$$

To calculate the space time, we replace the rate Equation 14.48 in the following expression:

$$\tau = \frac{V}{v_0} = C_{A0}\int_0^{X_A} \frac{dX_A}{(-r_A)} \tag{14.38}$$

The solution will be:

$$\tau = -\frac{1}{k}[(1 + \varepsilon_A)\ln(1-X_A) + \varepsilon_A X_A] \tag{14.43}$$

Then:

$$\tau = 6.7\ \text{s}$$

The space time is longer than the mean residence time due to the gas expansion in the reactor (variable volume system).

In the CSTR, the mean residence time will be:

$$\bar{t} = \frac{\tau}{(1 + \varepsilon_A X_A)} \qquad (14.28)$$

But space time in the CSTR is known

$$\frac{V}{v_0} = \tau = C_{A0} \frac{X_A}{(-r_A)} \bar{t} = \frac{X_A}{k(1-X_A)} \qquad (14.27)$$

Substituting the rate:

$$\bar{t} = \frac{X_A}{k(1-X_A)}$$

Thus,

$$X_A = 0.68$$

For the same mean residence time in the PFR and CSTR reactors, the conversion in the CSTR is approximately 21% lower.

Example

E14.12 The chlorination of propene:

$$C_3H_6 + Cl_2 \rightarrow CH_2 = CH - CH_2Cl + HCl$$

takes place in gas phase inside a PFR reactor of 5 m^3 at 300°C and 20 atm. One introduces 30% C_3H_6, 40% Cl_2, and balance in N_2. The total molar flow is 0.45 kmol/h. The rate constant is given:

$$k = 4.12 \times 10^{10} \exp{(-27,200/(RT))} \, (m^3/(kmol\,min))$$

Calculate the final conversion and the mean residence time.

Solution

The reactant C_3H_6 (A) is the limiting one and even in gas phase there is no volume variation, $\varepsilon_A = 0$ ($A + B \rightarrow R + S$). Based on the unit of rate constant, we have a

The mean residence time is 5 s.

(a) Calculate the conversion and space time.
(b) If the reaction is carried out in a CSTR under the same conditions, what is the conversion? Compare both cases.

Since it is a first-order reaction, the rate expression will be:

$$(-r_A) = kC_{A0}\frac{(1-X_A)}{(1 + \varepsilon_A X_A)} \tag{14.48}$$

The mean residence time will be:

$$\bar{t} = C_{A0}\int_0^{X_A} \frac{dX_A}{(1 + \varepsilon_A X_A)(-r_A)} \tag{14.47}$$

Where the rate is given by Equation 14.48. Substituting and integrating, we obtain:

$$\bar{t} = -\frac{1}{k}\ln(1-X_A)$$

The rate constant is calculated at 300°C (573 K):

$$k = 0.408 \text{ s}^{-1}$$

Therefore, if $\bar{t} = 5$ s, then:

$$X_A = 0.86$$

To calculate the space time, we replace the rate Equation 14.48 in the following expression:

$$\tau = \frac{V}{v_0} = C_{A0}\int_0^{X_A} \frac{dX_A}{(-r_A)} \tag{14.38}$$

The solution will be:

$$\tau = -\frac{1}{k}[(1 + \varepsilon_A)\ln(1-X_A) + \varepsilon_A X_A] \tag{14.43}$$

Then:

$$\tau = 6.7 \text{ s}$$

The space time is longer than the mean residence time due to the gas expansion in the reactor (variable volume system).

In the CSTR, the mean residence time will be:

$$\bar{t} = \frac{\tau}{(1 + \varepsilon_A X_A)} \tag{14.28}$$

But space time in the CSTR is known

$$\frac{V}{v_0} = \tau = C_{A0}\frac{X_A}{(-r_A)} \quad \bar{t} = \frac{X_A}{k\,(1-X_A)} \tag{14.27}$$

Substituting the rate:

$$\bar{t} = \frac{X_A}{k(1-X_A)}$$

Thus,

$$X_A = 0.68$$

For the same mean residence time in the PFR and CSTR reactors, the conversion in the CSTR is approximately 21% lower.

Example

E14.12 The chlorination of propene:

$$C_3H_6 + Cl_2 \rightarrow CH_2 = CH - CH_2Cl + HCl$$

takes place in gas phase inside a PFR reactor of $5\,m^3$ at $300°C$ and $20\,atm$. One introduces 30% C_3H_6, 40% Cl_2, and balance in N_2. The total molar flow is $0.45\,kmol/h$. The rate constant is given:

$$k = 4.12 \times 10^{10} \exp\left(-27,200/(RT)\right) (m^3/(kmol\,min))$$

Calculate the final conversion and the mean residence time.

Solution

The reactant C_3H_6 (A) is the limiting one and even in gas phase there is no volume variation, $\varepsilon_A = 0$ ($A + B \rightarrow R + S$). Based on the unit of rate constant, we have a second-order rate expression. One assumes an elementary reaction. The rate expression will be:

$$(-r_A) = kC_A C_B = kC_{A0}^2(1 - X_A)(M - X_A)$$

Substituting into PFR balance equation:

$$\tau = \frac{V}{v_0} = C_{A0} \int_0^{X_A} \frac{dX_A}{(-r_A)} \tag{14.38}$$

We obtain the following solution:

$$\tau = \frac{1}{kC_{A0}(M-1)} \ln \frac{(M-X_A)}{M(1-X_A)} \tag{14.49}$$

where:

$$M = \frac{C_{B0}}{C_{A0}}$$

The rate constant is calculated at 300°C (573 K):

$$k = 2.02 \, m^3/(kmol \, min)$$

The initial concentration can be calculated:

$$C_{A0} = 0.3(P/(RT)) = 0.127 \, mol/L = kmol/m^3$$

The ratio $M = \dfrac{C_{B0}}{C_{A0}}$

Thus,

$$M = 1.33$$

Therefore, the molar flow of A:

$$F_{A0} = y_{A0}F_0 = 0.3 \times 0.45 = 0.135 \, kmol/h$$

and the volumetric flow will be:

$$F_{A0} = C_{A0}.v_0 \quad v_0 = 1.063 \, m^3/h$$

Then:

$$\tau = 4.7 \, min$$

Substituting the values into expression 14.49, we obtain the conversion:

$$X_A = 0.66$$

Example

E14.13 The reaction takes place between two gases that mix at the entrance of the PFR reactor. The first gas A enters at a flow rate of $14.2 \, \text{m}^3/\text{min}$ and the second gas contains 50% B and balance in inert I, entering with a total flow rate of $7.1 \, \text{m}^3/\text{min}$. Both are instantaneously mixed at the entrance of the reactor at 86°C and 1 atm. The gases are ideal and the product obtained is R according to the stoichiometric reaction:

$$A + B \rightarrow R$$

The rate constant is given by the following expression:

$$\ln k_p = -\frac{5470}{T} + 12.5 (\text{mol}/(\text{atm}^2 \, \text{L} \, \text{min}))$$

Calculate the reactor volume operating isothermally to achieve 90% conversion. Calculate the molar flow and space velocity.

Solution

Taking into account 1 mol as reference, we obtain the following molar or volumetric fractions, with temperature and pressure constant:

$$y_{A0} = \frac{V_{10}}{v_0} = \frac{14.2}{21.5} = 0.67$$

$$y_{B0} = \frac{V_{20}}{v_0} = \frac{0.5 \times 7.1}{21.5} = 0.165$$

$$y_I = y_{B0} = 0.165$$

The component B is the limiting one. In this case, the rate expression as a function of product R or reactant B will be:

$$(-r_B) = k C_{B0}^2 \frac{(1-X_B)(M-X_B)}{(1 + \varepsilon_B X_B)^2} \tag{14.50}$$

The ideal PFR equation is:

$$\tau = \frac{V}{v_0} = C_{B0} \int_0^{X_B} \frac{dX_A}{(-r_B)} \tag{14.51}$$

Substituting the rate expression 14.50 into PFR equation we obtain the following solution:

$$\tau = \frac{1}{k C_{B0}} \left[\left(\frac{(1 + \varepsilon_B M)^2}{(M-1)} \right) \ln \frac{(M-X_B)}{M} - \frac{(1 + \varepsilon_B)^2}{(M-1)} \ln (1 - X_B) - \varepsilon_B^2 X_B \right] \tag{14.52}$$

Determining ε_B:

	A	B	R	Inert	Total
Initial	0.67	0.165	0	0.165	1.0
Final	0.50	0	0.165	0.165	0.83

$$\varepsilon_B = -0.17$$

Calculating $M = \dfrac{C_{A0}}{C_{B0}}$

where:

$$C_{B0} = \frac{y_{B0}}{RT} = \frac{0.165 \times 1.0}{0.082 \times 359} = 0.00564 \,(\text{mol/L})$$

and

$$C_{A0} = \frac{y_{A0}}{RT} = \frac{0.67 \times 1.0}{0.082 \times 359} = 0.0228 \,(\text{mol/L})$$

Thus,

$$M = \frac{C_{A0}}{C_{B0}} = 4.0$$

Calculating the rate constant at $T = 359\,\text{K}$:

$$\ln k_p = -\frac{5470}{T} + 12.5 = -2.736$$

$$k_p = 0.0648 \,(\text{mol}/(\text{atm}^2 \text{L min}))$$

Thus,

$$k = k_p(RT)^2 = 0.0648(0.082 \times 359)^2 = 56.1 \,\text{L}/(\text{mol min})$$

$$k = 56.1 \,\text{L}/(\text{mol min})$$

Substituting $X_B = 0.90$ (limiting), we obtain:

$$\tau\, K C_{B0} = 0.486$$

$$\tau = 1.537 \,\text{min}$$

The reactor volume will be:

$$V = \tau V_0 = 1.537 \times 21.3 = 32.7 \,\text{m}^3$$

$$V = 32.7 \,\text{m}^3$$

And the space velocity:

$$v_e = \frac{1}{\tau} = \frac{v_0}{V} = 0.65 \text{min}^{-1}$$

Example

ER14.14 Acetaldehyde, ethane, and other hydrocarbons are mixed and undergo two PFR reactors in parallel. The acetaldehyde decomposition into methane and carbon monoxide occurs preferentially at 520°C, but at 800°C ethane decomposes into ethylene and hydrogen. One introduces the mixture containing 9% acetaldehyde, 8% ethane, and stream as diluent (molar %). The other components are negligible. The first reactor works at 520°C and 1 atm and the second one at 800°C and 1.4 atm. To achieve 60% conversion, what should be the ratio between the volumetric flow at the entrance of the reactors, assuming that they have the same volume? The rate constants are given by:

Acetaldehyde: $k = 0.33$ L/(mol s) at 520°C
Ethane: $k = 1.535 \times 10^{14} \cdot \exp(-70{,}000/(RT))$ s^{-1}

Solution
The acetaldehyde decomposition occurs in the first reactor, since ethane decomposition is thermodynamically negligible at that operating condition. On the other hand, ethane decomposition takes place in the second reactor. These components should be included into the global balance.

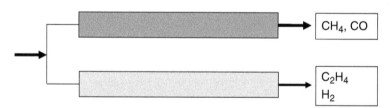

The kinetics of the reactions is known. The acetaldehyde decomposition is of second order at variable volume and the ethane decomposition is of first order according to the unit of the rate constants.

For the first reactor, we have the following integrated equation:

$$\tau = \frac{1}{kC_{A0}}\left[(1 + \varepsilon_A)^2 \frac{X_A}{(1 - X_A)} + \varepsilon_A^2 X_A + 2\varepsilon_A(1 + \varepsilon_A)\ln(1 - X_A)\right] \qquad (14.53)$$

Calculating ε_A: first reactor

$$CH_3CHO \rightarrow C_2H_4 + CO$$

	CH_3CHO	C_2H_6	CO	CH_4	H_2O	Total
Initial	0.09	0.08	0	0	0.83	1.0
Final	0	0.08	0.09	0.09	0.83	1.09

$\varepsilon_A = 0.09$

Calculating C_{A0}:

$$C_{A0} = \frac{y_{A0}}{RT} = \frac{0.9 \times 1.0}{0.082 \times 793} = 1.34 \times 10^{-3} (moles/L)$$

Substituting these values and conversion $X_A = 0.60$ into Equation 14.53, we obtain:

$$\tau_1 = 58.6 \, min$$

For the second reactor, the reaction is of first order and the equation will be:

$$C_2H_6 \rightarrow C_2H_4 + H_2 \tag{14.54}$$

$$\tau = -\frac{1}{k} [(1 + \varepsilon_A)\ln(1 - X_A) + \varepsilon_A X_A]$$

But the factor of expansion is:

	C_2H_6	CH_3CHO	C_2H_4	H_2	H_2O	Total
Initial	0.08	0.09	0	0	0.83	1.0
Final	0	0.09	0.08	0.08	0.83	1.08

$\varepsilon_A = 0.08$

We calculate $C_{A0} = 1.273 \times 10^{-3} \, mol/L$
We calculate k_2 at 800°C (1073 K):

$$k_2 = 0.685 \, s^{-1}$$

Substituting k_2 into Equation 14.54, we obtain for $X_A = 0.6$:

$$\tau_2 = 18 \, min$$

Since both PFR reactors have the same volume, we can calculate the ratio between the volumetric flow at the entrance of the reactors from the ratio between the space times:

$$\frac{v_{02}}{v_{01}} = 3.25$$

Example

E14.15 The mixture 15% C_2H_4, 55% Br_2, and inert as balance (wt. %) flows in a PFR reactor at 330°C and 1.5 atm. The feed flow rate is 10 m³/h. The reaction is reversible as follows:

$$C_2H_4 + Br_2 \underset{}{\overset{k_1}{\rightleftharpoons}} C_2H_4Br_2$$

The rate constants are:

$$k_1 = 500 \, m^3/(kmol \, h)$$

$$k_{-1} = 0.032 \, h^{-1}$$

Calculate the reactor volume to a conversion of 75% (limiting reactant).
(Data: Molecular weight: $C_2H_4 = 28$; $Br_2 = 80$; inert $= 44$)

Solution
Since the composition is given in weight percent, we have to calculate the number of moles at the entrance:

$$m = n \cdot M$$

To $C_2H_4 (A) \rightarrow m_A = n_A \cdot M_A = 0.15, \blacktriangleright \quad n_A = 5.3510^{-3} \, moles/h$

To $Br_2 (B) \rightarrow m_B = n_B \cdot M_B = 0.55, \blacktriangleright \quad n_B = 6.8710^{-3} \, moles/h$

To Inert $(I) \rightarrow m_I = n_I \cdot M_I = 0.30, \blacktriangleright \quad n_I = 6.8110^{-3} \, moles/h$

We will use the variable α that represents the extent of reaction and organize the following molar balance:

Component	Moles Entering	Moles Leaving	Molar Fraction – y_i
C_2H_4 (A)	0.00535	$0.00535 - \alpha$	$\dfrac{0.00535 - \alpha}{0.0191 - \alpha}$
Br_2 (B)	0.00687	$0.00687 - \alpha$	$\dfrac{0.00687 - \alpha}{0.0191 - \alpha}$
Inert (I)	0.00681	0.00681	0.00681
$C_2H_4 Br_2$ (R)	0	α	$\dfrac{\alpha}{0.0191 - \alpha}$
Total	0.0191	$0.0191 - \alpha$	

The partial pressures can be determined for each component:

$$p_i = y_i P$$

where P is the total pressure of the system.

$$r = \frac{k_1}{(RT)^2} p_A p_B = \frac{k_1 P^2}{(RT)^2} y_A y_B = \frac{k_1 P^2}{(RT)^2} \frac{(0.00535-\alpha)(0.00687-\alpha)}{(0.0191-\alpha)^2}$$

Since the reaction is reversible, one can calculate the equilibrium constant equaling the resulting rate to zero:

$$r = k_1 C_A C_B - k_{-1} C_R = 0$$

$$K_e = \frac{k_1}{k_{-1}} = 1.56 \times 10^4 \, (m^3/kmol)$$

The equilibrium constant is high and the forward rate constant is much higher than the reverse one, therefore, one may disregard the reversible reaction. Then, we assume a second-order forward reaction given by the following expression in terms of partial pressure:

$$r = \frac{k_1}{(RT)^2} p_A p_B = \frac{k_1 P^2}{(RT)^2} y_A y_B = \frac{k_1 P^2}{(RT)^2} \frac{(0.00535-\alpha)(0.00687-\alpha)}{(0.0191-\alpha)^2}$$

Substituting the rate expression into the PFR balance:

$$\tau = \frac{V}{v_0} = n_0 \int_0^\alpha \frac{d\alpha}{r} \tag{14.55}$$

where n_0 is the total initial number of moles.
Then:

$$\tau = \frac{V}{v_0} = \frac{(RT)^2}{n_0 P^2} \int_0^\alpha \frac{(0.0191-\alpha)^2 \, d\alpha}{(0.00535-\alpha)(0.00687-\alpha)}$$

Substituting the value of k_1, considering $n_0 = C_0 v_0$ and with the total initial concentration given:

$$\frac{k_1 P^2}{(RT)^2} = 0.46$$

with $P = 1.5$ atm and $T = 603$ K, we have:

$$X_A = \frac{n_{A0} - n_A}{n_{A0}} = \frac{\alpha}{n_{A0}} = 0.75$$

Since the conversion of the limiting reactant (A) is 75%,

$$\alpha = 0.0040 = 0.0040 \, mol/h$$

Therefore,

$$\alpha = 0.0040 \, \text{mol/h}$$

With the feed volumetric flow $v_0 = 10 \, \text{m}^3/\text{h}$, and integrating we obtain the reactor volume:

$$V = 4.06 \, \text{m}^3$$

14.4   IDEAL NONISOTHERMAL REACTORS

The kinetics of reactions has been studied for different reaction systems in isothermal reactors. The majority of reactions and processes are not isothermal, since the reactions are endothermic or exothermic. Depending on the extent of exothermicity or endothermicity, the thermal effects on conversion, selectivity, or yield are quite pronounced.

The reactions in liquid phase with low heat capacity can be performed in a reactor operating isothermally. However, reactions with high heat capacity in liquid or gas phase affect significantly the conversion, selectivity, yield, and/or the reactor features, as volume (especially volume).

The effect of temperature is observed on the reaction rate, because temperature affects the rate constant through the Arrhenius constant. The temperature influences significantly the rate constant due to the exponential term that contains the activation energy and temperature. The temperature effect is lower on the preexponential factor, which takes into account the collision between molecules, but it may become important in catalyzed reactions.

Thermodynamics show that the equilibrium conversion increases exponentially with increasing temperature in endothermic reactions but decreases in exothermic reactions.

Figure 14.12 shows the curves of equilibrium conversion X_{Ae} for endothermic and exothermic cases, as well as the conversion with increasing temperature until reaching equilibrium conversion.

The reactors can operate in three different ways:

(1) *Isothermally:* The temperature is constant and uniform in the reactor. If the reaction is endothermic, one needs to supply heat to keep it isothermal. If the reaction is exothermic one needs to remove heat to keep the temperature constant.
(2) *Nonisothermally:* In this case, the temperature varies throughout the reactor and also the conversion, depending on if the reaction is endothermic or exothermic.
(3) *Adiabatically:* The reactor remains insulated; no heat transfer occurs and both the temperature and conversion vary throughout the reactor. When the reaction is exothermic, a rigorous control is necessary because the temperature inside the reactor can exceed the desired value leading to explosion risks. Usually the oxidation reactions are highly exothermic.

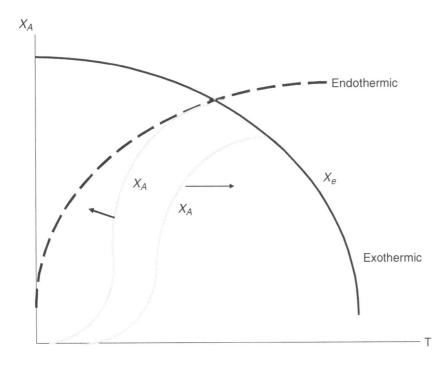

X_A

Endothermic

X_A

X_e

X_A

Exothermic

T

Figure 14.12 Equilibrium conversion for endothermic and exothermic reactions.

The profiles of temperature, conversion, and concentration of the reactants in nonisothermal reactions taking place in batch, tank, or tubular reactors are shown in Figure 14.13.

If the conversion varies with temperature, we have one more unknown variable, i.e., the temperature, and therefore we need to perform an energy balance.

Let us consider an open system according to the following scheme and Figure 14.14:

$$
\begin{Bmatrix} \text{energy entering} \\ \text{system due to} \\ \text{reactants flow} \end{Bmatrix} - \begin{Bmatrix} \text{energy leaving system} \\ \text{due to reactants} \\ \text{and products flow} \end{Bmatrix} + \begin{Bmatrix} \text{rate of generated heat} \\ \text{in the system due to} \\ \text{the chemical reaction} \end{Bmatrix}
$$

$$
+ \begin{Bmatrix} \text{rate of external} \\ \text{heat transfer} \end{Bmatrix} - \begin{Bmatrix} \text{work done by system} \\ \text{on its external} \\ \text{surrounding} \end{Bmatrix} = \begin{Bmatrix} \text{rate of accumulated} \\ \text{energy into the} \\ \text{system} \end{Bmatrix}
$$

If E represents the total energy of the system and F_j the molar flow of components, we have:

$$
\sum_j F_j E_j \big]_{\text{in}} - \sum_j F_j E_j \big]_{\text{out}} + Q_{\text{generated}} + Q_{\text{external}} - W_{\text{done}} = \frac{dE}{dt} \tag{14.56}
$$

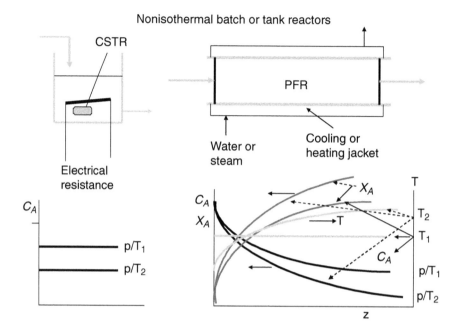

Figure 14.13 Distribution of concentration and temperature in nonisothermal reactors.

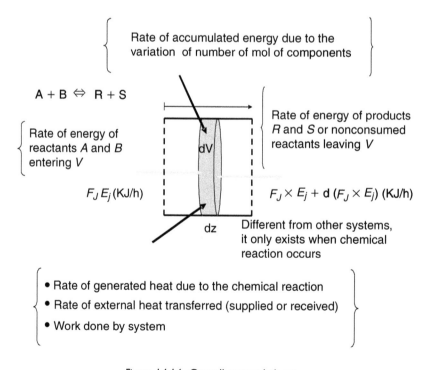

Figure 14.14 Overall energy balance.

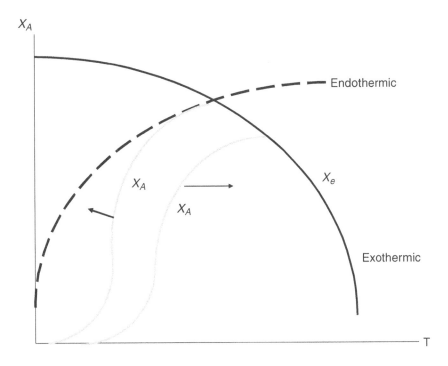

X_A

Endothermic

X_A

X_A

X_e

Exothermic

T

Figure 14.12 Equilibrium conversion for endothermic and exothermic reactions.

The profiles of temperature, conversion, and concentration of the reactants in nonisothermal reactions taking place in batch, tank, or tubular reactors are shown in Figure 14.13.

If the conversion varies with temperature, we have one more unknown variable, i.e., the temperature, and therefore we need to perform an energy balance.

Let us consider an open system according to the following scheme and Figure 14.14:

$$\left\{ \begin{array}{l} \text{energy entering} \\ \text{system due to} \\ \text{reactants flow} \end{array} \right\} - \left\{ \begin{array}{l} \text{energy leaving system} \\ \text{due to reactants} \\ \text{and products flow} \end{array} \right\} + \left\{ \begin{array}{l} \text{rate of generated heat} \\ \text{in the system due to} \\ \text{the chemical reaction} \end{array} \right\}$$

$$+ \left\{ \begin{array}{l} \text{rate of external} \\ \text{heat transfer} \end{array} \right\} - \left\{ \begin{array}{l} \text{work done by system} \\ \text{on its external} \\ \text{surrounding} \end{array} \right\} = \left\{ \begin{array}{l} \text{rate of accumulated} \\ \text{energy into the} \\ \text{system} \end{array} \right\}$$

If E represents the total energy of the system and F_j the molar flow of components, we have:

$$\sum_j F_j E_j \Big]_{in} - \sum_j F_j E_j \Big]_{out} + Q_{generated} + Q_{external} - W_{done} = \frac{dE}{dt} \qquad (14.56)$$

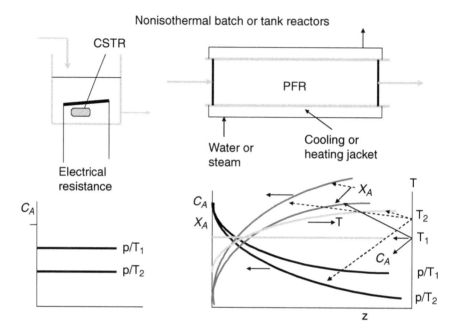

Figure 14.13 Distribution of concentration and temperature in nonisothermal reactors.

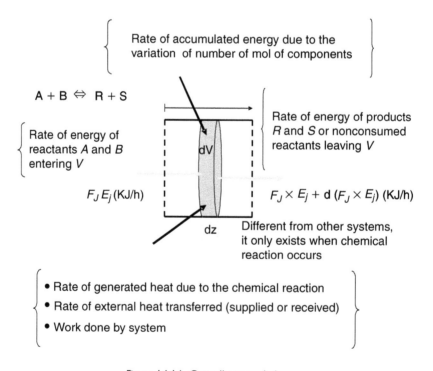

Figure 14.14 Overall energy balance.

where:

E_j = energy of each component
F_j = molar flow of each component
W = work done
Q = heat (J/mol)
t = time

The work done W includes the work due to the reactants flow and external work, which may be from a turbine or a stirrer. Usually this kind of work is negligible. But the work done by the flow depends on the pressure and molar volume. Therefore,

$$W_{done} = -\sum_j F_j (P.V_j)]_{\text{in}} + \sum_j F_j (P.V_j)]_{\text{out}} + W_{external} \tag{14.57}$$

where:

(PV_j) (Pa m^3/mol) = 1 J/mol
V_j = specific volume (m^3/mol)

Substituting Equation 14.57 into Equation 14.56:

$$\frac{dE}{dt} = -\sum_j F_j (E_j + P.V_j)]_{\text{in}} - \sum_j F_j (E_j + P.V_j)]_{\text{out}}$$
$$+ Q_{generated} + Q_{external} - W_{done} \tag{14.58}$$

However, the energy of the system consists of the internal energy (U_j), kinetic energy, and potential energy, i.e.,

$$E_j = U_j + \frac{v_j^2}{2} + gz_j \tag{14.59}$$

In conventional reactors, the kinetic and potential energies are negligible compared to the internal energy of the system.

From thermodynamics, one knows that the enthalpy of reaction in an open system is defined as a function of the internal energy and flow, therefore:

$$H_j = U_j + PV_j$$

Considering that the energy of the system is equal to the internal energy, $E_j = U_j$:

$$H_j = E_j + PV_j \tag{14.60}$$

Substituting Equation 14.60 into Equation 14.58:

$$\frac{dE}{dt} = \sum_j F_j H_j]_{\text{in}} - \sum_j F_j H_j]_{\text{out}} + Q_{generated} + Q_{external} - W_{external} \tag{14.61}$$

The heat generated due to the chemical reaction depends on the enthalpy of reaction and the reaction rate in a system with volume dV. Thus, we can express this generated heat as follows:

$$\dot{G}_{generated} = \Delta H_r \cdot r_j \, dV \tag{14.62}$$

where:

$\Delta H_r =$ total enthalpy of reaction:

$$\left(\sum H_{products} = \sum H_{reactants} \right)$$

and

$r_j =$ total reaction rate (mol/(L h))

According to thermodynamics, the enthalpy of each component depends on the temperature. As we have seen:

$$H_j = \int_{T_0}^{T} c_p \, dT \tag{14.63}$$

where c_{pj} is the specific heat of each component, reactant or product, which in turn depends on the temperature. There are two possibilities:

(1)   The specific heat c_{pj} does not vary with temperature within a certain range, which is quite normal for most systems.
(2)   The specific heat c_{pj} varies with temperature and it is calculated according to the following expression, assuming ideal gas systems:

$$c_{pj} = \alpha_j + \beta_j T + \gamma_j T^2 \tag{14.64}$$

where the constants α, β, and γ are known and tabulated for several compounds and mixtures.

There are two different cases that depend on the units. When the unit is expressed as cal/(g °C) one should use directly c_{pj}, but when it is expressed as cal/(mol K) one should use the nomenclature with bar, i.e., $\bar{c}_{pj}$.

Substituting Equation 14.64 into Equation 14.63, we obtain the variation of enthalpy with temperature.

If the system operates in steady state, the accumulation term is zero. In the transient system, the energy should be a function of the number of moles and the specific heat of each component. This term is different from the feed heat flow passing through the system. Here, we also have two cases:

(a) $$\frac{dE}{dt} = 0 \rightarrow \text{steady state regime} \tag{14.65}$$

(b)
$$\frac{dE}{dt} = \frac{d\left(\sum_j n_j \bar{c}_{pj} T\right)}{dt} \rightarrow \text{transient state regime} \qquad (14.66)$$

$$c_{pj} = \alpha_j + \beta_j T + \gamma_j T^2$$

Substituting the terms of Equations 14.62 and 14.66 into Equation 14.61 and neglecting the external work when compared with the other terms, we have:

For the transient regime:

b)
$$\frac{dE}{dt} = \frac{d\left(\sum_j n_j \bar{c}_{pj} T\right)}{dt} \qquad (14.67)$$

For the steady-state regime:

$$\sum_j F_j H_j\big]_{in} - \sum_j F_j H_j\big]_{out} + \Delta H_r r_j \, dV + Q_{external} = \frac{d\left(\sum_j n_j \bar{c}_{pj} T\right)}{dt} \qquad (14.68)$$

The enthalpies of components at the inlet and outlet of the system depend on the specific heat and temperature. Usually one assumes that heat capacities are independent of temperature, and therefore, the difference between the first two terms will be, considering Equation 14.63:

$$\sum_j F_j H_j\big]_{in} - \sum_j F_j H_j\big]_{out} + \Delta H_r r_j \, dV + Q_{external} = 0$$

Moreover, by the molar balance for the reactant A in a PFR reactor, we have:

$$-\sum_j F_j \bar{c}_{pj} (T - T_0)$$

$$(-r_A)dV = F_{A0} \, d\bar{X}_A$$

Substituting these two expressions into Equation 14.68 and integrating between T_0 and T, and $X_A = 0$ and X_A, we have:

$$\sum_j F_j \bar{c}_{pj} (T - T_0) + \Delta H_r F_{A0} X_A = Q_{external} \qquad (14.69)$$

Therefore:

$$Q_{sensible} + Q_{generated} = Q_{external}$$

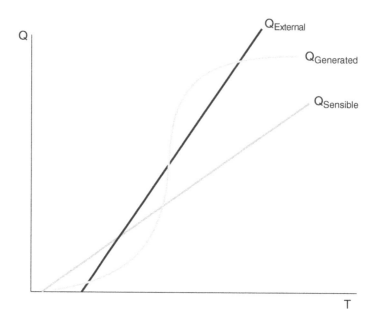

Figure 14.15   Heat as a function of temperature in nonisothermal reactors.

where:

$$\Delta H_r = \left(\sum_j H^0_{products} - \sum_j H^0_{reactants} \right) \tag{14.70}$$

ΔH_r is constant at the temperature range. If it varies with temperature, one should use Equation 14.63.

$\bar{c}_{pj}$ is the heat capacity and it is constant for each component. If it varies with temperature, one should use Equation 14.64.

The energy equation in terms of heat can be seen in Figure 14.15.

The generated heat varies exponentially since it contains the reaction term. In this term, the rate constant or Arrhenius constant varies exponentially with temperature. At first, the heat increases smoothly since the conversion is low, but with increasing temperature and conversion, the heat increases sharply and stabilizes as the reaction rate decreases at high conversions.

The sensible or flow heat varies linearly with temperature and depends only on the heat capacity of the components, which is approximately constant at this temperature range. However, the flow of each component varies with conversion.

Let us consider the irreversible gas-phase reaction:

$$aA + bB \rightarrow rR$$

Then:

$$\sum_j F_j \bar{c}_{pj} = F_A \bar{c}_{pA} + F_B \bar{c}_{pB} + F_R \bar{c}_{pR} \tag{14.71}$$

But by the relation:

$$\frac{F_{A0} - F_A}{a F_{A0}} = \frac{F_{B0} - F_B}{b F_{A0}} = \frac{F_R - F_{R0}}{r F_{A0}}$$

where:

$$X_A = \frac{F_{A0} - F_A}{F_{A0}}$$

We have:

$$F_A = F_{A0} \left(1 - X_A\right)$$

$$F_B = F_{A0} \left[M - \left(\frac{b}{a}\right) X_A \right]$$

$$F_R = F_{A0} \left(\frac{r}{a}\right) X_A$$

But $M = \dfrac{F_{B0}}{F_{A0}}$ and $F_{R0} = 0$. One assumes $a = b = r = 1$. Therefore:

$$\sum_j F_j \bar{c}_{pj} = F_{A0} [(1 - X_A) \bar{c}_{pA} + (M - X_A) \bar{c}_{pB} + X_A \bar{c}_{pR}]$$

where:

$\bar{c}_{pA}, \bar{c}_{pB}$ e $\bar{c}_{pR}$, are the specific heats of components expressed as $J/(molK)$

The external heat $Q_{external}$ depends on the heating (steam) or cooling fluid that flows through a cooling coil, jacket, or external heat exchanger.

Usually one considers the overall heat transfer coefficient, U which takes into account the heat transfer coefficients between the cooling/heating fluid and the fluid inside reactor, the transfer coefficient h_{fluid}, and the thermal conductivity k_c of the material. Therefore,

$$Q_{external} = U \cdot A_{area} \cdot (T - T_s) \tag{14.72}$$

where T is the temperature of the reaction system and T_s is the temperature of the external cooling/heating fluid. The external heat varies linearly with the temperature of the reaction system as shown in Figure 14.15.

Substituting Equations 14.72 and 14.71 into Equation 14.69, we have:

$$\sum_j F_j \bar{c}_{pj} (T - T_0) + \Delta H_r F_{A0} X_A = U A_s (T - T_s) \tag{14.73}$$

This is the energy balance equation for open systems with external heat transfer, thus nonisothermal. Along with the molar balance equation of a PFR (Equation 14.38) or a CSTR (Equation 14.27), we have a system with two equations and two unknown variable, T and X_A.

$$\tau = \frac{V}{v_0} = C_{A0} \int_0^{X_A} \frac{dX_A}{r} \tag{14.38}$$

or,

$$\frac{V}{v_0} = \tau = C_{A0} \frac{X_A}{(-r_A)} \tag{14.27}$$

where the rate,

$$(-r_A) = k_0 \exp(-E/(RT)) f(X_A)$$

The function $f(X_A)$ depends on the kinetics of reaction, which may be simple, multiple, complex, reversible, or irreversible, in systems at constant volume (liquids) or variable volume (gas).

14.4.1 Adiabatic continuous reactor

The isothermal and nonisothermal reactors are the most used when the reactions are exothermic or endothermic and one can provide or withdraw heat to keep the reactor isothermal (or not) and also determine how the temperature varies with the progress of the reaction. The adiabatic reactor is a particular case in which there is no heat exchange and the reactor is thermally insulated. The temperature and conversion vary differently within the reactor. The term Q_{external} is zero into Equation 14.69 or 14.73.

$$\sum_j F_j \bar{c}_{pj} (T - T_0) + \Delta H_r F_{A0} X_A = 0 \tag{14.74}$$

Then:

$$(T - T_0) = \frac{-\Delta H_r F_{A0} X_A}{\sum_j F_j \bar{c}_{pj}} \tag{14.75}$$

The temperature varies linearly with conversion. However, the term:

$$\sum_j F_j \bar{c}_{pj}$$

Depends on each component as seen in Equation 14.71. One may consider the enthalpy of reaction $-\Delta H_r$ and specific heat c_{pj} to be constant at the temperature range.

14.4.2 Nonadiabatic batch reactor

The temperature varies with the reaction time in the nonisothermal batch reactor. To perform the energy balance, we use the same energy balance equation 14.67, annulling the molar flow terms, but considering the variation of sensible heat with temperature and time. Then,

$$\Delta H_r r_j V + Q_{external} = \frac{d\left(\sum_j n_j \bar{c}_{pj} T\right)}{dt} \tag{14.76}$$

But considering that,

$$(-r_A) = -\frac{1}{V}\frac{dn_A}{dt} = \frac{n_{A0}}{V}\frac{dX_A}{dt}$$

And substituting the term $r_j V$ by the expression above, we obtain:

$$-\Delta H_r \frac{n_{A0}}{V}\frac{dX_A}{dt} + Q_{external} = \frac{\left(\sum_j n_j \bar{c}_{pj}.dT\right)}{dt}$$

As an integral equation:

$$\int_0^{X_A} \Delta H_R n_{A0} dX_A + \int_0^t Q_{external}\, dt = \int_{T_0}^T \sum_j n_j \bar{c}_{pj} dT \tag{14.77}$$

Integrating X_A between 0 and X_A, T between T_0 and T, being t the common variable.

Substituting the external heat as a function of the overall heat transfer coefficient:

$$UA_s (T - T_s) = \Delta H_r n_{A0} X_A + \sum_j n_j \bar{c}_{pj} (T - T_0) \tag{14.78}$$

This is the energy balance equation for a batch reactor, in which T is the temperature of reaction, T_s is the temperature of the cooling/heating coil, and T_0 is the initial temperature of the system. All other parameters have already been defined.

14.4.3 Adiabatic batch reactor

The adiabatic batch reactor is completely isolated; therefore, no external heat transfer occurs. One can determine the temperature variation as a function of the reactant conversion from Equation 14.78. Thus, we obtain the same expression 14.75,

$$(T - T_0) = \frac{-\Delta H_r n_{A0} X_A}{\sum_j n_j \bar{c}_{pj}} \tag{14.79}$$

where n_{A0} and n_j are, respectively, the initial number of mol of reactant A and number of moles of the components in the system, $\bar{c}_{pj}$ is the molar specific heat of the components (reactants or products).

Along with the molar or mass balance equation one can calculate the variation of conversion or temperature and determine the reaction time and volume of the adiabatic batch reactor.

$$t = C_{A0} \int_0^{X_A} \frac{dX_A}{V(-r_A)} \qquad (14.16)$$

where:

$$(-r_A) = k_0 \exp(-E/(RT))f(X_A)$$

14.4.4 Analysis of the thermal effects

In the continuous and batch reactors, some parameters allow to perform an analysis of the thermal effects on the reaction temperature of the systems.

Let us consider the energy equation for PFR or CSTR reactors, Equation 14.73, or for batch reactor, Equation 14.79.

$$\sum_j F_j \bar{c}_{pj}(T - T_0) + \Delta H_r F_{A0} X_A = UA_s(T - T_s) \qquad (14.73)$$

The heat transfer through the cooling/heating coil or jacket is due to the sensible heat. By definition, T_{f0} is the temperature of the fluid at the entrance of the exchanger, $\bar{c}_{pf}$ is the fluid specific heat and v_0 is the feed flow rate of the fluid, i.e.:

$$Q_{external} = UA_s(T - T_s)$$

and,

$$Q_{external} = \rho_f v_{f0} c_{pf}(T_f - T_{f0})$$

By equating the equations one can determine the temperature T_f and replacing it, one obtains the heat removed or supplied:

$$Q_{external} = U^* A_s(T - T_{f0})$$

where:

$$U^* = \frac{1}{\left(\dfrac{A_s}{\rho_f v_{f0} c_{pf}}\right) + \dfrac{1}{U}} \qquad (14.80)$$

This is the overall heat transfer coefficient, which takes into account the coefficient of the cooling/heating fluid. A_s is the surface area of heat transfer.

From Equation 14.73 and some rearrangements, one can calculate the temperature of the reaction system:

$$\frac{T}{T_0} = \frac{1 - \dfrac{UA_s}{\sum_j F_j \bar{c}_{pj}} \dfrac{T_s}{T_0} - \dfrac{\Delta H_r F_{A0} X_A}{\sum_j F_j \bar{c}_{pj} T_0}}{1 - \dfrac{UA_s}{\sum_j F_j \bar{c}_{pj}}} \qquad (14.81)$$

Let us consider different cases to analyze the effect of parameters on the temperature of the reaction system.

- If the heat capacity of the feed flow rate is high, i.e.:

$$\rho_f v_{f0} c_{pf} \to \infty \text{ or } \sum_j F_j \bar{c}_{pj} \to \infty$$

  The reaction system will be isothermal, i.e., $T \to T_0$
- If the heat capacity of removal or addition of heat is high due to the overall heat transfer coefficient or due to the large heat transfer area (coil or jacket), then,

$$UA_s \to \infty$$

  In this case, the system will be isothermal when $T \to T_0$
- On the other hand, if the heat capacity is very low due to the low overall heat transfer coefficient or heat transfer area, then:

$$UA_s \to 0$$

In this case, there will not be heat transfer and the reactor will be adiabatic. Therefore, the temperature varies linearly with conversion.

In other nonisothermal cases, the parameters have significant influence on the temperature profile, especially the parameters: flow rate, heat capacity, heat transfer coefficients, and heat transfer area.

Let us consider as an example the adiabatic CSTR and an irreversible first-order reaction at constant volume. The reaction rate is:

$$(-r_A) = k_0 \exp\left(-E/(RT)\right) C_{A0}(1 - X_A)$$

The Arrhenius constant:

$$k = k_0 \exp\left(-E/(RT)\right) = k_0 \exp[-\gamma / (T / T_0)]$$

where $\gamma = \dfrac{E}{RT_0}$ (activation parameter)

By the energy balance Equation 14.75:

$$\frac{T}{T_0} = 1 + \frac{(-\Delta H_r)F_{A0}}{\sum_j F_j \bar{c}_{pj}} X_A$$

Defining:

$$\beta = \frac{(-\Delta H_r)F_{A0}}{\sum_j F_j \bar{c}_{pj} T_0} \tag{14.82}$$

where β is a new parameter which we will call the **energy parameter**, since it has the enthalpy of reaction in the numerator and the sensible energy of heat flow in the denominator.

Therefore, the temperature in an adiabatic CSTR or PFR system varies as follows:

$$\frac{T}{T_0} = 1 + \beta X_A \tag{14.83}$$

In a CSTR, the space time will be:

$$\frac{V}{v_0} = \tau = \frac{X_A}{k_0 \exp(-\gamma/(T/T_0))(1 - X_A)} \tag{14.84}$$

Substituting T of Equation 14.83, we obtain X_A:

$$X_A = \frac{\tau \cdot k_0 \exp(-\gamma/(1 + \beta X_A))}{1 + \tau \cdot k_0 \exp(-\gamma/(1 + \beta X_A))} \tag{14.85}$$

The results are shown in Figure 14:16 for different values of parameters γ and β. The variable $\tau.k_0$, which is proportional to reactor volume for some feed conditions, is shown as a function of conversion X_A.

If the reactor operates isothermally ($\beta = 0$), the volume increases with conversion since $\tau \approx V$. Comparing to an exothermic reaction, the volume depends on β and γ. For the same value of $\beta = 0.2$ (exothermic) but different activation energies, $\gamma = 1.0$ and 5.0, there is a large volume variation. One verifies the same by keeping $\gamma = 5.0$ and varying parameter $\beta = -0.2$ to 0.5.

More the exothermic reaction ($\beta > 0$) the smaller the variation of volume if conversion is higher than 50%. On the other hand, for endothermic reactions ($\beta < 0$) the reactor volume is higher if compared to isothermal reactions ($\beta = 0$).

Similar calculation can be performed to PFR reactors, noting that for an adiabatic reactor we use the same energy balance equation, but the molar balance equation should be in the integral form. Thus, using the same rate expressions for an irreversible first-order reaction,

$$(-r_A) = k_0 \exp(-E/(RT)) C_{A0}(1 - X_A)$$

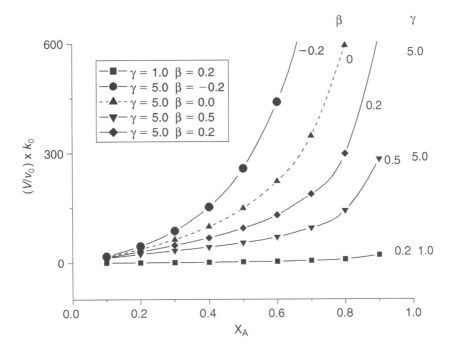

Figure 14.16 $(\tau \cdot k_0)$ as a function of conversion in nonisothermal CSTR reactor.

And the energy balance equation,

$$\frac{T}{T_0} = 1 + \beta X_A \tag{14.83}$$

Substituting the rate expression into molar balance equation, Equation 14.38, we obtain:

$$\tau = \frac{V}{v_0} = C_{A0} \int_0^{X_A} \frac{\mathrm{d}X_A}{k_0 \, \exp(-\gamma/(1 + \beta X_A)) \, (1 - X_A)} \tag{14.86}$$

Integrating different ranges of X_A, but keeping constant the energy parameters $\beta = 5.0$ and $\gamma = 0.5$ we obtain the curves shown in Figure 14.17. The volume of an adiabatic PFR reactor is significantly smaller than the volume of an isothermal reactor for the same reaction, especially if conversions are higher than 50%.

The comparison between CSTR and PFR reactors at isothermal and adiabatic conditions for the particular case $\beta = 0.5$ and $\gamma = 5.0$ is shown in Figure 14.18. One notes significant differences especially for conversions higher than 50%, which makes the adiabatic reactor to be very important.

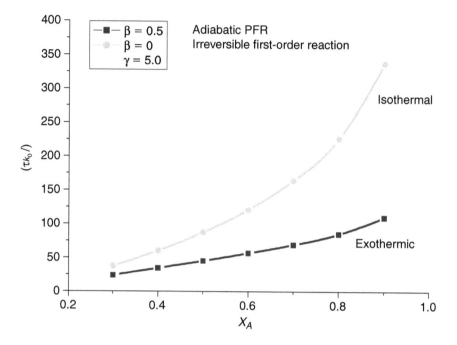

Figure 14.17 $(\tau \cdot k_0)$ as a function of conversion in nonisothermal PFR reactors.

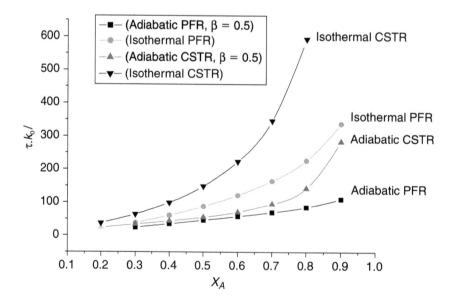

Figure 14.18 $(\tau \cdot k_0)$ as a function of conversion; comparison of nonisothermal reactors.

Example

E14.16 An irreversible reaction A → R is carried out in a nonisothermal CSTR reactor. Reactant is introduced into reactor at 130 kg/h at 20°C and the final conversion is 90%. The final temperature is 160°C. Calculate the heat required to reach this temperature and also the volume of the reactor operating isothermally at 160°C:

Rate constant:	$k = 2.61 \times 10^{14} e^{-(14570/T)} h^{-1}$
Enthalpy of reaction:	$\Delta H_R = 83$ cal/g
Specific heat:	$c_{PA} = c_{PB} = 0.5$ cal/(g · °C)
Density of the mixture:	$\rho = 0.9$ g/cm^3
Molecular weight of A:	250 g/mol

Solution
We calculate the CSTR volume for the isothermal condition at 160°C using the basic equation:

$$\frac{V}{v_0} = \tau = C_{A0} \frac{X_A}{(-r_A)} \tag{14.27}$$

The rate is of first-order, irreversible, and at constant volume.

$$(-r_A) = k_0 \underbrace{\exp(-E/(RT))}_{2.61 \times 10^{14} e^{-(14570/T)}} C_{A0}(1 - X_A)]$$

$$k = 0.635\underline{h}^{-1}$$

Calculating the volumetric flow:

$$\dot{G} = 130,000(\text{g/h}) = \rho \cdot v_0$$

$$v_0 = 0.144 \text{ m}^3/\text{h}$$

Substituting into Equation 14.87, we have:

$$\frac{V}{v_0} = \tau = \frac{X_A}{k(1 - X_A)} \tag{14.87}$$

$$V = 2.03 \text{ m}^3$$

Calculating the heat required to operate at 160°C, getting started at 20°C. Using the nonisothermal energy balance equation:

$$\sum_j F_j \bar{c}_{pj} (T - T_0) + \Delta H_r F_{A0} X_A = Q_{\text{external}} \tag{14.88}$$

where:

$$\sum_j F_j \bar{c}_{pj} = F_A \bar{c}_{pA} + F_R \bar{c}_{pR}$$

But:

$$F_A = F_{A0}(1 - X_A)$$

$$F_R = F_{A0} X_A$$

Therefore:

$$\sum_j F_j \bar{c}_{pj} = F_{A0}[(1 - X_A)\bar{c}_{pA} + X_A \bar{c}_{pR}](cal/(h \cdot {}^\circ C))$$

But $\bar{c}_{P_A} = \bar{c}_{P_R} \bar{c}_P$, then

$$\sum_j F_j \bar{c}_{pj} = F_{A0}[(1 - X_A) + X_A]\bar{c}_p = F_{A0}\bar{c}_p \qquad (14.89)$$

But,

$$F_{A0} = \frac{\dot{G}}{M} = \frac{130,000}{250} = 520(mol/h)$$

Or

$$F_{A0} \cdot M \cdot \bar{c}_p/M = G \times c_p = 130\,000 \times 0.5 = 6.5 \times 10^4\ (cal/(h \cdot K))$$

Equation 14.88 becomes:

$$(F_{A0}M) \cdot (\bar{c}_p/M)(T - T_0) + (\Delta H_r/M)(F_{A0}M)X_A = Q_{external} \qquad (14.90)$$

$$Q_{external} = G \cdot c_p(T - T_0) + \Delta H_r G \cdot X_A$$

$$Q_{external} = 1.88 \times 10^4\ kcal/h$$

Example

E14.17 A reversible second-order reaction was carried out in a nonisothermal CSTR and the rate as a function of temperature is known. The initial concentration is 8 kmol/m³. The volumetric flow rate is 1.415 m³/min and the initial temperature of the reactant is 50°C. The reaction temperature is 80°C and to keep it constant one utilizes a coil through which passes a fluid entering at 20°C and leaving at 80°C. Determine the reactor volume and the coil surface area to a conversion of 60% of the equilibrium conversion at 80°C. Some data are:

Specific heat of the mixture → $C_p = 1.864 \times \cdot 10^3$ J/(kg°C)
Enthalpy of reaction → $\Delta H_R = -1.50 \times 10^7$ J/kmol
Overall heat transfer coefficient → $U = 2.300$ J/(m² s°C)
Molecular weight of reactant → $M_A = 88$
Flow rate of fluid into the coil → $v_{c0} = 0.2$ m³/min
Reaction rate → $r = 1.11 \times \exp(12.5 - 5000/T)(1 - X_A)^2 - 2 \times \exp(25 - 10000/T)X_A^2$ – (lbmol/(ft³.min))

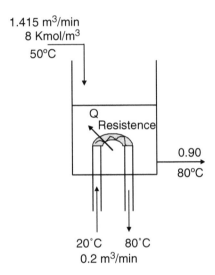

1.415 m³/min
8 Kmol/m³
50°C

Q
Resistence

0.90
80°C

20°C   80°C
0.2 m³/min

Solution

- *Equilibrium conversion at 80°C:*
   Using the rate expression and $T = 80°C$, we obtain:

$$r = 0.210 \times (1 - X_A)^2 - 0.0716X_A^2 = 0$$

$$X_{Ae} = 0.63$$

Since the conversion at the reactor outlet should be 60% of equilibrium conversion, then:

$$X_A = 0.378$$

- *Reactor volume:*

$$\frac{V}{v_0} = \tau = C_{A0}\frac{X_A}{r} \tag{14.27}$$

where the resulting rate will be equal to the reaction rate at 80°C:

$$r = 0.210 \ (1 - X_A)^2 - 0.0716 X_A^2$$

Substituting the rate expression and final conversion into Equation 14.27, we obtain:

$$r = 0.0708 \ \text{lbmol} \ / \ (\text{ft}^3 \ \text{min}) = 1.133 \ \text{kmol} \ / \ (\text{m}^3 \ \text{min})$$

With the volumetric flow rate $v_0 = 8$ m³/min:

$$V = 3.77 \ \text{m}^3$$

- Coil surface area:
  By the energy balance in the nonisothermal CSTR:

$$Q_{external} = \dot{G} \cdot c_p (T - T_0) + \Delta H_r \dot{G} X_A = U^* A_s (T - T_{s0})$$

$$F_{A0} = C_{A0}.v_0 = 1.415 \times 8 = 11.32 \ \text{kmol/min}$$

$$\dot{G} = F_{A0}.M = 11.32 \times 88 = 996.16 \ \text{kg/min}$$

$$c_p = 1.864 \times 10^3 \ \text{J/(kg K)}$$

$$60 U A_s = -8.48 \times 10^6$$

$$U A_s = 1.41 \times 10^5$$

$$A_s = 1.02 \ \text{m}^2$$

Example

E14.18 The irreversible second-order reaction $A + B \rightarrow 2R$ in liquid phase was carried out in an adiabatic CSTR of 1.5 L. The initial concentrations of A and B are respectively 10 and 40 mol/L, the volumetric flow at the reactor entrance is 0.2 L/min and the initial temperature is 17°C.

Determine the conversion and temperature. If the feed flow rate is 10 times lower, calculate the conversion and temperature. If the inlet temperature is 50°C and the inlet concentration 3.78 mol/L, calculate the conversion and temperature. Some data follow:

Specific heat of the mixture → $\bar{c}_p = 2.7$ cal/(g K)
Enthalpy of reaction → $\Delta H_R = -34,000$ cal/mol
Rate constant → $k = 33 \times 10^9 \exp(-20,000/(RT))$ (L/(mol min))

Solution

The reaction takes place in an adiabatic CSTR reactor in liquid phase. To calculate the conversion we use the CSTR equation:

$$\frac{V}{v_0} = \tau = \frac{X_A}{k(1 - X_A)} \tag{14.91}$$

Since the initial concentrations of A and B are respectively 10 and 40 mol/L, the reactant A is the limiting one.

$$(-r_A) = 33 \times 10^9 \exp(-20,000/(RT))C_{A0}^2(1 - X_A)(M - X_A)$$

where:

$$M = \frac{C_{B0}}{C_{A0}} = 4.0$$

Therefore:

$$\tau = C_{A0}\frac{X_A}{33 \times 10^9 \exp(-20,000/(RT))C_{A0}^2(1 - X_A)(M - X_A)} \tag{14.92}$$

There is temperature variation into the exponential term of the rate constant. Therefore, if the reaction occurs adiabatically, we should determine the temperature variation using the energy balance for the adiabatic reactor, whose solution was presented in Equation 14.83:

$$\frac{T}{T_0} = 1 + \beta X_A \tag{14.83}$$

where:

$$\beta = \frac{-\Delta H_r F_{A0}}{\sum_j F_j \bar{c}_{pj}} = \frac{-\Delta H_r C_{A0}}{\rho c_p T_0} = \frac{34,000 \times 10}{1070 \times 2.70 \times 290} \tag{14.93}$$

$$\beta = 0.407$$

Then:

$$T = T_0(1 + 0.407X_A) \tag{14.94}$$

We substitute T into Equation 14.92. To solve and make it explicit as a function of conversion, we can simplify the equation. Since $M = 4.0$, M is much higher than X_A. Therefore, we simplify the rate:

$$(-r_A) = 33 \times 10^9 \exp(-20,000/(RT))MC_{A0}^2(1 - X_A)$$

Thus, the kinetic expression is considered pseudo first order and by making it explicit as a function of conversion, we obtain an equation that has already been deduced (Equation 14.85).

$$X_A = \frac{\tau \cdot k_0 \exp(-\gamma/(1+\beta X_A))}{1+\tau \cdot k_0 \exp(-\gamma/(1+\beta X_A))} \tag{14.95}$$

Case 1

$\beta = 0.407$

$k_0 = 33 \times 10^9$

$\gamma = \dfrac{E}{RT_0} = 34.8$

$\tau = \dfrac{V}{v_0} = 7.5 \text{ min}$

Substituting these values into Equation 14.95 and solving, we obtain three solutions:

$X_A = 0.0085$ and $T = 18°C$

$X_A = 0.337$ and $T = 56.7°C$

$X_A = 0.994$ and $T = 134.3°C$

Case 2–If $v_0 = 0.02$ L/min

$\tau = \dfrac{V}{v_0} = \dfrac{1.5}{0.02} = 75 \text{ min}$

By Equation 14.94, we obtain:

$X_A = 1.0 \rightarrow$ total conversion

$k_0 M C_{A0} = 9.9 \times 10^{13}$

Case 3–If $v_0 = 0.2$ L/min, $T_0 = 50°C$ and $C_{A0} = 3.78$ mol/L, then:

$\beta = 0.138$

$\gamma = 31.2$

$X_A = 0.897$

$T = 90°C$

One observes that increasing the inlet temperature (T_0) and reducing the feed flow rate (v_0), thus increasing residence time, the conversion increases.

Example

E14.19 The gas-phase reaction of butadiene with ethylene $(A + B \rightarrow R)$ is irreversible and of second order and has been carried out in an adiabatic PFR. One feeds the reactor with 40% butadiene, 40% ethylene, and balance in inert (molar %) at 450°C and 1.25 atm.

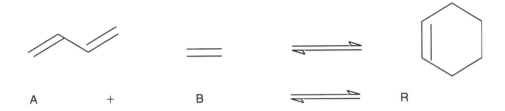

A   +   B   R

Determine the mean residence time and the PFR volume at isothermal conditions, as well as the reactor volume at the adiabatic condition for a final conversion of 10% in both cases, using the data:

Molar specific heat →

$\bar{c}_p$ C4H6 $= 36.8$ cal/(mol K)

$\bar{c}_p$ C2H4 $= 20.2$ cal/(mol K)

$\bar{c}_p$ C6H10 $= 59.5$ cal/(mol K)

- *Reaction heat* → $\Delta H_R = -30000\ Kcal/mol$
- *rate constant* → $k = 10^{7.5} \exp(-28{,}000/(RT))$ (L/(mol s))
- $v_0 = 0.05$ L/s

Solution
The rate expression to an irreversible second-order reaction with the same initial concentrations of reactants A and B will be:

$$(-r_A) = kC_{A0}^2 \frac{(1 - X_A)^2}{(1 + \varepsilon_A X_A)^2} \left(\frac{T}{T_0}\right)^2 \qquad (14.96)$$

The volume of an isothermal reactor can be determined by substituting the rate equation into the PFR molar balance, Equation 14.38:

Isothermal

$$\tau = \frac{V}{v_0} = C_{A0} \int_0^{X_A} \frac{dX_A}{r}$$

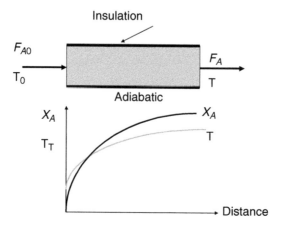

Figure E14.5 Profile of temperature and conversion.

Obtaining as solution:

$$\tau = \frac{1}{kC_{A0}}\left[(1+\varepsilon_A)^2\frac{X_A}{(1-X_A)} + \varepsilon_A^2 X_A + 2\varepsilon_A(1+\varepsilon_A)\ln(1-X_A)\right] \qquad (14.97)$$

Calculating ε_A:

	C_4H_6+	C_2H_4	C_6H_{10}	Inert	Total
Initial	0.4	0.4	0	0.2	1.0
Final	0	0	0.4	0.2	0.6

$\varepsilon_A = -0.4$

$$C_{A0} = y_{A0}\frac{P}{RT} = 8.43 \times 10^{-3}\,\text{mol/L}$$

$k = 0.1012\,\text{L/(mol s)}$

Therefore:

$\tau = 124.8\,\text{s}$

$V_{PFR} = 6.2\,\text{L}$

The mean residence time in a PFR is given by Equation 14.47, but with a correction when the reaction is not isothermal.

Isothermal

$$\bar{t} = C_{A0} \int_0^{X_A} \frac{dX_A}{(1 + \varepsilon_A X_A)(-r_A)} \tag{14.47}$$

Substituting the rate expression 14.95, we obtain:

$$\bar{t} = \int_0^{X_A} \frac{(1 + \varepsilon_A X_A)^2 dX_A}{k\, C_{A0}(1 + \varepsilon_A X_A)(1 - X_A)^2} \tag{14.48}$$

Integrating, we obtain:

$$\bar{t} = -\frac{1}{kC_{A0}} \left[\frac{X_A(1 + \varepsilon_A)}{(1 - X_A)} + \varepsilon_A \ln(1 - X_A) \right]$$

$$\bar{t} = 159.4s$$

Nonisothermal

The rate expression to an irreversible second-order reaction with the same initial concentrations of reactants A and B will be:

$$(-r_A) = kC_{A0}^2 \frac{(1 - X_A)^2}{(1 + \varepsilon_A X_A)^2} \left(\frac{T}{T_0} \right)^2$$

We determine the temperature performing the energy balance in an adiabatic system, according to the equation:

$$\frac{T}{T_0} = 1 + \beta X_A \tag{14.83}$$

where:

$$\beta = \frac{-\Delta H_r F_{A0}}{\sum_j F_j \bar{c}_{pj}}$$

where:

$$\sum_j F_j \bar{c}_{pj} = F_{A0}[(1 - X_A)\bar{c}_{pA} + (1 - X_A)\bar{c}_{pB} + X_A \bar{c}_{pR}]$$

$$\sum_j F_j \bar{c}_{pj} = F_{A0}(57 + 2.5X_A) \tag{14.99}$$

Therefore:

$$\beta = \frac{41.5}{(57 + 2.5X_A)}$$

$$\frac{T}{T_0} = 1 + \frac{41.5X_A}{(57 + 2.5X_A)} \tag{14.100}$$

The parameter γ is defined as a function of the activation energy:

$$\gamma = \frac{E}{RT_0} = 18.8$$

Substituting the expression 14.100 and 14.98 into Equation 14.38, we obtain:

$$\tau = \int_0^{X_A} \frac{dX_A}{10^{7.5} C_{A0} \frac{(1 - X_A)^2}{(1 + \varepsilon_A X_A)^2} \left((1 + \frac{41.5X_A}{(57 + 2.5X_A)})^2 \right) \times \exp\left(-18.8 / \left(1 + \frac{41.5\ X_A}{(57 + 2.5X_A)} \right) \right)} \tag{14.101}$$

where:

$$C_{A0} = y_{A0} \cdot \frac{P}{RT_0} = 8.43 \times 10^{-3}\,\text{mol/L}$$

Integrating:

$$\tau = 47.0\,\text{s}$$
$$V = 2.35\,\text{L}$$

Therefore, the nonisothermal reactor has a smaller volume.

Chapter 15

Specific reactors

The kinetics of reactions is specific for different reaction systems and processes and valid for isothermal and nonisothermal reactors. The effects of the kinetics on the conversion, selectivity, or yield depend on the reaction and may be quite pronounced. Liquid or gas phase reactions with high heat capacity can be performed in specific reactors, which operate isothermally or not. We will study the most common cases such as semibatch reactors, recycle reactors, fixed-bed reactors, and reactors with membranes.

15.1 SEMIBATCH REACTOR

The semibatch reactor (tank or tubular) contains a large amount of liquid reactant that continuously reacts on the addition of a second reactant (liquid or gas), which is instantly consumed. The kinetics is the same, but consumption increases over reaction time as the second component is added (Figure 15.1).

The semibatch reactors can be operated by three different manners:

1 The tank contains a liquid reactant B into which the reactant A (liquid or gas) is continuously added—isothermal system.

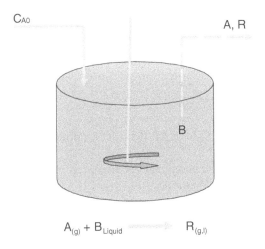

$$A_{(g)} + B_{Liquid} \longrightarrow R_{(g,l)}$$

Figure 15.1 Scheme of semibatch reactor.

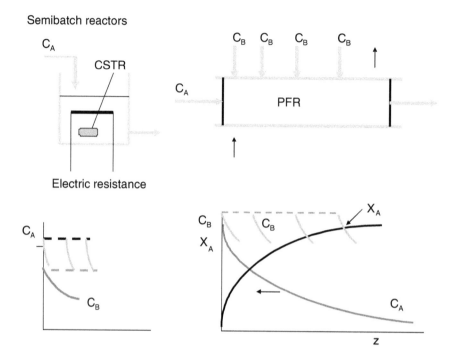

Figure 15.2 Concentration profiles in different cases.

2   The tank contains the reactants while the gas product is continuously formed.
3   The tank contains the liquid reactant while the gas reactant A is continually added. Nonreacted A is continually released.

The reactant concentration profiles have been represented in Figure 15.2 for semibatch reactor under isothermal conditions.

The concentration is uncertain because it varies over time. Consider a semiopen system according to the scheme shown in Figure 15.3.

F_A represents the molar flow of reactant A, then:

$$F_{A0} - F_A + (-r_A) \cdot V(t) = \frac{dn_A}{dt} \tag{15.1}$$

where $(-r_A)$ = reaction rate, F_A = molar flow of component A, and V = system volume which varies over time.

Since $F_{A0} = C_{A0} v_0$ and the reagent A is completely consumed during the reaction, there is no outlet flow of A. Therefore,

$$F_A = 0$$

$$C_{A0} v_0 - 0 + (-r_A) \cdot V(t) = \frac{d(C_A V)}{dt} = V \frac{dC_A}{dt} + C_A \frac{dV}{dt} \tag{15.2}$$

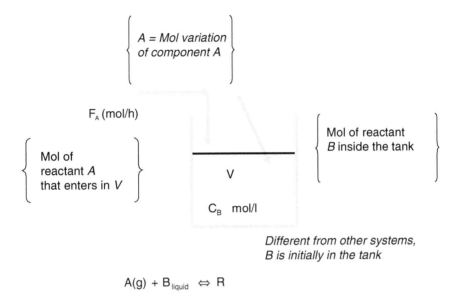

Figure 15.3 Molar mass balance in the semibatch reactor.

The volume variation over time caused by addition of reagent A into reagent B (present in the tank) is calculated by means of an overall mass balance taking into account the average density ρ in the system without chemical reaction. Thus:

$$\rho_0 v_0 - 0 + 0 = \frac{d(\rho V)}{dt} \tag{15.3}$$

Assuming that the system average density is constant, i.e.:

$$\rho = \rho_0 = \text{constant}$$

We obtain:

$$v_0 = \frac{dV}{dt} \tag{15.4}$$

With the boundary condition:

$$t = 0 \rightarrow V = V_0$$

We obtain the volume variation over time, i.e.:

$$V = V_0 + v_0 t \tag{15.5}$$

Or dividing by v_0, we obtain the following expression as a function space time:

$$\tau = \tau_0 + t \tag{15.6}$$

Substituting Equation 15.4 into Equation 15.2:

$$C_{A0}v_0 - C_A v_0 + (-r_A) \cdot V(t) = V \frac{dC_A}{dt} \tag{15.7}$$

But:

$$\frac{dC_A}{dt} = \frac{dC_A}{d\tau} \underbrace{\frac{d\tau}{dt}}_{\substack{=1 \\ eq15.6}} \tag{15.8}$$

Substituting Equation 15.8 into Equation 15.7, we obtain:

$$(C_{A0} - C_A)v_0 + (-r_A) \cdot V(t) = V \frac{dC_A}{d\tau} \tag{15.9}$$

Let us consider an irreversible second-order reaction:

$$A + B \rightarrow products$$

where the liquid reactant B with concentration C_B is present in the reactor of volume V and the reactant A is added at a constant rate. Therefore, the kinetics may be represented by:

$$(-r_A) = kC_A C_B \tag{15.10}$$

The concentration of reactant B is much higher than A and varies slightly with the addition of the reactant A, therefore one may assume that its variation is negligible with respect to A. In this case, one assumes a pseudo first-order reaction.

$$(-r_A) = kC_A C_{B0} = k^* C_A \tag{15.11}$$

where:

$$k^* = kC_{B0}$$

Substituting the rate expression 15.11 into Equation 15.9, we obtain:

$$(C_{A0} - C_A)v_0 - k^* C_A V(t) = V \frac{dC_A}{d\tau} \tag{15.12}$$

Rearranging:

$$\frac{dC_A}{d\tau} + \frac{(1 + \tau k^*)}{\tau} C_A = \frac{C_A}{\tau} \tag{15.13}$$

We solve this equation using the integrating factor $f_i = \exp(\int f(x)dx)$, with the initial condition: $\tau = \tau_0 \rightarrow C_A = C_{Ai}$.

C_{Ai} is the initial concentration of A in the tank before the reaction.

The solution will be:

$$\frac{C_A}{C_{A0}} = \frac{1}{(t+\tau_0)k^*} - \left(\left(\frac{1}{\tau_0 k^*}\right) - \frac{C_{Ai}}{C_{A0}}\right) \cdot \frac{\tau_0}{(t+\tau_0)} \cdot e^{-k^*t} \tag{15.14}$$

Example

E15.1 The irreversible reaction $A + B \rightarrow$ products, has the rate represented by a pseudo first-order reaction. This reaction takes place in a semibatch reactor of $10\,L$ and at $25°C$, some data follow:

$$t = 0 \rightarrow C_{A0} = 3 \times 10^{-4}\,\text{mol/L}$$
$$\rightarrow C_{Ai} = 5 \times 10^{-5}\,\text{mol/L}$$
$$\rightarrow v_0 = 2\,\text{L/min}$$
$$\rightarrow (-r_A) = 0.38\,\text{mol/(L min)}$$

Calculate the time required to reach the stabilized conversion.

Solution
By Equation 15.14, we calculate:

$$\tau_0 = \frac{V}{v_0} = 5\,\text{min}$$

With $k^* = 0.38$, we calculate:

$$\tau_0 k^* = 1.9$$

Substituting into Equation 15.14:

$$\frac{C_A}{C_{A0}} = \frac{1}{0.38(t+5)} - \left(\frac{1}{1.9} - \frac{5 \times 10^{-5}}{3 \times 10^{-4}}\right) \cdot \frac{5}{(t+5)} e^{0.38t}$$

By the definition of conversion:

$$X_A = 1 - \frac{C_A}{C_{A0}}$$

We have:

$$X_A = 1 - \frac{2.63}{(t+5)} + \frac{1.8}{(t+5)} e^{-0.38t} \tag{15.15}$$

t (min)	0	0.5	1.0	1.5	3.0	5.0	10.0	20.0	25.0
X_A		0.474	0.792	0.766	0.751	0.743	0.764	0.827	0.895	0.912

The results have been displayed in Figure E15.3.

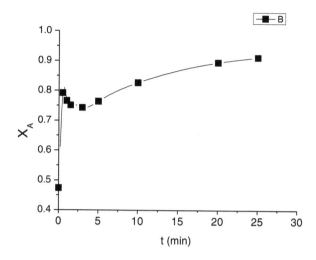

Figure E15.3 Conversion as function of time.

15.2   REACTOR WITH RECYCLE

The reactors with recycle are continuous and may be tanks or tubes. Their main feature is increasing productivity by returning part of unconverted reactants to the entrance of the reactor. For this reason, the reactant conversion increases successively and also the productivity with respect to the desired products. The recycle may also be applied in reactors in series or representing models of nonideal reactors, in which the recycle parameter indicates the deviation from ideal behavior. As limiting cases, we have ideal tank and tubular reactors representing perfect mixture when the recycle is too large, or plug flow reactor(PFR) when there is no recycle.

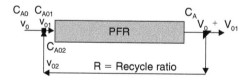

For PFR tubular reactor, Equation 14.38 is valid:

$$\tau = \frac{V}{v_0} = C_{A0} \int_{X_{A1}}^{X_A} \frac{dX_A}{r}$$ (15.16)

The lower limit in the reactor inlet changed to X_{A1}. The conversion at any position or at the outlet is X_A (Figure 15.4).

where:

$$C_A = C_{A0}(1 - X_A)$$

and:

$$C_{A1} = C_{A0}(1 - X_{A1})$$

To simplify, we assume an irreversible first-order reaction as follows: $A \rightarrow$ Products, whose kinetics at constant volume is represented by:

$$(-r_A) = kC_A = kC_{A0}(1 - X_A)$$

Substituting the rate expression into Equation 15.16 and integrating, we obtain:

$$\tau = -\frac{1}{k} \ln \frac{(1 - X_A)}{(1 - X_{A1})} \qquad (15.17)$$

Conversion X_{A1} is still unknown and maybe determined by the balance at the intersection(1)by calculating the inlet concentration C_{A1} or conversion X_{A1}. Thus, at the intersection (1), we have:

$$v_{01} = v_0 + v_{02}$$
$$F_{A01} = F_{A0} + F_{A02}$$
$$C_{A1} \cdot v_{01} = C_{A0}v_0 + C_A v_{02} \qquad (a)$$

Then, the concentration C_{A1} will be:

$$C_{A1} = \frac{C_{A0}v_0 + C_A v_{02}}{v_{01}} \qquad (15.18)$$

Substituting Equation (a) into 15.18 and considering a recycle ratio $R = \frac{v_{02}}{v_0}$, we have:

$$C_{A1} = C_{A0}\left(1 - \frac{R}{(R+1)}X_A\right) \qquad (15.19)$$

or substituting C_{A1} and C_{A0} by the respective conversions, we obtain directly:

$$X_{A1} = \frac{R}{(R+1)}X_A \qquad (15.20)$$

Substituting Equation 15.20 into Equation 15.17:

$$\tau_{recycle} = -\frac{1}{k} \ln \frac{(1+R)(1-X_A)}{(1+R(1-X_A))} \qquad (15.21)$$

This expression depends on the recycle ratio R and there are two particular cases:

1 If there is no recycle, $R = 0$, the reactor behaves like an ideal PFR.

$$\tau_{PFR} = -\frac{1}{k} \ln(1 - X_A) \qquad (15.22)$$

If the recycle is large ($R \to \infty$), there will be a mixing effect resulting in a solution corresponding to the ideal CSTR reactor. In this case, expanding the logarithmic function in series and neglecting the terms of higher order, one obtains:

$$\tau_{CSTR} = \frac{1}{k} \frac{X_A}{(1 - X_A)} \qquad (15.23)$$

Therefore, depending on the recycle ratio, one obtains solutions that may indicate an increase in the final conversion or productivity and may represent behavior of nonideal reactor. The recycle ratio would be a parameter that indicates the deviation from ideal behavior. It is equivalent to the average residence time, which also indicates the extent of deviation from the ideal behavior of a reactor.

Example

E15.2 A liquid phase reaction $A \to R + S$ takes place in a system consisting of a CSTR and a PFR reactor with recycle. One introduces1 kmol/m^3 of reagent A into the first reactor. Both CSTR and PFR operate isothermally at 300°C and atmospheric pressure. The volume of the first reactor is 0.086 m^3 and the inlet flow is 1.6 m^3/ks. The conversion at the outlet of the PFR is 90%.

It is a first-order reaction and the rate constant is $(-r_A) = 8 \, C_A$ kmol/(m^3 h).
Calculate the PFR volume considering a recycle ratio of $R = 1$.

Solution

Since we have a first-order reaction, the reaction rate is:

$$-r_A = 8C_A = 8C_{A0} (1 - X_A)$$

In the CSTR:

$$\tau_{CSTR} = \frac{1}{k} \frac{X'_{A1}}{(1 - X'_{A1})}$$

Substituting the values, we have:

$$\tau = \frac{V}{v_0} = \frac{0.083}{1.6} = 0.0518 \, ks = 0.0144 \, h$$

Then:

$$X'_{A1} = 0.103$$

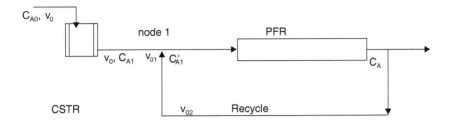

Balance at the intersection (1):

$$C_{A1}v_{01} = C_{A1}v_0 + C_A v_{02} \qquad \text{(a)}$$

$$v_{01} = v_0 + v_{02}$$

$$1 + R = \frac{v_{01}}{v_0}$$

$$v_{01} = v_0 (1 + R) \qquad \text{(b)}$$

where:

$$R = \frac{v_{02}}{v_0}$$

Substituting (b) into (a):

$$C_{A1}(1 + R) = C_{A1} + C_A R$$

Or as function of conversions:

$$1 - X_{A1} = \frac{(1 - X'_{A1}) + R(1 - X_A)}{(1 + R)}$$

Substituting $X'_{A1} = 0.103$ and $X_A = 0.9$:

$$X_{A1} = 0.501$$

For the second reactor (PFR) with recycle, we utilize Equation 15.17:

$$\tau = -\frac{1}{k} \ln \frac{(1 - X_A)}{(1 - X_{A1})}$$

$$\tau_{\text{PFR}} = 722 \text{ seg} = 0.722 \text{ ks}$$

where:

$$v_{01} = v_0(1 + R) = 3.2 \text{ m}^3/\text{ks}$$

We obtain:

$$V = \tau_{PFR} \cdot v_{01} = 0.722 \times 3.2 = 2.3 \, m^3$$

$$V = 2.3 \, m^3$$

15.3   PSEUDO-HOMOGENEOUS FIXED-BED REACTOR

In the ideal CSTR and PFR reactors in homogeneous phase, the reactants and products constitute a single phase. The homogeneous reaction takes place in a liquid or gas phase whose concentration varies throughout the PFR, while in the CSTR reactor the concentration is instantaneous.

In a specific case, when the reaction takes place on the catalyst present on the internal wall of the tube and the flow is uniform, the behavior is equal to a reaction in homogeneous phase.

If the catalyst consists of particles, one has a fixed or mobile stationary particles reactor. The flow is no longer uniform and the reaction is considered heterophasic. In this case, the catalyst consists of solid particles, but the reactants and products are gas or liquid and may flow through the particle bed in the concurrent or counter current direction. The reaction may be catalytic or noncatalytic and takes place on the particle surface and/or within the pores of the catalyst. In this case, the system is heterophasic and may be solid–gas type, solid–liquid, or solid–liquid–gas taking place in the respective reactors: fixed or mobile-bed, trickle bed, and slurry bed.

These reactors exhibit complex fluid dynamics. The velocity profile in a cross section is not uniform due to the influence of the wall and the porosity of the catalytic bed, causing the nonideality of the continuous reactor.

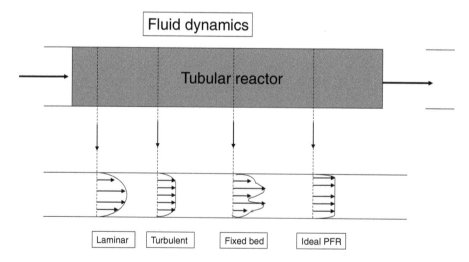

The solid particles have high specific surface areas due to high porosity (pore volume). Convection phenomena occur in fixed or mobile-bed reactor due to three

mechanisms: molecular diffusion, Knudsen diffusion within the particle pores, and surface diffusion, depending on the geometrical characteristics of the pores and particles.
Usually, the effective diffusion is determined combining these three phenomena.

Therefore, there are axial and radial dispersion effects due to mass and heat transfer
phenomena in isothermal and nonisothermal systems.

For mass transfer by convection, the mass flow is given by:

$$m = k_c \Delta C$$

where ΔC = concentration gradient (radial or axial) and k_c = mass transfer coefficient.

Fick's law describes the mass transfer by diffusion within the pores as:

$$\dot{m} = D_e \frac{dC}{dz}$$

where dC/dz = concentration gradient in the pore (cylinder) and D_e = effective gas
diffusion in the pore.

The Reynolds number is a flow characteristic, and if defined as a function of
particle diameter, it will be:

$$Re = \frac{u d_p}{v}$$

u = superficial gas velocity in the cross section
d_p = particle diameter
v = kinematic viscosity.

$$(15.24)$$

Moreover the Péclet number(Pé) characterizes radial and axial dispersions, which
represent the convective and diffusive coefficients in the pores (see Chapter 24):

$$\text{Pé} = \frac{u d_p}{D_e} = \frac{\text{Convective transport rate}}{\text{Diffusive transport rate}} \qquad (15.25)$$

where u = superficial gas velocity in the cross section, d_p = particle diameter, and D_e =
effective diffusivity.

The radial and axial dispersions generally exist in a particulate catalytic bed reactor, but under certain conditions one can neglect the radial gradients. One can take into
account the uniform velocity over a cross section (ideal PFR) when the ratio between
tube and particle diameters is higher than 30 and when the Reynolds number is high,
characterizing a quasi-turbulent flow, i.e., high superficial velocities (high flows). In
this case, Pé can be calculated according to the empirical equation (Aris,1969):

$$\text{Pé} = 1 + 19.4 \left[\left(\frac{d_p}{d_t} \right)^2 \right]^{-1} \qquad (15.26)$$

where d_t = tube diameter.

To $Pé = 10$ and ratio $(d_t/d_p) \geq 30$, the radial dispersion can be neglected under isothermal conditions. The superficial velocity is constant and the radial concentration is negligible.

To $Pé \approx 2$, the axial dispersion can be neglected if the ratio between the reactor length and diameter tube is higher than 50 $(L/d_t > 50)$.

To a very high $Pé$ number, one can consider a fixed-bed reactor as a pseudo-homogeneous one, therefore, a reactor with ideal behavior. In these conditions, its superficial velocity is high and/or the effective gas diffusion in the pores (D_e) is low. It means that if $(1/Pé) \to 0$, the gas residence time in the reactor tends to the average residence time of an ideal reactor according to Figure E15.6. Figure 15.6 shows $Pé$ values for different flows and the limits to achieve ideal conditions. The criterion $(1/Pé) = 0.002$ is used as a limit for a PFR in ideal conditions.

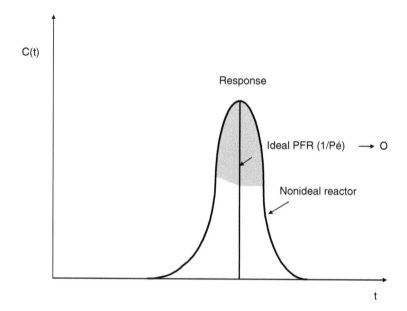

Figure E15.6 Concentration distribution in the reactor.

Therefore, the Péclet number is an important parameter to evaluate if mass transfer limitations take place, and the Reynolds number to verify if the reactor is in turbulent flow and consequently quasi-ideal conditions in which one may apply the basic equations for ideal reactors.

Example

E15.3 Hydrogenation of toluene was carried out in a tubular fixed-bed reactor containing a catalyst $(\rho = 2.3\ \mathrm{g/cm^3})$. The reaction is in gas phase operating isothermally at 600°C and pressure of 10 atm. One introduces the reactants as 20% toluene, 40% hydrogen, and balance in inert (%vol.) with a volumetric flow of 400 L/min. The benzene outlet flow is 10 mol/min. Calculate the catalyst mass in the reactor assuming that the kinetics is given by the following rate expression (Froment and Bischoff, 1979):

$$-r_T = \frac{1.4 \times 10^{-8} p_{H2} p_T}{(1 + 1.26 p_B + 1.01 p_T)} \frac{mol_{Toluene}}{g_{cat}\ s}$$

(15.27)

where B = benzene and T = toluene.

Additional data:
Kinematic viscosity: $v = 0.74 \times 10^{-6}$ m²/s
Tube diameter: $d_t = 2.54$ cm
Particle diameter: $d_p = 0.5$ cm
Diffusion coefficient $(H_2) = 0.1$ cm²/s.

Solution

T = 600°
P = 10 atm
ρ = 2.3 g/cm³

1   *Calculating the superficial velocity:*

$$A = \frac{\pi d^2}{4} = 5.06\ cm^2$$

$$u = \frac{v_0}{A} = \frac{4 \times 10^5}{5.06} = 7.9 \times 10^4\ cm/min = 13.1\ m/s$$

2   *Calculating the Reynolds number:*

$$Re = \frac{u d_p}{v} = \frac{13.1 \times 2.54}{0.74 \times 10^{-6}} = 4.52 \times 10^5$$

The Reynolds number is very high, characterizing a turbulent flow. One may consider a uniform flow in the cross section and also along the bed.

3   *Calculating the Péclet number:*

$$Pé = \frac{u d_p}{D_e} = \frac{7.9 \times 10^4 \times 0.5}{0.1 \times 60} = 6.58 \times 10^3$$

Therefore:

$$(1/Pé) = 1.5 \times 10^{-4}$$

The criterion for a reactor with ideal behavior is $1/Pé = 2 \times 10^{-3}$.

Therefore, the mass transfer limitations and diffusion effects are negligible.

The Reynolds and Péclet numbers indicate a pseudo-homogeneous reactor behavior, close to an ideal reactor. In such a condition, one can calculate the catalyst mass using the equations of an ideal PFR.

Therefore, by Equation 14.38:

$$\tau = \frac{W}{F_{T0}} = C_{A0} \int_0^{X_T} \frac{dX_T}{r} \tag{15.38}$$

The rate can be expressed as a function of toluene conversion assuming that:

$$p_T = C_T RT = C_{T0}(1 - X_T)RT$$

$$p_H = C_H RT = C_{T0}(M - X_T)RT$$

$$p_B = C_B RT = C_{T0}X_T RT$$

where:

$$C_{T0} = \frac{y_{T0}P}{RT} = \frac{0.2 \times 10}{RT} = \frac{2}{RT}$$

Substituting the partial pressures into Equation 15.27, we obtain:

$$M = \frac{C_{H0}}{C_{T0}} = 2$$

or

$$(-r_T) = \frac{5.6 \times 10^{-8}(1 - X_T)(2 - X_T)}{(1 + 2.52X_T + 2.02(1 - X_T))} \tag{15.28}$$

$$(-r_T) = \frac{3.36 \times 10^{-6}(1 - X_T)(2 - X_T)}{(3.02 + 0.5X_T)}$$

Substituting the rate expression 15.28 into Equation 14.38:

$$\frac{W}{F_{T0}} = \int_0^{X_T} \frac{(3.02 + 0.5X_T)dX_T}{3.36 \times 10^{-6}(1 - X_T)(2 - X_T)} \tag{15.29}$$

Calculating F_{T0}

$$F_{T0} = y_{T0}F_0 = C_{T0}v_0 = \frac{0.2 \times 10 \times 400}{0.082(273 + 600)} = 11.17 \, mol/min$$

Calculating the final conversion:

$$F_B = F_{T0} X_T$$

$$X_T = 0.89$$

Integrating Equation 15.29 and using the values above, we obtain:

$$W = 2.2 \times 10^4 \, kg_{cat}$$

15.4 MEMBRANE REACTORS

The membrane reactors enable one to separate the reaction products increasing the production capacity and also to shift the reaction beyond the equilibrium under the same conditions of pressure and temperature. This type of reactor consists of a permeable membrane with selective porous walls. The membrane should be able to withstand high temperatures and sintering without blocking fluid flow.

An example is the use of membrane reactors to generate and separate hydrogen to fuel cells.

Hydrogen can be obtained by steam reforming and autothermal reforming of methane which is represented by the following reactions:

Steam reforming: $CH_4 + H_2O \Leftrightarrow CO + 3H_2$ $\Delta H_{298K} = 206 \, kJ/mol$
Dry reforming: $CH_4 + CO_2 \Leftrightarrow 2CO + 2H_2$ $\Delta H_{298K} = 247 \, kJ/mol$
Partial oxidation of methane: $CH_4 + 1/2O_2 \Leftrightarrow CO + 2H_2$ $\Delta H_{298K} = -36 \, kJ/mol$
Water gas shift reaction: $CO + H_2O \leftrightarrow CO_2 + H_2$ $\Delta H_{298K} = -41 \, kJ/mol.$

Steam reforming and dry reforming are endothermic, but partial oxidation of methane and water gas shift reactions are exothermic. The autothermal process becomes exothermic and is thermodynamically favorable.

Two types of membrane reactors are shown in Figure 15.7:

1 Catalytic membrane (active porous wall) and
2 Noncatalytic membrane (inert).

In the first case, the membrane itself contains the catalyst, such as platinum dispersed on the membrane. The reaction takes place within the membrane and the products pass through well-defined pores.

In the second case, the membrane is porous and inert but the catalyst is placed inside the reactor where the reactions take place. One of the products is usually separated by passing through the inert porous membrane, the other passing directly to the exit of the reactor. Thereby, one increases the yield or selectivity to the desired product, also increasing the conversion since the reaction shifts toward products.

Let us consider the following reaction:

$$A + B \rightarrow R + S$$

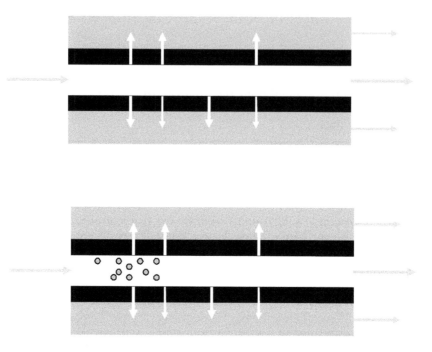

Figure 15.7 Scheme of membrane reactors.

where R is the product to be separated by the membrane.

By the molar balance of any component in the volume element dV, we have:

$$F_j - (F_j + dF_j) + \int r_j\, dV = \frac{dn_j}{dt} \tag{15.30}$$

Considering the steady state, we obtain generically (Equation 14.34):

$$\int dF_j = \int r_j\, dV$$

To the component A, we have (Equation 14.37):

$$\frac{V}{F_{A0}} = \int\limits_0^{X_A} \frac{dX_A}{(-r_A)} \tag{15.31}$$

The molar balance with respect to the product R to be separated by the membrane is:

$$F_R - (F_R + dF_R) + r_R\, dV - \dot{R}_R\, dV = 0 \tag{15.32}$$

Therefore:

$$dF_R = (r_R - \dot{R}_R)dV \tag{15.33}$$

The product formation rate is related to the reagent.

$$-r_A = r_R$$

And the rate R_R is defined as the molar flow passing through the membrane:

$$\dot{R}_R = k_c C_R \tag{15.34}$$

where k_c is the mass transfer coefficient, which depends on the fluid and membrane properties.

Example

E15.5 The gas phase reaction $A \overset{k}{\underset{k'}{\rightleftharpoons}} R$ is carried out in a membrane reactor and R is the product to be separated by the membrane. The reaction is reversible and of first order with $k = 0.05\,\text{s}^{-1}$. The reactor operates at 8.2 atm and isothermally at 227°C. The equilibrium constant at 227°C is 0.5. The inlet flow into reactor is 2 mol/min and its volume is 10 L. Calculate the outlet flow of R assuming $k_c = 0.3\,\text{s}^{-1}$.

Solution
We have a reversible and first-order reaction in gas phase. Therefore (Equation 3.19):

$$r = kC_A - k'C_R = kC_{A0}\left[(1 - X_A) - \frac{1}{K}X_A\right]$$

At equilibrium, the rate is zero. Thus:

$$X_{Ae} = \frac{K}{1 + K} = 0.33$$

The resulting rate as a function of the equilibrium conversion will be:

$$r = kC_{A0}\left(1 - \frac{X_A}{X_{Ae}}\right) \tag{15.35}$$

No separation

$$\tau = \frac{V}{v_0} = C_{A0}\int_0^{X_A} \frac{dX_A}{r} \tag{15.36}$$

Substituting the rate Equation 15.36, we have:

$$\tau = -\frac{1}{k}\ln\left(1 - \frac{X_A}{X_{Ae}}\right) \tag{15.37}$$

Calculating the initial concentration:

$$C_{A0} = \frac{P}{RT} = \frac{8.2}{0.082 \times 500} = 0.2 \, \text{mol/L}$$

Calculating the volumetric flow:

$$v_0 = \frac{F_{A0}}{C_{A0}} = \frac{2}{0.2} = 10 \, \text{L/min}$$

Calculating the space time:

$$\tau = \frac{V}{v_0} = 1.0 \, \text{min}$$

Then, with $k = 0.05 \, \text{s}^{-1}$ and Equation 15.37, we obtain:

$$\frac{X_A}{X_{Ae}} = 0.95$$

$$X_A = 0.316$$

With separation
The rate changes by including the mass transfer term.

$$r = kC_A - k'C_R - R_R = kC_A - k'C_R - k_c C_R$$

or

$$r = kC_{A0}\left[(1 - X_A) - \frac{1}{K}X_A\right] - k_c C_{A0} X_A$$

Rearranging:

$$r = kC_{A0}\left(1 - k^* \cdot \frac{X_A}{X_{Ae}}\right) \tag{15.38}$$

where:

$$k^* = (1 + k_c) = 1.3 \, \text{s}^{-1}$$

Substituting Equation 15.38 into Equation 15.36 and integrating, we obtain:

$$\tau = -\frac{X_{Ae}}{kk^*}\ln\left(1 - \frac{k^* X_A}{X_{Ae}}\right) \tag{15.39}$$

Data: $k = 0.05 \, \text{s}^{-1}$, $\tau = 1 \, \text{min} = 60 \, \text{s}$, we calculate a new conversion:

$$X_A = 0.248$$

Finally, we calculate the flow of R.

No separation

$$F_R = C_R v_0 = C_{A0} v_0 X_A = F_{A0} X_A = 2 \times 0.316 = 0.632 \, \text{mol/min}$$

With separation

$$F_R = F_{A0} X_A = 0.427 \, \text{mol/min}$$

The molar flow passing through the membrane will be:

$$R_R = k_c C_R = k_c C_{A0} X_A = 0.0644 \, \text{mol/(L s)}$$

Chapter 16

Comparison of reactors

The choice of the reactor is very important and should be carried out on technical basis. There are three main criteria to distinguish batch, tubular, and tank reactors:

1 Reactor volume
2 Productivity
3 Selectivity

For the first criterion, one compares the reactor volumes based on the average residence time for a given extent of reaction or final conversion. The average residence time depends on the reaction kinetics and therefore the reaction rate, which in turn depends on whether the reaction takes place at constant volume or variable volume. In a system at constant volume, one obtains directly a ratio between the volumes, because the average residence time is equal to space time which is defined as the ratio between reactor volume and inlet volumetric flow in the reactor. For the same conversion, the ratio between volumes is proportional. Since the average residence time in a PFR reactor is similar to the reaction time in a batch reactor, we may assume that they have similar behaviors and then we compare only the ideal tubular reactors (PFR — plug flow reactor) to the ideal tank reactors (CSTR—continuous stirred-tank reactor).

16.1 COMPARISON OF VOLUMES

To visualize the difference between the volumes and contact times in the reactors, we observe Figure 16.1, which shows the inverse of rate as a function of conversion. This curve represents the kinetic behavior of the reaction irrespective of the reactor type. The reaction rate decreases with increasing conversion, therefore the inverse of rate increases and is positive.

The basic equations for the reactors have already been seen as:

Batch reactor:

$$t = C_{A0} \int_0^{X_A} \frac{dX_A}{(-r_A)} \tag{16.1}$$

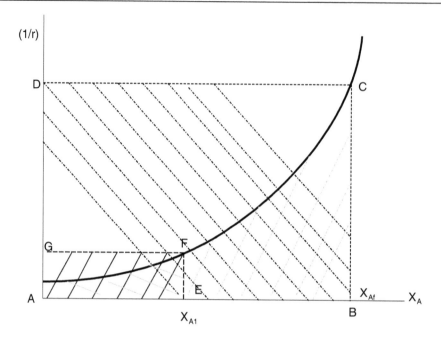

Figure 16.1 Inverse of reaction rate as a function of conversion.

Tank (CSTR):

$$\frac{V}{v_0} = \tau = C_{A0}\frac{X_A}{(-r_A)}$$

(16.2)

Tubular (PFR):

$$\tau = \frac{V}{v_0} = C_{A0}\int_0^{X_A}\frac{dX_A}{r}$$

(16.3)

Analyzing the equations of PFR, CSTR, and batch reactors represented by Equations 16.1–16.3, one notes that the area under the curve AC represents the integral in the PFR equation, whereas the rectangle area ABCD represents the CSTR equation. The rate related to the outlet concentration in the CSTR is equal to the average concentration in the tank, because perfect mixture is considered.

For a given constant temperature and reactions taking place at constant volume, these areas are directly proportional to the space time and also to the average residence time in the reactors, assuming the same inlet volumetric flow and the same final conversion. Therefore, the areas represent the reactors volume.

From Figure 16.1, one concludes that the CSTR volume is always greater than the PFR volume. The space time or average residence time (contact time) in the PFR is always smaller than in the CSTR. Since the reaction time in the batch reactor is equal to the contact time in the PFR, we can compare the batch reactor directly to the

PFR. If the reaction takes place at variable volume, then the average residence times are different from the space times and as a consequence, the volumes are not proportional.

Figure 16.1 illustrates that for low conversions, for example, X_{A1}, the CSTR and PFR volumes are similar, since the respective areas AEFG and under A′F curve are similar. As the desired conversion is decreased, the areas for the two reactor types approach the same value. Small reactors lead to small conversions and are useful for kinetic studies.

Let us consider some cases and compare CSTR and PFR volumes in order to verify the most significant comparison criteria.

16.1.1   Irreversible first-order reaction at constant volume

The reaction rate expression is:

$$(-r_A) = kC_A = kC_{A0}(1 - X_A)$$

Therefore, substituting the rate expression into PFR (16.3) and CSTR (16.2) equations and integrating, we obtain respectively:

$$\tau = -\frac{1}{k}\ln(1 - X_A) \tag{16.4}$$

$$\tau = \frac{1}{k}\frac{X_A}{(1 - X_A)} \tag{16.5}$$

For the same volumetric feed flow, the space times are directly proportional to the corresponding volumes. The ratio between CSTR and PFR volumes considering $\varepsilon_A = 0$,

$$\frac{V_{CSTR}}{V_{PFR}}$$

is shown in Figure 16.2 (curve 1) as a function of final conversion.

For conversions below 20%, the volumes have same order of magnitude, not varying more than 15%. Calculating the ratio,

$$\frac{V_{CSTR}}{V_{PFR}}$$

for $X_A = 0.20$, we obtain the value of 1.12. The CSTR volume is slightly larger than the PFR volume (i.e., 12%).

However, for conversions above 50%, the CSTR and PFR volumes differ significantly. For a conversion $X_A = 0.80$, this ratio is 2.5. Therefore, the CSTR volume is 2.5 times larger than the PFR volume.

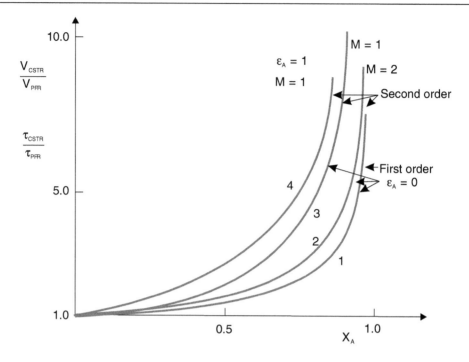

Figure 16.2 Comparison of reactors volume for different parameters.

16.1.2 Irreversible second-order reaction at constant volume

For an irreversible second-order reaction, the main parameter is the ratio between the initial concentrations:

$$M = \frac{C_{B0}}{C_{A0}}$$

The ratio between the CSTR and PFR volumes will be:

$$\frac{V_{CSTR}}{V_{PFR}} = \frac{\dfrac{X_A}{(1 - X_A)(M - X_A)}}{\dfrac{1}{(M-1)} \ln \dfrac{(M - X_A)}{M(1 - X_A)}} \tag{16.6}$$

The ratio of volumes as a function of conversion is shown in Figure 16.2 (curves 2 and 3). The influence of parameter M is significant, especially for high conversions.

The influence of the reaction order on the CSTR and PFR volumes is considerable for first- and second-order reactions, $M = 1$ and conversion $X_A = 0.80$. Taking into account a second-order kinetics, the CSTR volume is about five times larger than for a first order. On the other hand, this ratio is lower in the PFR, since the PFR volume is only about 2.5 times larger for a second-order kinetics, compared to a first-order and admitting kinetic constants of same order of magnitude.

16.1.3   Reactions at variable volume

The influence of volume variation during a reaction, where there is contraction or expansion of fluid, on the CSTR and PFR volumes may be expressed as a function of the parameter ε_A, for first- or second-order reactions. The effect of ε_A on the CSTR or PFR volume is depicted in Figure 16.2 (curve 4 with $M = 1$).

Increasing ε_A (when $\varepsilon_A > 0$, expansion), the ratio between the reactors volume,

$$\frac{V_{CSTR}}{V_{PFR}}$$

increases as well with respect to the reaction whose volume does not change ($\varepsilon_A = 0$), for the same final conversion. When contraction occurs ($\varepsilon_A < 0$), we have the opposite. For example, for an irreversible second-order reaction with $\varepsilon_A = 1$ and conversion $X_A = 0.80$, we obtain:

$$\frac{V_{CSTR}}{V_{PFR}} = 5.1$$

Examples

E16.1 Effect of parameter M

Butanol and monobutylphthalate (MBF) react in the presence of H_2SO_4 forming dibutylphthalate and water. Reaction can be carried out in CSTR or PFR reactor. The reactants are placed in two separate tanks containing 0.2 mol/L of MBF and 1 mol/L of butanol and fed to the reactor under 10 L/h and 30 L/h, respectively. The reactants are mixed before entering the reactor. The reaction rate constant is $k = 7.4 \times 10^{-2}$ L/(mol min). Calculate the CSTR and PFR volume separately and show the relation between the volumes when conversion is 70%.

Solution

Reaction rate expression: $-r_A = kC_A C_B$, where $A = $ Butanol and $B = $ MBF (reactants).

Total volumetric flow in the reactor entrance:

$$v_{01} + v_{02} = v_0 = 40 \, \text{L/h}$$

The concentrations in the reactor entrance:

$$C_{A01} \cdot v_{01} + C_{A02} \cdot v_{02} = C_{A0} \cdot v_0$$

With the previous data, we obtain:

$$C_{A0} = 5 \times 10^{-2} \, \text{mol/L}$$
$$C_{B0} = 7.5 \times 10^{-1} \, \text{mol/L}$$

442   Chemical reaction engineering

Considering reaction stoichiometry $1A:1B$, the reactant A is the limiting one. Therefore, the rate will be:

$$(- r_A) = kC_{A0}^2(1 - X_A)(M - X_A)$$

where:

$$M = \frac{C_{B0}}{C_{A0}} = 1.5$$

In the CSTR: for $X_A = 0.70$

$$\tau_{CSTR} = \frac{X_A}{kC_{A0}(1 - X_A)(M - X_A)}$$

Therefore, the CSTR volume, with $v_0 = 40$ L/h is:

$$V_{CSTR} = 520\,L$$

In the PFR:

$$\tau_{PFR} = \frac{1}{kC_{A0}(M - 1)} \ln \frac{(M - X_A)}{M(1 - X_A)} \qquad (14.40)$$

$$\tau_{PFR} = 311\ \text{min} = 5.2\ \text{h}$$

The PFR volume, with $v_0 = 40$ L/h is:

$$V_{PFR} = 207\,L$$

Then, the ratio between CSTR and PFR volumes will be:

$$\frac{V_{CSTR}}{V_{PFR}} = 2.5$$

E16.2 Effect of parameter ε_A
The gas-phase reaction $A + B \to R$ is carried out in a PFR and in a tank (CSTR) under isothermal condition at 550°C and 1 atm. The initial reactants' concentrations are equimolar and the inlet volumetric flow is 10 L/min. It is an irreversible second-order reaction and the specific reaction rate is given by:

$$k = 7 \times 10^4 \cdot e^{\frac{-9000}{T}}\ (L/(mol\ s))$$

Calculate the average residence time, space time, and reactor volume for a final conversion of 75% by carrying out the reaction in a PFR and CSTR separately. What is the ratio between their volumes?

Solution

The reaction rate as a function of conversion to an irreversible second-order reaction, with $C_{A0} = C_{B0}$ will be:

$$-r_A = kC_A^2 = \frac{C_{A0}^2(1 - X_A)^2}{(1 + \varepsilon_A X_A)^2}$$

And the average residence time in the PFR is as follows:

$$\bar{t} = C_{A0} \int_0^{X_A} \frac{dX_A}{(1 + \varepsilon_A X_A)(-r_A)}$$

Substituting $(-r_A)$ into the above equation and integrating, we obtain:

$$\bar{t} = \frac{1}{kC_{A0}} \left[\frac{1 + \varepsilon_A X_A}{1 - X_A} + \ln(1 - X_A) \right]$$

Calculating ε_A:

A	B	R	Inert	Total	
0.4	0.4	0	0.2	1.0	
0	0	0.4	0.2	0.6	$\varepsilon_A = -0.4$ (contraction)

Calculating C_{A0}:

$$C_{A0} = \frac{y_{A0}P}{RT} = \frac{0.4 \times 1}{0.082 \times 823} = 5.92 \times 10^{-3}$$

We obtain k by Arrhenius equation provided:

$$k = 0.712 \, L/(mol\,s)$$

Substituting the values into equation:

$$\bar{t} = 22.7 \, min$$

Space time can be determined by equation:

$$\tau = \frac{V}{v_0} = C_{A0} \int_0^{X_A} \frac{dX_A}{(-r_A)}$$

Substituting $(-r_A)$ and integrating:

$$\tau = \frac{1}{kC_{A0}} \left[(1 + \varepsilon_A)^2 \frac{X_A}{(1 - X_A)} + \varepsilon_A^2 X_A + 2\varepsilon_A(1 + \varepsilon_A) \ln(1 - X_A) \right]$$

Substituting $X_A = 0.70$, we obtain:

$\tau = 7.38\,\text{min}$

Therefore, the PFR volume will be:

$V_{\text{PFR}} = 73.8\,\text{L}$

In the CSTR, we obtain analogously:

$$\tau_{\text{CSTR}} = \frac{X_A(1 + \varepsilon_A X_A)^2}{k C_{A0}(1 - X_A)^2}$$

$\tau_{\text{CSTR}} = 15.9\,\text{min}$

That leads to a reactor volume of:

$V_{\text{CSTR}} = 159\,\text{L}$

Thus, the ratio between the different reactors is:

$$\frac{V_{\text{CSTR}}}{V_{\text{PFR}}} = 2.2$$

16.2 PRODUCTIVITY

The second criterion takes into account the yield and productivity for a given reactor volume.

Definition

The production rate in a system is defined as the number of moles of product, or converted reactant to main product, per unit of volume and time. It is an overall average rate, different from reaction rate defined previously, which is local.

Therefore, if α_A is the number of moles of reactant A converted to product over an average contact time or average residence time, one defines the productivity as (Hill, 1977):

$$W = \frac{\alpha_A}{\bar{t}} = \frac{C_{A0} \cdot X_A}{\bar{t}} \tag{16.7}$$

However, the average residence time previously defined depends on the system used: if at constant or at variable volume.

For the CSTR:

$$\bar{t}_{\text{CSTR}} = \frac{V}{v_f} = \frac{V}{v_0(1 + \varepsilon_A X_A)} = \frac{\tau_{\text{CSTR}}}{(1 + \varepsilon_A X_A)} \tag{16.8}$$

For the PFR:

$$\bar{t}_{PFR} = C_{A0} \cdot \int_0^{X_A} \frac{dX_A}{(1 + \varepsilon_A X_A)(-r_A)} \tag{16.9}$$

In both cases, the average contact times are different when the reaction volume (fluid volume) is constant or variable, depending on ε_A.

Then,

- If $\varepsilon_A = 0$ $\bar{t}_{CSTR} = \tau_{CSTR}$ and $\bar{t}_{PFR} = \tau_{PFR}$
- If $\varepsilon_A \neq 0$ $\bar{t}_{CSTR} \neq \tau_{CSTR}$ and $\bar{t}_{PFR} \neq \tau_{PFR}$
- In any case: $\bar{t}_{PFR} = t_{batch}$

Therefore, productivity (W) in both cases is given by the following equations:

CSTR:

$$W_{CSTR} = \frac{\alpha_A}{\bar{t}} = \frac{C_{A0} \cdot X_A}{\bar{t}} = (1 + \varepsilon_A X_A)(-r_A) \tag{16.10}$$

PFR:

$$W_{PFR} = \frac{\alpha_A}{\bar{t}} = \frac{C_{A0} \cdot X_A}{\bar{t}} = \frac{X_A}{\int_0^{X_A} \frac{dX_A}{(1 + \varepsilon_A X_A)(-r_A)}} \tag{16.11}$$

To illustrate, let us consider a reversible forward and reverse first-order reaction $A \leftrightarrow R$. The resulting rate as a function of conversion is known. Substituting it into Equations 16.10 and 16.11, we obtain the productivity in PFR and CSTR:

$$W_{PFR} = \frac{C_{A0} \cdot k(1 + N)X_A}{(N + X_{Ae}) \ln \left(1 - \frac{X_A}{X_{Ae}}\right)} \tag{16.12}$$

where:

$$N = \frac{C_{R0}}{C_{A0}}$$

And

$$W_{CSTR} = \frac{C_{A0} \cdot k(1 + N)X_{Ae}}{(N + X_{Ae})} \left(1 - \frac{X_A}{X_{Ae}}\right) \tag{16.13}$$

The productivity, or average production rate, depends on the final conversion and equilibrium conversion for a given temperature. Relating the CSTR and PFR rates, one obtains the relative productivity of the reactors. The average production rate is

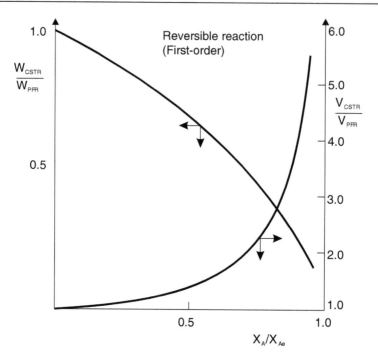

Figure 16.3 Comparison of productivity and reactors' volume.

inversely proportional to the reactor volume. The ratio between CSTR and PFR rates and the respective volumes ratio are shown in Figure 16.3.

The unitary production of the CSTR decreases significantly with respect to the PFR at high conversions, while the CSTR volume increases much more with respect to the PFR. For instance, at a conversion of 70% of equilibrium conversion, the CSTR productivity is 50% lower than the PFR, considering the same feed conditions. To increase the CSTR or PFR productivity, one can make use of reactors in series.

16.3 YIELD/SELECTIVITY

If more than one reaction takes place and a specific product or by-product is desired, the reactor choice and reaction conditions are very important, since the kinetics of each reaction depends on the temperature and reaction order.

Initially, let us define yield and selectivity for multiple reactions occurring simultaneously, for reactions in parallel, series, or mixed. When there is only one reaction step, the yield is confused with the conversion.

Considering the reaction system:

$$A + B \xrightarrow{k_1} R \text{ (desired product)}$$

$$A + B \xrightarrow{k_2} S \text{ (undesired product)}$$

The definitions follow (Hill, 1977):

Yield:

$$\varphi_A = \frac{\text{number of mol of } A \text{ transformed in desired product } (R)}{\text{number of mol of } A \text{ reacted (total)}} \qquad (16.14)$$

Operational Yield:

$$\varphi_o = \frac{\text{number of mol of } A \text{ transformed in desired product } (R)}{\text{total number of mol fed into reactor}} \qquad (16.15)$$

Selectivity:

$$S_A = \varphi_A \quad \text{(same definition of yield)}$$

or

$$S_A = \frac{\text{number of mol of } A \text{ transformed in desired product } (R)}{\text{number of mol of } A \text{ transformed in undesired product } (S)} \qquad (16.16)$$

Since the reactant A is present in both chemical reactions, we take A as a reference.

Let us consider the extent of reaction α_1 and α_2 (for first and second reactions, respectively). Then, by definition of yield φ_A or selectivity S_A, the mole number of A transformed in R at the time (t) will be α_1 (batch) or position (z) (continuous), and therefore,

$$\varphi_A = S_A = \frac{a\alpha_1}{r(n_{A0} - n_A)} \qquad (16.17)$$

where $(F_{A0} - F_A)$ is the molar flow of A reacted in both reactions.

If F_0 is the total molar flow fed into reactor (reactants and inert), the operational yield is as follows:

$$\phi_o = S_o = \frac{a\alpha_1}{r \cdot n_0} \qquad (16.18)$$

Similarly, the selectivity is defined considering α_2 as the molar flow of product S formed through the second reaction:

$$S_A = \frac{a_1\alpha_1}{a_2 \cdot \alpha_2} \qquad (16.19)$$

where a_1, a_2, and r are the stoichiometric coefficients of the respective compounds and reactions.

The yield and selectivity represent a local measure and depend on the feed and reactor type. In the CSTR, perfect mixing is obtained, which leads the local yield to

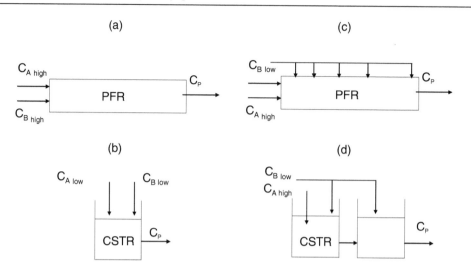

Figure 16.4 Reactor schemes—yield.

be equal to the final yield, i.e., since the reaction is instantaneous, the outlet flow has the same product concentration as the reactor.

Depending on the residence time or space velocity, the reactor is not suitable for multiple reactions where the by-products formed depending on the residence time in the reactor.

In the PFR, the concentration varies along the reactor and the yield as well (therefore we have local yields). Depending on the residence time or space velocity, the desired product formation can be facilitated; on the other hand, undesired by-products might also take place.

Comparing CSTR and PFR for the same conversion and space velocity, the yield in the PFR is higher, since the final A concentration is lower and R concentration is higher. The yields in the CSTR and PFR reactors are high always when the final product concentration is high, while the reactant concentration is low. Therefore, the higher the α_1 and $(n_{A0}-n_A)$ the higher the φ_A.

Depending on the initial concentration, one can design different feed into reactors in order to obtain a higher yield, as shown in Figure 16.4. With two reactants, A and B, both with low concentration, the CSTR is the best choice. However, if the concentrations are high, depending on the residence time, the yield in the PFR will be higher. If the concentrations have different levels, one can use the schemes (c) and (d) with intermittent addition in several positions in the PFR reactor or many CSTR in series.

16.4 OVERALL YIELD

As seen, the overall yield in a CSTR is equal to the local yield, since the reaction is instantaneous. On the other hand, the concentration in the PFR changes along the reactor, and therefore, the yield and selectivity are also local, varying along the reactor.

In the CSTR, the overall yield will be:

$$\Phi_A = \varphi_A = \frac{a\alpha_1}{r(C_{A0} - C_A)} \tag{16.20}$$

In the PFR, the concentration varies (dC_A) in a volume dV, and thus, the local yield in the differential form will be according to Equation 16.17:

$$d\varphi_A = -\frac{a \cdot d\alpha_1}{r \cdot dC_A} \tag{16.21}$$

where a and r are the stoichiometric coefficients of the chemical reaction.

Therefore, the final concentration of the desired product will be:

$$\alpha_1 = -\int_{C_{A0}}^{C_A} \frac{r}{a} \varphi_A \, dC_A \tag{16.22}$$

Substituting Equation 16.22 into overall yield Equation 16.20, we obtain the following expression for a system at constant volume (as a function of concentration):

$$\Phi_A = -\frac{1}{(C_{A0} - C_A)} \int_{C_{A0}}^{C_A} \varphi_A \, dC_A \tag{16.23}$$

where Φ_A is the overall yield.

16.4.1 Effect of reaction order

Let us consider the reactions in parallel:

$A \xrightarrow{k_1} R,$ where R is the desired product

$A \xrightarrow{k_2} S,$ S is the undesired product

The kinetic equations for systems at constant volume have a generic order, i.e.:

$$r_R = k_1 \cdot C_A^{a_1}$$
$$r_S = k_2 \cdot C_A^{a_2} \tag{16.24}$$

The number of moles of A transformed into R per unit of volume and time represents the formation rate of R, while the total number of moles of A reacted, under the same conditions, represents the consumption rate of A or formation rate of all products, then the local yield can be written as follows:

$$\varphi_A = \frac{r_R}{(r_R + r_S)} = \frac{r_R}{(-r_A)} \tag{16.25}$$

Substituting the rate expressions (Equation 16.24), we obtain:

$$\varphi_A = \frac{k_1 C_A^{a_1}}{(k_1 C_A^{a_1} + k_2 C_A^{a_2})} = \frac{r_P}{(-r_A)} \tag{16.26}$$

Rearranging:

$$\varphi_A = S_A = \frac{1}{\left[1 + \dfrac{k_2}{k_1} \cdot C_A^{(a_2 - a_1)}\right]} \tag{16.27}$$

The overall yield in the CSTR is equal to the local yield, Equation 16.27, while in the PFR the yield should be integrated, substituting Equation 16.27 into Equation 16.23, i.e.:

$$\Phi_A = -\frac{1}{(C_{A0} - C_A)} \int_{C_{A0}}^{C_A} \frac{1}{\left[1 + \dfrac{k_2}{k_1} \cdot C_A^{(a_2 - a_1)}\right]} dC_A \tag{16.28}$$

High yield is naturally desired and as the above equations show, they depend on the kinetics of each reaction, particularly the reaction order and kinetic constants for each step.

Analyzing the effect of the reaction order, having in mind that R is the desired product, by Equation 16.27, we note that if a_1 is higher than a_2, then the difference $(a_1 - a_2)$ is negative and therefore, the concentration appears in the denominator. In this case, to obtain a high yield, the local concentration should be high, i.e., low conversion so that $\varphi_A \to 1$. Otherwise, if $a_2 > a_1$, the concentration remains in the numerator and the local C_A concentration should be low, i.e., high conversion. For high conversions, the PFR reactor is suggested; while for low conversions, the CSTR is more appropriate. If a_1 is larger than a_2, the CSTR is the best solution.

16.4.2   Effects of kinetic constants

The yield also depends on the reaction rate constants or specific reaction rates (k_1 and k_2), and therefore depends on the reaction temperature. According to Arrhenius, the ratio between these constants is:

$$\frac{k_2}{k_1} = k_{021} e^{-(E_2 - E_1)/RT} \tag{16.29}$$

The effect of temperature depends on the activation energy of each reaction. If both energies are equal, the ratio depends only on the ratio between the frequency factors k_{021}. If $E_2 > E_1$, then we have the difference $(E_2 - E_1) = \Delta E > 0$, and thus the exponential term,

$$e^{-\Delta E/RT}$$

increases with increasing temperature and consequently k_2/k_1 increases. On the other hand, if $E_1 > E_2$, the exponential term decreases with increasing temperature and k_2/k_1 also decreases, indicating that the constant k_2 diminishes with respect to k_1. Consequently, an increase in temperature enhances the desired product (R) yield if $E_1 > E_2$. However, the yield is more sensitive to temperature variations if $E_2 > E_1$. In this case, the desired product yield is higher by decreasing the temperature.

E16.3 Let us consider a system with two irreversible decomposition reactions at constant volume. The first one is a second-order reaction while the other is a first-order reaction, as in the following rate expressions:

$$r_R = k_1 \cdot C_A^2$$
$$r_S = k_2 \cdot C_A \tag{16.30}$$

This reaction may be carried out in PFR and CSTR reactors separately. Calculate the local and overall yields as a function of outlet concentration in the reactor, considering the same space velocity and constant temperature. Obtain the yield and selectivity to the desired product R from the first reaction.

Solution

The local yield will be expressed by:

$$\varphi_A = S_A = \cfrac{1}{\left[1 + \cfrac{k_2}{k_1} \cdot \cfrac{1}{C_A}\right]} \tag{16.31}$$

We define $\chi^* = \frac{k_2}{k_1 C_{A0}}$ as a parameter that relates the two kinetic constants.

The CSTR overall yield is equal to the local yield and thus:

$$\Phi_{\text{CSTR}} = \cfrac{1}{\left[1 + \chi^* \cdot \cfrac{C_{A0}}{C_A}\right]}$$

The PFR overall yield is determined by substituting Equation 16.31 into Equation 16.23.

$$\Phi_{\text{PFR}} = 1 + \cfrac{\chi^*}{1 - \cfrac{C_A}{C_{A0}}} \ln \cfrac{\chi^* + \cfrac{C_A}{C_{A0}}}{1 + \chi^*} \tag{16.32}$$

The overall yield is a function of concentration and depends on the parameter χ^*. Figure 16.5 shows the PFR and CSTR overall yields for $\chi^* = 0.5$ and the ratio between CSTR and PFR overall yields for different values of χ^*. But $\chi^* = \frac{k_2}{k_1 C_{A0}}$ depends on the initial concentration C_{A0}.

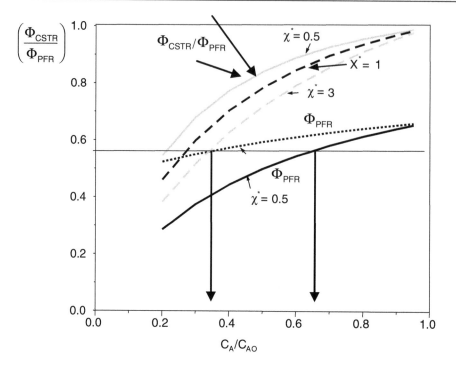

Figure 16.5 PFR and CSTR overall yield when $\chi^* = 0.5$ and ratio between CSTR and PFR overall yields Φ_{CSTR}/Φ_{PFR}.

At low conversions, yields in the CSTR and PFR are similar, but as the reactant A concentration decreases, or its conversion increases, the yields become quite different. However, the PFR yield is always higher than the CSTR.

Opposite to the reactors' volume comparison, in which the CSTR volume is always larger than the PFR, the yield in the PFR is higher than in the CSTR. The same analysis is valid for selectivity comparison between CSTR and PFR reactors.

For the selectivity with respect to the desired product, according to Equation 16.19, we obtain:

$$S_{local} = \frac{r_R}{r_S} = \frac{k_1 C_A^2}{k_2 C_A} = \frac{1}{\chi^*} \frac{C_A}{C_{A0}} \tag{16.33}$$

The ratio between CSTR and PFR selectivities is depicted in Figure 16.6.

The yield and selectivity (relative selectivity) are distinct. The yield ranges from 0 to 1.0 while the selectivity may be higher than 1, since it relates the formation rate of the desired product with the undesired one.

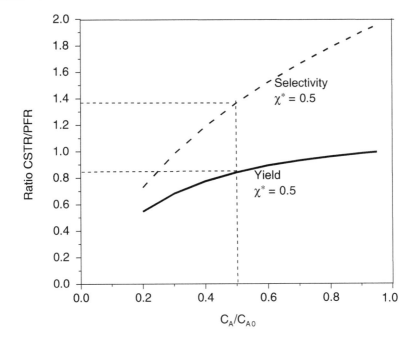

Figure 16.6 Relative selectivity and relative yield.

16.4.3   Presence of two reactants

When there is the simultaneous presence of a reactant—in this case B—in two parallel reactions, the yield is defined with respect to the reactant participating in both reactions. For instance,

$$A + B \xrightarrow{k_1} R \rightarrow \text{desired product}$$

$$B \xrightarrow{k_2} S \rightarrow \text{undesired product}$$

Assuming generic order for the product reaction rates:

$$r_R = k_1 C_A^{a_1} C_B^{b_1}$$
$$r_S = k_2 C_B^{b_2} \tag{16.34}$$

Since reactant B participates in both reactions and R is the desired product, the yield will be:

$$\varphi_B = \cfrac{1}{\left[1 + \cfrac{k_2}{k_1} \cfrac{C_B^{(b_2 - b_1)}}{C_A^{a_1}} \right]} \tag{16.35}$$

Analogously, if $(b_2 - b_1) > 0$, then it is necessary to have low concentrations of B at the reactor outlet. Therefore, a good scheme is a PFR reactor with A fed at the reactor inlet while B fed at several positions so that reagent B may be immediately consumed. This sequential B feeding allows complete consumption of reagent A until the end of the reactor.

Otherwise $(b_2 - b_1) < 0$, both reactants should leave the reactor at high concentration. A CSTR reactor is the best option.

Example

E16.4 The thermal decomposition of butane in butadiene takes place at 650°C and 1 atm. But then, butadiene cracks and polymerizes according to the following rates:

$$\text{Butadiene} \xrightarrow{k_1} \text{dimer (R)(desired product)}$$
$$\text{Butadiene} \xrightarrow{k_2} \text{HC (S)(undesired product)}$$

$$r_R = k_1 p_B^2$$
$$r_S = k_2 p_B \quad\quad\quad (16.36)$$

where:

$$\ln k_1 = \frac{25,000}{4.57T} + 8.063$$
$$\ln k_2 = \frac{30,000}{4.57T} + 7.24$$

The feed into reactor comprises 1 mol/L of hydrocarbons distributed in 72.5% butene and 27.5% butadiene. Determine the final conditions in the CSTR and PFR reactors, assuming an overall yield of 80% of the maximum yield of dimer.

Solution

Butadiene is the main reactant in the system of parallel reactions, in which the dimer (R) is the main product. The CSTR yield is given by Equation 16.34:

$$\Phi_{CSTR} = \frac{1}{\left[1 + \chi^* \cdot \dfrac{p_{B0}}{p_B}\right]}$$

where:

$$\chi^* = \frac{k_2}{k_1 p_{B0}}$$

At $650°C$, $k_2 = 1.14$ and $k_1 = 8.47$

$$p_{B0} = 0.275P = 0.275 \text{ atm}$$

Then,

$$\chi^* = \frac{k_2}{k_1 p_{B0}} = 0.49 \approx 0.50$$

The maximum yield in the PFR for $\chi^* = 0.50$ (Figure 16.5) is $\Phi_{PFR} = 0.67$. Assuming 80% of this maximum yield, then the overall yield will be $\Phi_{PFR} = 0.5$. Therefore, the outlet concentration in the CSTR and PFR is 0.65 and 0.35 and conversion 0.35 and 0.65, respectively.

E16.5 Let us consider an example involving two components in the first reaction. The rate expressions are known for both reactions.

$$A + B \xrightarrow{k_1} R \text{ (desired product)}$$
$$2B \xrightarrow{k_2} S \text{ (undesired product)}$$

$$r_R = k_1 C_A C_B$$
$$r_S = k_2 C_B^2$$

For 80% conversion of B, the CSTR yield is 85%. What is the PFR yield for the same conversion?

Solution

The consumption rate of B in both reactions is $(-r_B) = r_R + 2r_S$, therefore the local yield will be:

$$\varphi_B = \frac{1}{1 + 2\dfrac{k_2}{k_1}\dfrac{C_B}{C_A}}$$

By the stoichiometry: 1 mol of $A \rightarrow B/3$ mol. Then,

$$\Phi_{CSTR} = \frac{1}{1 + 18\dfrac{k_2}{k_1 C_{B0}}\dfrac{C_{B0}}{C_B}} = \frac{1}{1 + \chi^*\dfrac{C_{B0}}{C_B}}$$

where:

$$\chi^* = 18\frac{k_2}{k_1 C_{B0}}$$

For 80% conversion of B, the value $C_B/C_{B0} = 0.2$ and $\Phi_{CSTR} = 0.85$, then we obtain $\chi^* = 0.035$.

The PFR yield is given by:

$$\Phi_{PFR} = 1 + \frac{\chi^*}{1 + \dfrac{C_B}{C_{B0}}} \ln \frac{\chi^* + \dfrac{C_B}{C_{B0}}}{1 + \chi^*}$$

Considering the PFR conversion as 80%, we obtain:

$$\Phi_{PFR} = 0.934$$

16.5 REACTIONS IN SERIES

Reactions in series in a system at constant volume, where R is the desired product and S is the undesired one, are represented by:

$$A \xrightarrow{k_1} R \xrightarrow{k_2} S$$

For this particular case, we determine the concentration profiles C_A, C_R, and C_S (deduced in Section 6.1), which are dimensionless:

$$\varphi_A = e^{-\tau} \tag{16.37}$$

$$\varphi_R = \frac{1}{(\chi - 1)}(\varphi_A - \varphi_A^\chi) \tag{16.38}$$

$$\varphi_S = \frac{1}{(\chi - 1)}\varphi_A^\chi - \frac{\chi}{\chi - 1}\varphi_A \tag{16.39}$$

where:

$$\varphi_A = \frac{C_A}{C_{A0}}; \varphi_R = \frac{C_R}{C_{A0}}; \varphi_S = \frac{C_S}{C_{A0}}; \tau = k_1 t; \chi = \frac{k_2}{k_1}.$$

To determine the maximum yield of R, we calculate its maximum dimensionless concentration $\varphi_{R\,max}$ (according to Equation 6.13 or Equation 16.40) and the corresponding time.

$$\varphi_{R\,max} = \chi^{(\chi/1-\chi)} \tag{16.40}$$

The average residence time in the PFR is equal to the reaction time in the batch reactor. Therefore, the previous equations are valid for the batch and PFR. However, the yield and selectivity vary along the PFR reactor. The PFR overall yield, taking into account the outlet concentration, is:

$$\Phi_{PFR} = \frac{1}{(1-\varphi_A)} - \left[\frac{1}{(\chi-1)} (\varphi_A - \varphi_A^\chi) \right] \tag{16.41}$$

The maximum yield of R is determined by Equation 16.40. Thus,

$$\Phi_{PFR} = \frac{\varphi_{Rmax}}{(1-\varphi_A)} = \frac{1}{(\chi^{(\chi/\chi-1)} - \chi)} \tag{16.42}$$

For the CSTR, the residence time is different from the batch. Therefore, from the kinetics of reactions, one obtains the following equality for the same space time:

$$\tau = \frac{C_{A0} - C_A}{k_1 C_A} = \frac{C_R}{k_1 C_A - k_2 C_R} \tag{16.43}$$

We make it dimensionless and determine the concentration of R:

$$\varphi_R = \frac{\varphi_A(1-\varphi_A)}{\varphi_A + \chi(1-\varphi_A)} \tag{16.44}$$

The overall yield in the CSTR will be:

$$\Phi_{CSTR} = \frac{C_R}{C_{A0} - C_A} = \frac{1}{1 + \dfrac{\chi(1-\varphi_A)}{\varphi_A}} \tag{16.45}$$

The results are shown in Figure 16.7. The CSTR yield is always lower than the PFR and also depends on the χ parameter, which relates the reaction rate constants:

$$\frac{k_2}{k_1}$$

The difference is significant for high conversions of reactant A, i.e., for low values of its dimensionless concentration (φ_A). When the reaction rate constant for R formation is higher than its decomposition ($k_1 > k_2$), then PFR yield will be much higher in high conversions, decreasing significantly when ($k_2 > k_1$). Finally, the reaction rate constants depend on temperature and activation energy, which may influence in the reactor choice.

The maximum yield or selectivity is obtained for the maximum concentration of product R, by:

$$\Phi_{CSTRmax} = \frac{1}{1 + \chi^{1/2}} \tag{16.46}$$

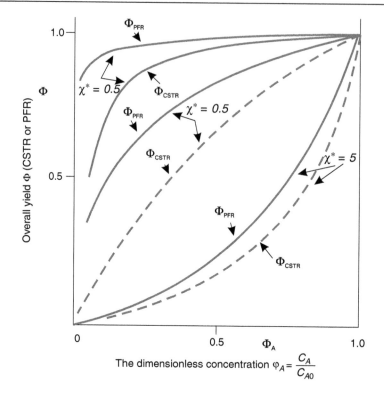

Figure 16.7 PFR and CSTR yields for reactions in series.

E16.6 The reaction in series $A \xrightarrow{k_1} R \xrightarrow{k_2} S$ is carried out in a CSTR. The kinetics is unknown but the yield is 68% and the conversion 35%. If the CSTR reactor volume is 300 L and the volumetric flow 150 L/min, calculate the PFR yield for this reaction and the PFR volume, considering the same feed.

Solution

The CSTR yield is given by Equation 16.45:

$$\Phi_{CSTR} = \frac{C_R}{C_{A0} - C_A} = \frac{1}{1 + \dfrac{\chi(1 - \varphi_A)}{\varphi_A}}$$

With $X_A = 0.35$, we obtain the dimensionless concentration of A, i.e., $\varphi_A = C_A/C_{A0} = 1 - X_A = 0.65$. Since we know $\Phi_{CSTR} = 0.68$, then we obtain $\chi = 0.874$.

The PFR yield can be calculated by Equation 16.42 for the same final conversion ($X_A = 0.35$), i.e., when the dimensionless concentration of A is $\varphi_A = 0.65$.

$$\Phi_{PFR} = \frac{1}{(\chi^{(\chi/\chi-1)} - \chi)}$$

$$\Phi_{PFR} = 0.822$$

To calculate the PFR volume, we should determine the constant k_1 which can be obtained through CSTR data. Thus,

$$\tau = \frac{V}{v_0} = \frac{C_{A0} - C_A}{k_1 C_A} = \frac{1 - \varphi_A}{k_1 \varphi_A}$$

Since $V_{CSTR} = 300$ L, $v_0 = 150$ L/min and the dimensionless concentration of A is $\varphi_A = 0.65$, we obtain:

$$k_1 = 0.27 \, \text{min}^{-1}$$

The PFR volume is calculated taking into account the PFR equation:

$$\tau_{PFR} = C_{A0} \int \frac{dX_A}{k_1 C_A} = -\frac{1}{k_1} \ln{(1 - X_A)}$$

Since $X_A = 0.35$

$$V_{PFR} = 240 \, \text{L}$$

Chapter 17

Combination of reactors

So far we have studied separately the continuous-flow stirred tank reactor (CSTR) and the plug flow reactor (PFR) each with their distinct characteristics. When comparing them, we have seen that for the same final conversion, the volume of the CSTR is always larger than that of the PFR, especially for high conversions. Additionally, the average residence time in the CSTR is also higher than that in the PFR. It is important to note, however, that the yield and selectivity are always higher in a PFR as compared to a CSTR.

The kinetics is the same for any system and the rate depends on concentration and temperature. The inverse of rate as a function of conversion is displayed again in Figure 17.1, this behavior depends on the reaction (reversible or irreversible) and also the order.

As discussed in Chapter 16, the basic equations show that the area under the kinetic curve ACDA' represents the integral of the balance equation in a PFR, whereas the rectangular area ACDF' represents the balance equation in a CSTR. Both represent the average residence time of the reactants in the reactors.

Figure 17.1 shows that it is possible to combine reactors and vary their residence times or volumes so that the final volume is equivalent or equal to the volume of a single reactor. The main advantage of combining reactors in series or parallel is to utilize less volume to yield the same efficiency, yield, selectivity, and final conversion as much as larger reactor yields. Two reactors in series are represented in Figure 17.1. If one assumes a PFR model for both reactors, the shaded areas under the kinetic curve ABEA' and BCDE represent the two PFRs in series. The conversion at the outlet of the first and second reactors is X_{A1} and X_{A2}, respectively. Note that the area is proportional to the volume of each reactor, and therefore the total volume is the sum of the volumes $V_1 + V_2 = V_{PFR}$.

Now consider two CSTR reactors in series. The hatched area ABEE' is proportional to the volume of the first one, while the area from BDCF to the second reactor. These volumes are distinct and their sum differs from that equivalent to a single CSTR that reaches the same final conversion, which is proportional to the area ACDF'. Therefore, the combination of two CSTR reactors in series results in a reduction in total reactor volume, when targeting the same final conversion. There will be different average residence times in each reactor, increasing the final yield. Note that the combination of multiple CSTR reactors in series tends to approach a single PFR reactor.

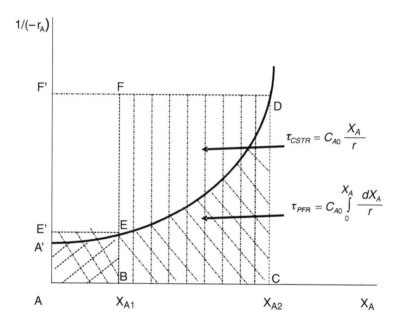

Figure 17.1 Inverse of reaction rate as a function of conversion.

Besides the series combination, reactors may be placed in parallel. Consequently, the final production increases with reactors of smaller volume compared to a larger single reactor. Various combinations are possible as illustrated in Figure 17.2.

Figure 17.2c and d shows diagrams of reactors in parallel. For this combination, it is essential that the conversion at the outlet of the reactors is the same. This leads to equal residence times in both reactors. If the outlet concentrations are different, then there will be mixture and the final concentration will be different from that desired.

The schemes (e) and (f) show separately the CSTR and PFR reactors in series. The residence time in each reactor is shorter and therefore the volumes are smaller. The sum of residence time or volume for PFR in series is equal to a single PFR with same final conversion. However, the sum of volume for CSTR in series is not the same for a single one with same final conversion.

The scheme (g) combining CSTR and PFR in parallel is unusual. This combination also requires equal conversions at the outlet of each reactor. Note, however, that the volumes are different from one another.

The schemes (a) and (b) involving CSTR and PFR in series are used when it is required to increase the final conversion. Note that the reaction rate is an important parameter in this analysis. In case (a) PFR + CSTR, first consider the PFR. As the conversion increases, the reaction rate decreases along the PFR. Therefore, it would take a long time to achieve the desired high conversion. Therefore, a CSTR is connected in series to complete this reaction. In case (b) CSTR + PFR, first consider the CSTR. Only a limited conversion may be obtained at the outlet of the CSTR and of course this is dependent on the space velocity of the reactor. In this case, a PFR is then connected in series to achieve higher conversions.

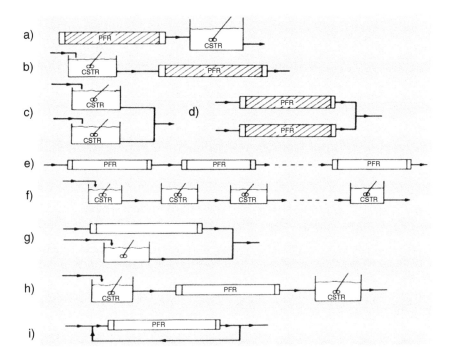

a) PFR — CSTR

b) CSTR — PFR

c) CSTR / CSTR

d) PFR / PFR

e) PFR — PFR — ---- — PFR

f) CSTR — CSTR — CSTR — ---- — CSTR

g)

h) CSTR — PFR — CSTR

i) PFR

Figure 17.2 Combination of reactors.

The scheme (h) is specific and aims to optimize the volume and yield. The conversion is low at the outlet of the first CSTR, but when a PFR is added in series, a higher final conversion is achieved. In order to complete the reaction, a CSTR with smaller volume is connected in series. The contact time or total residence time should be equal to the sum of the intermediate times (each reactor).

Finally, the scheme (i) shows a reactor with recycle (see Section 15.2) which aims to increase productivity by recycling unconverted reactants. The conversion at the outlet of the PFR depends on the recycle ratio, which is the ratio between the recycled flow and the inlet flow.

17.1 REACTORS IN SERIES

Let us consider two CSTR and PFR reactors in series as shown in Figure 17.2e and f or in Figure 17.3a. If initial concentration of A is C_{A0} and volumetric flow is v_0, then the initial molar flow is given by F_{A0}. At the exit of the first reactor, we have the concentration C_{A1} and subsequently decreasing concentrations, $C_{A_{i-1}}$ and $C_{A_{i-2}}$, until reaching the final concentration. In a system with constant or variable volume, one calculates the corresponding molar flows F_{Ai}. Conversion is defined with respect to the limiting reactant at the inlet of the first reactor such that the conversion varies between 0 and X_A, at the outlet of the last reactor. One should always take as reference the initial

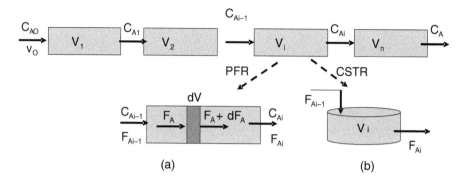

Figure 17.3 Combination of PFR and CSTR reactors in series.

concentration at the inlet of the first reactor and not in each reactor separately. The intermediate conversions must always be lower than the final conversion and should range from $X_{A_{i-1}}$ to X_{A_i}.

Constructing a molar balance with respect to the limiting reactant A in the ith reactor (with volume V_i), we obtain (Figure 17.3a):

$$F_A - (F_A + dF_A) + r_A \, dV_i = 0 \qquad (17.1)$$

Then,

$$-dF_A = r_A \, dV_i$$

Since conversion is defined as:

$$X_{Ai-1} = \frac{F_{A0} - F_{Ai-1}}{F_{A0}}; \quad X_A = \frac{F_{A0} - F_A}{F_{A0}}; \quad X_{Ai} = \frac{F_{A0} - F_{Ai}}{F_{A0}}$$

We obtain:

$$-dF_A = F_{A0} \, dX_A$$

and

$$F_{A0} \, dX_A = -r_A \, dV_i$$

Integrating, it follows:

$$\frac{V_i}{F_{A0}} = \int_{X_{Ai-1}}^{X_{Ai}} \frac{dX_A}{(-r_A)} \qquad (17.2)$$

On comparison with the molar balance in a single reactor, one observes a change in the integration limits. For n reactors in series, the total volume will be the sum of the intermediate volumes and the final conversion will be X_{An}. Thus,

$$\frac{V_{total}}{F_{A0}} = \frac{1}{F_{A0}}(V_1 + V_2 + \cdots + V_i + \cdots + V_n) \tag{17.3}$$

or

$$\int_0^{X_{An}} \frac{dX_A}{(-r_A)} = \left(\int_0^{X_{A1}} \frac{dX_A}{(-r_A)} + \int_{X_{A1}}^{X_{A2}} \frac{dX_A}{(-r_A)} + \cdots + \int_{X_{Ai-1}}^{X_{Ai}} \frac{dX_A}{(-r_A)} V_i + \cdots + \int_{X_{An-1}}^{X_{An}} \frac{dX_A}{(-r_A)} \right) \tag{17.4}$$

For CSTR reactors in series, we construct the molar balance in the ith reactor as in Figure 17.3b. Since the concentration in the tank is uniform and equal to the outlet concentration, we take directly the inlet and outlet molar flows.

$$F_{Ai-1} - F_{Ai} + r_A V_i = 0$$

Writing the molar flows as a function of conversion at the inlet and outlet of the ith reactor (note that conversion increases while the reactant concentration decreases), we obtain:

$$F_{A0}(X_{Ai} - X_{Ai-1}) + r_A V_i = 0 \tag{17.5}$$

or

$$\frac{V_i}{F_{A0}} = \frac{(X_{Ai} - X_{Ai-1})}{(-r_A)} \tag{17.6}$$

This expression applies to systems with constant or variable volume. When the system has constant volume, one can write the balance as a function of concentration according to Equation 17.7:

$$\frac{V_i}{v_0} = \tau = \frac{(C_{Ai-1} - C_{Ai})}{(-r_A)} \tag{17.7}$$

As mentioned before, the sum of the volumes of tanks (CSTR) in series is smaller than the volume of a single CSTR (see Figure 17.1). Besides, when the reactor volume is constant, the average residence time is equal to the space time $t_{CSTR} = \tau_{CSTR}$ for each reactor in series, but it is different for a single reactor. The reaction kinetics must be known and it is valid for any reactor in series or parallel.

The main goal is usually determining the required number of reactors in series or parallel to achieve the maximum desired conversion or productivity and to minimize the reactor volume.

It is possible to obtain analytical solutions for reactors in series when dealing with irreversible first- and second-order reactions at constant volume, as discussed below. For other cases, the solutions are complex and require computational methods.

17.1.1 Calculating the number of reactors in series to an irreversible first-order reaction

Let us consider a series of CSTR reactors of equal volumes. That means the average residence time and space time are the same in each reactor. Then,

$$\tau_i = \frac{V_i}{v_0}$$

If the volumetric flow v_0 is constant in a system at constant volume, we can claim that the total space time will be $\tau = n\tau_i$.

The rate for an irreversible first-order reaction will be:

$$(-r_A) = kC_A = kC_{Ao}(1 - X_A)$$

Then, for reactor i (Equation 17.6), we have the following expression:

$$\frac{V_i}{v_0} = C_{A0}\frac{(X_{Ai} - X_{Ai-1})}{(-r_A)} = \frac{(X_{Ai} - X_{Ai-1})}{k(1 - X_{Ai})} \tag{17.8}$$

or on rearranging:

$$(X_{Ai} - X_{Ai-1}) = \tau_i k(1 - X_{Ai})$$

Since the reactor volumes are equal and therefore t_i = constant, we obtain the following expression after adding the unity (1) on both sides of the equation:

$$(1 - X_{Ai-1}) = (1 + \tau_i k)(1 - X_{Ai})$$

The initial conversion is zero, i.e., $X_{A0} = 0$. Then,

$$(1 - X_{A1}) = \frac{1}{(1 + \tau_i k)} \tag{17.9}$$

Substituting into Equation 17.9, one obtains subsequently:

$$(1 - X_{A2}) = \frac{1}{(1 + \tau_i k)^2}$$

and

$$(1 - X_{An}) = \frac{1}{(1 + \tau_i k)^n} \tag{17.10}$$

or

$$(1 - X_{An}) = \frac{1}{\prod\limits_{1}^{n} (1 + \tau_i k)}$$

E17.1 Consider a cascade of CSTRs in series. Assuming the conversion at the outlet of the first reactor is 0.475 and of the last reactor is 0.99, calculate the number of CSTR reactors in series. Note that the kinetics has not been provided, but one may assume an irreversible first-order reaction.

Solution

Since the conversion in the first reactor is equal to 0.475, we can determine the term $(1 + \tau_i k)^{-1} = 0.525$ by means of Equation 17.9. From the information about final conversion in the last reactor, 0.99, we can calculate the intermediate conversions at the outlet of each reactor using Equation 17.10. Keep in mind that the conversion in the last reactor should not exceed 0.99. Then, it is possible to obtain the number of CSTR reactors required to achieve this conversion. Six reactors in series of equal volume are obtained.

17.1.2 Calculating the number of reactors in series for an irreversible second-order reaction

This solution is more difficult because the rate for an irreversible second-order reaction is given by:

$$(-r_A) = k C_A^2 = k C_{A0}^2 (1 - X_A)^2$$

Substituting into Equation 17.10, we obtain:

$$\frac{V_i}{v_0} = \tau_i = C_{A0} \frac{(X_{Ai} - X_{Ai-1})}{(-r_A)} = \frac{(X_{Ai} - X_{Ai-1})}{k C_{A0}(1 - X_A)^2} \tag{17.11}$$

This expression contains a quadratic term, whose solution has two roots. Let us instead use the inverse of the above method: start from the desired final conversion in the last reactor and calculate conversion or volume for the previous reactor. Note that the conversion in the first reactor should be positive. This method enables one to determine the volumes of the reactors in series using their inlet and outlet conversions. Therefore, rearranging Equation 17.11, we obtain:

$$(1 - X_{Ai-1}) = (1 - X_{Ai})[(1 + \tau_i k C_{A0})(1 - X_{Ai})] \tag{17.12}$$

E17.2 Considering n reactors in series and having the outlet conversions of the first and last reactors as 0.40 and 0.85, respectively, calculate the number of reactors even without the kinetic constants.

Solution

In the first reactor, we have conversion $X_{A1} = 0.40$. Substituting into Equation 17.12 for $i = 1$:

$$\tau_i k C_{A0} = 1.1$$

Considering similar reactor volumes, we have $t_i = t =$ constant. Starting from the last reactor whose conversion $X_{An} = 0.85$, we calculate the conversions in the previous reactors using Equation 17.12. The conversion at the outlet of the first reactor cannot be lower than 40%. Solving, we get:

$$X_{A_{n-7}} = 0.241 < 0.40$$

Therefore, we need $n = 7$ reactors in series with the same volume.

For more complex reactions, the solutions may be obtained using numerical methods.

17.1.3 Graphical solution

The graphical method is only useful for displaying a logical sequence during the determination of volumes and inlet and outlet conditions for reactors in series under isothermal conditions.

Both volumes and average residence times may be different. One should know the reaction rate, which is valid for any case as shown in Figure 17.4.

We start rearranging Equation 17.6 (CSTR):

$$(-r_{Ai}) = -\frac{1}{\tau_i}(C_{Ai-1} - C_{Ai}) \tag{17.13}$$

This expression represents a straight line that intersects the rate curve $(-r_{Ai})$, to give the outlet concentration of the following reactor. The slope is:

$$\left(-\frac{1}{\tau_i}\right)$$

Through Figure 17.4, if one knows the initial concentration and space time in the first reactor then it is possible to calculate directly the concentration at its outlet. This process may be repeated until the outlet concentration of the last reactor is determined. The points B, D, and F represent the concentrations at the outlet of reactors in series. Note that t_i may be different resulting in different volumes.

This method may be applied to PFR reactors in series as well as other different combinations of reactors.

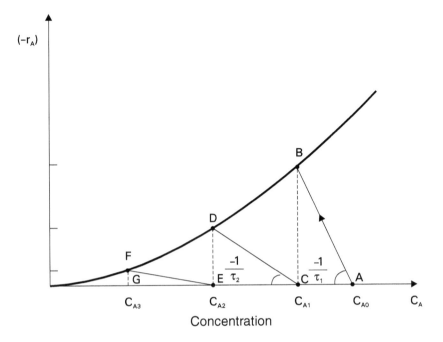

Figure 17.4 Graphical method – reactors in series.

17.2   REACTORS IN PARALLEL

Let us consider PFR or CSTR reactors in parallel as illustrated in Figure 17.2c and d and detailed in Figure 17.5, for both cases. The initial concentration, the volumetric flow, and volume may be different for the reactors. However, the main condition is that the outlet concentration or conversion for both reactors should be the same, which implies that the average residence time and space time must be the same as well.

Therefore, through molar balance we obtain the same expression of a single reactor, according to the following equations:

$$\text{PFR} \rightarrow \frac{V_i}{F_{A0i}} = \int_0^{X_A} \frac{dX_A}{(-r_A)} \tag{17.14}$$

$$\text{CSTR} \rightarrow \frac{V_i}{F_{A0i}} = \frac{(X_A)}{(-r_A)} \tag{17.15}$$

The aim is to determine the number of reactors in parallel necessary to achieve the desired maximum productivity as well as the reaction conditions for any kind of reaction.

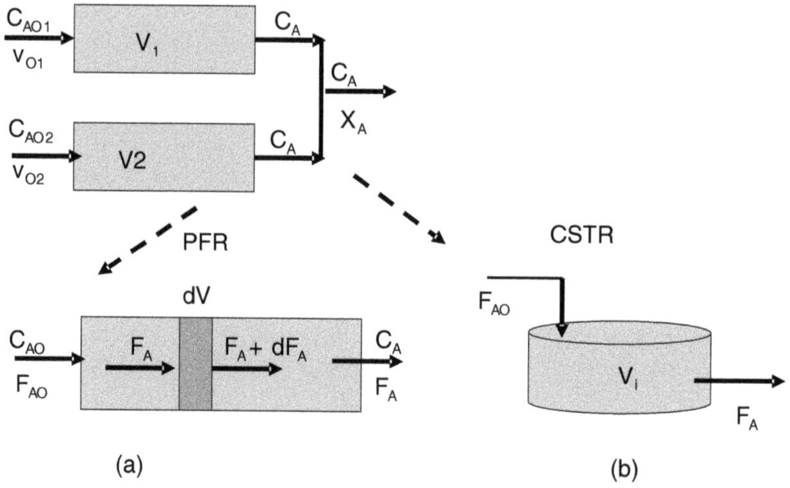

Figure 17.5 Combination of PFR and CSTR reactors in parallel.

It is possible to obtain analytical solutions for reactors in parallel to irreversible first- and second-order reactions at constant volume. For other cases, the solutions are complex and require computational methods.

17.3 PRODUCTION RATE IN REACTORS IN SERIES

By definition, *productivity* (W) is equal to the *production rate* at the outlet of the reactor, i.e., moles of reactant A converted into product over time. Both in CSTR or PFR, this production rate depends on the reaction kinetics and we take into account the *average residence time* in the reactor. For several reactors in series, we have different average residence times.

Figure 17.6 shows the productivity of CSTR and PFR reactors in series and separated.

The productivity presents higher values for PFR than for CSTR, but decreases with increasing conversion (X_{Ai}) for both reactors. By employing two CSTR in series, the productivity increases, and the more reactors in series the most it approaches a PFR.

17.4 YIELD AND SELECTIVITY IN REACTORS IN SERIES

The concept of yield in CSTR and PFR reactors has been studied in Section 16.3 and comparison showed that the yield in the PFR is always higher than that in the CSTR. The overall yield in the CSTR is equal to the local yield. On the other hand, the overall yield in the PFR is integrated along the reactor, according to Equations 16.20 and 16.23. For n reactors in series, the overall yield will be equal to the sum of yields in each reactor.

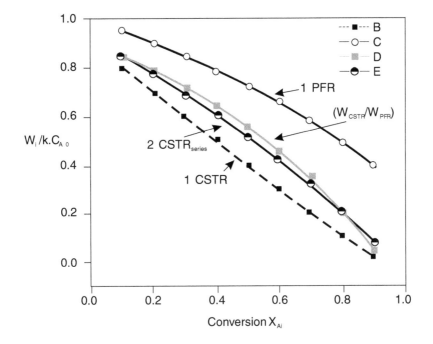

Figure 17.6 Productivity of reactors CSTR and PFR in series.

For n reactors in series:

- CSTR in series

$$\varphi_1(C_{A0} - C_{A1}) + \varphi_2(C_{A1} - C_{A2}) + \cdots + \varphi_n(C_{An-1} - C_{An}) = \Phi(C_{A0} - C_{An})$$

(17.16)

- PFR in series

$$\Phi_1(C_{A0} - C_{A1}) + \Phi_2(C_{A1} - C_{A2}) + \cdots + \Phi_n(C_{An-1} - C_{An}) = \Phi(C_{A0} - C_{An})$$

(17.17)

where:

$\varphi_i = $ local yield in each reactor i

$\Phi_i = $ overall yield in the PFR i

$\Phi = $ for n reactors in series

$C_{Ai} = $ inlet and outlet concentration of each reactor i.

$$\Phi_A = \varphi_A$$

(16.23)

The overall yield equation for n CSTR and PFR reactors in series or combined will be:

$$\Phi_A = -\frac{1}{(C_{A0} - C_A)} \int_{C_{A0}}^{C_A} \varphi_A dC_A$$

$$\Phi_n = \frac{\sum\limits_{1}^{n} \Phi_i(C_{Ai-1} - C_{Ai})}{(C_{A0} - C_{An})} \tag{17.18}$$

E17.3 From the kinetics of the reactions, one can calculate the yield of n reactors in series. Consider the reactions:

$$A + B \xrightarrow{k_1} P$$
$$A + B \xrightarrow{k_2} Q$$

whose rates are:

$$r_P = k_1 C_A^2 C_B$$

$$r_Q = k_1 C_A C_B^2$$

Solution

We calculate the local yield:

$$\varphi_A = \frac{r_P}{(-r_A)} = \frac{1}{\left(1 + \dfrac{k_2 C_B}{k_1 C_A}\right)} \tag{17.19}$$

The yield for n CSTR reactors in series will be, according to Equation 17.18:

$$\Phi_n = \frac{\sum\limits_{1}^{n} \dfrac{1}{\left(1 + \dfrac{k_2 C_B}{k_1 C_A}\right)}(C_{Ai-1} - C_{Ai})}{(C_{A0} - C_{An})} \tag{17.20}$$

For any kinetics, the local yield as a function of conversion can be represented by two cases (a) and (b) as displayed in Figure 17.7 (Denbigh, 1965).

In case (a), the yield decreases with conversion. The overall yield in the PFR is the integral (Equation 16.26) represented by the area under the curve (HF) to a final conversion (X_{Af}) at point C. On the other hand, the overall yield in the CSTR is represented by the rectangle (ACFG). By comparison, one observes that the yield in PFR is higher than that in the CSTR.

With two PFR reactors in series, aiming for the same final conversion (X_{Af}), one obtains the yields represented by the area under the curves (HD) and (DF), whose sum is equal to a single PFR that achieves the conversion X_{Af}. With two CSTR in series, their respective yields are represented by rectangular areas and the overall yield will be the sum of area of these two rectangles, leading to a higher yield when

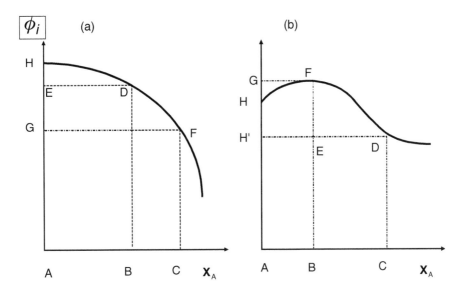

Figure 17.7 Yield as a function of conversion.

compared to a single CSTR. This, however, is still lower than the yield of two PFR in series.

In case (b), the yield curve shows a maximum F over the conversion at point B. In a PFR, the overall yield (area under the curve HFD) to a final conversion (X_{Af}) at C is higher than the yield in a CSTR for the same conditions (area of rectangle ACDH').

With two reactors in series we may have two situations:

1. Two PFR reactors in series, the first with outlet conversion in the maximum of the yield curve and the second connected in series (area under HF and FD curves).
2. The best combination will be a CSTR followed by a PFR. Therefore, the overall yield would be represented by the areas of the rectangle ABFG and under the curve FD. Any other combination will present a lower overall yield.

Example

E17.4 Ester is hydrolyzed with an excess of caustic soda. The ester solution is fed to the first reactor at a rate of 200 L/min and concentration 0.02 mol/L, while the caustic soda at 50 L/min and 1 mol/L. The reaction rate constant of this irreversible second-order reaction is 2 L/(mol min). Three CSTR reactors in series are utilized; the volume of the first is unknown, the second is 2200 L and the third is 800 L. Calculate the volume of the first reactor and the conversion at the outlet of each reactor. Assume final conversion of 95% and use the analytical method.

Solution

(a) Analytical method:

Since we have a second-order reaction, the rate is:

$$(-r_A) = kC_A C_B = kC_{Ao}^2(1 - X_A)(M - X_A) \qquad (17.21)$$

where:

$A =$ ester

$B =$ caustic soda

But caustic soda is very diluted and fed separately. Then, the new concentrations at the entrance of the first reactor are:

$$C_{A0} = 0.02 \times \frac{200}{250} = 0.016 \ (\text{mol/L})$$

$$C_{B0} = 1.0 \times \frac{50}{250} = 0.20 \ (\text{mol/L})$$

The concentration of caustic soda (B) is higher than ester and therefore the reactant A is limiting (considering a reaction stoichiometry 1:1 between ester and caustic soda). The ratio between the initial concentrations will be:

$$M = \frac{C_{B0}}{C_{A0}} = 12.5$$

In the rate equation, the value of M is much higher than the conversion X_A. In this case, one can neglect the second term and consider a pseudo first-order reaction.

$$(-r_A) = kC_A C_B = kMC_{A0}^2(1 - X_A) \qquad (17.22)$$

Then, substituting into Equation 17.7, we obtain an equation similar to Equation 17.9.

$$(1 - X_{Ai-1}) = (1 + \tau_i kMC_{A0})(1 - X_{Ai}) \qquad (17.23)$$

But,

$$\tau_i kMC_{A0} = \frac{V_3}{v_0}kMC_{A0} = \frac{800}{250} \times 2 \times 12.5 \times 0.016 = 1.28$$

Thus,

$$X_{Ai-1} = X_{A2} = 0.86$$

For the second reactor, we have its volume and the outlet conversion, then we can calculate:

$$\tau_2 kMC_{A0} = 3.52, \quad X_{A1} = 0.486$$

Since the total flow fed to the first reactor is 250 L/min, we obtain the reactor volume:

$$\tau_1 kMC_{A0} = 0.94; \quad \tau_1 = 2.35; \quad V_1 = 588 \ \text{L}$$

E17.5 One desires to process a reversible reaction $2A \leftrightarrow P + Q$ at constant volume in a CSTR reactor or several reactors in series. Pure reactant A is fed under flow of $3.5\,\mathrm{m^3/h}$ and initial concentration of $48\,\mathrm{kmol/m^3}$. The forward reaction rate constant is equal to $0.75\,\mathrm{m^3/(kmol\,h)}$ and the equilibrium constant $K = 16$. If the final conversion is desired to be 85% of the equilibrium conversion, calculate the volume of a single CSTR. If the capacity of the available reactors is only 5% of the calculated volume, how many CSTR reactors in series would be needed?

Solution

Since the reaction is reversible at constant volume, we have the rate equation as a function of conversion according to Equation 5.26, i.e.:

$$r = kC_{A0}^2 \left[(1 - X_A)^2 - \frac{1}{K}\frac{X_A^2}{4} \right] \tag{17.24}$$

The equilibrium conversion (X_{Ae}) for this forward and reverse second-order reaction is also given by:

$$K = \frac{X_{Ae}^2/4}{(1 - X_{Ae})^2} \tag{17.25}$$

Solving this last equation, one determines the equilibrium conversion. Therefore, from K provided, we obtain:

$$(4K - 1)X_{Ae}^2 - 8KX_{Ae} + 4K = 0 \tag{17.26}$$

Then,

$$X_{Ae} = 0.88$$

$$X'_{Ae} = 1.14$$

The outlet conversion in the reactor is desired to be 85% of the equilibrium conversion. Thus,

$$X_A = 0.75$$

Substituting the rate equation into CSTR Equation 17.6, we obtain:

$$\tau_{CSTR} = C_{A0}\frac{X_A}{(r)} = \frac{X_A}{kC_{A0}^2 \left[(1 - X_A)^2 - \frac{1}{K}\frac{X_A^2}{4} \right]} \tag{17.27}$$

With the values $K = 16$ and $X_A = 0.75$, we obtain the value of t. Then taking the flow $v_0 = 3.5\,\mathrm{m^3/h}$, we calculate the CSTR volume.

$$V = 1.35\,\mathrm{m^3}$$

Since the capacity of available reactors is only 5% of the calculated volume, each reactor has a volume of:

$$V = 0.067\,\text{m}^3$$

Then, the space time in each reactor (equal to the average residence time, since the volume is constant) will be:

$$\tau_i = 0.0193\,\text{h}$$

We assume the volumes are equal and then substitute the rate Equation 17.24 into Equation 17.6 to obtain the following expression, after some rearrangement:

$$\left(1 - \frac{X_{Ai-1}}{X_{Ae}}\right) = \left(1 - \frac{X_{Ai}}{X_{Ae}}\right)\left\{\left(1 + \tau_i k C_{A0}\frac{4K-1}{4K}X_{Ae}\right)\left[\left(\frac{X_A'}{X_{Ae}} - 1\right)\right.\right.$$
$$\left.\left. + \left(1 - \frac{X_{Ai}}{X_{Ae}}\right)\right]\right\} \tag{17.28}$$

Substituting the values known:

$$\left(1 + \tau_i k C_{A0}\frac{4K-1}{4K}X_{Ae}\right) = 0.604$$

Substituting the final conversion $X_A = 0.75$ and with the values $X_{Ae} = 0.88$ and $X_{Ae}' = 1.14$, we obtain successively:

$$X_{A1} = 0.188$$

$$X_{A2} = 0.472$$

$$X_{A3} = 0.631$$

$$X_4 = 0.729$$

Therefore, 4 CSTR reactors in series are necessary to reach the desired conversion.

E17.6 One desires to combine reactors in series based on the following graph that represents the kinetics of a reaction. Given the volumetric flow $v_0 = 1\,\text{m}^3/\text{h}$ and the initial concentration of the reactant $C_{A0} = 1 \, \text{kmol/m}^3$, choose the best combination of reactors and calculate their respective volume. The rate is given in mol/(Lh).

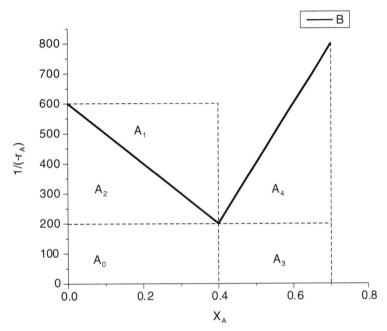

Graphical solution

Calculating the areas:

$$A_0 = 0.4 \times 200 = 80$$
$$A_1 = 0.4 \times 600 = 240$$
$$A_2 = 0.4 \times 200 + 0.4 \times (600 - 200)/2 = 160$$
$$A_3 = 0.3 \times 800 = 240$$
$$A_4 = 0.3 \times 200 + 0.3 \times (800 - 200)/2 = 150$$

Combination 1: CSTR + CSTR = $A_6 + A_3 = 320$
Combination 2: PFR + PFR = $A_2 + A_4 = 310$
Combination 3: CSTR + PFR = $A_0 + A_4 = 230$

The option that best satisfies the criterion of lowest volume is combination 3: CSTR + PFR. However, the correct combination is 2, satisfying the kinetics of reaction. In this case, the volumes would be:

$$V_{PFR1} = 160 \, \text{m}^3$$

$$V_{PFR2} = 150 \, \text{m}^3$$

E17.7 Two PFR reactors in parallel process the gas-phase reaction $A \xrightarrow{k} R + S$. The reactant A is introduced with inert 20%vol. at pressure 10atm and isothermal temperature 550°C. The average residence time in the reactor is 3.3 min. The volume of one reactor is double the other. Calculate the ratio between the inlet volumetric flows of the two reactors. The specific reaction rate can be estimated from the below equation. Calculate also the outlet conversion in the reactors.

$$\ln k = \frac{12,000}{T} + 10.6 \; (\text{L mol}^{-1} \text{s}^{-1})$$

Solution

The specific reaction rate is calculated at 550°C or 823 K using the expression provided:

$$k = 1.86 \times 10^{-2} \, \text{L/(mol s)}$$

The reaction is irreversible and of second order. Then, the corresponding rate as a function of conversion is:

$$(-r_A) = kC_A^2 = kC_{A0}^2 \frac{(1 - X_A)^2}{(1 + \varepsilon_A X_A)^2} \tag{17.29}$$

But the average residence time in a PFR is known, i.e.:

$$\bar{t} = C_{A0} \int_0^{X_A} \frac{dX_A}{(1 + \varepsilon_A X_A)(-r_A)} \tag{17.30}$$

Substituting Equation 17.29 into Equation 17.30 and integrating:

$$\bar{t} = \left[\frac{(1 - \varepsilon_A X_A)}{(1 - X_A)} + \varepsilon_A \ln(1 - X_A) \right] \tag{17.31}$$

Determining C_{A0}:

$$C_{A0} = \frac{p_{A0}}{RT} = \frac{y_{A0} P}{RT} = 1.35 \times 10^{-1} (\text{mol/L})$$

Calculating e_A:

A	R	S	I	Total
0.8	0	0	0.2	1.0
0	0.8	0.8	0.2	1.8

$$\varepsilon_A = 0.8$$

Substituting these values into Equation 17.31 and utilizing average residence time, we calculate the outlet conversion in the reactors:

$$X_A = 0.7$$

The final conversion in both reactors is equal, since the residence time in each one should be the same.

To calculate the space time, we take the general PFR equation:

$$\tau = C_{A0} \int_0^{X_A} \frac{dX_A}{(-r_A)} \qquad (17.32)$$

Substituting the rate Equation (17.29) and integrating, we obtain (Equation 14.45):

$$\tau = \frac{1}{kC_{A0}} \left[(1 + \varepsilon_A)^2 \cdot \frac{X_A}{(1 - X_A)} + \varepsilon_A^2 X_A + 2\varepsilon_A(1 + \varepsilon_A) \ln(1 - X_A) \right] \qquad (17.33)$$

Substituting the values known:

$$\tau = 829\,\text{s} = 13.7\,\text{min}$$

But

$$\tau_1 = \frac{V_1}{v_{01}} \qquad \tau_2 = \frac{V_2}{v_{02}} \qquad \tau_1 = \tau_2 \quad \text{and} \quad V_1 = 2V_2$$

Thus,

$$\frac{v_{01}}{v_{02}} = 2$$

E17.8 The reaction $A \xrightarrow{k} R$ was carried out in three PFR reactors in series. The reactant was fed at a flow rate of 100mol/min at 25°C, followed by heating until 100°C. Has a reaction taken place in the first reactor during this heating phase?

Then, the reactor operated adiabatically until a temperature of 650°C. When 650°C was reached, the reactor operated isothermally releasing heat. The final conversion was 85%.

Considerations and questions

1. The system comprises three PFR reactors in series, the first is nonisothermal, the second is adiabatic, and the third is isothermal. What are the inlet and outlet conversions in the second reactor (adiabatic)?
2. Calculate the volume of the isothermal reactor.

3.   Calculate the energy required to heat the first reactor and also the heat required to maintain the third reactor isothermal.

Data:

$$k = 4 \times 10^6\, e^{-8000/T}\, \text{min}^{-1}$$

$$\Delta H_R = -150\, \text{kJ/mol}$$

$$\overline{C}_{PA} = \overline{C}_{PR} = 180\, \text{J/(mol K)}$$

Heat exchange $= 300\, \text{J/(m}^2\text{s)}$
Heat exchange area $= 1\, \text{m}^2$

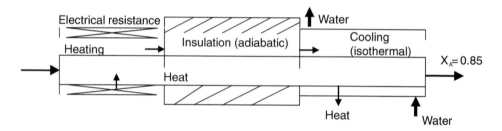

Solution

1. First reactor–Heating from 25°C to 100°C
The energy balance in PFR yields:

$$\sum F_i \overline{C}_{Pi}(T - T_0) + \Delta H_R F_{A0} X_A = Q = UA \times \Delta T \tag{17.34}$$

But

$$\sum F_i \overline{C}_{Pi} = F_A \overline{C}_{PA} + F_R \overline{C}_{PR} = F_{A0}(1 - X_A)\overline{C}_{PA} + F_{A0} X_A \overline{C}_{PR} \tag{17.35}$$

Since $\overline{C}_{PA} = \overline{C}_{PR} = 180\, J/mol$

$$\sum F_i \overline{C}_{Pi} = F_{A0} \overline{C}_P = 1.8 \times 10^4\, \text{J/min}$$

The second term:

$$\Delta H_R F_{A0} = 150 \times 10^3 \times 100 = 1.5 \times 10^7\, \text{J/min}$$

The third term:

$$Q = UA \cdot \Delta T = 300 \times 60 \times 1 \times 100 = 1.8 \times 10^6$$

$$Q = UA \times \Delta T = 300.60.1.100 = 1.8.10^6$$

Substituting the values into Equation 17.35, we calculate the outlet conversion in the first reactor. In this case,

$$X_{A1} = 0$$

Therefore, no conversion occurred during the heating.

In the second reactor, operated adiabatically, the temperature increases but does not exceed $650°C = 923K$.

With the energy balance in an adiabatic system, we deduce:

$$\frac{T}{T_0} = 1 + \beta \cdot X_A \tag{17.36}$$

where:

$$\beta = \frac{-\Delta H_R \cdot F_{A0}}{\sum F_i \cdot \overline{C_{Pi}} T_0} = \frac{1.5 \times 10^7}{1.8 \times 10^4 \times 393} = 2.12 \tag{17.37}$$

Then,

$$\frac{T}{T_0} = 1 + 2.12 X_A$$

If the maximum temperature is $650°C = 923$ K, we obtain:

$$\frac{T}{T_0} = 2.34$$

Therefore, the outlet conversion in the adiabatic reactor is:

$$X_{A2} = 0.636$$

The third reactor in series operates isothermally and its inlet and outlet conversions have been calculated and specified above. Then, by the molar balance in a PFR:

$$\tau = C_{A0} \int_0^{X_A} \frac{dX_A}{(-r_A)} \tag{17.38}$$

where:

$$(-r_A) = k \cdot C_A = k \cdot C_{A0}(1 - X_A)$$

Substituting the rate expression into Equation 17.38 and integrating between conversion limits in the third reactor:

$$\tau_3 = -\frac{1}{k} \left[\ln \left(\frac{1 - X_{A\text{final}}}{1 - X_{A2}} \right) \right] \tag{17.39}$$

Since the third reactor operates isothermally at 923 K, we need to calculate k (specific reaction rate):

$$k = 4.10^6 . e^{-8000/T}$$

Thus,

$$k = 0.688 \, \text{min}^{-1}$$

Therefore, with Equation 17.29, we calculate:

$$\tau_3 = 1.31 \, \text{min}$$

Obtaining the initial concentration:

$$F_{A0} = C_{A0} \cdot v_0 = 100$$

and

$$C_{A0} = \frac{P}{RT} = 4.09 \times 10^{-2}$$

and

$$v_0 = 2.44 \times 10^3 \, \text{L/min}$$

Therefore, the volume of the third reactor will be:

$$V_3 = 3.2 \, \text{m}^3$$

E17.9 The reaction $A \xrightarrow{k} R + S$ is carried out in a combined system of CSTR and PFR reactors in series with recycle. It takes place in liquid phase and the reaction rate follows: $(-r_A) = 8C_A$ $(\text{kmol m}^{-3} \, \text{h}^{-1})$. The feed into CSTR reactor consists in 1.6 m³/ks of reactant with concentration 1 kmol/m³. The CSTR volume is 0.086 m³ and the outlet conversion in the PFR is 90%. If the ratio recycle $R = 1$, calculate the PFR volume.

Solution

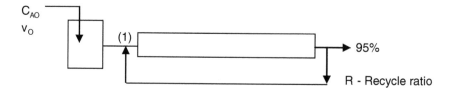

In the first reactor (a CSTR), we have:

$$\tau_1 = C_{A0} \frac{X'_{A1}}{(-r_A)} \tag{17.40}$$

In which the rate is given by:

$$(-r_A) = 8 \cdot C_A = 8C_{A0}(1 - X'_{A1}) \tag{17.41}$$

Since the CSTR reactor volume and flow are known, we obtain the space time and consequently, the outlet conversion X'_{A1}:

$$\tau_1 = 0.0144\,h = 51.8\,s$$

Substituting the above rate expression into Equation 17.40 yields:

$$X'_{A1} = 0.103$$

To calculate the inlet conversion or concentration in the PFR in series, which also has a recycle, we construct a molar balance in the intersection 1:

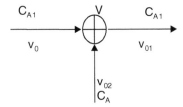

$$C'_{A1}v_0 + C_A v_{02} = C_{A1}v_{01} \tag{17.42}$$

$$v_0 + v_{02} = v_{01} \rightarrow v_{01} = v_0(1 + R) \tag{17.43}$$

where

$$R = \frac{v_{02}}{v_0} = 1 \tag{17.44}$$

Thus,

$$X_{A1} = 0.501$$

Then, for the PFR, we have the following Equation (15.17):

$$\tau_{PFR} = -\frac{1}{k} \ln \frac{(1 - X_A)}{(1 - X_{A1})} \tag{17.45}$$

Substituting the values $X_A = 0.90$ and $X_{A1} = 0.501$, we have:

$$\tau_{PFR} = 0.722\,ks = 722.7\,s$$

The reactor volume will be:

$$V = \tau \cdot v_{01} = 2.3\,m^3$$

E17.10 The reaction $A \xrightarrow{k} R + S$ takes place in a combined system of two PFR reactors in parallel where one reactor operates isothermally at 200°C and the other adiabatically, both under constant pressure of 2 atm. Pure reactant A is introduced at 10 mol/min in each reactor. The reaction rate constant is expressed as:

$$k = 8.19 \times 10^{15} \cdot e^{-34,222/(RT)} \ (\text{L mol}^{-1} \text{min}^{-1})$$

The average residence time in the isothermal reactor is 4 min. The other reactor operates adiabatically in parallel with the same flow entering at 200°C. Questions:

(a)   Calculate the volumes of both reactors.
(b)   Calculate the initial temperature in the adiabatic reactor.
(c)   What is the final conversion?

Data:

$$\overline{C}_{PA} = 170 \, \text{J/(mol K)}$$
$$\overline{C}_{PR} = 80 \, \text{J/(mol K)}$$
$$\overline{C}_{PS} = 90 \, \text{J/(mol K)}$$
$$\Delta H_R = -80 \, \text{kJ/mol}$$

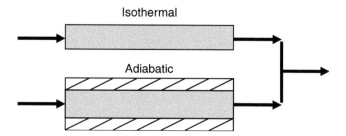

It is a gas-phase reaction and by unit of k it is a second-order reaction. Therefore, the rate as a function of conversion is:

$$(-r_A) = kC_A^2 = kC_{A0}^2 \frac{(1 - X_A)^2}{(1 + \varepsilon_A X_A)^2} \tag{17.46}$$

The average residence time in the isothermal reactor is:

$$\bar{t} = C_{A0} \int_0^{X_A} \frac{dX_A}{(1 + \varepsilon_A X_A)(-r_A)} \tag{17.47}$$

Substituting Equation 17.46 into Equation 17.47 and integrating, we obtain:

$$\bar{t} = \frac{1}{kC_{A0}} \left[\frac{(1 - \varepsilon_A X_A)}{(1 - X_A)} + \varepsilon_A \ln(1 - X_A) \right] \tag{17.48}$$

Calculating ε_A:

A	R	S	Total	
1	0	0	1	
0	1	1	2	$\rightarrow \varepsilon_A = 1$

Calculating C_{A0}:

$$C_{A0} = \frac{P}{RT} = 5.15 \times 10^{-2} (\text{mol/L})$$

Calculating the reaction rate constant, k, at 473 K:

$$k = 8.19 \times 10^{15} \cdot e^{-34,222/(RT)} \quad (\text{L mol}^{-1} \text{min}^{-1})$$

$$k = 1.106 \, \text{L mol}^{-1} \, \text{min}^{-1}$$

With the average residence time and above values, we determine the outlet conversion in the isothermal reactor:

$$X_A = 0.203$$

The space time can be calculated as follows:

$$\tau = \frac{1}{kC_{A0}}[(1 + \varepsilon_A)^2 \cdot \frac{X_A}{(1 - X_A)} + \varepsilon_A^2 X_A + 2\varepsilon_A(1 + \varepsilon_A)\ln(1 - X_A)]$$

Therefore,

$$\tau = 5.44 \, \text{min}$$

From the provided molar flow fed into reactors $F_{A0} = 10$ mol/min and the initial concentration, we calculate the volumetric flow. Then with space time, we calculate the isothermal reactor volume.

$$F_{A0} = C_{A0} \cdot v_0$$

Thus:

$$v_0 = 194 \, (\text{L/min})$$

and with:

$$\tau = \frac{V}{v_0}$$

$$V = 1.05 \, \text{m}^3$$

Since the average residence time in the reactors is equal, the outlet conversions in both reactors are the same and so, we should determine the initial temperature in the adiabatic reactor. The volume of the adiabatic reactor is the same.

By the energy balance in an adiabatic reactor, we have:

$$T = T_0 + \beta \cdot X_A$$

where:

$$\beta = \frac{-\Delta H_R \cdot F_{A0}}{\sum F_i \cdot \overline{C}_{Pi}}$$

where:

$$\sum F_i \overline{C}_{Pi} = F_A \overline{C}_{PA} + F_R \overline{C}_{PR} + F_s \overline{C}_{PS} = F_{A0}(1 - X_A)\overline{C}_{PA} + F_{A0}X_A\overline{C}_{PR} + F_{A0}X_A\overline{C}_{PS}$$

Then,

$$\sum F_i \overline{C}_{Pi} = F_{A0}[(1 - X_A)\overline{C}_{PA} + [\overline{C}_{PR} + \overline{C}_{PS})X_A = F_{A0}\overline{C}_{PA} = 170. \times F_{A0}$$

Therefore,

$$\beta = 470$$

And the inlet temperature in the adiabatic reactor will be:

$$T_0 = 377K = 105°C$$

E17.11 The local yield involving multiple reactions has been determined without previous knowledge about the kinetics, forming the desired product and several undesired co-products. The result is given by (Adapted from Denbigh, 1965):

$$\varphi = 0.6 + 2X_A - 5X_A^2$$

where X_A is the reactant conversion. With this result, one can calculate the yield of a reactor or several combined ones and find the best combination.

Solution

From the curve described by the local yield equation, we can observe a maximum for a given conversion. Deriving the above equation, we obtain:

$$\frac{d\varphi}{dX_A} = 0 = 2 - 10X_A = 0$$

Therefore, $X_{Amax} = 0.2$, as viewed in the following figure.

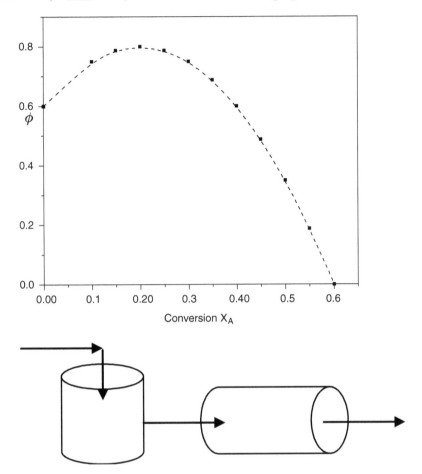

As already mentioned in the text, the maximum yield is obtained by considering two reactors in series, the first as a CSTR followed by a PFR, i.e.,

Thus, the overall yield in the first reactor (CSTR) will be equal to the local yield, when $X_{A1} = 0.2$.

$$\Phi_{CSTR} = \Phi_1 = 0.8$$

The overall yield in the PFR can be obtained by using Equation 16.23, i.e.:

$$\Phi_A = -\frac{1}{(C_{A0} - C_A)} \int_{C_{A0}}^{C_A} \varphi_A \, dC_A$$

where:

$$-dC_A = C_{A0} \, dX_A$$

Then,

$$-dC_A = C_{A0}\,dX_A$$

And,

$$\Phi_{PFR} = \frac{1}{(X_{Af} - X_{A1})} \int_{0.2}^{0.5} (0,6 + 2X_A - 5X_A^2)dX_A$$

Thus,

$$\Phi_2 = 0.65 = 65\%$$

Therefore, the overall yield will be calculated according to Equation 17.18:

$$\Phi_n = \frac{\displaystyle\sum_{1}^{n} \Phi_i(C_{Ai-1} - C_{Ai})}{(C_{A0} - C_{An})}$$

$$\Phi_n = \frac{\Phi_1(C_{A0} - C_{A1}) + \Phi_2(C_{A1} - C_{A2})}{C_{A0} - C_{A2}} = \frac{\Phi_1 X_{A1} + \Phi_2(X_{A2} - X_{A1})}{X_{A2}}$$

$$\Phi_n = 0.71 = 71\%$$

E17.12 The following reactions take place in an isothermal CSTR (Adapted from Fogler, 2000):

where *m*-xylene is the desired product while toluene the undesired one, some extra data:

$$r_{1M} = k_1[M][H_2]^{1/2}, \ k_1 = 55 \ (L/mol)^{0.5} \tag{17.49}$$

$$r_{2T} = k_2[X][H_2]^{1/2}, \ k_2 = 30 \ (L/mol)^{0.5} \tag{17.50}$$

For a space time of 30min, the maximum concentration of *m*-xylene is 3×10^{-3} and mesitylene is 2×10^{-2} mol/L at 700 K and 1 atm. Calculate the local yield in the PFR.

In short: M = Mesitylene

X = *m*-xylene
T = Toluene
CH_4 = Methane
H_2 = Hydrogen

$$M + H_2 \rightarrow X + CH_4$$
$$X + H_2 \rightarrow T + CH_4$$

From the molar balance of each component:

$$\frac{dF_H}{dV} = r_{1H} + r_{2H}$$

$$\frac{dF_X}{dV} = r_{1X} - r_{2X}$$

$$\frac{dF_M}{dV} = r_{1M}$$

$$\frac{dF_T}{dV} = r_{2T}$$

$$\frac{dF_{CH_4}}{dV} = r_{1CH_4} + r_{2CH_4}$$

For each reaction, we have:

$$-r_{1H_2} = -r_{1M} = r_{1X} = r_{1CH_4}$$

$$-r_{2H_2} = -r_{2X} = r_{2T} = r_{2CH_4}$$

By the definition of local yield, where X is the desired product and M reactant that has reacted:

$$\varphi_X = \frac{r_X}{r_M} = \frac{r_{1M} - r_{2T}}{r_{1M}} \tag{17.51}$$

But,

$$r_X = k_1[M] \times [H_2]^{1/2} - k_2[X] \times [H_2]^{1/2} = r_{1M} - r_{2T} \tag{17.52}$$

$$r_M = k_1[M] \times [H_2]^{1/2} = r_{1M} \tag{17.53}$$

Then, substituting Equations 17.52 and 17.53 into Equation 17.51:

$$\varphi_X = \frac{r_{1M} - r_{2T}}{r_{1M}} = \frac{k_1[M] \times [H_2]^{1/2} - k_2[X] \times [H_2]^{1/2}}{k_1[M] \times [H_2]^{1/2}}$$

Or

$$\varphi_X = 1 - \frac{k_2[X]}{k_1[M]} \tag{17.54}$$

Then, substituting the concentration values and constants, we obtain:

$$\varphi = 0.918$$

Chapter 18

Transport phenomena in heterogeneous systems

18.1 INTRAPARTICLE DIFFUSION LIMITATION—PORES

The presence of pores, for which the observed reaction rate is lower than the kinetically controlled intrinsic one, in the particles or pellets affects the reaction rate due to diffusion limitations. This intraparticle diffusion effect causes a concentration gradient within the pores. If diffusion is fast, then the concentration gradient is negligible.

Figure 18.1 shows a particle (pellet) with a pore through which the reactant A diffuses, reaches the active site, and reacts to form the product R, which follows the reverse path until it reaches the external film. Therefore, the observed rate should take into account the reaction and intraparticle diffusion, i.e.:

$$r_{obs} = r_{intrinsic}\eta \tag{18.1}$$

The intrinsic rate is defined by the kinetics on the pore surface or at the surface sites under the reaction conditions and η is called the effectiveness factor. For now, let us not consider external mass and heat transfer limitations.

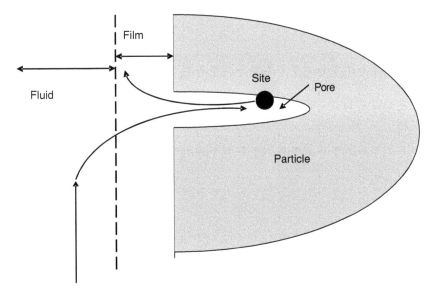

Figure 18.1 Intraparticle diffusion effects.

Initially, we define reactant A flow per unit surface area as:

$$F'(\text{mol/cm}^2/\text{h})$$

and consider a spherical particle model. Then, the molar flow of reactant (A) that passes through an element Δr equals:

$$F'_A\big|_r \cdot 4\pi r^2 - F'_A\big|_{r+dr} \cdot 4\pi r^2\big|_{r+dr} + r' \cdot A_i \cdot (4\pi r^{-2}\Delta r) = 0 \tag{18.2}$$

where:

$$F'_A\big|_r \text{ area} = F'_A\big|_r \cdot 4\pi r^2 \tag{18.3}$$

$$F'_A\big|_{r+dr} \cdot \text{area} = F'_A\big|_{r+dr} \cdot 4\pi r^2\big|_{r+dr} \tag{18.4}$$

The generated reaction rate in the volume element is:

$$R_A = r' \cdot A_i \cdot (4\pi r^{-2}\Delta r) \tag{18.5}$$

$$\text{mol/h} = (\text{mol/m}^2/\text{h})(\text{m}^2/\text{m}^3)(\text{m}^3)$$

where:
$A_i =$ internal surface area per volume; $(4\pi r^2 \Delta r) =$ volume element of the layer; $r =$ pore radius.
$r'_j =$ reaction rate with respect to component j/area.
   Taking the limit $\Delta r \to 0$ and dividing everything by $4\pi r^2 \Delta r$:

$$\frac{d\left(F'_A\big|_r \cdot r^2\right)}{dr} - r'_A \cdot A_i \cdot r^2 = 0 \tag{18.6}$$

Under equimolar conditions, the direct flow is equal to the counter-diffusion. By Fick's law, the diffusion flow of gas A through the pore will be:

$$F'_A\big|_r = -D_e \frac{dC_A}{dr} \tag{18.7}$$

Substituting Equation 18.7 into Equation 18.6, we have:

$$\frac{d\left(r^2 D_e \frac{dC_A}{dr}\right)}{dr} + r'_A \cdot A_i \cdot r^2 = 0 \tag{18.8}$$

The internal area A_i is unknown but the total surface area S_g and the solid particle density ρ_s can be measured. Then,

$$S_g = \frac{A_i}{\rho_s} \quad (\text{cm}^2/\text{g}_{\text{solid}}) \tag{18.9}$$

The consumption reaction rate of component j, r'_j (mol/h/m^2), depends on the reaction order. Assuming an irreversible reaction and a generic n-order:

$$(-r'_A) = k' C_A^n \tag{18.10}$$

Substituting Equations 18.10 and 18.9 into Equation 18.8, we have:

$$\frac{d\left(r^2 D_e \frac{dC_A}{dr}\right)}{dr} - k' C_A^n S_g \rho_s \cdot r^2 = 0 \tag{18.11}$$

where k' mol/(m^2 h) is the specific reaction rate or rate constant (per unit area) and D_e (m^2/s) is the effective diffusivity.

$$k' \left(\frac{mol}{m^2 \cdot h}\right)$$

Transforming the above equation, we obtain a general differential equation for the concentration profile:

$$\frac{d^2 C_A}{dr^2} + \frac{2}{r}\frac{dC_A}{dr} - \frac{k' C_A^n S_g \rho_s}{D_e} = 0 \tag{18.12}$$

This equation can be solved with the following boundary conditions:

$$\begin{aligned} r = 0 &\to C_A = finite \\ r = R &\to C_A = C_{As} \to concentration \mapsto \sup erficial \end{aligned} \tag{18.13}$$

Nondimensionalizing variables,

$$\lambda = \frac{r}{R} \text{ and } \varphi = \frac{C_A}{C_{As}}$$

We obtain:

$$\frac{d^2 \varphi}{d\lambda^2} + \frac{2}{\lambda}\frac{d\varphi}{d\lambda} - \underbrace{\frac{k' R^2 C_A^{n-1} S_g \rho_s}{D_e}}_{\Phi_n^2} \varphi^n = 0 \tag{18.14}$$

In this equation, the term Φ_n is called the *Thiele modulus*. It relates the surface reaction rate to the diffusion rate inside the pore, i.e.:

$$\Phi_n^2 = \frac{k' R C_A^n S_g \rho_s}{D_e \left(\frac{C_{As} - 0}{R}\right)} = \frac{reaction\ rate}{diffusion\ rate} \tag{18.15}$$

In the numerator, the most important parameter is the reaction rate constant so that the reaction rate is represented by $-r_A = k' C_A^n$, while the entire denominator represents the diffusion rate (R is porous or particle radius).

Thiele modulus is defined for any reaction order n. For a first-order reaction, we have:

$$\Phi_1 = R\sqrt{\frac{k'S_g\rho_s}{D_e}} \tag{18.16}$$

where:
k' is the specific reaction rate constant of a first-order reaction and its unit is $(cm^3/cm^2\ s)$
Total specific area S_g (cm^2/g) is measured by N_2 physisorption.
ρ_s (g/cm^3) is the solid density.
D_e is the diffusivity coefficient.

The Knudsen diffusivity and molecular diffusion coefficients of the components in gas phase are different and determined by empirical correlations or in most cases they are tabulated.

Thiele modulus presents limiting cases:

• When its value is high, the effective diffusion or diffusion rate is very low with respect to the reaction rate and consequently diffusion is the limiting step.
• The opposite (low Thiele modulus) represents a low reaction rate with respect to the diffusion rate and therefore the surface chemical reaction is the limiting step.

Equation 18.14 can be solved for the boundary conditions previously indicated, yielding the following solution:

$$\varphi = \frac{C_A}{C_{As}} = \frac{R}{r}\frac{\sinh(\Phi_1 \cdot r/R)}{\sinh \Phi_1} \tag{18.17}$$

This expression is valid for a first-order reaction. The concentration profiles are depicted in Figure 18.2 as a function of pore radius and they depend on the Thiele

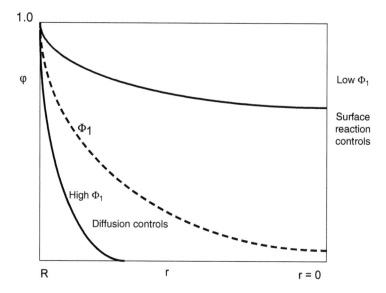

Figure 18.2 Concentration profile as a function of radius and Thiele modulus.

modulus. At the two extremes, a chemical reaction may be either diffusion limited or reaction limited. If the rate of diffusion through the particle is very low (presence of strong mass transfer limitation), the reaction effectively takes place at the external surface of the particle, i.e., $r \rightarrow R$. On the other hand, when the rate of diffusion through the particle is high, a concentration profile exists within particulate pores.

18.2   EFFECTIVENESS FACTOR

The effectiveness factor is a parameter that determines how much the intrinsic rate (kinetic control) is affected by diffusion limitations. Therefore, this factor measures the deviation from the real kinetics in the presence of diffusion phenomena. According to Equation 18.1, the effectiveness factor is:

$$\eta = \frac{r'_{obs}}{r'_{int}} \tag{18.18}$$

The intrinsic rate r'_{int} provides the surface reaction velocity controlled by the kinetic control regime, i.e.,

$$r'_{int} = k' C_{As} S_g \rho_s \frac{4\pi R^3}{3} \tag{18.19}$$

Since the reaction is irreversible and first-order with respect to A, the typical units for k' and r'_{int} are indicated in parentheses:

$$k' (cm^3/cm^2\,s); \quad r'_{int} (mol/s)$$

The observed rate is described by the diffusion rate within the pore, Equation 18.3, Then,

$$F'_A\big|_r \text{ area} = F'_A\big|_r \cdot 4\pi r^2$$

Along with Equation 18.7, we have:

$$F'_A\big|_r \text{ area} = F'_A\big|_r \cdot 4\pi r^2$$

Then,

$$r'_{obs} = (-r'_A) = -4\pi r^2 D_e \frac{dC_A}{dr}\bigg|_{r=R} = -4\pi R C_{As} D_e \frac{d\varphi}{d\lambda}\bigg|_{\lambda=1} \quad (mol/s) \tag{18.20}$$

where $x = r/R$.
   Deriving Equation 18.17, we have:

$$\frac{d\varphi}{d\lambda}\bigg|_{\lambda=1} = \Phi_1 \coth \Phi_1 - 1 \tag{18.21}$$

Substituting into Equation 18.20:

$$(-r'_A) = 4\pi R D_e C_{As}(\Phi_1 \coth \Phi_1 - 1) \,(\text{mol/s}) \tag{18.22}$$

Then substituting Equations 18.22 and 18.19 into Equation 18.18, we obtain:

$$\eta = \frac{r'_{obs}}{r'_{int}} = \frac{4\pi R D_e C_{As}(\Phi_1 \coth \Phi_1 - 1)}{k' C_{As} S_g \rho_s \frac{4\pi R^3}{3}}$$

Along with Equation 18.16, we have:

$$\eta = \frac{3(\Phi_1 \coth \Phi_1 - 1)}{\Phi_1^2} \tag{18.23}$$

The effectiveness factor is plotted as a function of Thiele modulus (logarithmic scale) in Figure 18.3. This curve is similar to that mentioned in several books.

It is observed that when Thiele modulus Φ_1 is low (order of magnitude 0.5), the effectiveness factor η is almost equal to 1 indicating that the observed reaction rate is not diffusion limited. On the other hand, the effectiveness factor varies linearly with Thiele modulus for Φ_1 values that range from 2 to 10, indicating that the observed reaction rate is diffusion limited. For Φ_1 values over 10, the effectiveness factor η has a more significant deviation, indicating a strong diffusion effect inside pores of the spherical particle.

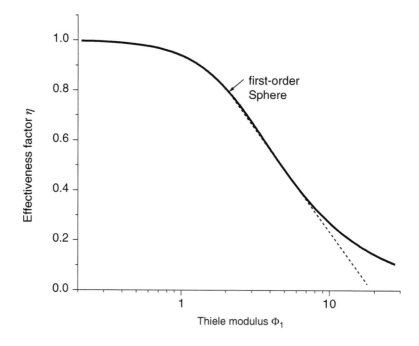

Figure 18.3 Effectiveness factor as a function of Thiele modulus.

For values of Φ_1 ranging from 2 to 10, we can admit a linear variation of the effectiveness factor along Thiele modulus, i.e. (coth $\Phi_1 \to 0$), then:

$$\eta = \frac{3}{\Phi_n} \tag{18.24}$$

Thiele modulus is defined for a generic-order reaction.

Considering an irreversible first-order reaction, then according to Equation 18.16, Thiele modulus is expressed as:

$$\Phi_1 = R\sqrt{\frac{k'S_g\rho_s}{D_e}}$$

Substituting into Equation 18.24, we obtain:

$$\eta = \frac{3}{R}\sqrt{\frac{D_e}{k'S_g\rho_s}} \tag{18.25}$$

In this case, one can determine directly the observed rate substituting Equation 18.25 into rate equation:

$$r'_{obs} = (-r'_A) = \frac{3}{R}\sqrt{D_e k'S_g\rho_s}\,C_{As}(\text{mol/cm}^3/\text{s}) \tag{18.26}$$

The observed rate (r'_{obs}) increases by decreasing particle or pore radius, and increases with temperature since D_e and $k\prime$ increase with temperature T and surface area.

To a generic order, we have:

$$\Phi^2 = \frac{k'R^2 C_A^{n-1}S_g\rho_s}{D_e} \tag{18.27}$$

$$\eta = \left(\frac{2}{n+1}\right)^{1/2} \cdot \frac{3}{\Phi_n} \tag{18.28}$$

In the sequence, substituting Equation 18.27 into Equation 18.28, we have:

$$\eta = \left(\frac{2}{n+1}\right)^{1/2} \cdot \frac{3}{R}\sqrt{\frac{D_e}{k'_n S_g\rho_s}}\,C_{As}^{(1-n)/2} \tag{18.29}$$

This expression is valid for generic-order reactions with large diffusion effects and catalyzed by spherical particles. It is observed that the effectiveness factor in isothermal conditions does not exceed the unit.

On the other hand, the effectiveness factor varies significantly with temperature. Therefore, for nonisothermal conditions, especially in the exothermic reactions, it will vary since the temperature varies within the pores or particles due to the temperature gradient caused by the chemical reaction. It is then necessary to construct an energy balance that takes the specific geometry of the catalyst particle into account (sphere, cylinder). It is important to remember that the reaction rate constant depends on temperature. The solution for the energy balance leads to an energy parameter defined as:

$$\beta = \frac{C_{As}(-\Delta H_R)D_e}{k_t T_s} = \frac{\Delta T_m}{T_s} \tag{18.30}$$

where:
$(-\Delta H_R) =$ reaction heat (kcal/mol)
$k_t =$ thermal conductivity $T_s =$ temperature at the surface of pore or particle
$\Delta T_m =$ maximum difference of temperature between the external surface and pore.
    The β energy parameter is, by convention, positive if the reaction is exothermic, and negative if endothermic.
    For these conditions, we obtain:

$$\frac{T_m - T_\infty}{T_\infty} = \frac{(-\Delta H_R)D_e}{k_e T_\infty}(C_\infty - C_m) \tag{18.31}$$

where C_m corresponds to the concentration at the maximum temperature.
    When Thiele modulus is high, C_m is effectively zero and the maximum temperature difference between particle center and its external surface is equal to:

$$T_m - T_\infty = \frac{(-\Delta H_R)D_e}{k_e}(C_\infty) \tag{18.32}$$

    The effectiveness factor for nonisothermal reactions as a function of Thiele modulus can be generically represented by Figure 18.4. Isothermal systems have $\beta = 0$. Considering an exothermic reaction with $(-\Delta H_R) = 20$ kcal/mol; thermal conductivity $k_t = 2.0 \times 10^{-3}$ cal/($°$C cm s), $D_e = 0.1$ cm^2/s and external concentration $C_\infty = 1.0 \times 10^{-4}$ mol/cm^3, one obtains a difference of temperature of 100$°$C.
    Indeed, the experimental values may be lower, but calculations indicate that if we do not take into account the temperature gradient, errors can be significant. Figure 4.18 shows the behavior of effectiveness factor when there is temperature gradient in the particles.
    In exothermic reactions, when $\beta > 0$, the effectiveness factor exceeds unity. This happens because the increase in the reaction rate caused by the increase of temperature within the particle, is higher than the decrease in concentration caused by the concentration gradient within the pores. Furthermore, for low Thiele modulus values, one obtains two or more corresponding effectiveness factor values, indicating the existence of at least three conditions where the generated heat inside the particle is equal to the removed heat; one condition is metastable and the other two conditions indicate that the rate is limited by chemical reaction at low temperatures and by diffusion effects within the pores at high temperatures, corresponding to high values of β.

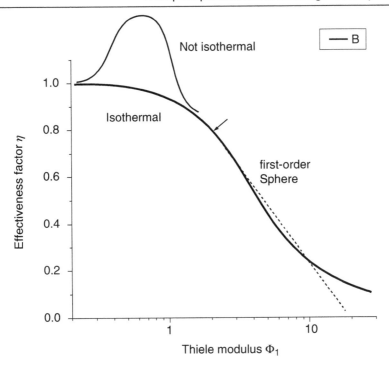

Figure 18.4 Effectiveness factor for nonisothermal reactions.

18.3 EFFECTS OF INTRAPARTICLE DIFFUSION ON THE EXPERIMENTAL PARAMETERS

In the presence of intraparticle diffusion effects, the observed rate differs from the surface intrinsic rate which is controlled only by the reaction kinetics. Therefore, the rate must be corrected by the effectiveness factor η.

Under strong diffusion effects, the effectiveness factor is inversely proportional to Thiele modulus. When the reaction has a generic n-order, one should take care in interpreting the experimental results, since, according to Equation 18.29, the reaction order is not the same as the intrinsic rate. The observed rate is as follows:

$$r'_{obs} = r'_{int} \cdot \eta$$

Substituting the effectiveness factor as a function of Thiele modulus, the factor is inversely proportional to Thiele modulus in the presence of strong diffusion effects. Therefore, from Equation 18.29, we obtain:

$$r'_{obs} = k_n C^n_{As} \cdot \eta = k'_n \left(\frac{2}{n+1} \right)^{1/2} \cdot \frac{3}{\Phi_n} C^n_{As}$$

$$r'_{obs} = \frac{3}{R} \left(\frac{2}{n+1} \right)^{1/2} \cdot \sqrt{\frac{D_e}{S_g \rho_s}} (k'_n)^{1/2} C^{(1+n)/2}_{As} \tag{18.33}$$

or

$$r'_{obs} = \frac{3}{R}\left(\frac{2}{n+1}\right)^{1/2} \cdot \sqrt{\frac{D_e}{S_g \rho_s}}(k'_0)^{1/2}(e^{-E/(RT)})^{1/2}C_{As}^{(1+n)/2}$$ (18.34)

The observed rate is defined as:

$$r'_{obs} = k'_{obs}C_{As}^{n'}$$ (18.35)

Therefore, we have a measured reaction order n' (apparent order) which is obtained by equating Equations 18.34 and 18.35:

$$n' = \frac{n+1}{2}$$ (18.36)

Consequently, a zero-order reaction will have an apparent order equals to ½, a first-order reaction will present the same apparent order and a second-order reaction, an apparent order of 3/2. Therefore, the order of reaction is apparent.

The reaction rate constant and activation energy are also apparent in the presence of intraparticle diffusion effects. Through Arrhenius equation, we have an apparent activation energy measured, i.e.:

$$k'_{obs} = k'_0\, e^{-E_{ap}/(RT)}$$ (7.7)

Comparing k'_{obs} in Equation 7.22 with the terms present in Equation 18.34, we obtain that:

$$E_{ap} = \frac{E}{2}$$

And the apparent frequency factor (k'_{0ap}) is:

$$k_{ap} = \frac{3}{R}\left(\frac{2}{n+1}\right)^{1/2} \cdot \sqrt{\frac{D_e}{S_g \rho_s}}(k'_0)^{1/2} \; (\mathrm{cm^3/\,cm^2/s})$$ (18.37)

The apparent frequency factor contains several parameters and is proportional to:

$$\sqrt{\frac{D_e}{S_g \rho_s}}$$

which suggests that the higher the diffusivity coefficient, the higher the apparent frequency factor. On the other hand, the higher the specific surface area, the lower the apparent frequency factor.

There are several approaches and criteria to determine the presence of diffusion limitations and their influence on the catalytic activity. Most of them are specific to a kind of reaction and under limited conditions. A very useful approach is Weisz–Prater criterion which is defined as the ratio between the observed rate and the diffusion rate (Froment and Bischoff, 1979), i.e.:

$$\Psi_{WP} = \frac{r'_{obs}}{r_d} = \frac{r'_{observed}}{r'_{int}} \cdot \frac{r'_{int}}{r'_{diffusion}} = \eta \cdot \Phi_n^2 \tag{18.38}$$

$$\Psi_{WP} = \frac{r'_{obs}}{r_d} = \frac{r' S_g \rho_s R^2}{D_e C_{As}} \tag{18.39}$$

But with Equation 18.23, we obtain:

$$\eta = \frac{3(\Phi_1 \coth \Phi_1 - 1)}{\Phi_1^2}$$

Then,

$$\Psi_{WP} = 3(\Phi_1 \coth \Phi_1 - 1) \tag{18.40}$$

for which Thiele modulus is known, but also depends on the reaction order. Equation 18.40 is applied to first-order reactions.

When $\Psi_{WP} < 1$, there is no mass transfer limitation, otherwise when $\Psi_{WP} > 1$, there are strong diffusion limitations.

Example

E18.1 The reaction $(CH_3)_2 C=CH_2 + H_2O \rightarrow (CH_3)_3COH$ with excess of water is done in a fixed-bed reactor, using a spherical catalyst with radius 0.213 cm and density of 2 g/cm³. The reaction is carried out at 100°C and 1 atm, it is reversible and the final conversion is 80% of equilibrium conversion. The equilibrium constant at 100°C corresponds to 16.6. The reaction takes place isothermally, the reaction rate is 1.1×10^{-5} mol/(s g_{cat}) and external concentration of the limiting reactant is equal to 1.65×10^{-5} mol/cm³. The mean effective diffusivity is 2×10^{-2} cm²/s. Determine the effectiveness factor.

Solution
One assumes a reversible first-order forward and reverse reaction, since the solution is quite dilute, i.e.:

$$A \leftrightarrow R$$

Then, the rate will be:

$$r'' = kC_A - k'C_R = k\left[C_A - \frac{1}{K}C_R\right] \tag{18.41}$$

At equilibrium, the resulting rate is zero, therefore:

$$K = \frac{C_{Re}}{C_{Ae}} = \frac{X_{Ae}}{1 - X_{Ae}} = 16.6$$

The equilibrium conversion will be:

$$X_{Ae} = 0.94$$

and

$$X_A = 0.75$$

Through Equation 18.41, one determines the rate as a function of conversion. The observed rate will be:

$$r''_{obs} = k'' \cdot C_{A0} \frac{K}{K + 1}(X_{Ae} - X_A)\eta = k^* C_{A0}(X_{Ae} - X_A)\eta \qquad (18.42)$$

Calculation of Thiele modulus:

$$\Phi_1 = R\sqrt{\frac{k''_{obs}\rho_s}{D_e}} = R\sqrt{\frac{r''_{obs}\rho_s}{(X_{Ae} - X_A)C_{A0}D_e\eta} \frac{(mol/g/s)(g/cm^3)}{(mol/\ cm^3)(cm^2/s)}} \qquad (18.43)$$

$$\Phi_1 = 0.213\sqrt{\frac{1.1 \times 10^{-5} \times 2}{1.65 \times 10^{-5} \times (0.94 - 0.75) \times 0.02 \times \eta}}$$

$$\Phi_1 = 0.213\sqrt{\frac{350.88}{\eta}} \qquad (18.44)$$

But the effectiveness factor as a function of the Thiele modulus is equal to:

$$\eta = \frac{3(\Phi_1 \coth \Phi_1 - 1)}{\Phi_1^2} = \frac{3}{\Phi_1}\left(\frac{1}{\tanh \Phi_1} - \frac{1}{\Phi_1}\right) \qquad (18.45)$$

This system is solved by iteration, arbitrating a value for η and calculating Φ_1 until finding the correct value of the effectiveness factor. By interaction, we start with $\eta = 0.70$ and obtain, after three iterations using Equations 18.44 and 18.45, the following values:

$$\Phi_1 = 6.30$$
$$\eta = 0.40$$

E18.2 Consider the previous problem, but now a new experiment has been performed using a catalyst with spherical particle of diameter 0.13 cm. The observed rate was 0.8×10^{-5} mol/(s g_{cat}). The other values remain the same. Determine the particle size necessary to eliminate the diffusion effects.

Solution

We use the Weisz–Prater criterion, Equation 18.40, along with Equation 18.39:

$$\Psi_{WP} = 3(\Phi_1 \coth \Phi_1 - 1) \tag{18.46}$$

$$\Psi_{WP} = \frac{r'_{obs}}{r_d} = \frac{r'' \rho_s R^2}{D_e C_{As}} \tag{18.47}$$

Equating the above two equations and relating them for two experiments, considering D_e, C_{As} as given previously, we obtain:

$$\frac{R_1^2 \, r''_{obs}\big|_1}{R_2^2 \, r''_{obs}\big|_2} = \frac{\Psi_{11}}{\Psi_{12}} \tag{18.48}$$

Substituting radius and rates:

$$\frac{\Psi_{11}}{\Psi_{12}} = 10.57$$

Then, $\Psi_{12} = \frac{\Psi_{11}}{10.6}$
But,

$$\Psi_{11} = \frac{r'_{obs}}{r_d} = \frac{r' S_g \rho_s R^2}{D_e C_{As}} = 3.02$$

Therefore,

$$\Psi_{12} = \frac{\Psi_{11}}{10.6} = 0.28$$

Thus, there are no diffusion limitations when the particle radius is smaller than that informed (spherical particle of diameter 0.13 cm).

18.4 EXTERNAL MASS TRANSFER AND INTRAPARTICLE DIFFUSION LIMITATIONS

In a catalytic reactor, the fluid flows through the catalyst particles and may face a resistance caused by the concentration gradient between the bulk fluid and the external particle surface. This resistance (interparticle or external mass transfer limitation) must be added to intraparticle diffusion limitation.

The overall rate should consider these phenomena along with the chemical reaction. Figure 18.5 shows the concentration profiles resulting from different phenomena in a gas–solid system or liquid–solid containing porous particles.

The external concentration C_{A0} is known (it is the bulk concentration), but a concentration gradient ranging from C_A to C_{As} may exist within the thin film around the particle, C_{As} being concentration at the external surface of the particle or at the

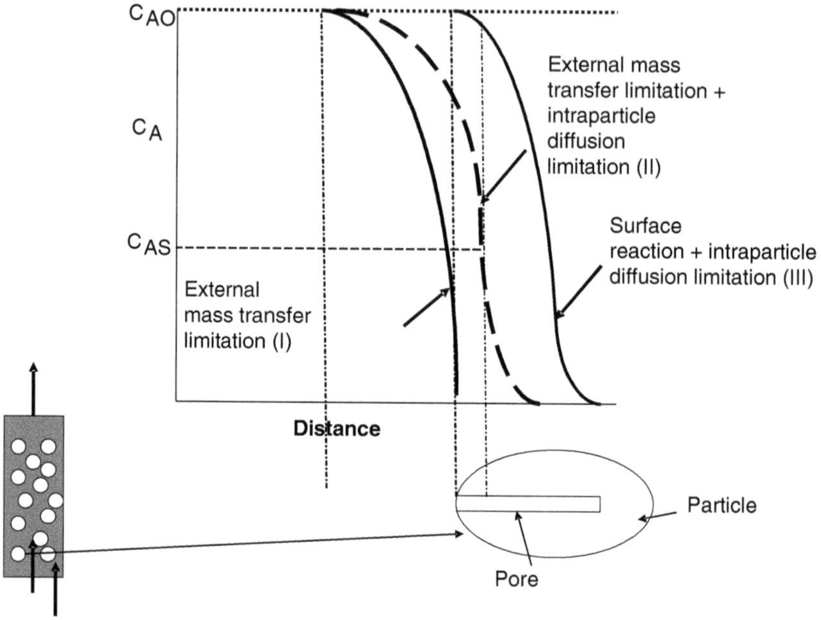

Figure 18.5 External and internal mass transfer limitations.

pore entrance. Due to the external mass transfer resistance there is a concentration gradient (curve I, see Figure 18.5). Therefore, the mass transfer rate in this region is:

$$\dot{R}_A = k_m a_m (C_A - C_{As}) \tag{18.49}$$

where
R''_A = mass transfer rate (mol/g/s)
k_m is the mass transfer constant and a_m is the gas/solid interface area.

Curve II shows a lower concentration gradient outside the particle, compared to curve I, but followed by a concentration gradient inside the pore due to the intraparticle diffusion limitation. On the other hand, curve III does not have influence of external limitations, representing the reaction rate in the presence of pore-diffusion resistance as shown in the following equation:

$$r'' = k'S \cdot f(C_{As}, T) \cdot \eta (\text{mol/g/s}) \tag{18.50}$$

where:
r'' = rate by mass of solid (mol/g/s)
η = effectiveness factor
k'' = rate constant by surface area.

The effectiveness factor depends on the Thiele modulus defined by Equation 18.13.

To determine the overall reaction rate in the PFR reactor, one starts with:

$$\frac{W}{F} = \int_0^{X_A} \frac{dX_A}{(-r_A'')} \tag{18.51}$$

Consider a particular case of irreversible first-order reaction (reactant A as limiting). Therefore, the rate will be:

$$(r_A'') = k'S_g C_{As} = k'S_g C_{A0}(1 - X_{As})\eta \tag{18.52}$$

At steady state, the mass transfer rate and reaction rate in the presence of diffusion effects are equal. Then,

$$k_m a_m (C_A - C_{As}) = k'S_g C_{As}\eta \tag{18.53}$$

As a function of the limiting reactant conversion:

$$k_m a_m (X_{As} - X_A) = k'S_g C_{A0}(1 - X_{As}) \times \eta \tag{18.54}$$

We determine $(1 - X_A)$ to replace it into rate equation, since the surface concentration is unknown. Then,

$$(1 - X_{As}) = \frac{(1 - X_A)}{1 + \dfrac{k'S_g C_{A0}\eta}{k_m a_m}} \tag{18.55}$$

Substituting into rate Equation (18.52), we have:

$$(-r_A'') = k'S_g C_{A0}(1 - X_A)\frac{\eta}{1 + \dfrac{k'S_g C_{A0}\eta}{k_m a_m}} \tag{18.56}$$

Defining a new factor that comprises both external mass transfer and intraparticle diffusion limitations (Fogler, 2000):

$$\Omega = \frac{\eta}{1 + \dfrac{k'S_g \eta}{k_m a_m}} \tag{18.57}$$

The rate equation can be expressed as:

$$(-r_A') = k'S_g C_{A0}(1 - X_A)\Omega \tag{18.58}$$

The reaction rate constant is temperature dependent. For nonisothermal systems, one should take into account the energy balance. Moreover, catalyst properties such as the specific surface area and the pore volume must be known. In this system, a spherical particle has been considered.

It is very important to know the mass transfer constant (k_m) and the gas/solid interface area (a_m). This information can be obtained from the fluid properties and characteristic numbers for flow such as Schmidt and Sherwood numbers. It is challenging to determine the interface area, but usually one tries to obtain $k_m a_m$ together and not separately.

E18.3 An irreversible reaction is done in a PFR reactor containing catalyst with spherical particles of 3 mm diameter. Pure reactant A is fed under flow of 1000 L/s, pressure of 1 atm, and constant temperature of 300°C. The conversion at the outlet of the reactor is supposed to be 80%. Calculate the catalyst mass that must be placed in the reactor. Additional data are provided as follows (Fogler, 2000):

$$k' = 0.07 \,(\text{cm}^3/\text{cm}^2/\text{s}) \quad k_m a_m = 4.0 \,(\text{cm}^3/\text{s/g})$$

$$D_e = 0.01 \,(\text{cm}^2/\text{s}) \qquad\qquad \rho_s = 2 \,(\text{g/cm}^3)$$

$$S_g = 3.10^5 \,(\text{cm}^2/\text{g})$$

Solution
The rate will be:

$$(-r_A'') = k' S_g C_{A0}(1 - X_A)\Omega$$

For a first-order reaction: $(-r_A'') = k' C_A \,(\text{mol g}^{-1}\,\text{s}^{-1})$
The molar balance equation is given by Equation 18.51, i.e.:

$$\frac{W}{F} = \int_0^{X_A} \frac{dX_A}{(-r_A'')}$$

Substituting the rate into Equation 18.51 and integrating, we obtain:

$$\frac{W}{F} = -\frac{1}{k^* \times C_{A0}} \ln(1 - X_A)$$

Considering $k^* = k' S_g \Omega$
We still need to calculate Ω using Equation 18.57:

$$\Omega = \frac{\eta}{1 + \dfrac{k' S_g \eta}{k_m a_m}}$$

Initially, we need to determine the effectiveness factor η. From Thiele modulus, we know that:

$$\Phi_1 = R\sqrt{\frac{k' S_g \rho_s}{D_e}} = 0.3\sqrt{\frac{7.10^{-5} \times 3.10^5 \times 2}{0.01}} = 19.4$$

Thiele modulus value reveals the presence of strong diffusion effects and therefore, the effectiveness factor varies linearly with Thiele modulus according to the equation:

$$\eta = \frac{3}{\Phi_1} = \frac{3}{19.4} = 0.154$$

From that we can determine the overall mass transfer limitation effect:

$$\Omega = \frac{\eta}{1 + \dfrac{k'S_g\eta}{k_m a_m}} = \frac{0.154}{1 + \dfrac{7.10^{-5} \times 3.10^5 \times 0.154}{4.0}} = 0.085$$

Then,

$$k^* = k'S_g\Omega = 1.785 (\text{cm}^3/\text{g/s})$$

$$C_{A0} = \frac{P}{RT} = 2.12 \times 10^{-2} \text{ mol/L}$$

Therefore,

$$\frac{W}{F} = -\frac{1}{k * C_{A0}} \ln(1 - X_A) = 42.1$$

But

$$F_{A0} = C_{A0} \cdot v_0 = 21.2 \text{ (mol/ min)}$$

Finally, we calculate the mass:

$$W = 896\,\text{g} = 0.89\,\text{kg}$$

In the absence of external mass transfer limitation:

$$k^* = k'S_g\eta = 7.10^{-5} \times 3.10^5 \times 0.154 = 3.23 (\text{cm}^3/\text{g/s})$$

Therefore,

$$\frac{W}{F} = -\frac{1}{k * C_{A0}} \ln (1 - X_A) = 23.4$$

$$W = 479\,\text{g} = 0.479\,\text{kg}$$

Part V

Deactivation

Chapter 19

Catalyst deactivation

Catalyst deactivation occurs over time during a catalytic process and it is a phenomenon that invariably takes place in most industrial processes. Catalyst deactivation may occur via four main phenomena: coke deposition which leads to the blocking of pores and active sites, poisoning via metals (S, As), sintering, and the loss of catalytic active phase.

Supported metal catalysts are commonly used in industrial processes. Examples include nickel at high concentration and low dispersion and palladium at a low metal concentration and high dispersion. Both catalysts enable high conversion and selectivity in reforming reactions, gasoline pyrolysis, and hydrogenation processes. However, these catalysts are prone to deactivation due to coke formation and presence of gums in the load. Coke is deposited preferentially on the metal sites during hydrogenation reactions while gum is deposited on both metal sites and support resulting in rapid catalyst. Thus, it is important to investigate catalyst deactivation on an industrial scale concerning the extent of deactivation, as well as explore possible strategies for regeneration.

To do so, one scales down the process to a laboratory scale. This entails using the same space velocity, pressure, and temperature but a scaled down catalyst mass. Furthermore, to evaluate the catalysts under kinetic control regime one determines the appropriate conditions to minimize heat and mass transfer effects. Comparative evaluations can then be made between the results obtained in laboratory and industry.

Stability is a key criterion when determining the most appropriate catalyst for industrial use. Among the deactivation phenomena, coke deposition and poisoning may be reversible but sintering is usually irreversible.

Sintering is a thermodynamically favorable process which decreases the catalyst surface area. Although the catalyst solid nature and the environment to which the catalyst is exposed influence the process, the sintering phenomenon is basically dominated by temperature. The minimum sintering temperature can be estimated from the melting temperature of the solid ($T_{melting}$) expressed in Kelvin:

$$\alpha_v = 0.3 T_{melting} \tag{19.1}$$

The melting temperature of nickel is 1726 K. Therefore, according to Equation 19.1, nickel sintering may be carried out above 517 K.

Coke is carbonaceous species deficient in H_2 which may deposit on the catalyst surface. Carbon often chemisorbs reversibly. Thus, the catalyst can be regenerated by removing the carbon by means of some oxidant such as O_2, H_2O, or even CO_2. However, the high temperatures necessary for the carbon oxidation can lead to the metal sintering in some cases, which could decrease catalytic activity for further reuse. Schematically:

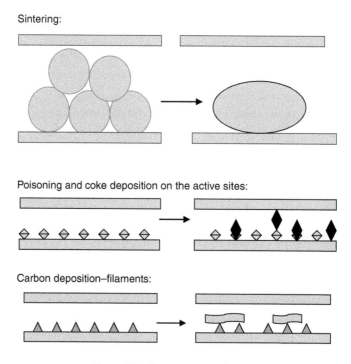

Figure 19.1 Deactivation schemes.

Figure 19.2 displays an interesting result for the ethanol conversion versus time in catalytic tests at different temperatures on supported nickel catalysts, i.e., samples tested isothermally over time at 723, 773, 873, and 973 K.

According to Figure 19.2, the initial activity of the catalyst increases with temperature and for the samples tested at 773 and 873 K the deactivation is very high. Deposited carbon was measured by thermogravimetric analysis (TGA), i.e., by oxidation following the mass loss. Figure 19.3 displays the total amount of mass loss attributed to carbon, obtained by TGA, for the different samples tested in the ethanol conversion at 723, 773, 873, and 973 K. TGA analysis revealed that the amount of carbon deposited on the samples ranged from 20 to 45 wt%, suggesting pore clogging and active sites covered by carbon, causing the constant deactivation of the catalysts.

Figure 19.4 displays a picture obtained by scanning electron microscopy (SEM) for a nickel catalyst deactivated. A large amount of carbon filaments and agglomerates of coke can be easily identified on the catalyst surface. These filaments do not cause a direct deactivation since they consist of α-type carbon. They do, however, lead to reactor clogging.

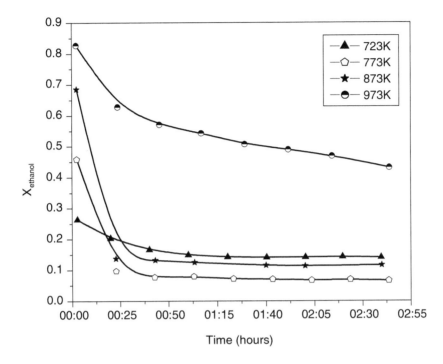

Figure 19.2 Ethanol conversion as a function of time at different temperatures of reaction (isothermal). Conditions: GHSV = 200 mL$_{EtOH}$/(min g$_{cat}$), H$_2$O/EtOH = 3, v_0 = 100 mL/min.

19.1 KINETICS OF DEACTIVATION

The observed rate varies with time and the catalyst activity decreases with all the variables of the system constant. The activity a is defined as the ratio between the observed reaction rate and the initial rate (when the catalyst is virgin), i.e.:

$$a = \frac{(-r''_A)_{obs}}{(-r''_{A0})} \qquad (19.2)$$

where:

a represents activity over the course of reaction
$(-r''_A)_{obs}$ is the observed reaction rate
$(-r''_{A0})$ is the reaction rate of the fresh catalyst.

We take a simple case assuming an irreversible first-order reaction as:

$$A \xrightarrow{k} R$$

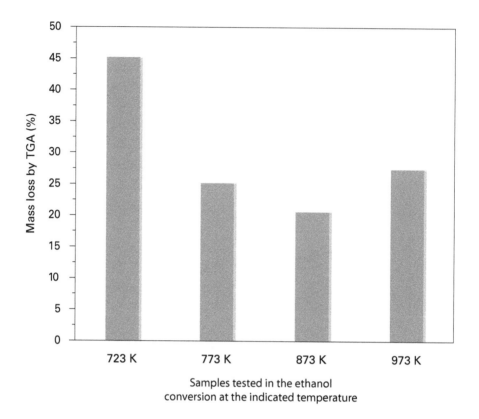

Figure 19.3 Mass loss obtained by TGA analysis for supported nickel catalysts.

It is assumed that the first-order kinetics, however, considering that the catalyst particles are porous, there may be diffusion effects. This initial rate or reaction rate of the virgin catalyst can be represented by:

$$\left(-r''_{A0}\right) = k'' \cdot a \cdot S_g \cdot C_A \qquad (19.3)$$

where η is the effectiveness factor which takes into account the diffusion effects. The reaction rate constant k' represents the specific reaction rate, depends on the temperature and usually is expressed by cubic centimeters per square centimeters per second (defined per unit surface area). The reactant concentration in the gas phase varies across the length in the continuous reactor and with time in the batch reactor. Therefore, the observed rate will be:

$$\left(-r''_A\right)_{obs} = k' \cdot a \cdot S_g \cdot C_A \cdot \eta \qquad (19.4)$$

The reaction rate is given by mass of catalyst, and in this case, the specific surface area of the particles S_g must be known.

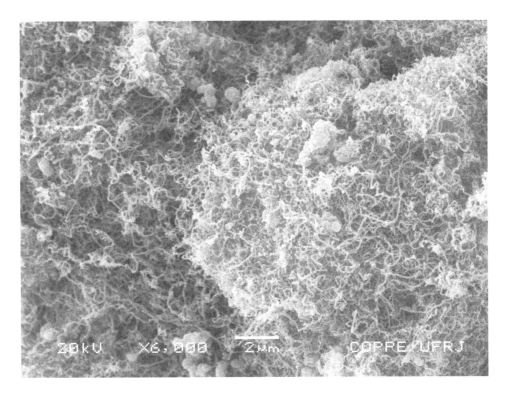

Figure 19.4 Picture of nickel catalyst performed in test at 723 K.

The deactivation rate varies with time and can have a generic order:

$$r_d = -\frac{da}{dt} = k_d \cdot a^n \tag{19.5}$$

Generally, it is a first-order reaction rate but it can also be of fractional order, depending on the deactivation mechanism and whether or not it satisfies power–law kinetics. We assume the simplest case, i.e., first order:

$$r_d = -\frac{da}{dt} = k_d \cdot a \tag{19.6}$$

where k_d is the deactivation constant that satisfies the Arrhenius equation.

$$k_d = k_d(T_0)e^{-(E_d/(RT))} \tag{19.7}$$

and k_0 is the pre-exponential factor.

Integrating Equation 19.6, one obtains the variation of activity with time at a constant temperature, i.e.:

$$a = e^{-k_d t} \tag{19.8}$$

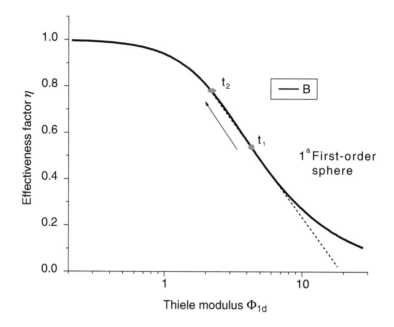

Figure 19.5   Modified effectiveness factor.

Finally substituting Equation 19.8 into Equation 19.4, we obtain:

$$\left(-r_A''\right)_{obs} = k'S_g \cdot C_A \cdot \eta \cdot e^{-k_d t} \tag{19.9}$$

This rate depends on particle size (pellet) and in this case spherical particles. If there is no mass transfer limitations (diffusion effects), the effectiveness factor $\eta = 1$. However, if diffusion effects take place this factor decreases significantly in a manner that is dependent on the Thiele modulus. The effectiveness factor varies according to Equation 18.24:

$$\eta = \frac{3}{\Phi_1} \tag{18.24}$$

Thiele modulus Φ_1 must be corrected as it varies with the activity a.

$$\Phi_{1d} = R\sqrt{\frac{k' \cdot a \cdot S_g \cdot \rho_s}{D_e}} \tag{19.10}$$

Therefore, the effectiveness factor, also depicted in Figure 18.4, varies with Thiele modulus as illustrated in Figure 19.5.

These expressions change for irreversible n-order reactions (n different from 1), as well as for nonspherical models.

19.2   DEACTIVATION IN PFR OR CSTR REACTOR

Starting from Equation 14.38 of PFR reactor:

$$\tau = C_{A0} \int_0^{X_A} \frac{dX_A}{(-r'_A)_{obs}} \tag{14.38}$$

where τ is the space time.

The observed rate as a function of conversion is given by Equation 19.4, which also depends on the effectiveness factor and activity:

$$(-r''_A)_{obs} = k' \cdot a \cdot S_g \cdot C_{A0}(1 - X_A) \cdot \eta$$

Substituting this expression into Equation 14.38 and using Equation 18.24, we obtain the following expression for a PFR reactor with diffusion effects:

$$-\ln(1 - X_A) = \frac{3 \cdot k' \cdot S_g \cdot \tau}{\Phi_1} e^{(-k_d t/2)} \tag{19.11}$$

An expression for CSTR reactor is also derived from Equation 14.27:

$$\frac{X_A}{1 - X_A} = \frac{3 \cdot k' \cdot S_g \cdot \tau_{CSTR}}{\Phi_1} e^{(-k_d t/2)} \tag{19.12}$$

In these cases, one can monitor conversion versus time and determine the deactivation rate as well as the deactivation rate constant. Figure 19.2 shows an example of conversion varying with reaction time in the ethanol reforming reaction.

Example

E19.1 An irreversible reaction takes place in a CSTR containing spherical particles catalyst of 24 mm diameter. Pure reactant A is introduced at pressure of 1 atm and constant temperature of 300°C with a space time of 4000 (g × s/m^3). Conversion was measured with time as shown in the table (Fogler, 2000). Calculate the deactivation rate with and without diffusion, knowing the additional data:

$$D_e = 2 \times 10^{-8} \text{ m}^2/\text{s}; S_g = 30 \text{ m}^2/\text{kg}; \rho_s = 1500 \text{ g/m}^3$$

Solution

Let us first consider the case in the absence of diffusion effects:

We start from Equation 19.12 with $\Phi_1 = 1$ and take the natural log of both sides of the equation:

$$\ln\left(\frac{X_A}{1 - X_A}\right) = \ln(k' \cdot S_g \cdot \tau_{CSTR}) - k_d t$$

Plotting a graph:

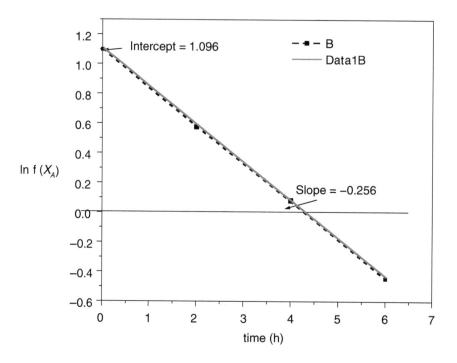

Figure 19.6 Deactivation rate.

Therefore, the slope provides $k_d = 0.256\,\text{h}^{-1}$ and the intercept allows calculating $\ln(k'S_g\tau_{CSTR}) = 1.096$.

Then, we calculate the value of k' using $S_g = 30\,\text{m}^2/\text{g}$ and $\tau = 4000\,(\text{g s/m}^3)$:

$$k' = 2.493 \times 10^{-5}\,\text{m}^3/(\text{m}^2\,\text{s})$$

We calculate the Thiele modulus to verify if there are diffusion effects in the data above.

$$\Phi_1 = R\sqrt{\frac{k' \cdot S_g \rho_s}{D_e}}$$

$$\Phi_1 = 2.4 \times 10^{-3}\sqrt{\frac{2.493 \times 10^{-5} \times 30 \times 1500}{2.0 \times 10^{-8}}} = 18$$

We note strong diffusion effects.

Therefore, now we admit the rate with mass transfer limitations.

We start from Equation 19.12

$$\frac{X_A}{1 - X_A} = \frac{3 \cdot k' \cdot S_g \cdot \tau_{CSTR}}{\Phi_1}e^{-(k_d t/2)}$$

Table 19.1 Conversion as a function of time.

t (h)	0	2	4	6
X_A	0.75	0.64	0.52	0.39

Taking the natural log of both sides of the equation, we obtain:

$$\ln\left(\frac{X_A}{1-X_A}\right) = \ln(k^*) - \frac{k_d}{2}t$$

Using the experimental data of Table 19.1, we obtain the same graph. Thus,

$k_d = 0.52\ \text{h}^{-1}$
$\ln k^* = 1.096$
$k^* = 3.0$

But:

$$k* = \frac{3 \cdot k' \cdot S_g \cdot \tau_{CSTR}}{\Phi_1}$$

And:

$$\Phi_1 = R\sqrt{\frac{k \cdot S_g \rho_s}{D_e}}$$

Consequently:

$$k' = \frac{R^2 \cdot \rho_s}{S_g \cdot \tau^2 \cdot D_e} = 9 \times 10^{-4}\ (\text{m}^3/\text{m}^2/\text{s})$$

Therefore, the new value of the constant k' is:

$k'' = 2.7 \times 10^{-2}\ \text{m}^3/(\text{kg s})$

$k' = 9 \times 10^{-4}\ (\text{m}^3/\text{m}^2/\text{s})$

Then we calculate again the Thiele modulus using the data above.

$$\Phi_1 = R\sqrt{\frac{k' \cdot S_g \rho_s}{D_e}}$$

$$\Phi_1 = 2.4 \times 10^{-3}\sqrt{\frac{9 \times 10^{-4} \times 30 \times 1500}{2.0 \times 10^{-8}}} = 108$$

Thus, the correct value of the constant is:

$$k' = 1.5 \times 10^{-4} \ (m^3/m^2/s)$$

And we can calculate the effectiveness factor, i.e.:

$$\eta = \frac{3}{\Phi_1} = \frac{3}{108} = 0.009$$

Depending on the activity, the Thiele modulus changes as Equation 19.9:

$$\Phi_{1d} = R\sqrt{\frac{k' \cdot a \cdot S_g \rho_s}{D_e}} = 108\sqrt{a}$$

And the effectiveness factor will be:

$$\eta = \frac{3}{\Phi_{1d}} = \frac{3}{108\sqrt{a}}$$

Finally, the conversion varies with time and activity will be:

In the absence of mass transfer limitations:

$$\frac{X_A}{1-X_A} = k' \cdot S_g \cdot \tau_{CSTR} \ e^{-(0.256t)} = 2.99 \ e^{-(0.256t)}$$

In the presence of mass transfer limitations:

$$\frac{X_A}{1-X_A} = \frac{3 \cdot k' \cdot S_g \cdot \tau_{CSTR} \cdot a}{\Phi_1} = 0.5\sqrt{e^{-0.52t}}$$

Therefore,

$$\frac{X_A}{1-X_A} = 0.5\sqrt{e^{-0.52t}}$$

To verify the data, we have Figure 19.7.

19.3 FORCED DEACTIVATION

Catalyst deactivation may be simulated by carrying out forced deactivation experiments at high temperatures. Under some conditions, there is a drastic loss of activity. Next, we try to determine the regeneration conditions.

Forced deactivation experiments may be performed in two ways:

1. Inducing the sintering of the catalytic material at high temperatures in order to verify the extent of deactivation during a short period of time.

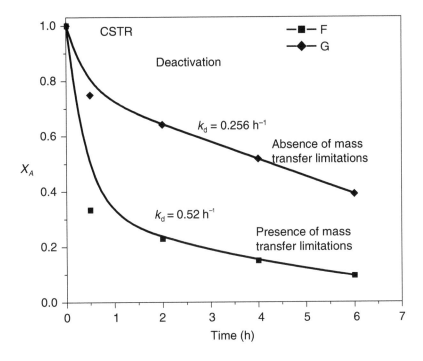

Figure 19.7 Diffusion effects in the deactivation.

Figure 19.7 Diffusion effects in the deactivation.

2. Using a high concentration of the main component in the feed in order to increase cracking and coke formation.

In these experiments, catalytic conversion is monitored over reaction time. Temperature is gradually increased but later the initial condition is taken up. The catalytic material may then be investigated by TGA to quantify the coke formed.

Example

E19.2 Let us take as an example the hydrogenation of pyrolysis gasoline which contains as main products styrene, isoprene, dicyclopentadiene, and gums in the load. Catalytic experiments were performed in a steel reactor with ascendant flow and catalyst bed volume of $27 \, cm^3$. The operating parameters were set according to the plant and the space velocity is $4.5 \, h^{-1}$. Several α-alumina-supported nickel and palladium catalysts were tested. After proceeding to pretreatment and reduction used in the industry, tests were performed at four temperatures, 60°C, 80°C, 100°C, and 120°C and then the initial condition (60°C) was taken up. After this step, the temperature was increased to 350°C for 3 h with N_2 flow with real and synthetic load, forcing the catalysts deactivation. The feed flow was reduced to half that used in the catalytic test, increasing the residence time in the reactor and forcing catalyst deactivation. Next, the deactivated

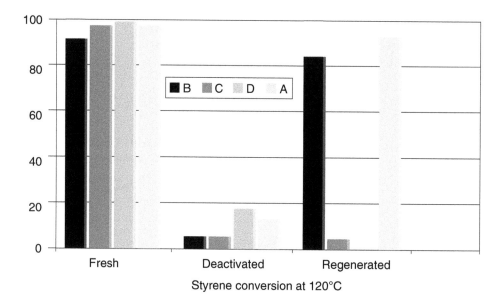

Figure 19.8 Styrene conversion at 120°C as a function of the sample tested.

catalysts were tested at 60°C and 120°C under the same initial operating conditions of the virgin catalyst. Finally, regeneration tests were performed to verify the regeneration degree or permanent deactivation.

Solution
Table 19.1 shows results of styrene conversion obtained from experiments in which the catalysts were subjected to forced deactivation at 120°C and regeneration at 60°C, 100°C, and 120°C. For comparison, the results obtained in step 1 with synthetic load are shown as well.

One can observe a considerable deactivation with respect to step 1 using synthetic load, especially the catalyst C which practically had not deactivated previously (around 4%) after hard treatment to 350°C. The other catalysts A and B, though both have presented an increased deactivation degree from 65% to 94% and from 54% to 88%, respectively, one observes recovery in the activity of these catalysts after the regeneration step.

The results of activity (styrene conversion) at 120°C in the hydrogenation of pyrolysis gasoline are shown in Figure 19.8 for selected samples.

For all of the catalysts, the real load (gasoline) led to a stronger deactivation than employing the synthetic load.

The regeneration step was not enough to recover the catalytic activity of the catalyst C. For catalysts A and B, despite the high deactivation, there was a reasonable recovery of the initial activity, reaching 90% of conversion after regeneration.

Table 19.2 Styrene conversion data for hydrogenation of pyrolysis gasoline and synthetic load using commercial catalysts after forced deactivation and regeneration procedures

Sample/Feed	Styrene Conversion (%)					
	After Deactivation* 120°C	Deactivation Degree (%)	After Regeneration			
			60°C	100°C	120°C	60°C
B/gasoline	5.2	94.3	51.7	77.1	87.1	60.8
B/synthetic load	34.5	65.2	42.8	90.9	98.5	–
C/gasoline	5.4	94.4	6.6	8.6	3.9	–
C/synthetic load	96.5	3.6	78.1	98.9	99.3	–
A/gasoline	11.6	88.1	52.6	82.8	93.8	83.5
A/synthetic load	45.7	54.2	97.2	–	99.7	–

*At 350°C.

Example

E19.3 Based on the experimental results above, data from previous example and assuming forced deactivation in the absence of mass transfer limitations, we have calculated the conversion over reaction time. From Equation 19.12 and data in Table 19.2, we have also calculated the deactivation rate constants assuming CSTR-type behavior.

$$\frac{X_A}{1-X_A} = 2.99 \ e^{-(k_d t)}$$

The estimated reaction time was $t = 5$ h. Therefore, we have calculated the constant k_d using the styrene conversion given in Table 19.2:

$$X_A = 0.35 \rightarrow\rightarrow k_d(60°C) = 0.35$$

After deactivation at 350°C and new activity test at 120°C for 5 h:

$$X_A = 0.05 \rightarrow k_d(120°C) = 0.77$$

Substituting these values into Equation 19.11 and using the same conversion data of the previous problem, we vary X_A with reaction time and obtain Figure 19.9.

The results reveal that after 6 h, natural deactivation led to a reactant conversion of 0.38. For this same time frame, forced deactivation led to a significantly lower conversion of approximately 0.02. Extrapolating the curve of natural deactivating (non-forced), the time required for complete deactivation would be 11.5 h.

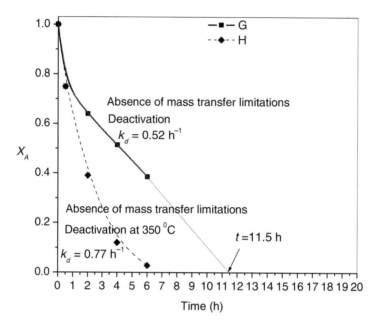

Figure 19.9 Deactivation of catalysts–deactivation rate constants.

19.4   CATALYST REGENERATION

The example in the earlier section shows a catalyst may be sufficiently regenerated after deactivation. However, this regeneration depends on the type of deactivation. For example, if sintering is the cause of deactivation, it may be difficult to regenerate catalytic material as a result of particle agglomeration and the formation of aluminates.

If the deactivation is due to the poisoning of active sites caused by deposition of compounds coming in the feed, its removal will depend on the type of poisoning. Deactivation due to gums may be reversible depending on the type of regeneration method. Sulfur compounds or metals, however, are irreversibly adsorbed. Regeneration is possible when deactivation is caused by cracking and coke deposition on particles or inside pores. Carbon often chemisorbs reversibly. Thus, the catalyst can be regenerated by removing the carbon by means of some oxidant such as O_2, H_2O, or CO_2. However, in some cases, high temperatures for the carbon oxidation can lead to the metal sintering. Oxidation or combustion in the presence of steam allows removing coke.

The catalyst regeneration process is not simple and involves many parameters that are directly related to deactivation. It is essential to understand the deactivation mechanisms in order to choose the appropriate regeneration conditions. Textural properties as surface area and pore volume are affected by the regeneration procedures as well as metal surface phases or acidic properties of the support. Therefore, a regeneration process will depend on the history of the catalytic process, the feed, the type of process, and its variables such as pressure and temperature, which are the main deactivation factors. The regeneration makes sense when the deactivation can be reversed or phases can be recovered. Each situation must be evaluated separately.

The parameters that influence the regeneration process comprise heating rate in the coke combustion process, regeneration time, and gas flowing during regeneration. The combustion rate is very critical and the ignition temperature causes uncontrollable increase in temperature, since combustion reaction is highly exothermic. Heat dissipation is an alternative to avoid sudden heating due to ignition, which is significant when the atomic ratio H/C is high. Finally, the time for coke removal is an important parameter and several regeneration cycles can be carried out not affecting the process at all.

Carbon is burnt in the process and the following reactions take place:

$C + O_2 \rightarrow CO_2$	$\Delta H^0_{298K} = -406.4 \ kJ/mol$
$2C + O_2 \rightarrow 2CO$	$\Delta H^0_{298K} = -246.3 \ kJ/mol$
$CO + 1/2O_2 \rightarrow CO_2$	$\Delta H^0_{298K} = -567.3 \ kJ/mol$
$C + 2H_2 \rightarrow CH_4$	$\Delta H^0_{298K} = -83.8 \ kJ/mol$
$C + H_2O \rightarrow CO + H_2$	$\Delta H^0_{298K} = 118.5 \ kJ/mol$

These reactions are exothermic and thermodynamically favorable, except the last. Most regeneration techniques are based on coke combustion increasing the temperature.

The regeneration can be evaluated by differential scanning calorimetry (DSC), temperature programmed oxidation (TPO), and catalytic tests.

19.4.1 Differential scanning calorimetry

A first evaluation of the catalytic performance is usually carried out by using DSC. Samples are heated at a rate of 10 K/min under synthetic air flow (50 mL/min).

Enthalpy of reaction (ΔH) is calculated using the following correlation:

$$\Delta H = \frac{K \cdot A}{m} [cal/g]$$

where:

$K =$ dimensionless constant (1.06)

$m =$ mass of sample (mg)

$A =$ quantity of heat (mcal) corresponding to the area under DSC curve ((mcal/s) s).

By the Kissinger method (Boudart), one can determine the activation energy (E) via DSC curves obtained under three different heating rates (β) at least. Once the rates (β) are known, as well as temperatures corresponding to the maximum on each peak (T_C), it is possible to establish the following linear relationship:

$$\ln\left(\frac{\beta}{T_C^2}\right) = -\frac{E}{R} \cdot \frac{1}{T_C} + \ln\left(\frac{A \cdot R}{E}\right) \tag{19.13}$$

where R is the universal gas constant and A is the pre-exponential factor of Arrhenius equation. Therefore, the energy E is determined by the slope.

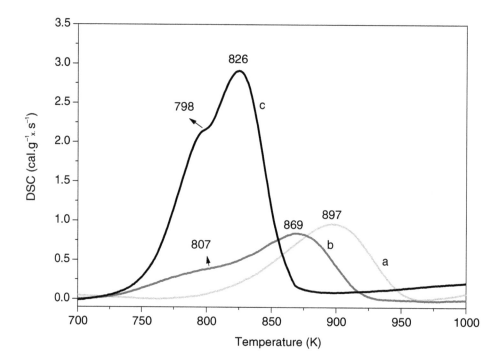

Figure 19.10 DSC curves of samples: (a) Al_2O_3, (b) catalyst A, (c) catalyst B. Conditions: 10 K/min, 21% O_2/N_2.

DSC curves of samples for different cases are shown in Figure 19.10.

The temperature corresponding to the maximum in the DSC peak is a parameter widely used in the literature for the evaluation of the catalytic performance. This parameter is often known as combustion temperature and represented by T_C.

19.4.2 Temperature programmed oxidation

The compounds from carbon oxidation are quantified by a mass spectrometer coupled to the catalytic test unit. The sample is flowed by 5% O_2/He under v_0 flow and heating rate of 10 K/min. The combustion is monitored by mass spectrometer as exemplified in Figure 19.11.

19.4.3 Catalytic evaluation

A fixed-bed reactor is usually utilized. Temperature is ranged during the tests but total pressure and space velocity remain constant. The experiments are performed continuously for 30 days.

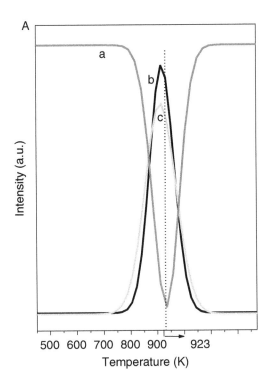

Figure 19.11   Oxidation profiles of samples: (a) O_2, (b) CO, (c) CO_2. Conditions: $10\,K\,min^{-1}$, 5% O_2/He no water added.

19.5   KINETIC STUDY OF REGENERATION

The most widely used model is a gas–solid reaction, assuming a carbon spherical particle. The most common cases are the regeneration and diesel soot combustion. In this model, carbon combustion is admitted and a cinder layer remains. Therefore, there is gas diffusion through the cinder layer and reaction on the carbon particle surface, which moves inward until total consumption. CO_2 is formed during combustion and must diffuse through the cinder in the opposite direction, as shown in Figure 19.12.

Oxygen (O_2) diffuses through the cinder layer, where no reaction takes place, reaches the interface, oxidizes carbon and then CO_2 just formed diffuses through the opposite direction crossing the cinder layer. Considering a carbon particle radius R_0 and interface radius r_c, we obtain for each mol of O_2 (reactant A) the diffusion rate:

$$N_A = -D_e \frac{dC_A}{dr} \ mol/(cm^2\ s) \tag{19.14}$$

The molar balance on a surface element $4\pi r^2$ between r and $r + dr$, where no reaction takes place will be:

$$N_A . 4\pi r^2 \big|_r - N_A \cdot 4\pi r^2 \big|_{r+dr} = 0$$

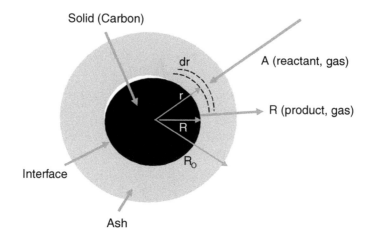

Figure 19.12 Uniform model of solid consumption (carbon consumption).

Then:

$$\frac{d}{dr}(N_A \cdot r^2) = 0 \tag{19.15}$$

Substituting into Equation 19.14:

$$\frac{d}{dr}\left(-D_e \cdot r^2 \frac{dC_A}{dr}\right) = 0 \tag{19.16}$$

where D_e is the diffusion coefficient (cm^2/s).
   With the boundary condition:

On the external surface: → $r = R_0$ → $C_A = C_{A0}$
At the interface: → $r = R(t)$ → $C_A = 0$

Solving Equation 19.13, we obtain (Equation 10.38):

$$\frac{C_{O_2}}{C_{O_{20}}} = \varphi = \frac{\dfrac{1}{R} - \dfrac{1}{r}}{\dfrac{1}{R} - \dfrac{1}{R_0}} \tag{10.38}$$

The concentration profile of the reactant A (O_2) along the carbon particle radius is shown in Figure 19.13:
   Therefore, the reactant A molar flow (N_A) at the interface will be:

$$N_A = -D\frac{dC_A}{dr}\bigg|_R = -\frac{DC_{A0}}{\left(\dfrac{1}{R} - \dfrac{1}{R_0}\right)r^2} \tag{19.18}$$

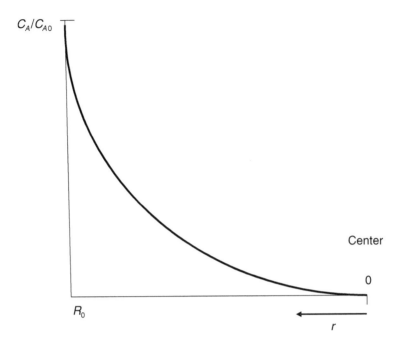

Figure 19.13 Oxygen concentration profile ($A = O_2$) along the solid radius (carbon radius).

19.5.1   Balance with respect to solid (carbon)

Generated rate = accumulated rate

$$r_s'' \cdot 4\pi r^2 = \frac{d\left(\frac{4}{3}\pi R^3 \rho\varepsilon\right)}{dt}$$

$$\frac{dR}{dt} = \frac{r_s''}{\rho\varepsilon} \tag{19.19}$$

where

ε = fraction of solid (carbon)
ρ = density of solid (carbon).

The reaction rate (carbon consumption) at the interface is equal to the reactant A molar flow that reaches the interface:

$$-r_s'' = N_A|_{r=R} = \frac{DC_{A0}}{\left(R - \dfrac{R^2}{R_0}\right)} \tag{19.20}$$

Substituting this rate expression into Equation 19.19:

$$-\frac{dR}{dt} = \frac{DC_{A0}}{\rho\varepsilon}\frac{1}{\left(R-\dfrac{R^2}{R_0}\right)} \tag{19.21}$$

Integrating for $t=0 \rightarrow r=R_0$, we have (Equation 10.42) the time for solid consumption (t):

$$t = \frac{\rho R_0^2 \varepsilon}{6DC_{A0}}\left[1-3\left(\frac{R}{R_0}\right)^2+2\left(\frac{R}{R_0}\right)^3\right] \tag{19.22}$$

Defining the solid S conversion (carbon conversion) as X_S:

$$X_S = \frac{V_{S0}-V_S}{V_{S0}} = \frac{4/3\pi R_0^3 - 4/3\pi R^3}{4/3\pi R_0^3} = 1-\left(\frac{R}{R_0}\right)^3$$

where S is the spherical solid (spherical carbon particle), R is the radius of the solid particle which is consumed along time and R_0 is the radius of particle in initial conditions.

The time for complete solid consumption (τ) is reached when $R=0$. Substituting this condition into Equation 19.22, we obtain:

$$\tau = \frac{\rho R_0^2 \varepsilon}{6DC_{A0}} \tag{19.23}$$

Then we can rearrange Equation 19.22 in order to obtain the following expression for the reaction time (solid consumption):

$$\frac{t}{\tau} = \left[1-3(1-X_S)^{2/3}+2(1-X_S)\right] \tag{19.24}$$

19.5.2 Particular case

The chemical reaction at the unreacted layer is rate determining. The rate will be a first-order rate in this case, i.e.:

$$r_s'' = k_s C_A$$

Substituting in Equation 19.19, we have:

$$\frac{dR}{dt} = \frac{r_s''}{\rho\varepsilon} = \frac{k_s C_{A0}}{\rho\varepsilon}$$

Integrating in between the boundaries R_0 and R:

$$t = \frac{\rho R_0 \varepsilon}{k_s C_{A0}}\left[1-\left(\frac{R}{R_0}\right)\right]$$

As a function of conversion:

$$\frac{t}{\tau} = \left[1 - (1 - X_S)^{1/3}\right]$$

(19.25)

Graphically:

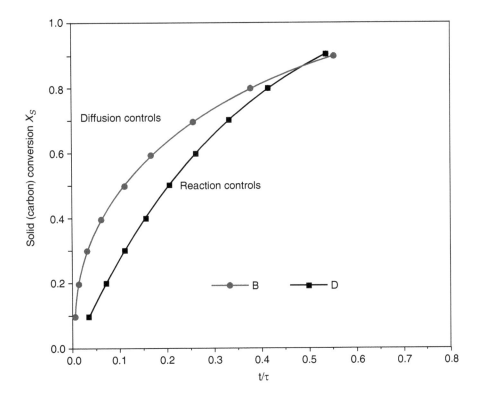

Figure 19.14 Solid conversion as a function of time.

Chapter 20

Exercises reactors and heterogeneous reactors

20.1 SOLUTIONS TO EXERCISES: REACTORS

SE.1 A gas phase reaction $A + B \xrightarrow{k} R$ is carried out in a CSTR and PFR separately at a rate of 100 mol/min where the temperature is constant at 150°C and the pressure is 10 atm. The reactant B is three times more than A in the feed reactor. The rate constant is given by:

$$\ln k = -\frac{5400}{T} + 12.5 \, (\text{min}^{-1})$$

If the average residence time in the PFR is 3 min and assuming equal values of conversion on outputs of each reactor, find the volumes and molar flow for both reactors, compare them.

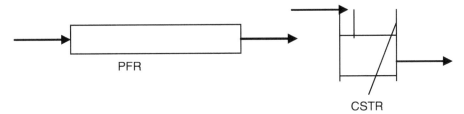

PFR

CSTR

Solution
The ratio between reactants, $B/A = 3$, indicates that there is a large dilution. Therefore, A is limiting. Since k is given in min^{-1}, one can conclude that it is a first-order reaction, but with variable volume. Thus:

$$-r_A = kC_A = C_{A0}\frac{(1 - X_A)}{1 + \varepsilon_A X_A} \tag{20.1}$$

Given the average residence time,

$$\bar{t} = C_{A0} \int_0^{X_A} \frac{\mathrm{d}X_A}{(1 + \varepsilon_A X_A)(-r_A)} \tag{20.2}$$

Replacing the rate expression, we get:

$$\bar{t} = -\frac{1}{k}\ln(1 - X_A) \tag{20.3}$$

The value of the rate constant can be calculated at 150°C (reaction temperature), which is:

$$\ln k = -\frac{5400}{T} + 12.5$$

where:
$k = 0.649\,\text{min}^{-1}$
As $\bar{t} = 3\,\text{min}$
$X_A = 0.857$

To calculate the volume, one has the PFR equation:

$$\tau = \frac{V}{v_0} = C_{A0}\int_0^{X_A}\frac{dX_A}{(-r_A)}$$

Substituting $(-r_A)$, we obtain the following solution:

$$\tau = -\frac{1}{k}[(1 + \varepsilon_A)\ln(1 - X_A) + \varepsilon_A X_A] \tag{20.4}$$

Calculating ε_A:

	A	B	R	Total
Initial	I	3	0	4
Final	0	2	I	3

$\varepsilon_A = -0.25$

Substituting these values into Equation 20.4, we get:

$\tau = 2.58\,\text{min}$

But

$F_0 = 100\,mol/min$

However:

$F_{A0} + F_{B0} = 100$
$F_{A0} + 3F_{A0} = 100$
$F_{A0} = 25\,\text{mol/min}$

Calculation of concentration:

$$C_0 = \frac{P}{RT} = 2.88 \times 10^{-1}\,\text{mol/L}$$

For v_0:

$$v_0 = \frac{F_0}{C_0} = 346\,\text{L/min}$$

Therefore, the volume of PFR reactor is:

$$V = 0.89\,\text{m}^3$$

In the CSTR reactor, assuming the same conversion, we use the equation:

$$\frac{V}{v_0} = \tau = C_{A0}\frac{X_A}{(-r_A)}$$

Replacing the rate and substituting these values for the same conversion, we obtain:

$$\tau = 4.71\,\text{min}$$

With the same input flow, we get:

$$V = 1.63\,\text{m}^3$$

This gives us a ratio of:

$$\frac{V_{\text{CSTR}}}{V_{\text{PFR}}} = 1.8$$

SE.2 The reaction $A \xrightarrow{k} R + S$ is irreversible and first order. It is conducted in a PFR with 50 tubes, each with "½" in diameter and 1.0 m of height. 200 kg/h of reactant A (MW = 80 g/gmol) with 30% inert is introduced at a pressure of 50 bar at 500°C. The output conversion is 80%. Calculate the average residence time.

Solution
The reaction rate will be:

$$-r_A = kC_A = C_{A0}\frac{(1-X_A)}{1+\varepsilon_A X_A} \tag{20.5}$$

Substituting that rate in PFR equation, we get:

$$\tau = -\frac{1}{k}[(1+\varepsilon_A)\ln(1-X_A)+\varepsilon_A X_A] \tag{20.6}$$

To determine τ, we need to calculate both the volume and flow.

As the total mass flow of G is given, we have for each tube:

$$G_1 = \frac{G}{50} = 4 \, \text{kg/h}$$

The molar flow would be:

$$F_{A0} = \frac{G_1}{M_A} = \frac{4}{80} = 0.05 \, \text{kmol / h} = 50 \, \text{mol/h}$$

Calculation of initial concentration:

$$C_{A0} = \frac{P_{A0}}{RT} = \frac{0.7 \times 50}{0.082 \times 773} = 5.52 \times 10^{-1} \, \text{mol/L}$$

Hence:

$$F_{A0} = C_{A0} v_0$$

With the values of flow and concentration, we have:

$$v_0 = \frac{F_{A0}}{C_{A0}} = 90.5 \, \text{L/h} = 1.5 \, \text{L/min}$$

Calculating the volume of the tube:

$$V = \frac{\pi d^2}{4} \cdot L = 126 \, \text{cm}^3$$

Hence:

$$\tau = \frac{V}{v_0} = 0.084 \, \text{min} = 5.04 \, \text{s} \tag{20.7}$$

Calculating ε_A:

	A	R	S	Inert	Total
Initial	0.7	0	0	0.3	1
Final	0	0.7	0.7	0.3	1.7

$$\varepsilon_A = 0.7$$

With these values, we calculate the reaction rate constant using Equation 20.6:

$$k = 35 \, \text{min}^{-1}$$

The average residence time can be determined by Equation 20.3 (PFR equation):

$$\bar{t} = -\frac{1}{k}\ln(1 - X_A)$$

$$\bar{t} = 2.76\,\text{s}$$

SE.3 The reaction $A + B \xrightarrow{k} R$ is carried out in the liquid-phase reactor. The components A and B are introduced separately at 4 mol/L and 5 L/min, respectively, and they are mixed together at the inlet to the reactor. This mixture goes into the reactor at 27°C. There are two types of reactors to choose from:

1   A CSTR with 200 L, which can operate at 77°C or 0°C.
2   A PFR with 800 L, which can only operate at 27°C.

It is known that the reaction is elementary and $k = 0.07$ L/(mol min) at 27°C and the activation energy is equal to 20 kcal/mol.

What reactor would you recommend and under what conditions? Justify.

If the reaction was performed in a batch reactor with 200 L and the final conversion was equal to the PFR, calculate the time at 77°C and 0°C, knowing that the initial concentrations are equal to 1 mol/L.

Solution

A:5L/min, 4 (mol/L)

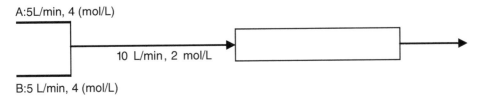

10 L/min, 2 mol/L

B:5 L/min, 4 (mol/L)

Since the initial concentrations are equal at the inlet, we have a second-order irreversible reaction, where the rate is given by the equation:

$$(-r_A) = kC_A^2 = kC_{A0}^2(1 - X_A)^2 \tag{20.8}$$

For the CSTR:

$$\frac{V}{v_0} = \tau = C_{A0}\frac{X_A}{(-r_A)} \tag{20.9}$$

Substituting Equation 20.8 into Equation 20.9, we get:

$$\frac{V}{v_0} = \tau = C_{A0}\frac{X_A}{kC_{A0}^2(1 - X_A)^2} \tag{20.10}$$

The value of k is solved at 77°C and 0°C. But at 27°C, we have k and, therefore, k_0 will be:

$$k = k_0 e^{-(E/RT)}$$

$$0.07 = k_0 e^{-(20,000/1.98 \times 300)}$$

$$k_0 = 8.51 \times 10^{14}$$

At 77°C:

$$k_{350} = 13.8 \text{ (L/mol min)}$$

At 0°C:

$$k_{273} = 1.79 \times 10^{-3} \text{(L/mol min)}$$

But the space-time will be:

$$\tau = \frac{V}{v_0} = \frac{200}{10} = 20 \text{ min}$$

Substituting these values into Equation 20.10:

$$20 = \frac{X_A}{kC_{A0}(1 - X_A)^2}$$

At 77°C, we get:

$$k_{350} = 13.8 \text{ (L/mol min)}$$

The solution is:

$$2X_A^2 - 2X_A + 1 = 0$$

Therefore:

$X_A > 1$ is impossible.

At 0°C:

$$k_{273} = 1.79 \times 10^{-3} \text{(L/mol min)}$$

We find:

$$X_A = 0.1$$

For the PFR:

Calculating:

$$\tau = \frac{V}{v_0} = \frac{800}{10} = 80 \, \text{min}$$

With $k = 0.07$ L/(mol min) at 27°C, we get:

$$\tau = \frac{V}{v_0} = C_{A0} \int\limits_{0}^{X_A} \frac{dX_A}{r} \tag{20.11}$$

Rearranging Equation 20.8:

$$\tau = \frac{1}{kC_{A0}} \frac{X_A}{(1 - X_A)} \tag{20.12}$$

In the condition of 27°C, we obtain:

$$X_A = 0.9$$

This is the best solution.

In the Batch reactor:

The time required would be calculated with a conversion of 90%. With $C_{A0} = C_{B0} = 1$ mol/L and k at 77°C and 0°C, we have:

$$t = C_{A0} \int\limits_{0}^{X_A} \frac{dX_A}{(-r_A)} \tag{20.13}$$

Substituting Equation 20.8 into Equation 20.13, we have (Equation 20.12):

$$t = \frac{1}{kC_{A0}} \frac{X_A}{(1 - X_A)}$$

Therefore, at 77°C will be:

$$k_{350} = 13.8 \, \text{(L/mol min)}$$

We get:

$$t = 0.65 \, \text{min}$$

At 0°C:

$$k_{273} = 1.79 \times 10^{-3} (\text{L/mol min})$$

$$t = 5000 \text{ min} = 83 \text{ h} = 3 \text{ days}$$

SE.4 A polymerization reaction of a monomer is conducted in a CSTR. The monomer (with initiator) is introduced at a rate of 12 L/min. The concentrations of the monomer and initiator are 3 and 0.01 mol/L, respectively. The rate of polymerization is known as:

$$(-r_M) = k_p[M] \sqrt{\frac{2\gamma k_0[I_2]}{k_t}} \tag{20.14}$$

where
$k_p = 10^{-2} \text{ s}^{-1}$
$k_0 = 10^{-2} \text{ s}^{-1}$
$k_t = 5 \times 10^{-7} \text{ mol/(L|s)}$
Calculate the volume of the reactor to a conversion of 80%.

Solution
The reaction is first order, according to Equation 20.14. Depending on the conversion, we have:

$$(-r_M) = k_p[M] \sqrt{\frac{2\gamma k_0[I_2]}{k_t}} = k^*[M] = k^*[M_0](1 - X_M) \tag{20.15}$$

The CSTR equation is:

$$\frac{V}{v_0} = \tau = C_{M0} \frac{X_M}{(-r_M)} \tag{20.16}$$

Substituting Equation 20.15 into Equation 20.13, we obtain:

$$\frac{V}{v_0} = \tau = [M_0] \frac{X_M}{k^*[M_0](1 - X_M)} \tag{20.17}$$

But substituting these values in k^*:

$$k^* = k_p \sqrt{\frac{2\gamma k_0[I_2]}{k_t}} = 10^{-2} \sqrt{\frac{2 \times 10^{-3} \times 10^{-2}}{5 \times 10^{-7}}} = s^{-1} \sqrt{\frac{s^{-1} \text{mol L}^{-1}}{\text{mol L}^{-1} s^{-1}}}$$

$$k^* = 10^{-2} \sqrt{\frac{2 \times 10^{-3} \times 10^{-2}}{5 \times 10^{-7}}} = 6.32 \times 10^{-2}$$

And substituting the values obtained from X and 80% conversion into Equation 20.17, we obtain:

$$\frac{V}{v_0} = \tau = 63.2\,\text{s} = 1.053\,\text{min}$$

Therefore, for a flow of 12 L/min:

$$V = 12.6\,L$$

SE.5 A second-order irreversible reaction $A \longrightarrow 2R$ is carried out in gas phase in a PFR reactor. A reactant ($MW = 40$) with 50% by weight and the rest with an inert ($MW_{inert} = 20$) are introduced into the PFR. It is desired to produce 20 kmol/h of a product R, knowing that the final conversion was 35%. Calculate the volume of the reactor.

Data: rate constant $= 500\,\text{m}^3/(\text{kmol ks})$; pressure $= 4.7$ atm; temperature $= 60°C$ (constant).

Solution
The reaction has second-order kinetics,

$$-r_A = kC_A^2 = kC_{A0}^2 \frac{(1 - X_A)^2}{(1 + \varepsilon_A X_A)^2} \tag{20.18}$$

Substituting Equation 20.18 into PFR equation gives:

$$\tau = \frac{V}{v_0} = C_{A0} \int_0^{X_A} \frac{dX_A}{r} \tag{20.19}$$

$$\tau = \frac{1}{kC_{A0}} \left[(1 + \varepsilon_A)^2 \frac{X_A}{(1 - X_A)} + \varepsilon_A^2 X_A + 2\varepsilon_A(1 + \varepsilon_A)\ln(1 - X_A) \right]$$

Calculation basis: 1.0 g

$$n_A = \frac{m_A}{M_A} = \frac{0.5}{40} = 0.0125\,\text{mol}$$

$$n_I = \frac{m_I}{M_I} = \frac{0.5}{20} = 0.025\,\text{mol}$$

Calculation of ε_A:

	A	2R	Inert	Total
Initial	0.0125	0	0.025	0.0373
Final	0	0.025	0.025	0.05

$\varepsilon_A = 0.33$

Calculation of initial concentration:

$$C_{A0} = \frac{y_{A0}P}{RT} = \frac{0.33 \times 4.7}{0.082 \times 333} = 5.68 \times 10^{-2} \text{ mol/L}(\text{kmol/m}^3)$$

Substituting these values into Equation 20.19, we get:

$$\tau = 0.041 \text{ ks} = 41.15 \text{ s} = 0.01143 \text{ h}$$

From R output stream, we determine the flow rate:

$$F_R = C_R v_0 = 2C_{A0}v_0 X_A$$
$$20 = 2C_{A0}v_0 X_A$$
$$v_0 = 503 \text{ m}^3/\text{h}$$

Thus, the volume of the reactor is:

$$V = \tau v_0 = 0.01143 \times 503 = 5.75 \text{ m}^3$$

SE.6 The following figure shows the rate behavior as a function of conversion. One can suggest the following cases:

- Two PFRs in series.
- One PFR and one CSTR in series.
- Two CSTRs in series.

What is the reaction scheme resulting in lower total volume?
Units and values:
Rate constant is mol/(L min)
Feed stream is $v_0 = 10 \text{ L/min}$
The initial concentration is $C_{A0} = 1 \text{ mol/L}$.

Solution
We calculate the areas by considering that the volume is proportional to the area.
PFR:

$$\tau = \frac{V}{v_0} = C_{A0} \int_0^{X_A} \frac{dX_A}{r} = C_{A0} \cdot \text{area (integral)}$$

CSTR:

$$\frac{V}{v_0} = \tau = C_{A0} \frac{X_A}{(-r_A)} = C_{A0} \cdot \text{area (rectangle area)}$$

Calculating separately for X_A from 0 to 0.5, which is the maximum from reverse rate and from 0.5 to 0.8, one can calculate the areas separately.

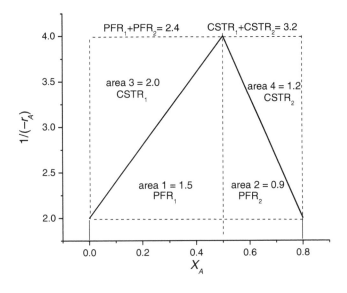

Figure 20.1

For the first case: admitting PFR_1 in series with PFR_2, one obtains a total area of 2.4.

For the second case: PFR_1 in series with $CSTR_2$, one obtains a total area of 2.7.

For the third case: $CSTR_1$ in series with $CSTR_2$, one obtains a total area of 3.2.

The total volume with a smaller area is that when we have a PFR_1 in series with PFR_2.

The most suitable system has a total volume: $V_{TOTAL} = 24\,L$.

SE.7 Dibutyl phthalate (DBP) is produced from mono-n-butyl phthalate (MBP) with butanol in liquid phase and catalyzed with H_2SO_4 in a CSTR reactor, according to the following reaction:

The reactants are in separate tanks, with 2 mol/L of MBP and 1 mol/L butanol, and they are fed at a rate of 10 and 30 L/h, respectively. Both are mixed before entering the reactor. The specific rate is equal to 4.4×10^{-2} L/(mol h). Calculate the volume of the reactor to a conversion of 70% of limiting reactant.

MBF

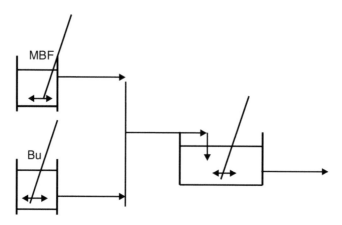

The reaction is reversible and occurs in the liquid phase. It is assumed as elementary reaction and therefore the reaction rate is directly proportional to the concentrations, i.e.:

$$(-r_A) = k C_A C_B = k C_{A0}^2 (1 - X_A)(M - X_A) \tag{20.20}$$

where:

$A = $ MBP and $B = $ Butanol

Calculation of inlet flow rates and concentrations of the reactor:

$v_{01} + v_{02} = v_0 = 10 + 30 = 40$ L/h

$C_{A01} v_{01} = C_{A0} v_0$
$2 \times 10 = C_{A0} \cdot 40 \longrightarrow C_{A0} = 5 \times 10^{-1}$ mol/L

$C_{B01} v_{01} = C_{B0} v_0$
$1 \times 30 = C_{B0} \cdot 40 \longrightarrow C_{B0} = 7.5 \times 10^{-1}$ mol/L

Therefore, A is the limiting reactant, and:

$$M = \frac{C_{B0}}{C_{A0}} = 1.5$$

For the CSTR:

$$\frac{V}{v_0} = \tau = C_{A0} \frac{X_A}{(-r_A)} \tag{20.21}$$

Substituting Equation 20.20 into Equation 20.21:

$$\frac{V}{v_0} = \tau = C_{A0} \frac{X_A}{k C_{A0}^2 (1 - X_A)(M - X_A)}$$

And substituting calculated values for a conversion of 70%, we find:

$$\tau = 19.7\,h$$

$$V = v_0 \tau = 788\,L$$

SE.8 An irreversible reaction must be carried separately in PFR and CSTR reactors. Calculate the volumes of each reactor. The kinetics data are provided and the feed flow is $2.0\,m^3/min$. Its pressure is 1 atm at 300°C. The reactant is introduced with 30% inert into the reactor.

X_A	0	0.2	0.4	0.6
$(-r_A)$ mol/(L s)	0.01	0.005	0.002	0.001

Solution
To solve this problem, plot $[1/(-r_A)]$ versus X_A from the below figure.

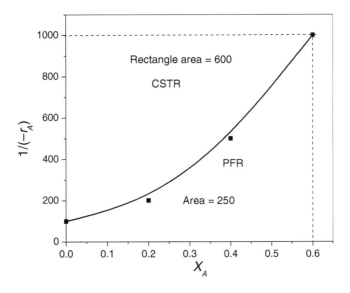

Figure 20.2

$$C_{A0} = \frac{0.7 \times 1}{0.082 \times 573} = 1.49 \times 10^{-2}(mol/L)$$

For the PFR:

$$\tau = \frac{V}{v_0} = C_{A0} \int_0^{X_A} \frac{dX_A}{r} = C_{A0} \cdot \text{area (integral)} = 3.72\,\text{s}$$

For the CSTR:

$$\frac{V}{v_0} = \tau = C_{A0} \frac{X_A}{(-r_A)} = C_{A0} \cdot \text{area} \left(\text{retangle area}\right) = 8.9\,\text{s}$$

The volumes are:

$V_{\text{PFR}} = 7.4\,\text{L}$
$V_{\text{CSTR}} = 17.8\,\text{L}$

The relationship is:

$$\frac{V_{\text{CSTR}}}{V_{\text{PFR}}} = 2.4$$

SE.9 For the irreversible reaction, $A \xrightarrow{k} 2R$, it is carried out in two CSTR reactors in series in liquid phase, with the temperature of second at 120°C. The inlet stream flowing for A is 20 kmol/ks and outlet flow rate of the last reactor is R, 28 kmol/ks. It is known that the rate constant is 1.5 m³/(kmol ks) at 120°C and the activation energy is 84 kJ/mol. The average residence time in the first and second reactor is 2.75 ks. The concentration of reactant is 1 kmol/m³. The two reactors operate isothermally but with different temperatures. Calculate the conversions, the temperature of the first reactor, and the volumes thereof.

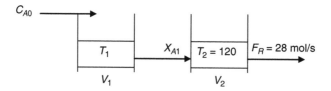

Solution
For the rate constant unit, we have a second-order reaction. Thus, the rate expression is:

$$-r_A = kC_A^2 = kC_{A0}^2(1 - X_A)^2 \quad \text{kmol/(m}^3\,\text{ks)} \tag{20.22}$$

For the Arrhenius equation for any temperature, we calculate k_0:

$$k = k_0 e^{-(84,000/(8.27 \times 393))} = 1.5$$

Hence:

$$k_0 = 2.46 \times 10^{11}$$

$$k = 2.46 \times 10^{11} e^{-(84,000/(8.27T))} \tag{20.23}$$

For the conversion in the second reactor: it is known that

$$\frac{F_{A0} - F_A}{F_{A0}} = \frac{F_R}{2F_{A0}} = X_A$$

Then:

$$X_{A2} = 0.7$$
$$F_{A2} = 6.0$$

For the CSTR equation:

$$\bar{t} = \frac{(F_{A1} - F_{A2})/v_0}{(-r_A)} \tag{20.24}$$

But:

$$v_0 = \frac{20}{1} = 20\,\mathrm{m^3/ks}$$

The rate at second reactor can be calculated as:

$$-r_A = kC_A^2 = kC_{A0}^2(1 - X_A)^2 = 1.5 \times 1 \times (1 - 0.7)^2$$
$$= 0.135\,(\mathrm{kmol/\,m^3\,ks}) \tag{20.25}$$

Substituting for Equation 20.24 from Equation 20.25 and after considering that the average residence time is 2.75 ks, we have:

$$F_{A1} = 13.4$$

The conversion at the output of the first reactor:

$$\frac{F_{A0} - F_{A1}}{F_{A0}} = X_{A1} = 0.33$$

Since $\bar{t} = \tau$, we have:

$$V_2 = \tau v_0 = 54\,\mathrm{m^3}$$

By considering that the rate at the first reactor outlet is the same, the calculated temperature in the first reactor as analogous to Equation 20.24:

$$(-r_A)_1 = 0.12$$

Then:

$$-r_{A1} = kC_A^2 = kC_{A0}^2(1 - X_{A1}) = k \times 1 \times (1 - 0.33)^2 = 0.12 \, \text{kmol/(m}^3 \, \text{ks)}$$

Hence:

$$k = 0.267 \, \text{m}^3/(\text{kmol ks})$$

Substituting into Equation 20.23, we get:

$$k = 2.46 \times 10^{11} e^{-(84,000/(8.27.T))} = 0.267$$

It takes the temperature in the first reactor:

$$T = 367 = 94°C$$

Thus, the reactor volume of the first reactor can be calculated:

$$\frac{V}{v_0} = C_{A0} \frac{X_{A1}}{(-r_A)} = 2,75$$

Thus

$$V = 55 \, \text{m}^3$$

SE.10 Two reactors in parallel conduct a reaction in the gas phase, $A \xrightarrow{k} R + S$. The reactant A is introduced separately with 20% (v) of inert species with a pressure of 10 atm and at 550°C. It is known that the average residence time in the reactor is 3.33 min. The volume of one reactor is double that of the other one. Calculate the ratio of inflow stream in the reactors, given:

$$\ln k = -\frac{12000}{T} + 10,6 \, \text{L/(mol s)}$$

Find the conversions on the outputs of the reactors.

Solution

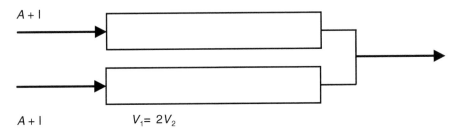

$$A + I$$

$$V_1 = 2V_2$$

To calculate the rate constant at 550°C

$$\ln k = -\frac{12,000}{T} + 10.6 \tag{20.26}$$

$$k = 1.86 \times 10^{-2} (\text{L/mol s})$$

With the unit of k, we have a second-order reaction:

$$-r_A = kC_A^2 = kC_{A0}^2 \frac{(1 - X_A)^2}{(1 + \varepsilon_A X_A)^2} \tag{20.27}$$

The mean residence time:

$$\bar{t} = C_{A0} \int_0^{X_A} \frac{dX_A}{(1 + \varepsilon_A X_A)(-r_A)} \tag{20.28}$$

Substituting Equation 20.27 into Equation 20.28, we have:

$$\bar{t} = \frac{1}{kC_{A0}} \left[\left[\frac{(1 - \varepsilon_A X_A)}{(1 - X_A)} \right] + \varepsilon_A \ln(1 - X_A) \right] \tag{20.29}$$

Initial concentration:

$$C_{A0} = \frac{0.8.10}{0.082.723} = 1.35 \times 10^{-1} \, (\text{mol/L})$$

For ε:

	A	R	S	Inert	Total
Initial	0.8	0	0	0.2	1.0
Final	0	0.8	0.8	0.2	1.8

$$\varepsilon_A = 0.8$$

Substituting the values into Equation 20.29, we get:

$$X_A = 0.7$$

This is the same conversion at the output of the two reactors.
Calculation of space-time and the volumes:

$$\tau = \frac{1}{kC_{A0}}\left[(1+\varepsilon_A)^2\frac{X_A}{(1-X_A)} + \varepsilon_A^2 X_A + 2\varepsilon_A(1+\varepsilon_A)\ln(1-X_A)\right]$$

$$\tau = 829\,s = 13.7\,min$$

The volume ratio at the input is different in the two reactors, because $V_1 = V_2$ and the residence time in both must be the same.
But:

$$\tau_1 = \frac{V_1}{v_{01}}, \quad \tau_2 = \frac{V_2}{v_{02}}, \quad \tau_1 = \tau_2, \text{ and } V_1 = 2V_2, \text{ we get:}$$

$$\frac{v_{01}}{v_{02}} = 2$$

SE.11 A dimerization reaction $2M\xrightarrow{k}D$ is carried out in a PFR with adiabatic operation (liquid-phase system). The reactant is introduced at a rate of $1.3\,m^3/ks$ and at a temperature of 312 K. Calculate the volume and temperature at the outlet of the reactor to a conversion of 70%.
Data:

- The enthalpy at 300 K: $-42\,kJ/mol$
- The heat exchange coefficient: 200 J/mol
- The initial concentration: $16\,kmol/m^3$
- The reaction rate constant: $k = 2.7 \times 10^{11}\,e^{-(12,185/T)}\,m^3/(mol\,ks)$

Solution

$$2A\xrightarrow{k}R$$

We have a second-order reaction.
Hence:

$$(-r_A) = kC_A^2 = kC_{A0}^2(1-X_A)^2 \tag{20.30}$$

For the PFR:

$$\tau = \frac{V}{v_0} = C_{A0} \int_0^{X_A} \frac{dX_A}{-r_A}$$

(20.31)

Substituting Equation 20.30 into Equation 20.31, we get:

$$\tau_{PFR} = \frac{V}{v_0} = C_{A0} \int_0^{X_A} \frac{dX_A}{kC_{A0}^2(1 - X_A)^2}$$

(20.32)

Note that k is not constant and varies with the temperature according to the Arrhenius equation. Equation 20.32 cannot be integrated.

To correlate the temperature with the conversion, we need the energy balance under adiabatic conditions. This was deduced and we obtain, from the energy balance:

$$\sum_j F_j \bar{c}_{pj}(T - T_0) + \Delta H_r F_{A0} X_A = Q_{ext} = 0$$

(20.33)

$$\frac{T}{T_0} = 1 + \beta X_A$$

(20.34)

where:

$$\beta = \frac{-\Delta H_r F_{A0}}{\sum_j F_j \bar{c}_{pj} T_0}$$

(20.35)

$$\sum_j F_j \bar{c}_{pj} = F_A \bar{c}_{pA} + F_R \bar{c}_{pR}$$

But:

$$\frac{F_{A0} - F_A}{2F_{A0}} = \frac{F_R}{F_{A0}} = \frac{X_A}{2}$$

Substituting these values, we get:

$$\sum_j F_j \bar{c}_{pj} = F_{A0}(1 - X_A)\bar{c}_{pA} + F_{A0}\bar{c}_{pR}\frac{X_A}{2}$$

As

$$\bar{c}_{pA} = \bar{c}_{pR} = \bar{c}_p = 200 \ (\text{J/mol K})$$

Thus:

$$\sum_j F_j \bar{c}_{pj} = F_{A0} \bar{c}_p \left[(1 - X_A) + \frac{X_A}{2} \right] = \frac{F_{A0} \bar{c}_p}{2} (2 - X_A) \tag{20.36}$$

Substituting these values, we get:

$$\sum_j F_j \bar{c}_{pj} = 100 F_{A0} (2 - X_A) \tag{20.37}$$

Substituting Equation 20.37 into Equation 20.35:

$$\beta = \frac{-\Delta H_r F_{A0}}{\sum_j F_j \bar{c}_{pj} T_0} = \frac{42,000 F_{A0}}{F_{A0} 100 (2 - X_A) 312} = \frac{1.34}{(2 - X_A)} \tag{20.38}$$

Therefore, from Equation 20.34:

$$\frac{T}{T_0} = 1 + \beta X_A = 1 + \frac{1.34 X_A}{(2 - X_A)} \tag{20.39}$$

But, according to the Arrhenius equation:

$$k = 2 \times 7 \times 10^{11} e^{-(12,185 /T)} = 2 \times 7 \times 10^{11} e^{-[39/(T/T_0)]} \tag{20.40}$$

Substituting Equation 20.40 into Equation 20.32:

$$\tau_{PFR} = \frac{V}{v_0} = C_{A0} \int_0^{X_A} \frac{dX_A}{k C_{A0}^2 (1 - X_A)^2}$$

Thus:

$$\tau_{PFR} = \frac{V_{PFR}}{1.3} = \int_0^{X_A} \frac{dX_A}{4.32 \times 10^{12} e^{-[39/(T/T_0)]}(1 - X_A)^2} \tag{20.41}$$

which are solved together with Equation 20.39:

$$\frac{T}{T_0} = 1 + \frac{1.34 X_A}{(2 - X_A)}$$

For a conversion of 70%, the exit temperature will be equal to 538 K, or

$$\frac{T}{T_0} = 1.72$$
$$T = 538 \, K = 265°C$$

Integrating Equation 20.41, one obtains the following graphic solution.

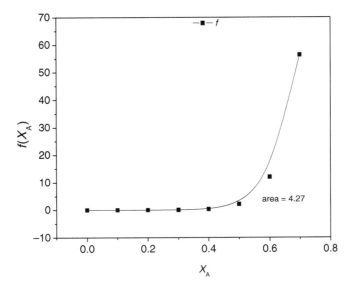

Figure 20.3

The PFR volume is:

$$V_{PFR} = 1.3 \times area = 5.55 \, m^3$$

SE.12 The reaction below is conducted in an adiabatic process:

$$H_3C-\underset{\underset{CH_2}{\|}}{CH} \quad + \quad Cl-Cl \quad \longrightarrow H_2C=\underset{\underset{CH_2Cl}{\backslash}}{CH} \quad + \quad HCl$$

The input stream is 0.45 kmol/h, with 30% C_3H_6, 40% Cl_2 and 30% inert (N_2). Its pressure is 2 atm at 30°C. The rate constant is:

$$k = 4.12 \times 10^3 \, e^{-(27,200/RT)} \quad (m^3/\text{kmol min})$$

Additional data:

$$\Delta H^0_R = -2.67 \, \text{kJ/mol}$$

Molar specific heat capacity in J/mol:

$$C_3H_6 = 1.7; Cl_2 = 0.5; C_3H_5Cl = 1.1; HCl = 4.0; N_2 = 1.04 \, \text{J/mol}$$

Calculate the rate constant.
Calculate the input stream.
Calculate the volume of the PFR.

Solution

It is a second-order reaction and its volume does not change.
Thus:

$$(-r_A) = kC_A C_B = kC_{A0}^2(1 - X_A)(M - X_A) \tag{20.42}$$

where:

$$M = \frac{C_{B0}}{C_{A0}} = 1.33$$

Using PFR equation:

$$\tau = \frac{V}{v_0} = C_{A0} \int_0^{X_A} \frac{dX_A}{-r_A} \tag{20.43}$$

Substituting Equation 20.42 into Equation 20.43, we get:

$$\tau = \frac{V}{v_0} = C_{A0} \int_0^{X_A} \frac{dX_A}{kC_{A0}^2(1 - X_A)(M - X_A)} \tag{20.44}$$

This equation cannot be integrated, since k varies with temperature.
With the energy balance, we have:

$$\frac{T}{T_0} = 1 + \frac{(-\Delta H_r)\, F_{A0}}{\sum_j F_j \bar{c}_{pj} T_0} X_A \tag{20.45}$$

But

$$\sum_j F_j \bar{c}_{pj} = F_A \bar{c}_{pA} + F_B \bar{c}_{pBA} + F_R \bar{c}_{pR} + F_s \bar{c}_{pS} + F_I \bar{c}_{pI} \tag{20.46}$$

However:

$$F_A = F_{A0}(1 - X_A)$$
$$F_B = F_{A0}(M - X_A)$$
$$F_R = F_{A0} X_A$$
$$F_S = F_{A0} X_A$$

Substituting these values into Equation 20.46, we have:

$$\sum_j F_j \bar{c}_{pj} = F_{A0}\left[(1 - X_A)\bar{c}_{pA} + (M - X_A)\bar{c}_{pBA} + \bar{c}_{pR}X_A + \bar{c}_{pS}X_A + \frac{F_I}{F_{A0}}\bar{c}_{pI}\right]$$

$$\sum_j F_j \bar{c}_{pj} = F_{A0}[3.4 + 2.9X_A] \tag{20.47}$$

Substituting Equation 20.47 into Equation 20.45:

$$\frac{T}{T_0} = 1 + \frac{(-\Delta H_r)F_{A0}}{F_{A0}[3.4 + 2.9X_A]T_0}X_A \tag{20.48}$$

Thus:

$$\frac{T}{T_0} = 1 + \frac{8.8X_A}{[3.4 + 2.9X_A]} \tag{20.49}$$

From Arrhenius equation:

$$k = 4.12 \times 10^3\, e^{-(27,200/RT)} = 4.12 \times 10^3\, e^{-(27,200/8.3 \times 303(T/T_0))}$$

$$k = 4.12 \times 10^3\, e^{-(10.8/(T/T_0))} \tag{20.50}$$

Substituting Equation 20.50 into Equation 20.44, we have:

$$\tau = \frac{V}{v_0} = \int_0^{X_A} \frac{dX_A}{C_{A0}4.12 \times 10^3\, e^{-(10.8/(T/T_0))}(1 - X_A)(M - X_A)} \tag{20.51}$$

The integral is solved with Equation 20.49.
Additional data:

$$C_{A0} = \frac{0.3 \times 2}{0.082 \times 303} = 2.41 \times 10^{-2}(\text{kmol/m}^3)$$

From the inflow, we have:

$$F_{A0} = C_{A0}v_0 = y_{A0}F_0 = 0.3 \times 0.45 = 0.175C_{A0}v_0$$

$$v_0 = 5.6\,\text{m}^3/\text{h} = 336\,\text{m}^3/\text{min}$$

The volume to a final conversion of 70% is:

$$V = 11.9\,\text{m}^3$$

The volume for 50% is:

$$V = 11.7\,\text{m}^3$$

The temperature as a function of conversion (Figure 20.5).

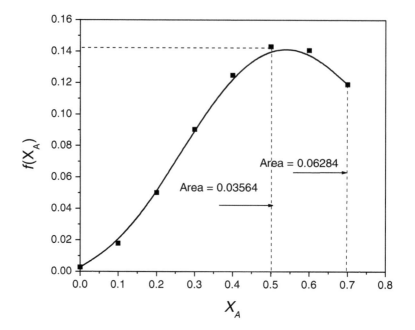

Figure 20.4

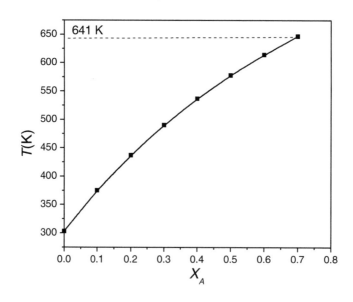

Figure 20.5

The temperature corresponding to 70% is 641 K or 368°C.

SE.13 Propylene reacts with water to produce ethylene glycol. A reactant with excess water is introduced into a PFR at 27°C and at a rate of 5 L/min with concentration of 1 mol/L. The reaction rate constant is $k = 1.69 \times 10^5 \exp(-41,800/RT)(\text{min}^{-1})$, E (J/mol). The reactor is adiabatic, but the maximum temperature cannot exceed a value of 400 K, then it operates isothermally until the final conversion of 90%. Is it possible to set the temperature at 300°C? Explain your answer. What would the temperature be if the conversion was half the output of the reactor? Calculate the volumes of adiabatic and isothermal sections of the system.

Data:

$$\Delta H_R^0 = -4.7 \times 10^4 \text{ kJ/mol}$$

$$\bar{c}_{p(\text{propylene})} = 45.0 \text{ J/(mol k)}$$

$$\bar{c}_{p(\text{water})} = 23.0 \text{ J/(mol k)}$$

$$\bar{c}_{p(\text{glycol})} = 59.0 \text{ J/(mol k)}$$

→	Adiabatic		Isothermal	→

Solution:
The reaction is a pseudo first order.

The energy balance

$$\frac{T}{T_0} = 1 + \frac{(-\Delta H_r) F_{A0}}{\sum_j F_j \bar{c}_{pj} T_0} X_A \qquad (20.52)$$

$$\sum_j F_j \bar{c}_{pj} = F_A \bar{c}_{pA} + F_B \bar{c}_{pBA} + F_R \bar{c}_{pR}$$

$$\sum_j F_j \bar{c}_{pj} = F_{A0}\left[(1 - X_A)\bar{c}_{pA} + \frac{F_{B0}}{F_{A0}}\bar{c}_{pBA} + \bar{c}_{pR}X_A\right]$$

Substituting the values we obtain:

$$\sum_j F_j \bar{c}_{pj} = F_{A0}[277 + 14.3X_A] \qquad (20.53)$$

Substituting Equation 20.53 into Equation 20.52

$$\frac{T}{T_0} = 1 + \frac{156.6 X_A}{[227 + 14.3 X_A]} \qquad (20.54)$$

To a temperature of 300°C (573 K), with the initial temperature at 300 K, we get:

$$\frac{573}{300} = 1.91 = 1 + \frac{156.6 X_A}{[227 + 14.3 X_A]}$$

Thus, we obtain a value of $X_A > 1.0$, which is impossible.
To a maximum temperature of 400 K

$$X_A = 0.607$$

This is the conversion at the outlet of adiabatic reactor.

If $X_A = 0.45$

Thus,

$$T = 390 \, K = 117°C$$

Calculation of volumes:
For the isothermal reactor,

$$\tau = \frac{V}{v_0} = C_{A0} \int_{0.6}^{0.9} \frac{dX_A}{r} \qquad (20.55)$$

Being a pseudo-first-order reaction

$$(-r_A) = k C_A = k C_{A0}(1 - X_A) \qquad (20.56)$$

Substituting Equation 20.56 into Equation 20.55 and integrating, we get

$$\tau_2 = -\frac{1}{k*} \ln(1 - X_A)|_{0,6}^{0,9}$$

Where

$$k = 1.69 \times 10^5 e^{-(41800/8.3 \times 400)} = 0.575$$

And integrating, we have

$$\tau_2 = 2.41 \, min$$

$$V_{Isothermal} = 12.0 \, L$$

For the adiabatic reactor

$$\tau_1 = \frac{V}{v_0} = \int_0^{X_A} \frac{dX_A}{1.698 \times 10^5 e^{-(10.8/(T/T_0))}(1 - X_A)} \tag{20.57}$$

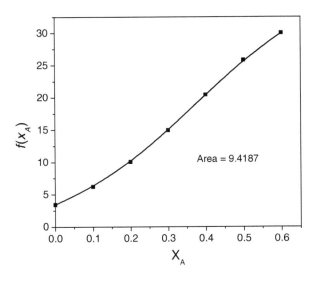

Area = 9.4187

Figure 20.6

Therefore,

$$\tau_1 = 9.41$$

$$V_1 = 47\,L$$

SE.14 An experiment was performed in a certain laboratory using a catalytic reactor, obtaining the following table data to a conversion of 5% at 400 K.

Experiment	$r_0 \times 10^4$ (mol/L s)	P_{CH} (atm)	P_{CH-} (atm)	P_{H2O} (atm)
1	7.1	1	1	1
2	7.11	5	1	1
3	7.62	10	1	1
4	7.82	2	5	1
5	7.11	2	10	1
6	5.36	3	0	5
7	5.08	3	0	10
8	6.20	1	10	10
9	0	0	5	8
10	7.0	3	3	3

The reaction is: cyclohexanol → cyclohexane + water

$$A \xrightarrow{k} R+S$$

Propose a rate that can present adsorption, desorption, and reaction. Try solving analyzing from table above and calculate the rate constants.

Solution

Admitted a reaction such as

$$A \xrightarrow{k} R+S$$

Assuming a model of Langmuir–Hishelwood, we have a generic rate expression, assuming that all components are adsorbed

$$r = \frac{kK_A p_A}{1 + K_A p_A + K_R p_R + K_S p_S} \tag{20.58}$$

Analyzing experiments 1, 2, and 3 where only the partial pressure of component A changes.

1	7.10			
2	7.11	5		
3	7.62	10		

It is observed that the pressure of compound A of experiments 1 and 2 increases five times, but the rate is virtually constant. The same is true for the experiments 1 and 3, where the pressure increases 10 times and the rate is constant. This means that the rate is not directly proportional to the concentration of A, but it falls due to the terms of adsorption–desorption in the denominator, which are significant. Therefore, it is adsorbed.

Now let us look at the products. Observing the table, set up the partial pressures of A (cyclohexanol) and R(cyclohexane).

6	5.36	3	0	5
7	5.08	3	0	10
10	7.0	3	3	3

It is noted that for fixed pressures of A and R, from the experiments 6 and 7, the pressure of S varies two times and its rate is virtually constant. The same result occurs with the experiments 7 and 10; the pressure of S decreases by about three times and the rate is in the same order of magnitude. This suggests that the S component (water), which should be in the denominator, does not influence significantly on the rate. Thus, it is not adsorbed.

Consider analogously to the influence of the product R (cyclohexene). Thus, we observe the experiences 4 and 5.

4	7.82	2	5	1
5	7.11	2	10	1

It is observed from experiments 4 and 5 that an increase of two times the pressure of cyclohexene (R) does not significantly affect the denominator, indicating that it is not adsorbed, and you can consider it as negligible.

Therefore, the rate can be represented:

$$r = \frac{kK_A p_A}{1 + K_A p_A} \tag{20.59}$$

Rearranging its terms, we have

$$\frac{1}{kK_A} + \frac{1}{k}p_A = \frac{p_A}{r} \tag{20.60}$$

Becoming $C_A = \frac{p_A}{RT}$ and considering $T = 400\,K$, we obtain the following values of the table and graph below.

Experiment	$r_0 \times 10^4$ (mol/L s)	C_A (mol/L)	C_A/r_0
1	7.1	0.030488	4.29E+01
2	7.11	0.152439	2.15E+02
3	7.62	0.304878	4.00E+02
8	6.20	0.030488	4.92E+01
10	7.00	0.091463	1.31E+02

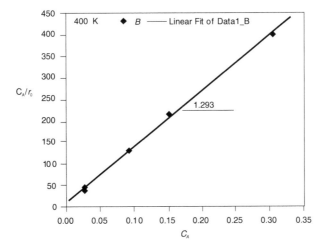

Figure 20.7

The slope is 1.293. Therefore,

$$\frac{1}{k} = 1293$$

Thus,

$$k = 0.522 \text{ min}^{-1}$$

We found the linear coefficient as 9.9.

$$\frac{1}{kK_A} = 9.9$$

and consequently,

$$K_A = 0.193$$

SE.15 The cumene cracking over a silica–alumina catalyst at 950 °C is described as $A \xrightarrow{k} R + S$, where $A =$ cumene, $R=$benzene, and $S=$propylene.

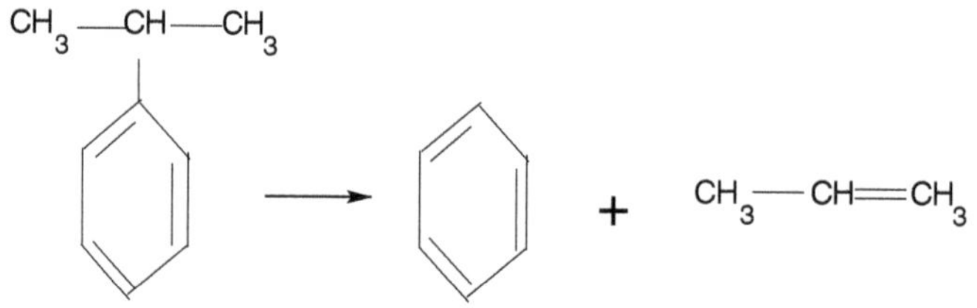

The reaction is irreversible and both the reactant and products can be adsorbed. Experiments were performed to determine the initial rate as a function of the total pressure, as shown in Figure 20.8.

Determine the reaction rate constant and the adsorption constant, as in the previous case.

Calculate the desorption rate constant of benzene (R), considering the collision theory, with temperature at 950°C and knowing that the specific area of the active sites is equal to $1.4 \times 10^{-15} \text{ cm}^2$, with $s_0 = 1$. Desorption constant was determined by transition state theory (TST), with $v = 10$ and $E = 1.8 \times 10^4$ J/mol.

Data:

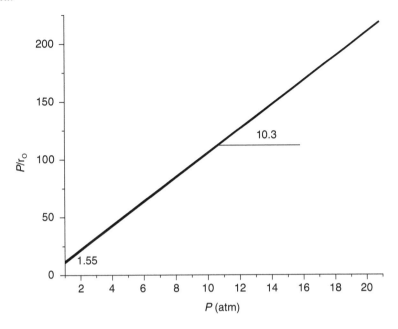

Figure 20.8 Initial rate as a function of the total pressure

Solution
It is assumed that all components are adsorbed and the reaction is first order, irreversible.

Thus,

$$r = \frac{kK_A p_A}{1 + K_A p_A + K_R p_R + K_S p_S}$$

For low conversions and initial rate r_0, one can assume that the partial pressures of R and S are small and insignificant. Therefore,

$$r_0 = \frac{kK_A p_A}{1 + K_A p_A}$$

From Figure 20.1, we obtain the linear and angular coefficients, 1.55 and 10.3, respectively.

$$k = 0.097 \, \text{min}^{-1}$$

$$K_A = 6.64$$

From the TST, we know that

$$K_{R(\text{benzene})} = \frac{k_{\text{ads}}}{k_{\text{des}}}$$

But

$$k_{ads} = \frac{s_0}{4} \left(\frac{8RT}{\pi M}\right)^{1/2} \pi d^2$$

$$k_{ads} = \frac{1}{4} \left(\frac{8 \times 8.31 \times 1223 \times 1 \times 10^7}{3.14 \times 78}\right)^{1/2} \times 1.7 \times 10^{-15} = 2.44 \times 10^{-11}$$

The desorption equilibrium constant:

$$k_{des} = 10\, e^{-(1.8 \times 10^4/T)} = 4.05 \times 10^{-6}$$

Therefore, the adsorption–desorption equilibrium constant will be

$$K_{R(benzene)} = \frac{k_{ads}}{k_{des}} = 6.03 \times 10^{-6}$$

The result shows that the equilibrium constant $K_{R(benzene)} \ll K_A$ can be neglected.

SE.16 Considering the same reaction to the previous question, but it is performed in a PFR, with 40 g of catalyst and operating at 550 °C and 10 atm. In this reaction is introduced 5 mol/min of reactant with 20% inert. The catalyst is spherical and has a radius of 1 cm, but it is porous and has an area of the 50 m²/g. The effective diffusion coefficient is 0.23 cm²/s. The density of the catalyst is 2.3 g/cm³. Determine the effectiveness factor and apparent activation energy. Check the rate ratio with and without diffusion.

Solution
The rate equation was determined in the previous problem, with

$$r_0 = \frac{kK_A p_A}{1 + K_A p_A}$$

As K_A, we have $K_A p_A > 1$. One can simplify the rate constant as

$$r_0 = \frac{kK_A p_A}{1 + K_A p_A} = \frac{kK_A p_A}{K_A p_A} = kp_A$$

In this case, a first-order reaction is assumed whose solution is

$$-r_A = kC_A = C_{A0}\frac{(1 - X_A)}{1 + \varepsilon_A X_A}$$

Substituting into PFR equation (Equation 14.37) and integrating it, we get

$$\frac{W}{F_{A0}} = -\frac{1}{k}[(1 + \varepsilon_A)\ln(1 - X_A) + \varepsilon_A X_A]$$

Calculating ε:

	A	R	S	Inert	Total
Initial	0.8	0	0	0.2	1.0
Final	0	0.8	0.8	0.2	1.8

$\varepsilon_A = 0.8$

$$\frac{40}{5} = -\frac{1}{0.067}[(1+0.8)\ln(1-X_A) + 0.8X_A]$$

$$0.536 = -[1.8\ln(1-X_A) + 0.8X_A]$$

Therefore, the conversion or concentration output will be:

$$X_A = 0.38$$

$$C_A = C_{A0}\frac{(1-X_A)}{1+\varepsilon_A X_A} = 0.064\,(\text{mol/L})$$

where

$$C_{A0} = \frac{0.8 \times 10}{0.082 \times 723} = 1.35 \times 10^{-1}(\text{mol/L})$$

From the diffusion rate (Equation 18.33):

$$r'_{obs} = \sqrt{\frac{2}{n+1}}\left(\frac{3}{r}\sqrt{\frac{D_e k}{\rho \times S_g}}.C_A^{(n+1)/2}\right)$$

For $n=1$ and $r=1$ and substituting the data we obtain the effective reaction rate at 550°C:

$$r_e = 7.4 \times 10^{-4}\,(\text{mol/L min})$$

The intrinsic rate

$$r = kC_A = 4.28 \times 10^{-3}$$

The effectiveness factor can be determined from the ratio

$$\eta = \frac{r_e}{r_{int}} = 0.17$$

The apparent energy, E_e, is

$$E_e = \frac{E}{2} = 900\,\text{J/mol}$$

Conclusion: there are strong diffusion effects.

SE.17 A reaction in the gas phase $A \xrightarrow{k} R + S$ is processed in two reactors, PFR operating in parallel, the first one operates isothermally at 2 atm and 200°C and the other adiabatically. The reactant A (pure) at a rate of 10 mol/min is introduced separately, with

$$k = 8.19 \times 10^{15} e^{-(34,222/RT)} \text{ (L/mol min).}$$

The average residence time in the isothermal reactor is 4 min, operating at 200 °C. The parallel reactor operates adiabatically with the same flow.
Calculate the volumes of isothermal and adiabatic reactors.
Calculate the inlet temperature of the adiabatic reactor.
What is the output conversion?
Are the volumes equal?
Data for the adiabatic reactor:

$\bar{C}_{pA} = 170$ (J/mol K)
$\bar{C}_{pR} = 80$ (J/mol K)
$\bar{C}_{pS} = 90$ (J/mol K)
$\Delta H_R^0 = -80$ (KJ/mol)

Solution

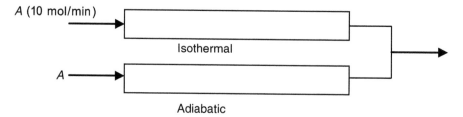

Because of the unit provided above, we have a second-order reaction. Thus,

$$(-r_A) = kC_{A0}^2 \frac{(1 - X_A)^2}{(1 + \varepsilon_A X_A)^2} \tag{20.61}$$

The average residence time in the first reactor

$$\bar{t} = C_{A0} \int_0^{X_A} \frac{dX_A}{(1 + \varepsilon_A X_A)(-r_A)} \tag{20.62}$$

Substituting Equation 20.61 into Equation 20.62, we get

$$\bar{t} = \frac{1}{kC_{A0}} \left[\varepsilon_A \ln(1 - X_A) + \frac{(1 - \varepsilon_A)X_A}{(1 - X_A)} \right]$$

Calculating ε:

	A	R	S	Total
Initial	I	0	0	1.0
Final	0	I	I	2.0

$\varepsilon_A = 1.0$

$$C_{A0} = \frac{2}{0.082 \times 473} = 5.15 \times 10^{-2} (\text{mol/L})$$

Calculating k (200°C)

$$k = 8.19 \times 10^{15} e^{-(34,222/1.98 \times 473)} = 1.106 \ (\text{L/mol min})$$

Since the average residence time of 4 min and with values obtained above, we solve the output conversion, which will be the same for both reactors. Thus,

$$X_A = 0.203$$

Thus, the space-time to the isothermal reactor is

$$\tau = \frac{1}{kC_{A0}} \left[(1 + \varepsilon_A)^2 \frac{X_A}{(1 - X_A)} + \varepsilon_A^2 X_A + 2\varepsilon_A (1 + \varepsilon_A) \ln (1 - X_A) \right]$$

$$\tau = 5.5 \ min$$

If the inlet stream is given, we can calculate v_0; therefore,

$$F_{A0} = 10 = C_{A0} v_0$$

Thus,

$$v_0 = 194 \ (\text{L/min})$$

The volume of the isothermal reactor is

$$V_1 = 1.05 \ m^3$$

In the case of the adiabatic reactor, with the same initial temperature and the final conversion provided, one can find the outlet temperature using the energy balance:

$$\frac{T}{T_0} = 1 + \frac{(-\Delta H_r)F_{A0}}{\sum_j F_j \bar{c}_{pj} T_0} X_A$$

But

$$\sum_j F_j \bar{c}_{pj} = F_A \bar{c}_{pA} + F_R \bar{c}_{pR} + F_S \bar{c}_{pS}$$

$$\sum_j F_j \bar{c}_{pj} = F_{A0}[(1 - X_A)\bar{c}_{pA} + (\bar{c}_{pR} + \bar{c}_{pS})X_A]$$

Thus,

$$\sum_j F_j \bar{c}_{pj} = 170 F_{A0}$$

and

$$T = T_0 + \frac{80,000 X_A}{170} = T_0 + 470 X_A$$

The outlet temperature must be the same, for the conversion does not change. Therefore, inlet temperature is estimated as:

$$473 = T_0 + 470 X_A$$

$$T_0 = 377\,\text{K} = 104.4\,°\text{C}$$

The space-time must be equal to the first reactor. Therefore, we calculate the initial concentration and volume flow at the entrance of the second PFR adiabatic reactor.

$$C_{A0} = \int_0^{X_{A=0.203}} \frac{(1 + \varepsilon_A X_A)^2 \, dX_A}{\tau_2 \times 8.19 \times 10^{15}\, e^{-45.8/(T/T_0)}(1 - X_A)^2}$$

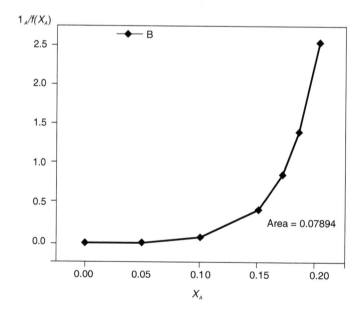

Figure 20.9

That is,

$C_{A0} = 0.07894$

For the pressure, we have

$P = C_{A0}RT = 0.07894 \times 0.082 \times 377 = 2.43 \, \text{atm}$

$V_2 = \tau_2 \cdot v_0 = V_1$

and

$V_2 = V_1$

SE.18 Two reactions are processed in an isothermal PFR differential isothermal at 700 K:

where *m*-xylene is the desired product and toluene is the undesired product, the following rate constants are known:

$$r_{1M} = k_1[M] \cdot [H_2]^{1/2} \quad k_1 = 3.42 \, [(\text{L/mol})^{0.5} \cdot \text{h}] \tag{20.63}$$

$$r_{2T} = k_2[X] \cdot [H_2]^{1/2} \quad k_2 = 1.87 \, [(\text{L/mol})^{0.5} \cdot \text{h}] \tag{20.64}$$

The indices 1 and 2 represent the reactions 1 and 2, respectively.

When the space-time is 0.5 h, the maximum concentration of X is 0.313 × 10^{-2} mol/L and the concentration of M is 7.6 × 10^{-3} mol/L, with $[M_0] = 0.02$ mol/L. Calculate the overall yield of the PFR using the maximum production of X.

Solution

$$-r_{1H} = -r_{1M} = r_{1X} = r_{1Me} \tag{20.65}$$

$$-r_{2H} = -r_{2X} = r_{2T} = r_{2Me} \tag{20.66}$$

The yield:

$$\phi = \frac{r_X}{r_M} \tag{20.67}$$

But

$$r_X = k_1[M] \cdot [H_2]^{1/2} - k_2[X] \cdot [H_2]^{1/2} = r_{1M} - r_{2T}$$

$$r_M = k_1[M] \cdot [H_2]^{1/2} = r_{1M}$$

Thus:

$$\phi_X = \frac{r_{1M} - r_{2T}}{r_{1M}} \tag{20.68}$$

$$\phi_X = \frac{r_{1M} - r_{2T}}{r_{1M}} = \frac{k_1[M] \cdot [H_2]^{1/2} - k_2[X] \cdot [H_2]^{1/2}}{k_1[M] \cdot [H_2]^{1/2}}$$

or

$$\phi_X = 1 - \frac{k_2[X]}{k_1[M]} \tag{20.69}$$

Therefore:

$$\phi_X = 1 - \frac{1.67 \times 3 \times 13 \times 10^{-3}}{3.42 \times 7.6 \times 10^{-3}} = 0.79$$

SE.19 The dehydrogenation of cyclohexane is conducted in a PFR, according to reaction:

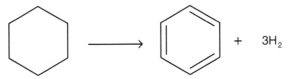

Cyclohexane and hydrogen are introduced at 8.0 and 2.0 mol/s, respectively. The reaction is first order and irreversible at 435°C and 1.5 atm. What would be the reactor volume in this system to achieve a conversion of 70%?

Solution
The following reaction $A \xrightarrow{k} R + 3S$ occurs in the gas phase. The total flow inlet is 10 mol/s and the rate constant can be represented by:

$$-r_A = kC_A = k\frac{C_{A0}(1-X_A)}{(1+\varepsilon_A X_A)}$$

Calculating ε:

$$\tau = -\frac{1}{k}[(1+\varepsilon_A)\ln(1-X_A) + \varepsilon_A X_A]$$

	A	R	S	Total
Initial	8	0	2	10
Final	0	8	26	34

$\varepsilon_A = 2.4$

$$C_{A0} = \frac{0.8 \times 1.48}{0.082 \times 708} = 2.04 \times 10^{-2}\,(\text{mol/L})$$

Thus:

$$v_0 = 3.92\,L/s$$

With these calculated values, we conclude that:

$$\tau = 3.83\,s$$

$$V = 15\,L$$

SE.20 People get sick, probably, as shown in the reaction mechanism below (Fogler, 2000):

Get sick: $H \xrightarrow{k_1} D$

In contact with other sick patients: $H + D \xrightarrow{k_2} 2D$

The patient is cured: $D \xrightarrow{k_3} H$

Patient dies (irreversible): $D \xrightarrow{k_4} M$

Questions:
Determine the death rate.
Determine the rate of people getting sick.
What is the half-life?
Data:

$$k_1 = 10^{-5}h^{-1} \quad k_2 = 10^{-7}h^{-1}$$

$$k_3 = 5 \times 10^{-6}h^{-1} \text{ and } k_4 = 10^{-7}h^{-1}$$

Solution

The reaction rates that correspond to each step are:

$$r_M = k_4[D] \tag{20.70}$$

$$r_D = k_1[H] - k_2[D][H] + 2k_2[D][H] - k_3[D] - k_4[D] = 0 \tag{20.71}$$

or

$$r_D = k_1[H] + k_2[D][H] - (k_3 + k_4)[D] = 0 \tag{20.72}$$

Neglecting the first term, the initial rate, because it is small if compared to others, yields:

$$[H] = \frac{k_3 + k_4}{k_2} \tag{20.73}$$

Furthermore:

$$r_H = k_3[D] - k_2[D][H] = 0 \tag{20.74}$$

Adding Equations 20.70 and 20.72, we get:

$$[D] = \frac{k_1}{k_4}[H] \tag{20.75}$$

Therefore, substituting Equations 20.73 and 20.75 into Equation 20.70:

$$r_M = k_1 \frac{k_3 + k_4}{k_2} \tag{20.76}$$

Substituting the values of the rate constants, one obtains:

$r_M = 5.10 \times 10^{-4} \, 1/h$

The rate expression of the [H] is:

$$-r_H = k_1[H] + k_2[D][H] - k_3[D] \qquad (20.77)$$

Substituting Equation 20.75 into Equation 20.77 and rearranging it, we obtain:

$$-r_H = \frac{k_1}{k_4}(k_2 - k_3 + k_4)[H]$$

Substituting the values obtained previously, the rate of people getting sick is:

$$-r_H = r_D = 4.8 \times 10^{-4}[H] \qquad (20.78)$$

Then, it is a first-order reaction, whose solution is:

$$-\ln(1 - X_A) = kt \qquad (20.79)$$

Calculating X_A:

$$X_A = \frac{[H_0] - [H]}{[H_0]}$$

The half-time is:

$$t_{1/2} = \frac{0.693}{k} = \frac{0.693}{4.8 \times 10^{-4}} = 1444 \, h = 60 \, days$$

SE.21 The reaction $3A \xrightarrow{k} 2R + S$ is conducted in a batch reactor. The reaction is exothermic. The reactant is heated up to 400°C, but after reaching this value it should operate adiabatically. During the heating period, there was obtained a conversion of 10%. What is the time required for converting the remaining up to 70%? Data:

-   Reactor volume = 1 m³
-   Mass = 950 kg/m³
-   Initial concentration = 10.2 kmol/m³
-   $\Delta H = -25,000 \, kcal/kmol$
-   $\ln k = -\frac{10,000}{RT} + 5 \, [(m^3/kmol^2/s)]$

Solution
The reaction is irreversible and second order. However, during heating, the initial concentrations were:

$$C_{A1} = C_{A0}(1 - X_A) = 10.2(1 - 0.10) = 9.18 \, kmol/L$$

$$\frac{C_{A0} - C_{A1}}{3} = \frac{C_{R0} + C_{R1}}{2} = \frac{C_{S0} + C_{S1}}{1}$$

$$\frac{10.2 - 9.18}{3} = 0.34 = \frac{0 + C_{R1}}{2} = \frac{0 + C_{S1}}{1}$$

Therefore, the new stage is:

$$C_{R1} = 0.68$$
$$C_{S1} = 0.34$$

It is noted that these initial concentrations are practically very small relative to the initial concentration of A and therefore we despise.
The rate expression is:

$$(-r_A) = kC_{A0}^2(1 - X_A)^2$$

In the batch reactor, we have:

$$t = C_{A0} \int_0^{X_A} \frac{dX_A}{(-r_A)}$$

Substituting the rate expression:

$$t = \int_0^{X_A} \frac{dX_A}{kC_{A0}(1 - X_A)^2} \tag{20.80}$$

In this case, k is not constant, but depends on temperature. For the energy balance in adiabatic system, we get:

$$\frac{T}{T_0} = 1 + \frac{-\Delta H_r C_{A0} X_A}{m\bar{c}_p T_0}$$

$$\frac{T}{T_0} = 1 + \frac{25{,}000 \times 9.18 X_A}{950 \times 0.59 \times 673} = 1 + 0.608 X_A \tag{20.81}$$

But

$$k = 1.48 \times 10^2 \, e^{-(10{,}000/RT)} = 1.48 \times 10^2 \, e^{-(7.5/(T/T_0))} \tag{20.82}$$

Substituting Equations 20.82 and 20.81 into Equation 20.80 (see the following figure) and integrating it, we have:

$$t = \int_0^{X_A} \frac{dX_A}{1.35 \times 10^3 \, e^{-(7.5/(T/T_0))}(1 - X_A)^2}$$

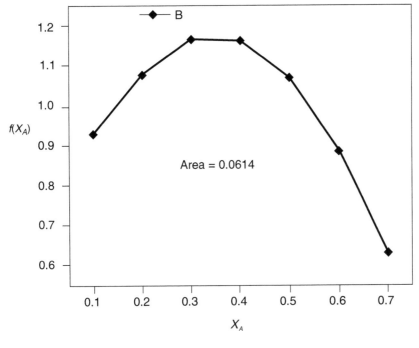

Figure 20.10

The time was $0.0614\,\text{h} = 3.68\,\text{min}$.

SE.22 The rate expression of the reaction is represented by:

$$-r_A = \frac{C_A}{(1 + C_A)^2} \tag{20.83}$$

It is conducted in a CSTR reactor where 13 mol/L of reactant at a rate of 0.2 L/s is introduced.

What is the concentration of A at a maximum rate achieved by the system?

What is the volume of CSTR at a maximum rate achieved by the system?

Solution

Assuming the reactor CSTR, we have:

$$\frac{V}{v_0} = \tau = C_{A0}\frac{X_A}{(-r_A)} = \frac{C_{A0} - C_A}{(-r_A)}$$

or in another way:

$$-\frac{1}{\tau} = \frac{(-r_A)}{C_{A0} - C_A}$$

This is represented in the figure below.

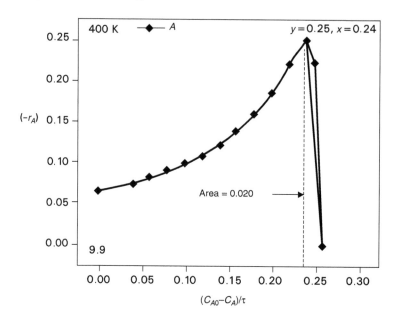

Figure 20.11

From the figure, we obtain the maximum rate, $(-r_A) = 0.25$.
Substituting into Equation 20.83 and neglecting the square term, we obtain:

$$C_A = 0.166$$

From the Figure, we have:

$$\frac{C_{AO} - C_A}{\tau} = 0.24$$

Thus:

$$\tau = 53.4 \text{ s and } V = 10.7 \text{ L}$$

SE.23 A gas phase reaction $A \xrightarrow{k} R$ is carried out in an adiabatic fixed-bed reactor and it is first-order reaction at the surface, $r_{As} = k'C_{As}$. The reactor is fed with a pure reactant (A) at 100 L/s and 300 K. The Reynolds and Sherwood numbers are known. The initial concentration is 1 M. Calculate the mass of catalyst to achieve a conversion of 60%. Given:

$$Sh = 100Re^{1/2}, \; Re = \frac{d_p u}{v};$$

$\upsilon = 0.02\,\text{cm}^2/\text{s};\ d_p = 0.1\,\text{cm};\ u = 10\,\text{cm/s};\ D_e = 0.01\,\text{cm}^2/\text{s}$
$\bar{c}_{PA} = \bar{c}_{PR} = 25\,\text{cal/ mol K};\ \rho = 2\,\text{g/cm}^3$
$k'_{300} = 0.01\,\text{cm}^3/\text{g}_{\text{cat}}\,\text{s};\ E = 4000\,\text{cal/mol}$
$a = 60\,\text{cm}^2/\text{g}_{\text{cat}};\ \Delta H = -10,000\,\text{cal/mol}$

Solution

There is no contraction or expansion of gas, $\varepsilon = 0$.
Reynolds number:

$$Re = \frac{d_p u}{\upsilon} = \frac{0.1 \times 10}{0.02} = 50$$

Sherwood number:

$$Sh = 100 Re^{1/2} = 707$$

Thus:

$$h_m = \frac{Sh \cdot D_e}{d_p} = 70.7$$

And the mass transfer coefficient

$$k_m = h_m a$$
$$k_m = 4242\,\text{cm}^3/\text{g}_{\text{cat}}\,\text{s}$$

The reaction rate constant is:

$$k' = k'_0\,e^{-(E/RT)} = 0.01$$

Thus:

$$k'_0 = 8.2\,\text{cm}^3/\text{g}_{\text{cat}}\text{s}$$
$$k' = 8.2\,e^{-(2013/T)}$$

In the adiabatic reactor, we have:

$$\frac{T}{T_0} = 1 + \frac{(-\Delta H_r)\,F_{A0}}{\sum_j F_j \bar{c}_{pj} T_0} X_A \tag{20.84}$$

where:

$$\sum_j F_j \bar{c}_{pj} = F_{A0}[\,(1 - X_A)\,\bar{c}_{pA} + (\bar{c}_{pR} + \bar{c}_{pS})X_A\,]$$

$$\sum_j F_j \bar{c}_{pj} = 100 F_{A0}$$

Thus:

$$T = T_0 + 100X_A$$

The simplified rate expression is:

$$r'_{AS} = \frac{k_m k'}{k_m + k'} C_{A0}(1 - X_A) \tag{20.85}$$

By mass balance:

$$W = F_{A0} \int_0^{X_A} \frac{dX_A}{r'_{As}} \tag{20.86}$$

Substituting Equations 20.85 and 20.84 into Equation 20.86 and integrating it, one obtains Figure 20.2.

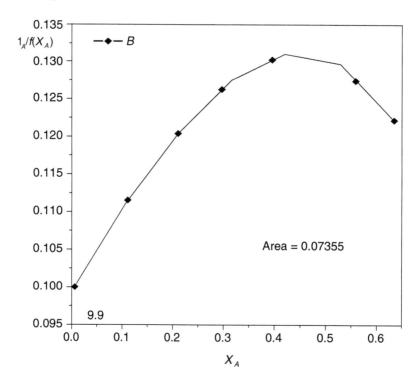

Figure 20.12

The area is 0.0732 and $v_0 = 100$ L/s, we get:

$$W = 7.3\,g$$

SE.24 Two parallel reactions are known, whose rates are given below:

$$A + 2B \xrightarrow{k_1} P$$
$$A \xrightarrow{k_2} S$$

where:

$$r_P = 0.14 C_A C_B$$
$$r_S = 0.07 C_A$$

The molar flow P in the output of reactor is 1.5 mol/min, with the initial concentrations of A and B equal to 2 and 1 mol/L, respectively. The inlet flow is 10 L/min.

Calculate the yield and selectivity in a CSTR.

Calculate the total yield in a PFR. Compare your results.

Solution

The rate of conversion of A is:

$$-r_A = r_P + r_S = k_1 C_A C_B + k_2 C_A$$

Thus, the yield is:

$$\phi_A = \frac{r_P}{-r_A} = \frac{k_1 C_A C_B}{k_1 C_A C_B + k_2 C_A} = \frac{1}{1 + \frac{k_2 C_A}{k_1 C_A C_B}} \qquad (20.87)$$

Knowing the data:

$$F_P = 1.5 = C_P v_0 \quad F_P = 1.5 = C_P v_0 = C_P \times 10$$

we get:

$$C_{Pf} = 0.15 \text{ mol/L}$$

	A	B	P
$t = 0$	C_{A0}	C_{B0}	0
$t \to \infty$	C_A	C_B	C_P

But

$$\frac{C_{A0} - C_A}{1} = \frac{C_{B0} - C_B}{2} = \frac{C_P}{1}$$

Thus:

$$C_P = \frac{C_{B0} - C_B}{2}$$

or

$$C_{Bf} = C_{B0} - 2C_P = 0.7$$

Therefore:

$$\phi_A = \frac{1}{1 + \frac{k_2 C_A}{k_{1A} C_B}} = 0.58$$

For the PFR (from Equation 16.23):

$$\Phi_A = -\frac{1}{C_{A0} - C_{Af}} \int \phi_A dC_A$$

or

$$\Phi_A = -\frac{1}{C_{A0} - C_{Af}} \int \frac{1}{1 + \frac{k_2}{k_1 C_B}} dC_A$$

where:

$$-dC_A = -\frac{dC_B}{2}$$

$$\Phi_A = -\frac{1}{C_{A0} - C_{Af}} \int \frac{1}{1 + \frac{k_2}{k_1 C_B}} \left(-\frac{dC_B}{2} \right) \tag{20.88}$$

$$C_{A0} - C_{Af} = 0.15$$

with:

$$C_{A0} - C_{Af} = 0.15$$

Integrating Equation 20.88, we have:

$$\Phi_A = -\frac{1}{2(C_{A0} - C_{Af})} \left(x - \ln\left(\frac{a - x}{a} \right) \right) \tag{20.89}$$

Substituting the values $a = k_2/k_1$ and $x = 0.7$ into Equation 20.89:

$$\Phi_A = 0.58$$

SE.25 An irreversible reaction $A + B \xrightarrow{k} R$ is conducted in an adiabatic reactor. The reactants are fed in equimolar concentrations to 27°C and 20 L/s.

- Calculate the volume of PFR to achieve 85% conversion.
- What is the temperature maximum input to the boiling point of the liquid not exceeding 550 K?
- Plot spatial-time versus conversion.
- What would be the amount of heat removed to keep the isothermal reaction, such that the external temperature does not exceed 50°C?

Data

$$k = 1.93 \times 10^5 \, e^{-(10,000/RT)}$$

$$H_A^0(273\,K) = -20\,kcal/mol$$

$E = 10\,(kcal/mol)$ $\qquad H_B^0(273\,K) = -15\,kcal/mol$

$C_{A0} = 0.1\,(kmol/m^3)$ $\quad H_R^0(273\,K) = -41kcal/mol$

$C_{PA} = C_{PB} = 15\,cal/mol\,K$

$C_{PR} = 30\,cal/mol\,K$

$\dfrac{UA}{\rho} = 20$

Solution

The rate

$$(-r_A) = kC_A C_B = C_{A0}^2(1 - X_A)^2 \tag{20.90}$$

Because $M = 1$
Calculation of ΔH_r^0:

$$\Delta H_r^0 = H_R^0 - (H_A^0 + H_B^0)$$

But $\Delta C_P = C_{PR} - (C_{PA} + C_{PB}) = 0$

Thus,

$$\Delta H_r^0 = -6000\,cal/mol$$

The energy balance is:

$$\frac{T}{T_0} = 1 + \frac{(-\Delta H_r)\,F_{A0}}{\sum_j F_j \bar{c}_{pj} T_0} X_A$$

Where,

$$\sum_j F_j \bar{c}_{pj} = F_{A0}[(1 - X_A)\bar{c}_{pA} + \bar{c}_{pB}(1 - X_A) + \bar{c}_{pR}X_A] =$$
$$F_{A0}[(1 - X_A)(\bar{c}_{pA} + \bar{c}_{pB}) + \bar{c}_{pS}X_A] = 30F_{A0}$$

Thus:

$$T = 300 + \frac{6000}{30}X_A = 300 + 200X_A$$

and

$$\frac{T}{T_0} = 1 + \frac{6000}{30 \times 300}X_A = 1 + 0.666X_A \qquad (20.91)$$

Substituting Equations 20.90 and 20.91 into PFR equation, we get:

$$V = v_0 \int_0^{X_A} \frac{dX_A}{1.93 \times 10^5 \, e^{-(16.8/(T/T_0))} C_{A0}(1 - X_A)^2} \qquad (20.92)$$

From the solution of the above equation, we obtain Figures 20.13 and 20.14.

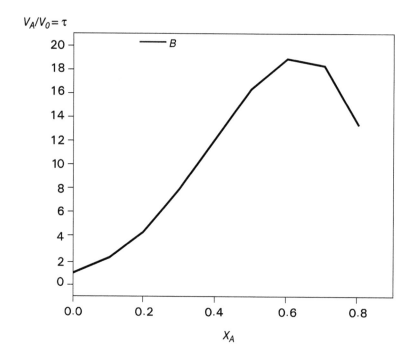

Figure 20.13 Solution from Equation 20.92.

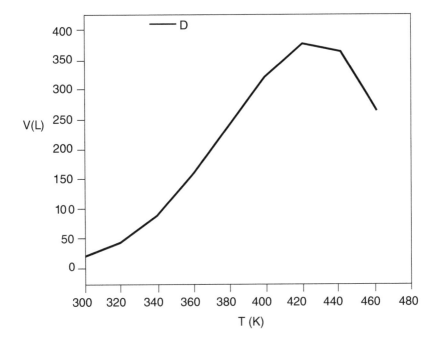

Figure 20.14 Solution from Equation 20.90.

For a conversion of 85%, we have:

$$V = 304\,L \approx 0.3\,m^3$$

When $T = 550\,K$ and $X_A = 0.85$, the initial temperature will be:

$$T_0 = 380\,K = 107°C$$

Calculation of heat removed to keep the isothermal reactor, such that the external temperature does not exceed 50°C.
The energy balance is:

$$\frac{dQ}{dV} = UA(T_a - T) \tag{20.93}$$

where:

$$T_a = 50°C = 323\,K$$

But, by Equation 20.91, the temperature varies with X_A.

$$\frac{T}{T_0} = 1 + 0.666X_A$$

And from equation 20.92 in differential form we have:

$$dV = f(X_A)\,dX_A \tag{20.94}$$

However, from Equation 20.93:

$$dQ = UA(T_a - T)\,dV$$

Therefore, substituting equations 20.91. 20.94 and 20.92, and integrating we obtain:

Interposing,

$$\frac{Q}{\rho} = \frac{UA}{\rho}v_0 \int_0^{X_A} \frac{(T_a - T)dX_A}{1.93 \times 10^5 e^{-(16.8/(T/T_0))}C_{A0}(1 - X_A)^2} \tag{20.95}$$

The solution is shown in Figure 20.15.

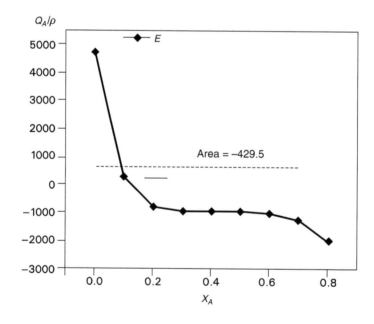

Figure 20.15 Solution from Equation 20.95.

Thus,

$$\frac{Q}{\rho} = -429(\text{m}^3 \cdot \text{J}/\text{Kg} \cdot \text{s})$$

SE.26 An irreversible reaction in vapor phase $A \rightarrow R + S$ is carried out in an adiabatic reactor of 500 L. The reactant A is fed into the reactor with inert 50% and a volumetric flow rate of 10 mol/min with a total pressure of 40 atm, the inlet temperature is 1100 K.

(a)   Calculate the conversion as a function of volume.
(b)   If the inlet temperature increases or decreases by 200°C, what will happen?
(c)   Calculate the outlet temperature of the adiabatic reactor to a conversion of 80%.
(d)   Calculate the heat required under nonisothermal conditions.

Data:
  Adiabatic reactor
  Irreversible gas-phase reaction $A \rightarrow R + S$
  $F_0 = 10$ mol/min
  $P = 2$ atm
  $T_{in} = 1100$ K
  $k = \exp(34.34 - 34,222/T)$ L/(mol min)
  $C_{PI} = 200$ J/(mol k)
  $C_{PA} = 170$ J/(mol k)
  $C_{PR} = 80$ J/(mol k)
  $C_{PS} = 90$ J/(mol k)
  $\Delta H_R^0 = 80$ kJ/mol

Solution
(a) Calculation of conversion
  Kinetics: according to the unit of the rate constant, we consider it as a second-order reaction:

$$-r_A = \frac{kC_{A0}^2(1 - X_A)^2}{(1 + \varepsilon_A X_A)^2} \tag{20.96}$$

$$\tau = \frac{V}{v_0} = C_{A0} \int_0^{X_A} \frac{dX_A}{(-r_A)} \tag{20.97}$$

By the energy balance in a reactor adiabatic, k varies with temperature, then:

$$\frac{T}{T_0} = 1 + \frac{(-\Delta H_r)F_{A0}}{\sum_j F_j \bar{c}_{pj} T_0} \times X_A$$

$$\sum_j F_j \bar{c}_{pj} = F_{A0}[(1 - X_A)\bar{c}_{pA} + \bar{c}_{pR}X_A + \bar{c}_{pS}X_A + F_I c_{pI}]$$

$$= \sum_j F_j \bar{c}_{pj} = 3700F_{A0}$$

Substituting these values, we have:

$$\beta = \frac{-80,000 \times 10}{3700 \times 100} X_A = -0.196 X_A$$

where:

$$F_{A0} = 0.5 F_0$$
$$F_I = 0.5 F_0$$

Thus:

$$\frac{T}{T_0} = 1 - 0.19 X_A \tag{20.98}$$

Replacing this variable into rate constant equation:

$$k = \exp\left(34.34 - \frac{31.1}{T/T_o}\right)$$

$$k = \exp\left(34.34 - \frac{31.1}{1 - 0.19 X_A}\right) \tag{20.99}$$

Rearranging the equations and substituting them into Equation 20.97, we get:

$$\tau = \frac{V}{v_0} = C_{A0} \int_0^{X_A} \frac{(1 + \varepsilon_A X_A) \, dX_A}{k C_{A0}^2 (1 - X_A)^2} \tag{20.100}$$

Replacing equation 20.99 in 20.100 we obtain:

$$V = \frac{v_0}{C_{A0}} \int_0^{X_A} \frac{(1 + \varepsilon_A X_A)^2 \, dX_A}{\exp\left(34.34 - \frac{31.1}{1 - 0.19 X_A}\right)(1 - X_A)^2} \tag{20.101}$$

Thus, we need to calculate the concentration and flow.

$$C_{A0} = \frac{p_{Ao}}{RT} = 0.22$$

$$v_0 = \frac{F_{Ao}}{C_{A0}} = 45.4$$

with:

$$\varepsilon_A = 0.5$$

Substituting the values into Equation 20.101:

$$V = 2.06 \times 10^2 \int_0^{X_A} \frac{(1 + \varepsilon_A X_A)^2 \, dX_A}{\exp\left(34.34 - \frac{31.1}{1-0.19X_A}\right)(1 - X_A)^2}$$

Plotting the conversion versus volume, we have:

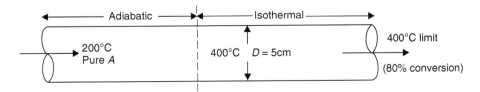

Figure 20.16

(b) What will happen with the increase or decrease in temperature?

If the inlet temperature increase by 200°C, $T = 1300$ K, the conversion increases considerably for this reaction volume, $\tau \times C_{A0}$ getting $X_A \cong 0.7$. In case of decreasing the inlet temperature 200°C, $T = 900$ K, the conversion decreases drastically and can even be considered that there will be no reaction.

(c) Calculation of the exit temperature for a conversion of 80%.

$$\frac{T}{T_0} = 1 + 0.19 X_A$$

For $X_A = 0.80$:

$$T = 1,000 (1 + 0.19 \times 0.8) = 1267K$$

$$T = 994°C$$

(e) Calculation of heat

$$Q = 3700(1267 - 1100) + 80,000 \times 10 \times 0.8 = 6.17 \times 10^5 + 6.40 \times 10^{-5}$$

$$Q = 1.25 \times 10^6 \text{ J/min}$$

SE.27 The reaction $A + B \rightarrow C$ is carried out adiabatically in a constant volume (batch reactor). The reactants A and B are introduced at a temperature of 100°C and initial concentrations of 0.1 and 0.125 mol/dm³, respectively. The rate equation is given by:

(a) Describe the mass balances for all compounds as well as the energy balance for the system in transient state.
(b) Determine the concentration and temperature at steady state.
(c) Plot the temperature and the concentrations of species as a function of time.

Data:

$k_1 (373\,\text{K}) = 2 \times 10^{-3}\,\text{s}^{-1}$
$k_2 (373\,\text{K}) = 2 \times 10^{-5}\,\text{s}^{-1}$
$\Delta H^0 (298\,\text{K})_A = -40.000\,\text{J/mol}$
$E_1 = 100\,\text{kJ/mol}$
$E_2 = 150\,\text{kJ/mol}$
$C_{PA} = 25\,\text{J/(mol K)}$
$C_{PB} = 25\,\text{J/(mol K)}$
$C_{PC} = 40\,\text{J/(mol K)}$

Solution

(a) Mass balance

$$\text{input} - \text{output} + \text{generation} = \text{accumulation} \tag{20.102}$$

$$n_{A0} = \frac{dX_A}{dt} = -r_A \times V \tag{20.103}$$

The rate of system:

$$r = k_1 C_A^{0.5} C_B^{0.5} - k_2 C_C$$

But:

$$C_A = C_{A0}(1 - X_A)$$
$$C_B = C_{B0}(1 - X_A) = C_{A0}(M - X_A) \text{ where } M = 1.25$$
$$C_C = C_{C0} + C_{A0}X_A$$

Therefore:

$$r = k_1 C_{A0}\sqrt{(1 - X_A)(m - X_A)} - k_2 C_{A0}X_A$$

and knowing that:

$$\frac{dX_A}{dt} = \frac{-r_A \times V}{n_{A0}} = \frac{-r_A}{C_{A0}}$$

we have:

$$\frac{dX_A}{dt} = k_1 C_{A0}\sqrt{(1 - X_A)(M - X_A)} - k_2 C_{A0}X_A \tag{20.104}$$

But k is a function of T, which is a function of t. Thus:

$$k_1 = 2.02 \times 10^{11} e^{-(100,000/8.314T)} \tag{20.105}$$

$$k_2 = 3.05 \times 10^{16} e^{-(150,000/8.314T)} \tag{20.106}$$

Adiabatic energy balance

$$\text{input} - \text{output} + \text{generation} = \text{accumulation}$$

$$\frac{dT}{dt} = \frac{(-\Delta H_r)\, n_{A0}}{\sum_j n_j \bar{c}_{pj}} \frac{dX_A}{dt} = 1 + \frac{40,000 X_A}{25(1 - X_A) + 25(1.25 - X_A)40} \frac{dX_A}{dt} \qquad (20.107)$$

Solving the integral of the above equation:

$$\frac{T}{T_0} = 1 + \frac{40 \times X_A}{(1.25 - X_A)} \qquad (20.108)$$

Rearranging the equations and substituting them into Equation 20.104, we get:

$$\frac{dX_A}{dt} = 2.02 \times 10^{11} e^{-(100,000/8.314T)} \times \sqrt{(1 - X_A)(1.25 - X_A)}$$
$$-3.05 \times 10^{16} e^{-(150,000/8.314T)} \times X_A \qquad (20.109)$$

(a) Calculating the equilibrium conversion when $\frac{dX_A}{dt} = 0$, we conclude that:

$$X_{Ae} = 0.25$$

Substituting this value in equation 20.108 we get:

$$T = 562K$$

Finally, substituting the value of X_{Ae} for finding the concentrations of reactants, we get:

$$C_{Ae} = 0.0747 \,\text{mol/L}$$
$$C_{Be} = 0.0933 \,\text{mol/L}$$
$$C_{Ce} = 0.0254 \,\text{mol/L}$$

(b) From Equation 20.109, we obtain the profiles concentration and temperature conversion:

$$\frac{dX_A}{dt} = 2.02 \times 10^{11} e^{-(100,000/8.314T)} \times \sqrt{(1 - X_A)(1.25 - X_A)}$$
$$-3.05 \times 10^{16} e^{-(150,000/8.314T)} \times X_A$$

Rearranging the above equation with 20.107:

$$X_A(0) = 0$$
$$T(0) = 373$$

and solving this equation with computational resources, we arrive at solutions whose graphs are shown below (b1, b2, and b3).

Note that the time to steady state is very small.

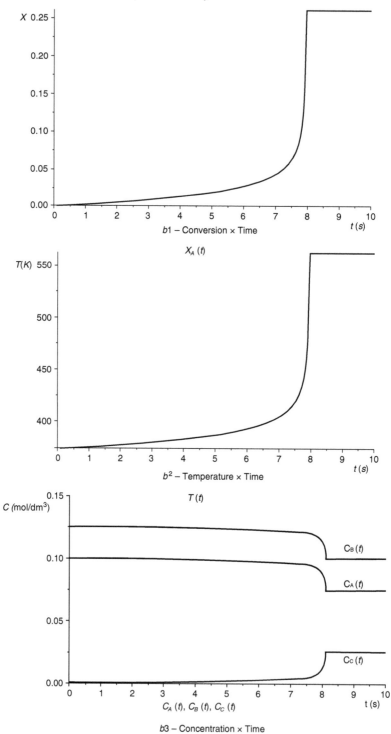

b1 – Conversion × Time

$X_A(t)$

b2 – Temperature × Time

$T(t)$

$C_A(t)$, $C_B(t)$, $C_C(t)$

b3 – Concentration × Time

SE.28 The endothermic reaction $A \rightarrow R$ is performed in three stages using a set of CSTR reactors (see the scheme below). It is desired a production (R) of 0.95×10^{-3} kmol/s to a conversion of 95%. The three reactors of equal volume operate at 50°C. The concentration of A is 1 kmol/m³ at 75°C. The three reactors are heated by condensing steam at 100°C in coils. Calculate the volume of the reactor and heat exchange area in each tank.

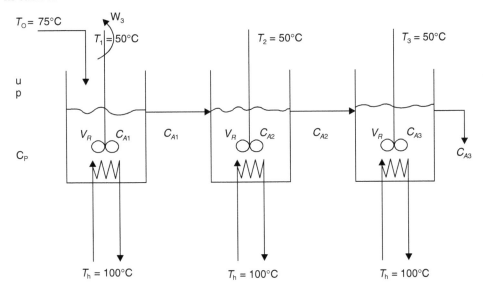

Data:

The reaction rate constant at 50°C, $k(50°C) = 4 \times 10^{-3}$ s⁻¹

Density of mixture, $\rho = 1000$ kg/m³

Specific heat of mixture, $C_p = 4 \times 10^3$ J/(kg °C)

Overall coefficient of heat transfer, $U = 1500$ J/(m² °C s)

Heat of reaction, $H_A = 1.5 \times 10^8$ J/kmol

Molecular weight of $A = 100$ kg/kmol

$F_R = 0.95 \times 10^{-3}$ kmol/s

$X_{A3} = 0.95$

$C_{A01} = 1$ kmol/m³

$U = 1500$ J/m² s°C

$k = 4 \times 10^{-3}$ s⁻¹ (first-order)

$\Delta H_R (A) = 1.5 \times 10^8$ J/kmol

Solution

→ Calculating the volume of the reactors:

– 3° reactor:

$$(1 - X_{An}) = \frac{1}{(1 + \tau \times k)^n}$$

$$(1 - 0.95) = \frac{1}{(1 + \tau \times k)^3}$$

$$\tau_1 = \nu_2 = \tau_3 = \tau$$

$$\tau = 428.61 \text{ s}$$

→ Conversions from each reactor:
− 2° reactor:

$$X_{A2} = 0.8643$$

− 1° reactor:

$$X_{A1} = 0.6316$$

→ Flow

$$F_R = F_{A0} X_A$$
$$0.95 \times 10^{-3} = F_{A0} \cdot \times 0.95$$
$$F_{A0} = 10^{-3} \text{ kmol/s}$$
Since $F_{A0} = C_{A0} \nu_0$

$$\nu_0 = 10^{-3} \text{ m}^3/\text{s} = 10 \text{ L/s}$$
$$\therefore \quad V = \tau \times V_0 \quad V_1 = V_2 = V_3 = V$$
$$V = 428.61 \times 10^{-3} \text{ m}^3 = 428 \text{ L}$$

→ Area of heat exchange
Energy balance:

1° reactor:

$$Q = \frac{F_{A0} X_A \times \Delta H_R}{V} + F_{A0} c_p (T - T_0) \overline{M}$$

$$Q_1 = -104.74 \text{ kJ/s}$$

2° reactor:

$$Q_2 = -34.905 \text{ kJ/s}$$

3° reactor:

$$Q_3 = -12.855 \text{ kJ/s}$$
$$Q = UA(T - T_s)$$
$$A_1 = 1.39 \text{ m}^2$$
$$A_2 = 0.47 \text{ m}^2$$
$$A_3 = 0.172 \text{ m}^2$$

SE.29 The reaction in liquid phase $A \rightarrow B$ is conducted at 85°C in a CSTR with a volume of 0.2 m³. The coolant temperature of the reactor is 0°C and the coefficient of heat transfer is 120 W/(m² K). The reactant A with concentration of 2.0 mol/L (pure) is introduced at a temperature of 40°C, its feed flow is 90 kg/min.

Data:

Specific heat of solution: 2 J/(g K)
Density of solution: 0.9 kg/dm³
Heat reaction: −250 J/g
MW_A: 90 g/mol
The reaction rate constant: 1.1 min⁻¹ (40°C) and 3.4 min⁻¹ (50°C)

(a) Calculate concentrations of A and B in the stationary state and the area of heat exchange.

(b) Determine the critical values of the thermal exchange area below which the reaction will undergo an uncontrolled temperature and then the reactor will explode.

Solution

(a) Data: $F_{A0} = 90,000$ g/min; $C_{A0} = 2$ mol/1; $MW_A = 90$ g/mol; $U = 120$ J/s mol K; $V = 200$ dm³; $k_1 = 1.1$ min⁻¹ (313 K); $k_2 = 3.4$ min⁻¹ (323 K); $T_s = 273$ K; $C_p = 2$ J/(g K).

$$C_{A0} = 2 \text{ (mol/g)} \times 90 \text{ (g/mol)} = 180 \text{ g/dm}^3$$

$$\tau = V/v_0 = 200 \text{ (dm}^3)/500 \text{ (dm}^3/\text{min}) = 0.4 \text{ min}$$

From $k = k_0 \times \exp(-E/RT)$, we have:

$$\ln(k_2/k_1) = (E/8.31) \times (1/323) - (1/313) \rightarrow$$

Thus:

$$E = 94,829 \text{ J/(mol K)}$$
$$\rightarrow k_0 = 7.49 \times 10^{15}$$

For $T = 358$ K: $k = 107.5$ min⁻¹

(b) For an irreversible reaction, the rate will be:

$$-r_A = kC_A \tag{20.110}$$

From the CSTR equation:

$$\tau = \frac{C_{A0} - C_A}{(-r_A)} \tag{20.111}$$

and substituting Equation 20.110 into Equation 20.111:

$$C_A = C_{A0} + k\tau \tag{20.112}$$

At 358 K, we have $C_A = 4.1$ g/dm³(final concentration of A) and $C_B = 176.3$ g/dm³ (final concentration of B)

Energy balance:

$$\sum_j F_j \bar{c}_{pj} (T - T_0) + \Delta H_r F_{A0} X_A = Q_{\text{external}} \tag{20.113}$$

Thus:

1 $Q = UA(T_s - T) = 120$ (J/(s mol K)) $\times$ 60 (s/min) $\times A$ $(273 - 358) = 7200A$
   $(273 - 358) = 6.12 \times 10^5 \times A$
2 $F_i C_{pi} (T - T_{i0}) = 90,000 \times 2 \times (358 - 313) = 8.1 \times 10^6$.
3 $(-\Delta H_R)(r_A)V = 250 \times 107.5 \times 4.1 \times 200 = 2.2 \times 10^7$

Then, we have:

$A = 22.77$ m²

(b) From Equation 20.113 (see above), we have:

$F_{A0} X_A = -r_A V$; but $-r_A = k_0 \times exp(-E/RT) \times C_A$

Since C_p is independent of temperature (for this case):

Thus, $(-r_A)V(\Delta H_R) = F_{A0} C_p + UAT$

Differentiating with respect to T, we have:

$$\frac{E \times \Delta H_r(-r_A)}{RT^2} = UA \tag{20.114}$$

Dividing Equation 20.114 by Equation 20.115, we get:

$$\frac{RT^2}{E} = \frac{F_{A0} C_p}{UA} + (T - T_s) \tag{20.115}$$

To an uncontrolled situation:

$$\frac{RT^2}{E} > \frac{F_{A0} C_p}{UA} + (T - T_s) \tag{20.116}$$

We obtain $A < 0.34$ m²
That is, the minimum heat exchange area is 0.34 m².

SE.30 The reaction in liquid phase $A \rightarrow B$ is performed in a tubular reactor at constant pressure of 202.6 kPa. The load consists of 600 kmol/ks A pure (specific volume = 0056 m³/kmol) at 200°C. The heat of the reaction at 200°C is -15 kJ/mol and the heat capacities of both compounds are equal to 42 J/(mol.K). The rate constant is given by:

$$k = 110 + 0.8(T - 473)$$

$$k = [ks^{-1}]; \quad T = [K]$$

The reactor must operate adiabatically. However, the maximum temperature permitted is 400°C. Above this temperature, some undesirable subproducts are formed. Calculate the minimum length of the reactor to obtain a conversion of 80%. What is the heat transfer rate in the cooling section of the reactor?

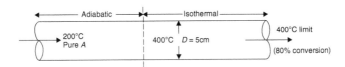

Solution:

$$P = 202.6 \, kPa = 2.02 \, atm$$
$$F_{A0} = 600 \, mol/s$$
$$v_e = 0.056 \, m^3/kmol$$
$$T = 473 \, K$$

$$\Delta H^0_{R(200°C)} = -15 \, kJ/mol \quad C_{PA} = C_{PB} = 42 \, J(/mol \, K)$$

$$(-r_A) = kC_{A0}(1 - X_A) \tag{20.117}$$

$$k = 110 + 0.8 \, (T - 473) \quad k = [ks^{-1}]; \quad T = [K]$$

$$T - T_0 = \frac{-\Delta H_r \, F_{A0} \, X_A}{\sum\limits_j F_j \bar{c}_{pj}} \tag{20.118}$$

Substituting Equation 20.118 into Equation 20.117:
    Then,

$$T - T_0 = \frac{-\Delta H_r F_{A0} X_A}{F_{A0}[(1 - X_A) + X_A] \times \bar{c}_{pA}}$$

Therefore,

$$200 = \frac{15,000 \, X_A}{42}$$

$$X_A = 0.56$$

$$T = T_0(1 + \beta X_A) \tag{20.119}$$

$$T = 473(1 + 0.76X_A)$$

Substituting into rate constant equation:

$$k = 110 + 0.8[473(1 + 0.76\,X_A) - 473]$$

To find the volume in the tubular reactor at constant pressure:

$$\tau_{PFR} = \frac{V}{v_0} = \int_0^{X_A} \frac{dX_A}{[110 + 0.8\,(359.4 \times X_A)](1 - X_A)}$$

Thus $\tau = 3.6\ s$

$$\tau = \frac{V}{v_0} \rightarrow v_0 = VF_{A0} = 0.056 \times 600$$

$$v_0 = 33.6 \times 10^{-3}\ \text{mol/s}$$

$$V = 3.6 \times 33.6 \times 10^{-3}$$

$$V = 121\ L \rightarrow \text{adiabatic volume}$$

To find the volume in isothermal tubular reactor (400°C):

$$\tau_{PFR} = \frac{V}{v_0} = \int_0^{X_A} \frac{dX_A}{k(1 - X_A)}$$

where:

$$k = 270\ \text{k s}^{-1}$$

Thus:

$$\tau = 2.9\ s$$

$$V = 97.4\ L \rightarrow \text{isothermal volume}$$

Total volume is:

$$V_t = V_{ad} + V_{isot}$$

$$V_t = 218.4\ L$$

$$L = 111.3\ m$$

Calculating the rate of heat transfer (isothermal): $(T_e = T_s = 673\ K)$
Substituting in Q:

$$Q = (-15,000 \times 600 \times (0.8 - 0.56))$$

$$Q = 2.16 \times 10^6\ J/s$$

SE.31 The production of styrene from vinylacetylene is an irreversible reaction of the type:

$$A \rightarrow R$$

(a)   Determine the conversion at the outlet of the PFR to the inlet temperature of 675 K. Plot the temperature and conversion according to the length of the reactor.

(b)   For a variation of the inlet temperature, plot of the conversion reaction as a function of temperature input.

(c)   Vary the inlet temperature of the heat exchanger and find the maximum inlet temperature, such that avoid the thermal runaway.

Data

$C_{A0} = 1 \, \text{mol/L}$
$F_{A0} = 5 \, \text{mol/s}$
$\Delta H_R = -231 - 0.012(T - 298) \, \text{kJ/mol}$
$\bar{c}_{PA} = \bar{c}_{PR} = 0.122 \, \text{kJ/(mol k)}$
$k = 1.48 \times 10^{11} \exp(-19124/T) \, \text{L/mol s}$
$T_0 = 675 \, \text{K}$
$T_a = 700 \, \text{K}$
$U \times A = 5 \, \text{kJ/s dm}^3/\text{K}$

Solution

$$A \rightarrow R$$

From PFR, we have:

$$\frac{dX_A}{dV} = \frac{-r_A}{F_{A0}}$$

The unit of k indicates a second-order reaction, so the rate will be:

$$-r_A = kC_A^2$$

where $C_A = C_{A0}(1 - X_A)$
Energy balance:

$$\frac{dT}{dV} = \frac{U \times A \, (T - T_0) + (-r_A)(-\Delta H_r)}{F_{A0}\bar{c}_{pA}}$$

Using the integration package DASSL (Petzolo, 1989) and implementing a routine in FORTRAN 90, we obtain the profiles in Figures 20.17 and 20.18.

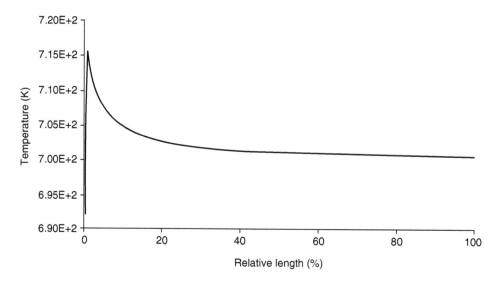

Figure 20.17 Temperature versus relative length.

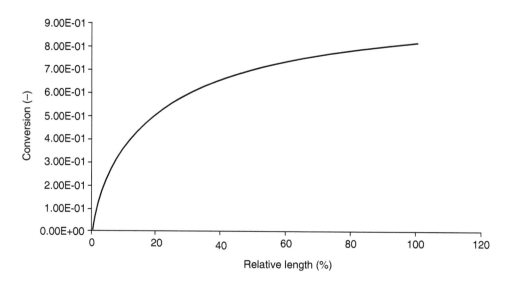

Figure 20.18 Conversion versus relative length.

Thus, the conversion obtained was 0.81 (81%).
It was assumed value of 100 dm³ to V_f.

(a)   Figure 20.19 shows conversion × temperature
(b)   From Figure 20.19, we get $T = 740\,K$.

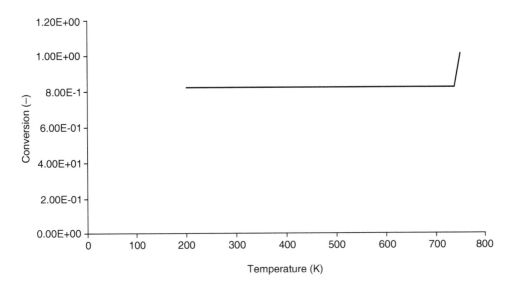

Figure 20.19 Conversion versus temperature.

SE.32 The fermentation of glucose to ethanol is conducted in a batch reactor. Calculate the concentration of cells, the substrate, and the growth rate versus time.

Data:

$C_{p*} = 93\,g/L$; $K_d = 0.01\,l/h$; $C_{c0} = 1$; $M = 0.031\,/h$; $Y_{cs} = 0.08$;
$C_{s0} = 250$; $\mu_{max} = 0.331/h$; $Y_{ps} = 0.45$; $n = 0.52$; $K_s = 1.7\,g/L$;
$Y_{pc} = 5.6$

Batch reactor:
Rate equation:

$$r_g = \mu_{max}\left(1 - \frac{C_p}{C_{p*}}\right)^n \frac{C_c \cdot C_s}{K_s + C_s}$$
$$r_d = K_d \cdot C_c$$
$$r_{smc} = M \cdot C_c$$
$$(-r_s) = r_g \cdot \frac{1}{Y_{cs}} + M \cdot C_c$$
$$r_p = Y_{pc} \cdot r_g$$

Mass balance
Cells:

$$\frac{d}{dt}C_c = r_g - r_d$$

Substrate:

$$\frac{d}{dt}C_s = \frac{r_g}{Y_{cs}} - r_{smc}$$

Product:

$$\frac{d}{dt}C_p = Y_{pc} \cdot r_g$$

Combining equations we get:

$$\frac{d}{dt}C_c(t) = \mu_{max} \cdot \left(1 - \frac{Cp(t)}{Cp^*}\right)^n \cdot \frac{C_c(t) \cdot C_s(t)}{K_s + C_s(t)} - K_d \cdot C_c(t)$$

$$\frac{d}{dt}Cs(t) = \frac{-1}{Ycs} \cdot \mu_{max} \cdot \left(1 - \frac{Cp(t)}{Cp^*}\right)^n \cdot \frac{Cc(t) \cdot Cs(t)}{Ks + Cs(t)} - M \cdot Cc(t)$$

$$\frac{d}{dt}Cp(t) = Ypc \left[\mu_{max} \cdot \left(1 - \frac{Cp(t)}{Cp^*}\right)^n \cdot \frac{Cc(t) \cdot Cs(t)}{Ks + Cs(t)}\right]$$

Solution
Initial conditions:

$$C_c(0) = 1.0$$

$$C_s(0) = 250 \cdot \frac{g}{L}$$

$$C_p(0) = 0$$

Concentration of the product (g/L) with time (h)

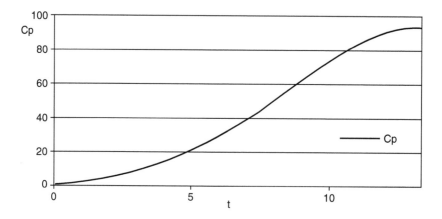

Figure 20.20 C_p(g/L) × t (h).

Concentration of the substrate (g/L) with time (h)

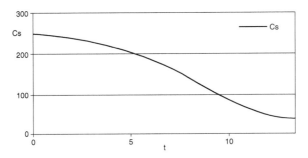

Figure 20.21   C$_s$(g/L) × t (h).

Concentration of the cells (g/L) with time (h)

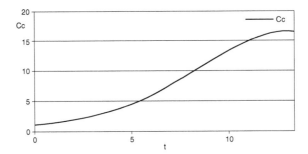

Figure 20.22   C$_c$(g/L) × t (h).

Rate of cell growth (g/L) with time (h)

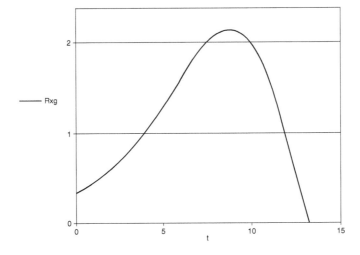

Figure 20.23   R$_c$(g/L) × t (h).

SE.33 The glycerol is made from grains using specific cells, substrate, and products.

$$\mu = \mu_0 \frac{[S] + k_2[ES][S]}{k_S + [S] + \frac{[S]^2}{K_{SE}}} \left(1 - \frac{[P]}{[P*]}\right) * e^{-K_{PE}}[P] \tag{20.120}$$

The rate equation for the substrate and cell are presented below.

$$\mu = \mu_m \frac{[S]}{k_S + [S] + \frac{[S]^2}{K_{SE}}}$$

$$\frac{dC}{dt} = \mu[C]$$

$$\frac{d[P]}{dt} = (\alpha\mu + \beta)[C]$$

$$\frac{d[S]}{dt} = \frac{1}{Y_{p/s}} \frac{d[P]}{dt}$$

$$r_g = \mu_m \frac{[S][C]}{k_S + [S]} * k_i$$

$$k_i = \left(1 - \frac{[P]}{[P*]}\right)^n$$

where:

$K_S = 0.018$ mg/mL
$K_{SE} = 11.84$ mg/mL
$K_{PE} = 0.06$ mg/mL
$[P*] = 32.4$ mg/mol

Determine the concentration profile versus time for cells, substrate and product in a batch reactor.

Additional data:
$C_0 = 10^{-8}$ mg/mL
$\beta = -0.147$
$Y_{PS} = 1.33$
$S_0 = 50$ mg/mL
$m_0 = 0.25$ Lmg/(mL s)

Solution

The computer program MATLAB was used to solve the differential equations for the concentrations versus batch time, according to Figure 20.20.

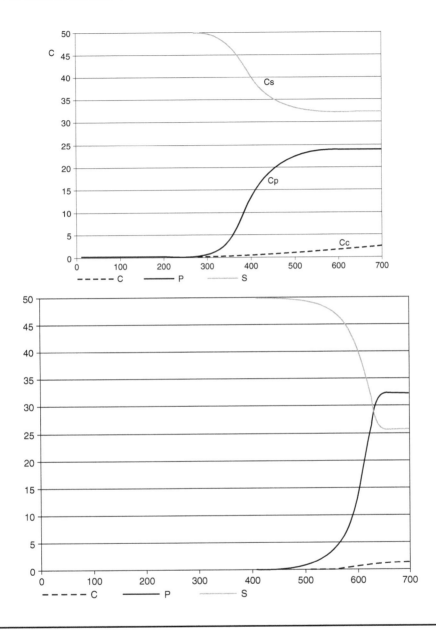

ER34. The reduction of NO is performed over a pellet catalyst. 2% of NO and 98% of air are introduced at a rate of 10^{-6} m^3/s into the tubular reactor. The process temperature is 1173 K at a pressure of 101 kPa. The chemical reaction is a first-order reaction.

$$-r_{NO} = k'C_{NO} \qquad (20.121)$$

where:

$$k' = kSg$$

(a)   Calculate the Thiele modulus and consider the intra particle diffusion with surface reaction.
(b)   Calculate the mass required to achieve a reduction of NO to 0.4%.

Data:
$D_e = 2.66 \times 10^{-8} \ m^2/s$
$D_{AB} = 2.0 \times 10^{-8} \ m^2/s$
$v = 1.53 \times 10^{-8} \ m^2/s$
Pore radius, $r_P = 1.3 \times 10^{-9} \ m$
Porosity, $\varepsilon_b = 0.5$
Pellet radius, $R_P = 3 \times 10^{-3} \ m$
Particle density $\rho_p = 1.4 \times 10^6 \ g/m^3$.

Solution

(a)   It is a first-order reaction, according to Equation 20.121.
       NO conversion:

$$X_A = \frac{C_{A0} - C_{Ab}}{C_{A0}} = 0.996$$

The reactor is a fixed bed having an internal diameter of 2″. Thus, according to mass balance for the PFR, we have:

$$\frac{dC_A}{dz} = \frac{\Omega.k.S_g.\rho_b.C_A}{U} \tag{20.122}$$

We can make the above equation based on the mass of the solid w, as follows:

$$\frac{dC_A}{dW} = \frac{\Omega.k.S_g.C_A}{U} \tag{20.123}$$

where:

$$\Omega = \frac{\eta}{\left(1 + \frac{\eta k S_g \rho_b}{k_c.a}\right)} \tag{20.124}$$

Thus, we integrate Equation 20.123 to find the mass equation.

$$W = \left[-\frac{v_0}{\Omega k S_g} \ln(1 - X_A)\right] \tag{20.125}$$

Thiele modulus and effectiveness factor:
From the definition of Thiele modulus (ϕ_1):

$$\Phi_1 = R\sqrt{\frac{k'S_g\rho_p}{D_e}} \tag{20.126}$$

The density of the pellet (ρ_p) can be calculated from the equation of the average pore radius in the pore cross section.
To relate the gas density (ρ_g), the following relationship can be used:

$$\rho_g = \rho_p(1 - e_b)$$

Thus, from Equation 20.126, we solve:

$$\phi_1 = 14.9$$

It is noted that this result is high. Therefore, the rate of diffusion is small in relation to reaction rate and hence the diffusion is the limiting step.
For the effectiveness factor, η:

$$\eta = \frac{3(\Phi_1 \coth \Phi_1 - 1)}{\Phi_1^2} \tag{20.127}$$

which is:

$$\eta = \frac{3}{\Phi_n} = \frac{3}{14.9} = 0.2$$

(b)   To find the mass, we need to calculate the coefficient of mass transfer:

$$K_c = \frac{(1 - \varepsilon_b)D_{AB}}{\varepsilon_b} \times Sh \tag{20.128}$$

Sherwood number:

$$Sh = Re^{1/2}Sc^{1/3} \tag{20.129}$$

where:
Re = Reynolds number
Sc = Schmidt number.

We calculate the cross-sectional area of the fixed bed ($d_t = 2''$):

$$A_s = \frac{\pi d^2}{4} = 2.03 \times 10^{-3}m^2$$

Thus:

$$u = \frac{v_0}{A_s} = 4.93 \times 10^4 m/s$$

Reynolds number:

$$Re = \frac{u.d_p}{(1 - \varepsilon_b)} = 386.7$$

Schmidt number:

$$Sc = \frac{v}{D_{AB}} = 0.765$$

Therefore, we have the Sherwood number:

$$Sh = Re^{1/2} Sc^{1/3} = 18$$

Now, we are able to calculate the mass transfer coefficient (k_c):

$$K_c = \frac{(1 - \varepsilon_b) D_{AB}}{\varepsilon_b} \times Sh = 6.0 \times 10^{-5}$$

Calculation of the external area per unit volume of reactor (a_c):

$$a_c = \frac{6 (1 - \varepsilon_b)}{d_p} = 500 m^2/m^3$$

Calculating the overall effectiveness factor (Ω):

$$\Omega = 0.063\, m$$

Calculation of mass for 99.8% conversion:

$$W = 421\, g$$

SE.35 Ethanol (B) diluted in water (2–3% w/w) is oxidized to acetic acid in the presence of oxygen (A) pure bubbled to 10 atm in catalyst pellets (Pd/Al$_2$O$_3$). This reaction occurs at 30°C in a slurry bed reactor, according to the following reaction (Sato et al., 1972).

$$O_2(g) + CH_3CH_2OH \rightarrow CH_3COOH + H_2O$$

The reaction rate is given by:

$$-r'_A = k'C_A, \tag{20.130}$$

Determine the conversion of ethanol in this process.

Data:
$k' = 1.77 \times 10^{-5}\, m^3/(kg\ s)$
$D_P = 10^{-4}\, m,$
$Ps = 1800\, kg/m^3,$

$$D_{ef} = 4.16 \times 10^{-10} \, m^3/s$$

Mass transfer:

to O_2

$$K_{Ac} = 4 \times 10^{-4} \, m/s \text{ to } O_2$$

Reactor:

Volume $= 5.0 \, m$(height) $\cdot \, 0.1 \, m^2$ (cross section)
Volume fraction: $f_g = 0.05; f_l = 0.75; f_s = 0.2$

Feed:

$v_g = 0.01 \, m^3/s; H_A = 86,000 \, Pa \, m^3/mol$ (Henry constant)
$v_t = 2 \times 10^{-4} \, m^3/s; C_{B0} = 400 \, mol/m^3$

Solution:

A slurry bed reactor is a reactor for multiphase flow in which the reactant gas dissolves in the liquid and then both reactants diffuse or move toward the surface of the catalyst for the reaction to occur. In this way, the resistances of transfer through the liquid/gas interface and the surface of the solid are present in the rate equation. One of the main advantages of this type of reactor is the control of temperature and heat recovery. Moreover, they can operate continuously or in batch.

Concentration of A:

$$C_A = \frac{P}{H} = 11.77 \, mol/m^3$$

Therefore, by comparing the concentrations of A and B, there is an excess of B ($C_B = 400 \, mol/m^3$).

Calculating a_c (spheres):

$$a_c = \frac{6f_s}{d_p} = 1.2 \times 10^4 m^2_{cat}/m^3$$

$$K_c a_c = 4.8 m^3_{liq}/m^3_{cat}$$

Thiele modulus:

$$\Phi_1 = L \times \sqrt{\frac{(n+1) \, k' \rho_p C_b^{(n-1)}}{2D_e}} = 0.145$$

where (spheres)

$$L = \frac{d_p}{6} = 1.6 \times 10^{-5}$$

Calculating the effectiveness factor (ε):

$$\eta = \frac{3}{\Phi} = 20.6$$

For the first-order reaction, $n = 1$, we have:

$$-r'_A = k'_{ap} C_A \qquad (20.131)$$

where:

$$\frac{1}{\frac{1}{k'_{ap}}} = \frac{1}{\frac{1}{k_{Ag}a_i} + \frac{1}{k_{Al}a_l} + \frac{1}{k_{Ac}a_c} + \frac{1}{k' \times \eta \rho_p \varepsilon_p}} \qquad (20.132)$$

Substituting the values, we obtain:

$$k'_{ap_c} = 0.102$$

Conversion of ethanol:

Assuming that oxygen consumption is proportional to ethanol reacted to stream the PFR reactor, we have:

$$\tau = -\frac{1}{k'_{ap}} \ln(1 - X_A) \qquad (20.133)$$

where X_A is the ethanol conversion. But we know that: $\tau = 5s$ and substituting it into Equation 20.133:

$$1 - X_A = e^{-k_{ap}\,t} = e^{-0.102 \times 5} = 0.60$$

Therefore:

$$X_A = 0.40$$

SE.36 A series of experiments were conducted using different sizes of particles that were milled to verify the effect of diffusion. The reaction is first order and irreversible. The concentration on the surface is equal to 2×10^{-4} mol/cm^3. The table below shows the rates observed in these experiments.

Sphere diameter (cm)	0.25	0.075	0.025	0.0075
$r_{obs(mol/h\,cm^3)}$	0.22	0.70	1.60	2.40

Data:

$$\rho = 2\,g/cm^3, S_g = 20\,m^2/g, D_e = 2 \times 10^{-3}\,cm^2/g$$

Determine the "real" rate of these experiments.
Determine the effectiveness factor.
Is there any change to a second-order reaction and irreversible for this case? Verify.

Solution

$$r_{AS} = \kappa C_{As}$$

The observed rate is:

$$r''_{obs} = \eta \times (-r''_A) \tag{20.134}$$

where

$$\eta = \frac{3}{\Phi} \tag{20.135}$$

$$\Phi_1 = R\sqrt{\frac{k'C_{A0}^{(n-1)}S_g\rho_p}{D_e}} \tag{20.136}$$

Substituting Equation 20.135 into Equation 20.136:

$$\eta = \frac{3}{R}\sqrt{\frac{D_e}{k'S_g\rho_p}} \tag{20.137}$$

Substituting Equation 20.137 into Equation 20.134:

$$r''_{obs} = \frac{3}{R}(k')^{1/2}\sqrt{\frac{D_e}{S_g\rho_p}} \tag{20.138}$$

The calculated values are shown in the table and Figure 20.24.

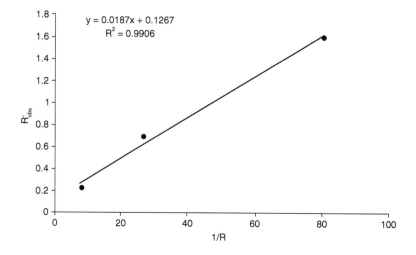

Figure 20.24　Rate versus radius.

Diameter	Radius	1/R	r''_{abs} (mol/h cm³)
0.25	0.125	8	0.22
0.075	0.0375	26.67	0.7
0.025	0.0125	80	1.6
0.0075	0.00375	266.67	2.4

From Figure 20.24, we found a slope of 0.0187. Thus:

$$0.0187 = \frac{3}{R}(k')^{1/2}\sqrt{\frac{D_e}{S_g \rho_p}}$$

$$C_{As} = 2 \times 10^{-4} mol/cm^3$$

Substituting the values, we get:

$$\kappa = 0.11 \ (cm^3/mol \ s)^2 \qquad\qquad (20.139)$$

For spherical particles and first-order reaction:

$$n = 1$$

$$D_e < 0.9 \left(\frac{C_{As}}{R^2 r_{obs} \rho_p}\right)$$

→ Assuming a second-order reaction and irreversible:

$$-r'_{As} = r''_{obs} = kC_{As}^2 \eta \qquad\qquad (20.140)$$

From Thiele modulus:

$$\Phi = R \sqrt{\frac{k' C_{As}^{(n-1)} S_g \rho_p}{D_e}} \qquad (20.141)$$

And $n = 2$, we obtain the following equation:

$$\Phi = R \sqrt{\frac{k' C_{As} S_g \rho_p}{D_e}} \qquad (20.142)$$

Thus, effectiveness factor will be:

$$\eta = \frac{3}{\Phi} \qquad (20.143)$$

Rearranging the above equations, we find:

$$r''_{obs} = 3 (C_{As})^{3/2} \sqrt{\frac{k D_e}{S_g \rho_p}} \qquad (20.144)$$

As the slope is equal to 0.0187, thus:

$$k = 2203 \left(\frac{S_g \rho_p}{D_e} \right)$$

We obtain:

$$k = 9.3 \times 10^{-2} \ (\text{cm}^3 / \text{mol s})^2$$

→ Calculating the effectiveness factor, η

$1/d_p$	r'' (mol/h cm³)
4	0.22
13.33	0.7
40	1.6
133.33	2.4

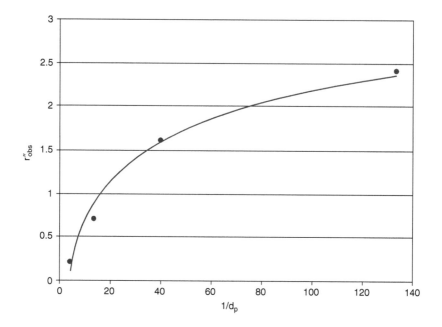

Figure 20.25 Rate versus $1/d_p$.

From Figure 20.25, $r''_{abs} = 2.4$:

$1/d_p$	$r \ (mol/h \ cm^3)$	N
4	0.22	0.092
13.33	0.7	0.29
40	1.6	0.67
133.33	2.4	1

SE.37 The data below were obtained in a slurry bed reactor, where hydrogenating a compound A is made. The table gives the solubility of H_2 in the liquid mixture (mol/dm^3) based on the mass of catalyst (g/dm^3) and the reaction rate $(mol/(dm^3 \times min))$.

Particle	$S/(-r_L) \ (min)$	$1/m \ (dm^3/g)$
A	4.2	0.01
A	7.5	0.02
B	1.5	0.01
B	2.5	0.03
B	3.0	0.04

What size particles have the lowest effectiveness factor?

If we use the catalyst A in the reactor with a concentration of $50\ g/dm^3$, there is an increased rate. Thus, if we use the catalyst B, minimum amount of catalyst can be used such that the diffusional resistance is 50% of the total resistance?

Solution

It is considered first-order reaction.

Reaction rate equation: considering the steady state, we get:

$$\dot{R}_A\ (\text{bubble}) = \dot{R}_A\ (\text{surface}) = \dot{R}_A\ (\text{solid}) =$$

where:

$\dot{R}_A$ (bubble): transfer rate from the bubble

$\dot{R}_A$ (surface): transfer rate of catalyst

$\dot{R}_A$ (solid): reaction rate in the catalyst

Thus:

$$\frac{S}{(-r_i)} = \left(\frac{1}{k_b a_b} + \frac{1}{m} \left(\frac{1}{k_c a_p} + \frac{1}{k\eta} \right) \right)$$

$$r_b = \frac{1}{k_b a_b} \quad : \text{resistance to gas absorption}$$

$$r_c = \frac{1}{k_c a_c} \quad : \text{transport resistance to the surface of the particle}$$

$$r_r = \frac{1}{k\eta} \quad : \text{Diffusion resistance and reaction}$$

We consider the H_2 concentration equal to solubility. Figure 20.26 shows $S/-r_L$ versus $1/m$ for the particles A and B:

$$\frac{S}{(-r_i)} = r_b + \frac{1}{m} r_{cr}$$

where $r_{cr} = r_c + r_r$

Particle A:

$$r_b = 0.9\ min$$

$$r_{cr} = 300 \frac{g}{dm^3\ min}$$

Particle B:

$$r_b = 1.0\ min$$

$$r_{cr} = 50 \frac{g}{dm^3\ min}$$

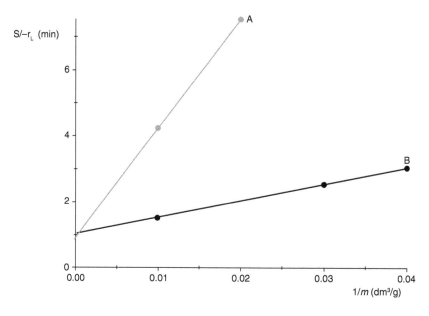

Figure 20.26

The values shown for r_b should be equal for both particles as they depend on the bubble but not the catalyst, the difference between the values found can be attributed to experimental errors.

From the curves, one can conclude that the particle diameter A is larger than that of the particle B.

We also know that for small diameters of the particles, the efficiency is equal to 1. While for larger particle sizes, the efficiency is inversely proportional to the Thiele modulus:

Thus, we conclude that the particle A generates the lowest effectiveness.

In fact, if we consider the transport resistance to surface (r_c) negligible:

Particle A:

$$r_{cr} = \frac{1}{k\eta_A} = 330 \Rightarrow \eta_A = \frac{1}{330 \times k}$$

Particle B:

$$r_{cr} = \frac{1}{k\eta_B} = 50 \Rightarrow \eta_B = \frac{1}{50 \times k}$$

And by comparison:

$$\frac{\eta_B}{\eta_A} = 6.6$$

$$\eta_A < \eta_B$$

For the concentration of catalyst 50 g/dm^3 (particle A):

$$\frac{1}{m} = 0.02\frac{dm^3}{g}$$

From the above figure, we get:

$$\frac{S}{(-r_i)} = 7.5$$

Remembering that for the particle A, $r_b = 0.9$ min, and the total resistance, $r_T = S/-r_L$

% of r_b in the total resistance $\quad \frac{r_b}{r_t} = \frac{0.9}{7.5} = 15$

% diffusion resistance $\quad \frac{r_t - r_b}{r_t} = 88$

We have: $r_T - r_b = 6.6$ min

Thus, we see that the diffusion resistance is the limiting factor. Therefore, the measure that increases the dispersion of gas will not be effective.

Comparing with the particle B:

$$\frac{1}{m} = 0.02\frac{dm^3}{g} \Rightarrow \frac{S}{(-r_i)} = r_t = 2$$

and $r_b = 1.0$ min

Diffusion resistance: $r_T - r_b = 1.0$ min

We can conclude that the particle A has a greater diffusional limitation. For a diffusional resistance with 50% of total resistance, we will have:

$$\frac{\frac{S}{(-r_i)} - r_b}{\frac{S}{(-r_i)}} = 50 \Rightarrow \frac{S}{(-r_i)} = 2 \times r_b = 2 \text{ min}$$

$$\Rightarrow m = 50g/dm^3$$

Then, the minimum load is 50 g/dm^3 (larger loads generate s/—r_L and therefore higher percentages).

20.2 EXERCISES PROPOSED: REACTORS

EP.1 The following reaction $A + B \xrightarrow{k} R$ is carried out in a batch reactor at 227°C and 10.3 atm, its initial composition is 33.3% of A and 66.6% of B. The following data were obtained in the laboratory:

$(-r_A)$ mol/(L s)	0.010	0.005	0.002	0.001
X_A	0	0.2	0.4	0.6

Calculate the volume of a PFR with a conversion of 30% A at an input stream at 2 m^3/min.

Calculate the volume of the CSTR under the same conditions.

Compare the volume to higher conversions, such as 60% and 80%.

Plot the curve of reaction rate and conversion as a function of volume.

If the density of the catalyst is 0.8 g/cm³, calculate the ratio rate/mass? Calculate the amount of mass required to achieve the same conversions of the item 3.

Is it possible to determine the average residence time in each reactor? Calculate and compare the space-time values.

EP.2 Summarize the residence times to the batch system in CSTR and PFR, considering the different cases:

- Constant volume
- Liquid phase
- Vapor phase

Consider the following reaction $A \xrightarrow{k} 3S$, whose rate is $(-r_A) = kC_A^2$, introducing the reactant with 50% inert at a pressure of 1 atm at 300 K.

EP.3 Two PFR reactors are in parallel at a volume of 50 and 30 L, respectively. These reactors must produce a certain quantity of products according to the following reaction $A \xrightarrow{k} R + S$ in gaseous phase. The rate constant is 0.15 min⁻¹ at 50°C. The average residence time in the reactor is 5 min.

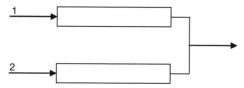

(a) Calculate the space-time in each reactor and flows of entry and exit.
(b) Calculate the molar flow output.

EP.4 Propylene glycol (C₃H₈O₂) is produced from propylene oxide (C₃H₆O) by liquid phase hydrolysis with excess water in the presence of a small concentration of sulfuric acid as the catalyst, and its reaction rate constant is $k = 1.69 \times 10^6 \exp(-41,800/RT)(\text{min}^{-1})$, with E (J/mol). It is known that the concentration of water is 10 times larger. The reactant is introduced into the CSTR reactor with a volume of 5 L at 27°C (isothermal) and the initial concentration of propylene is 2.0 mol/L. This reactor is placed in series with a PFR reactor and operates adiabatically, its final conversion is 70%. Calculate the exit temperature and the volume of PFR and the intermediate conversion.

$$\Delta H_R^0 = -4.7 \times 10^4 \, kJ/mol$$
$$\bar{c}_{p \text{ propylene}} = 45 (J/mol \times K)$$
$$\bar{c}_{p \text{ water}} = 23 (J/mol \times K)$$
$$\bar{c}_{p \text{ glycol}} = 59 (J/mol \times K)$$

Data:

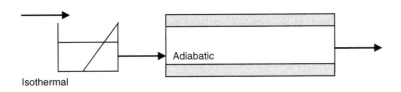

Isothermal   Adiabatic

EP.5 For an irreversible reaction such as $A \xrightarrow{k} R$ has a rate equation:
     The initial concentration of the reactant A (pure) is equal to 10 mol/L.
     Plot the reaction rate as a function of conversion.
     Calculate the volumes of two reactors in series, CSTR and PFR. The intermediate conversion is 50% and the final conversion was 90%.
     Justify your answer.

EP.6 A parallel reaction scheme is described as below:

$$A \xrightarrow{k_1} 2R$$
$$A \xrightarrow{k_2} S$$

     Both reactions are first-order and irreversible. It is desired to obtain the R product at a conversion of A equal to 70%. Calculate the rate constant and the selectivity, knowing that the space-time is 1 min. It is assumed $k_2 = 0.012 \, s^{-1}$.

EP.7 An irreversible reaction $A \xrightarrow{k} R$ occurs in the liquid phase batch reactor of volume 5 m^3. The reaction rate constant is given:
     One can describe this reaction as follows:
     The pure reactant is introduced at 20°C and heated in inlet tube to 55°C. There was some conversion?
     The reactor operates adiabatically, but it cannot exceed the temperature of 95°C and the conversion cannot be greater than 90%. Calculate the outlet temperature.
     After exiting the reactor, the product is cooled until it reaches 45°C. Calculate the conversion obtained.
     Calculate the final time of reaction.

Data:

Overall coefficient of heat transfer during heating: 1.360 W/(m²°C)
Overall coefficient of heat transfer during cooling: 1.180 W/(m²°C)
Heat of reaction: 1670 kJ/kg
Specific heat: 4.2×10^6 J/(m³°C)
Molecular mass: 80
Initial concentration: 1 kmol/m³

EP.8 The dehydrogenation of ethanol was conducted in an integral reactor at 275°C. The following experimental data were obtained in the laboratory.

X	0.118	0.196	0.262	0.339	0.446	0.454	0.524	0.59	0.60
P (atm)	1	1	1	1	1	1	1	1	1
W/F (kgh/kmol)	0.2	0.4	0.6	0.88	1.53	1.6	2.66	4.22	4.54
X	0.14	0.2	0.25	0.286	0.352	0.14	0.196	0.235	0.271
P (atm)	3	3	3	3	3	4	4	4	4
W/F (kgh/kmol)	0.2	0.4	0.6	0.88	1.6	0.2	0.4	0.6	0.88
X	0.32	0.112	0.163	0.194	0.214	0.254	0.1	0.148	0.175
P (atm)	4	7	7	7	7	7	10	10	10
W/F (kgh/kmol)	1.6	0.2	0.4	0.6	0.88	1.6	0.2	0.4	0.6
X	0.188	0.229							
P (atm)	10	10							
W/F (kgh/kmol)	0.88	1.6							

The equilibrium constant is 0.589. The system consists of an ethanol/water azeotrope containing 13.5% (molar base) of water. The water does not adsorb. Estimate the parameters of adsorption–desorption using the conversion as variable regression. Demonstrate what the best model.

EP.9 The reaction $A \xrightarrow{k} R$ is carried out in liquid phase in a PFR reactor at a pressure of 202 kPa. The reactant A has 600 kmol/ks and its specific volume is 0.056 m³/kmol. The heat of reaction at 200°C is 15 kJ/mol and the heat capacities of both components are equal to 42 J/(mol K). $k = 110 + 0.8(T - 473)$ The reaction rate constant is given by (k s⁻¹).

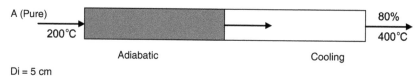

A (Pure)
200°C
Adiabatic   Cooling
80%
400°C

Di = 5 cm

The reactor must operate adiabatically. However, the maximum temperature permitted is 400°C. Above this temperature, unwanted byproducts are formed. Thus, calculate:

(a) The length of the reactor for this process reaches a conversion of 80%.
(b) The rate of heat transfer in the cooling of the reactor.

EP.10 It is desired to obtain a certain product in a PFR reactor from the following reaction $2A \xrightarrow{k} R$. Calculate the conversion at the reactor outlet when the inlet temperature is equal to 675 K.

By varying the input temperature of the heat exchanger, calculate the maximum inlet temperature, such that there is no "thermal runaway." Compare the results obtained using adiabatic reactors. Introducing inert and knowing that $F_1 = 3F_{A0}$, demonstrate how the conversion varies with the inlet temperature.

Data:

$C_{A0} = 1$ mol/L
$F_{A0} = 5$ mol/s
$UA = 5$ kJ/s dm^3/K
C_p (inert) $= 100$ J/(mol°C)
Heat of reaction: $-231 - 0.012$ $(T-298)$ kJ/mol
The specific heat of A: 0.122 kJ/(mol K)
Reaction rate constant: $k = 1.48 \times 10^{11}$ exp$(19{,}124/T)$ L/(mol s)
Inlet temperature: 675 K
Outlet temperature: 700 K

EP.11 The reaction $2A \xrightarrow{k} R$ occurs in gas phase, where:

$$k = 10.33 \exp\left[-\frac{E}{R}\left(\frac{1}{450} - \frac{1}{T}\right)\right] s^{-1}$$

The heat of the system obeys the following relationship:

$$\frac{Ua}{\rho} = 0.8 \text{ J/s kg K}$$

The flow of the heat exchanger is sufficiently high to maintain the temperature at 50°C. The component A (pure) is introduced at a rate of 5.42 mol/s and with a concentration of 0.27 mol/L. The catalyst is fed with reactant at 450 K and the coefficient of heat exchange between the gas and the catalyst is infinite. The heat capacity of the solid is 100 J/(kg K). The deactivation is given by the following equation:

$$k_d = 0.01 \exp\left[7000\left(\frac{1}{450} - \frac{1}{T}\right)\right] s^{-1}$$

Calculate the mass of catalyst that produces a greater conversion:

EP.12 A fuel cell system aims to produce 6.5×10^7 kW h/month. It is fed with hydrogen at a rate of 1.5×10^3 kg/h. This hydrogen is generated by steam reforming with the 15% of alumina-supported nickel catalyst ($MgAl_2O_4$). The main reactions are described below:

1   Steam-reforming:

$$CH_4 + H_2O \longleftrightarrow CO + 3H_2$$

2   Shift reaction:

$$CO + H_2O \longleftrightarrow CO_2 + H_2$$

3   Reverse reaction:

$$CO_2 + 3H_2 \longleftrightarrow CH_4 + H_2O$$

The rate expressions are known:

$$r_1 = \frac{k_1}{p_{H_2}^{2.5}} \left(p_{CH_4} \cdot p_{H_2O} - \frac{p_{H_2}^3 \cdot p_{CO}}{K_1} \right) / (DEN)^2$$

$$r_2 = \frac{k_2}{p_{H_2}} \left(p_{CO} \cdot p_{H_2O} - \frac{p_{H_2} \cdot p_{CO_2}}{K_2} \right) / (DEN)^2$$

$$r_3 = \frac{k_3}{p_{H_2}^{3.5}} \left(p_{CH_4} \cdot p_{H_2O}^2 - \frac{p_{H_2}^4 \cdot p_{CO_2}}{K_3} \right) / (DEN)^2$$

with

$$DEN = \left(1 + K_{CO}p_{CO} + K_{H_2}p_{H_2} + K_{CH_4}p_{CH_4} + K_{H_2O}\frac{p_{H_2}}{p_{H_2O}} \right)$$

The rate and equilibrium constant can be obtained as a function of the reference temperature (T_r):

$$k_i = k_{i,T_r} \exp\left[-\frac{E_i}{R} \left(\frac{1}{T} - \frac{1}{T_r} \right) \right] s^{-1}$$

$$K_j = K_{j,T_r} \exp\left[-\frac{\Delta H_j}{R} \left(\frac{1}{T} - \frac{1}{T_r} \right) \right]$$

where:
  i = reactions 1, 2, 3
  j = CO, H_2, CH_4, H_2O

The data in the table below were obtained by Xu and Froment.[8] (FONTE)

Reaction	$k_{i,Tr}$ $(T_r = 648\,K)$	E_i (kJ/mol)	ΔH_j (kJ/mol) at 948 K	ΔH_j (kJ/mol) at 298 K
1	1.842×10^{-4}	240.1	224	206.1
2	7.558	67.13	−37.3	−41.15
3	2.193×10^{-5}	243.9	187.5	164.9

$[k_1] = [k_3] = $ kmol bar$^{1/2}$/(kg € h) and $[k_2] = $ kmol/(kg h bar)

Molecule	K_{j},T_r	ΔH_j^0	T_r (K)
CO	40.91	−70.65	648
H_2	0.0296	−82.90	648
CH_4	0.1791	−38.28	823
H_2O	0.4152	88.68	823

$[K_{CH4}] = [k_{CO}] = [k_{H2}] = $ bar^{-1} and $[K_{H2O}] = $ dimensionless

Calculate the mass of catalyst required to obtain H_2 at 900°C. This reactor is fed with natural gas (98% CH_4) at a feed flow rate of 9.0×10^3 kg/h and excess steams, where H_2O/CH_4 is 3:1 at a total pressure of 10 bar.

Plot the composition along the reactor (based on the mass of the catalyst).

Do the same calculation for the temperatures of 600°C, 700°C, 800°C, and 1000°C.

Chapter 21

Multiphase reacting systems

The most catalytic or noncatalytic processes involving reactions in multiphase systems. Such processes include heat and mass transfer and other diffusion phenomena. The applications of these processes are diverse and its reactors have their own characteristics, which depends on the type of process. For example, the hydrogenation of vegetable oils is conducted in a liquid phase slurry bed reactor, where the catalyst is in suspension, the flow of gaseous hydrogen keeps the particles in suspension. This type of reaction occurs in the gas–liquid–solid interface.

As a second example, we have the Fischer–Tropsch synthesis. This process allows producing liquid hydrocarbons of high molecular mass with long carbon-chain molecules (such as gasoline, kerosene, diesel, and lubricant) from synthesis gas $(CO+H_2)$. This process takes place in slurry bed reactors, where the reactant gases flow through a liquid mixture inert.

The types of reactors commonly used in multiphase processes are represented in Figure 21.1.

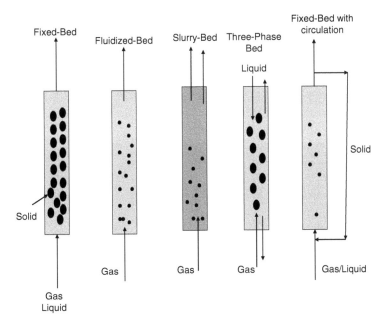

Figure 21.1 Examples of multiphase reactors.

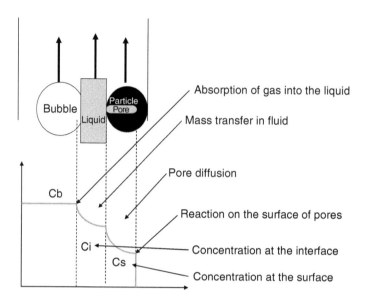

Figure 21.2 Concentration profiles in the different steps.

For fluidized bed and slurry bed, the reactant gases keep small particles in suspension and the reaction occurs in the gas–solid interface. In this set of reactors, the fluid cracking catalyst (FCC) is the most known. While the three-phase reactor is similar to the fixed bed, where gas and liquid act in concurrent or countercurrent flow, the chemical reaction occurs in the gas–liquid–solid interface. For example, the catalytic hydrotreating of heavy oils, which aims to the elimination of nitrogenous compounds, sulfur compounds, and oxygenates from petroleum. Among others, one can highlight the biological applications.

The multiphase reactions are characterized from the following mechanisms:

Gas absorption in the liquid phase as bubbles
Gas diffusion in the liquid phase
Diffusion from liquid to solid surface
Diffusion in porous systems
Reactions at the liquid–solid or gas–solid interface.

These steps are described in Figure 21.2.

1   Concentration in the gas–liquid interface:
In general, for low concentrations, then Henry's law applies.

$$C_i = H \times pi \tag{21.1}$$

where H is Henry's constant and i is the gas–liquid interface and p_i is the partial pressure.

2   Rate of mass transfer from a gas phase to liquid phase:

$$R_A = k_b a_b (C_i - C_b) \tag{21.2}$$

where:
C_i = concentration of A in the gas–liquid interface (mol/cm^3)
C_b = concentration of A in the gas phase or global concentration
$k_b a_b$ = represents the product of the mass transfer constant in the gas phase (global) and gas–liquid interface area (cm^3/s).

3   The rate of mass transfer of gas (A) in the liquid to the solid surface:

$$\dot{R}_A = k_l a_l m (C_b - C_s) \tag{21.3}$$

where:
C_s = concentration of A in the solid surface
C_b = concentration of A in the gas–liquid interface
k_l = mass transfer coefficient (cm/s)
a_l = external surface area of the particle at the liquid–solid interface (cm^2/g)
m = mass of solid particles (g).

4   Diffusion/reaction in porous catalysts:
     The reaction rate on the surface of the pores, by considering the diffusion and represented by the effectiveness factor is

$$(-r'_A) = (-r'_{As}) \times \eta \tag{21.4}$$

where:
η = effectiveness factor
$(-r'_{As})$ = intrinsic rate on the surface.

The rate per unit volume is:

$$\dot{R}_A = m \cdot \eta (-r'_{As}) \tag{21.5}$$

Considering an irreversible first-order reaction on the pores' surface, one obtain:

$$(-r'_{As}) = k'' \times C_{As} \tag{21.6}$$

Where
C_{As} (mol/L) and k'' (L/g · s)
In steady state, the transfer rates are equal.

$$\dot{R}_{A(\text{bubble})} = \dot{R}_{A(\text{liquid})} = \dot{R}_{A(\text{solid})}$$

Thus,

$$k_b a_b (C_i - C_b) = k_l a_l m (C_b - C_s) = m\eta \left(k'' C_{As}\right)$$

Rearranging the equations as a function of concentrations, we obtain:

$$\left.\begin{array}{l} \dfrac{\dot{R}_A}{k_b a_b} = (C_i - C_b) \\[2mm] \dfrac{\dot{R}_A}{k_l a_l m} = (C_b - C_s) \\[2mm] \dfrac{\dot{R}_A}{m\eta} = C_s \end{array}\right\} + \tag{21.7}$$

$$\overline{\left(\dfrac{1}{k_b a_b} + \dfrac{1}{m}\left(\dfrac{1}{k_l a_l} + \dfrac{1}{k''\eta}\right)\right) = \dfrac{C_i}{\dot{R}_A}}$$

Equation 21.7 includes the mass transfer, pore diffusion, and chemical reaction. Figure 21.3 allows observing the effect of each step, represented by Equation 21.7. As measure parameters, one has the inverse of the mass on the abscissa and the inverse of the global rate on the ordinate.

It is observed that the resistance to mass transfer in the gas–liquid interface is independent of the mass of the solid (case a). On the other hand, both the mass transfer in the liquid phase to the interface as diffusion and the reaction in the pores depend on the mass of the solid (case b) and particle size. The inverse of the rate $\dot{R}_A$ represents the resistance to flow mass in the reaction during the passage of reactant gas from the gas phase to the solid surface.

However, the interface area liquid/solid, a_l, and the effectiveness factor are dependent on the particle diameter. From Figure 21.3, it can be seen that the overall resistance varies linearly with the inverse of the solid mass. This demonstrates the dependence on particle diameter. Thus, the resistance to the rate decreases with the reduction of particle diameter (cases c and b).

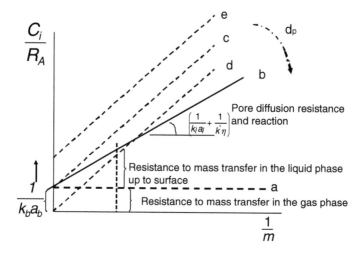

Figure 21.3 Profiles of mass resistance in each stage.

The case d shows that the resistance to mass transfer in the bubble is almost zero, this indicates that the size of gas bubbles is very small. On the other hand, the larger the bubble diameter greater the resistance to mass transfer (case e).

If we observe separately the effects of each step, we can obtain from Equation 21.7:

$$\frac{1}{\dot{R}_{lr}} = \left(\frac{1}{k_l a_l} + \frac{1}{k''\eta}\right) \tag{21.8}$$

If the particle diameter is small, the effectiveness factor is practically independent of Thiele modulus, i.e., $\eta = 1$, as we have seen earlier. In this case, the pore diffusion does not affect the rate and the resistance is due to the chemical reaction surface, which is the limiting step of the process. It is known that the product $k_l a_l$ depends on the particle diameter and diffusion, which are usually represented by Sherwood number. Thus,

$$Sh = \frac{k_l d_p}{D}$$

This means that if the particle diameter is small, the resistance to mass transfer in liquid phase becomes negligible. On the other hand, when analyzing the same expressions for particle diameters is larger, we see that the effectiveness factor decreases and the Sherwood number increases. Knowing that the effectiveness factor is inversely proportional to the Thiele modulus and Sherwood number is directly proportional to the diameter of the particle, we obtain such combined effects. Therefore:

For,

$$\eta = \frac{3}{\Phi}$$

Then,

$$\frac{1}{\dot{R}_r} = \left(\frac{1}{k''\eta}\right) \approx d_p \tag{21.9}$$

Note that with increasing the particle diameter the resistance increases significantly. Figure 21.4 shows a general behavior of the resistance as a function of particle

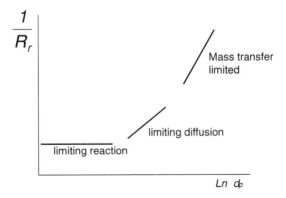

Figure 21.4 Limiting steps (Adapted from Fogler (2000)).

diameter. Small values of the diameters show that the reaction is independent, becoming a limiting step of the process. With increasing particle diameter, the diffusion becomes limiting. Thus, the mass transfer controls the system.

Example

E21.1 Hydrogenation of particular oil is performed in a liquid phase catalytic reactor (plug flow reactor (PFR)) containing catalytic particles (pellets—spherical diameters) of 1 cm. The external concentration is 1 kmol/L and on the particle surface is 0.1 kmol/L at a superficial velocity of 0.1 m/s. Verify if there are mass effects. There will be a change if the particle diameter is equal to 0.5 cm? Neglect the effects in diffusive pores (Fogler, 2000). Additional data:

$$\text{Viscosity kinematic} = 0.5 \times 10^{-6} \text{ m}^2/\text{s}$$
$$\text{Diffusivity} = 10^{-10} \text{ m}^2/\text{s}$$
$$Sh = 2 + 0.6\, Re^{0.5} Sc^{0.33}$$

where Re = Reynolds number, Sc = Schmidt number
Supface area = 100 m²/g

Solution
Rate of mass transfer: Equation 21.3

$$R_A = k_1 a_{ls}(C_b - C_s) \text{ (mol/L s)}$$

To calculate the mass transfer coefficient, we compute the Sherwood number from Re and Sc.

$$Sh = 2 + 0.6 Re^{0.5} Sc^{1/3} \tag{21.10}$$

$$Re_1 = \frac{d_p u}{v} = \frac{1 \times 10^{-2} \times 0.1}{0.5 \times 10^{-6}} = 2000$$

$$Sc = \frac{v}{D_e} = \frac{0.5 \times 10^{-6}}{1 \times 10^{-10}} = 5000$$

$$Sh = 2 + 0.6 Re^{0.5} \cdot Sc^{1/3} = 2.0 + 0.6 \cdot \sqrt{2000} \cdot 5000^{0.33} = 460$$

$$Sh = \frac{(k_1 a_1) d_p}{D_e} = 460$$

Then,

$$(k_1 a_1) = 460 \frac{D_e}{d_p} = 4.6 \times 10^{-6} \text{ (L/L s)}$$

The rate of reaction is:

$$\dot{R}_A = k_1 a_{ls} \times S_g(C_b - C_s) = 4.6 \times 10^{-6} \times 100 \times 0.9 \times 1000 = 4.14 \cdot 10^{-1} (\text{mol/L} \cdot \text{s})$$

Thus,

$$\dot{R}_A = 4.14 \cdot 10^{-1} \ (\text{mol/L} \cdot \text{s})$$

If the particle diameter is equal to 0.5 cm:

$$Re_l = \frac{d_p u}{\nu} = \frac{1 \times 10^{-1} \times 0.005}{0.5 \times 10^{-6}} = 1000$$

$$Sh = \frac{(k_l a_l) d_p}{D_e} = 325$$

$$(k_l a_l) = 6.5 \times 10^{-6} \ (\text{m/s})$$

$$R_A = 5.85 \times 10^{-1} \ (\text{mol/s})$$

E21.2 The hydrogenation of lanoline is conducted in a slurry bed reactor by using two types of catalyst, A and B. The reaction rates were measured at different mass functions, according to the table below. From these data, determine which catalyst that has the lowest effectiveness factor is. Using the catalyst A with 50 g/L, the reaction rate will increase if the gas distribution works best (Fogler, 2000)?

Catalyst	$S/(-r_A)$ (min)	$(1/m)$ (L/g)
C_1	4.2	0.01
C_2	7.5	0.02
C_3	1.5	0.01
C_4	2.5	0.03
C_5	3.0	0.04

where:
S is the solubility of H_2 in liquid (mol/L)
$(-r_A)$ is the rate of reaction of A (mol/L min).

Solution
From Equation 21.7 and considering a first-order reaction, we obtain:

$$\left(\frac{1}{k_b a_b} + \frac{1}{m} \left(\frac{1}{k_l a_l} + \frac{1}{k'' \eta} \right) \right) = \frac{S}{(-r_A)} \tag{21.11}$$

And defining,

$$r_b = \frac{1}{k_b a_b}$$

is the resistance to absorption in the gaseous phase;

$$r_1 = \frac{1}{k_1 a_1}$$

is the resistance to mass transfer in the liquid specific up to surface;

$$r_r = \frac{1}{k'\eta}$$

is the specific resistance to diffusion in the pores of the particle.

$$r_{lr} = r_1 + r_r$$

$r_{ep} = r_e + r_p$, it is specific resistance combined from internal and external diffusion:
From Equation 21.11, we plot

$$\frac{S}{(-r_A)}$$

as function of the inverse of mass. The slope will be the resistance combined at internal and external diffusion. The ratio between the resistance to gas absorption and the resistance to diffusion into the particles and subsequent reaction for mass m will be:

$$\frac{r_b}{r_{ep}} = \frac{Linear\ coeficient}{Angular\ coeficient} \qquad (21.12)$$

As the adsorption in the gas phase is independent of particle size, the intersection remains constant. With decreasing particle size, both the effectiveness factor and the coefficient of mass transfer increase. Consequently, the combined resistance to mass transfer and diffusion decrease (r_{ep}). This is shown in Figure 21.5 for both cases.
The angular coefficient for the experiments (C_1 and C_2) is 330 and the linear coefficient is equal to 0.9. Whereas for the second set of experiments (C_3, C_4, and C_5), the slope is equal to 50 and the linear coefficient is 1.
Thus,

$$r_{ep} = r_1 + r_r = 330 \text{ for } C_1 \text{ and } C_2$$

and,

$$r_{ep} = 50 \text{ for } C_3, C_4, \text{ and } C_5$$

As angular coefficient for the set C_1 and C_2 is six times larger than C_3, C_4, and C_5, further the angular coefficient inversely proportional to the effectiveness factor, it is concluded that the set, C_1 and C_2, has a lower effectiveness factor.

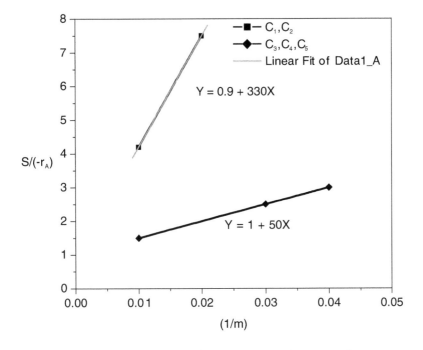

Figure 21.5 The influence of reaction rates and mass transfer.

From Equation 21.12,

(a)   For the group C_1 and C_2:

$$\frac{r_b}{r_{ep}} = \frac{0.9}{330} = 2.72 \times 10^{-3}$$

(b)   For the group C_3, C_4, and C_5:

$$\frac{r_b}{r_{ep}} = \frac{1.0}{50} = 2.0 \times 10^{-2}$$

It is concluded that the resistance to diffusion and reaction to both sets are much greater than the resistance to absorption.
Since,

$$r_b = \frac{1}{k_b a_b}$$

Thus, r_b increases for decreasing a_b.

E21.3 The hydrogenation of a diol is performed in a slurry bed reactor by adding 0.500 g of catalyst powder into reactor and the initial concentration of the diol is 2.5 mol/cm^3. The reaction is first order with respect to both H$_2$ and diol H$_2$ is bubbled through distributor at 1 atm and 35°C. The concentration of H$_2$ in equilibrium conditions is 0.01 mol/cm^3 and the constant specific reaction rate is 4.8×10^{-5} (cm^3/(cm$^2 \times$ s)). The flow rate of catalyst is 0.1 kg/m^3, the particle diameter is 0.01 cm, and its density is equal to 1.5 g/cm^3. Calculate the overall reaction rate. The pore diffusion can be neglected? Verify if the mass transfer in the pores and the external transfer can be negligible (Adapted from Fogler, 2000).

Data: $k_b a_b = 0.3$ cm^3/s and $k_l = 0.005$ cm/s, $D_e = 10^{-5}$ cm^2/s, $S_g = 25$ m^2/g

Solution
Calculate the area of the particles

$$a_{l_c} = \frac{area}{mass} = \frac{6}{d_p \rho_p} = \frac{6}{0,01.1.5} = 400 \, \text{cm}^2/\text{g}$$

Then,

$$k_l \cdot a_{lc} = 0.005 \times 400 = 2.0 (\text{cm}^3/\text{g} \cdot \text{s})$$

Using equation 18.24 we calculate the effectivess factor:

$$\eta = \frac{3}{\Phi} = \frac{6}{d_p} \sqrt{\frac{D_e}{k' S_g \rho_s}} = \frac{6}{0,01} \sqrt{\frac{10^{-5}}{4.8 \times 10^{-5} \times 25 \times 10^4 \times 1.5}} = 0.447$$

There are strong diffusion effects
   Thus,

$$k'' \cdot \eta = 4.8 \times 10^{-5} \times 25 \times 10^4 \times 0.447 = 5.36 (\text{cm}^3/\text{g} \cdot \text{s})$$

To calculate the overall rate, we use Equation 21.7:

$$\left(\frac{1}{k_b a_b} + \frac{1}{m} \left(\frac{1}{k_l a_l} + \frac{1}{k'' \eta} \right) \right) = \frac{C_i}{R_A}$$

And substituting the values into the above equation, we obtain:

$$\left(\frac{1}{0.3} + \frac{1}{m} \left(\frac{1}{2.0} + \frac{1}{5.36} \right) \right) = \frac{C_i}{R_A}$$

Thus,

$$\left(3.3 + \frac{1}{m} \times 0.68 \right) = \frac{C_i}{R_A}$$

For the mass of 0.5 g, we get:

$$a = 3.3$$

$$b = 0.68 \, s$$

Therefore,

$$\frac{C_i}{\dot{R}_A} = 4.66$$

And

$$\dot{R}_A = \frac{C_i}{4.66} = \frac{0.01}{4.66} = 2.14 \cdot 10^{-3} (\text{mol/cm}^3 \cdot s)$$

E21.4 The hydrogenation of $CH_3–CO–CH_3$ is conducted in a slurry bed reactor by using a Raney–Ni catalyst with a particle diameter of 0.5 mm. Hydrogen (pure) has a flow rate of 0.04 m³/s and the liquid into the reactor at a flow rate of 10^{-2} m³/s. Calculate the mass of catalyst to a conversion of 50%. The rate expression is known. Additional details are provided below.

Data:

Henry's constant of the $H_2 = 0.37$ (atm m³/kmol)
    Liquid concentration = 1000 kmol/m³

$$H_2 + CH_3 - CO - CH_3 \rightarrow CH_3 - CHOH - CH_3$$

$$(A + B \rightarrow R)$$

$$(-r_A'') = r_B = k'' C_{H_2}^{0.5} \ (\text{kmol/(kg s)})$$

where:

$$k'' = 10 \left(\frac{m^3}{kg\,s}\right) \left(\frac{kmol}{m^3}\right)^{0.5}$$

Diffusion coefficients:

$$D_e = 8 \cdot 10^{-10} \, m^2/s$$

Area: $A = 0.01 \, m^2$
Bed height $= 5.0 \, m$

Solution

Initially, calculate the concentration of H_2 to verify the condition of thermodynamics. From Henry's law:

$$C_{A(H_2)} = \frac{P_A}{H_A} = \frac{1}{0.37} = 2.7 \, (\text{kmol/m}^3)$$

It is observed that the concentration of the liquid (B) is much larger than H_2. Therefore, the fluid is very dilute and the rate depends on the concentration of H_2.

From Equation 21.7, we get:

$$\dot{R}''_A = \frac{C_A}{\left(\frac{1}{k_b a_b} + \frac{1}{m} \left(\frac{1}{k_l a_l} + \frac{1}{k'' \eta} \right) \right)}$$

The continuous stirred-tank reactor (CSTR) reactor is used to calculate the volume of the reactor. Thus,

$$\tau = C_{A0} \frac{X_A}{R_A}$$

We need to determine the different parameters of mass transfer and diffusion.

- Calculation of the effectiveness factor, before we determine the Thiele modulus. As the reaction order is 0.5, we obtain:

$$\Phi_n = R \sqrt{\frac{k'' \rho_s C_A^{n-1}}{D_e}} = 5 \times 10^4 \sqrt{\frac{10 \times 0.6 \times (2.7)^{-0.5}}{8 \times 10^{-10}}} = 33.7$$

$$\eta = \left(\frac{2}{n+1} \right)^{1/2} \times \frac{3}{\Phi_n} = \sqrt{\frac{2}{1.5}} \times \frac{3}{33.7} = 0.102$$

Therefore, there is a strong diffusion effect.

Calculation of k_l is conducted from the Sherwood number. We use an empirical relation known, such that:

$Sh = 0.6 Re^{0.5} Sc^{1/3}$

$Re = $ Reynolds number

$Sc = $ Schmidt number.

But,

$$u_l = \frac{\bar{q}_l}{A} = \frac{0.01}{0.01} = 1 \text{ m/s}$$

where:

$u_l = $ superficial velocity

$\bar{q}_l = $ flow of the liquid phase (m^3/s).

Thus,

$$Re_l = \frac{d_p u}{\nu} = \frac{5 \times 10^{-4} \times 1.0}{0.5 \times 10^{-8}} = 10^5$$

$$Sc = \frac{\nu}{D_e} = \frac{0.5 \times 10^{-8}}{8 \times 10^{-10}} = 6.25$$

$$Sh = 0.6 Re^{0.5} Sc^{1/3} = 0.6 \times \sqrt{10^5} \times 6.25^{0.33} = 349$$

We find the Sherwood number:

$$Sh = \frac{k_l d_p}{D_e} = 349$$

Thus

$$k_l = 5.54 \times 10^{-4} (m/m^3 \, s)$$

We do not know the interfacial area. Furthermore, it is difficult to determine the interfacial area. Then, we can assume a value $a_l = 1000 \, m^2/kg$:
    Thus,

$$k_l a_l = 5.54 \times 10^{-1} (m^3/kg \, m^3 \, s)$$

It is assumed that the mass transfer in the gas phase is negligible,

$$k_b a_b = \text{large value}$$

Thus, we calculate the overall reaction rate,

$$\dot{R}_A'' = \frac{C_A}{\left(\frac{1}{k_b a_b} + \frac{1}{m} \left(\frac{1}{k_l a_l} + \frac{1}{k'' \eta} \right) \right)}$$

and,

$$\left(\frac{1}{m} \left(\frac{1}{k_l a_l} + \frac{1}{k'' \eta} \right) \right) = \frac{1}{m} \left(\frac{1}{5.54 \times 10^{-1}} + \frac{1}{10 \times 0.102} \right) = \frac{1}{m} (1.8 + 0.98) = \frac{2.77}{m}$$

For CSTR reactor,

$$\tau = C_{A0} \frac{X_A}{\dot{R}_A}$$

The total flow is $\overline{q_0} = \overline{q_l} + \overline{q_g} = 0.05 \, m^3/s$
    For the volume, $V_r = A.h = 0.01.5 = 0.05 \, m^3$ One obtain the space time as $\tau = 1 \, s$
with $X_A = 0.5$

We get,

$$\tau = 1 = C_{A0} \frac{X_A}{\overline{R}_A} = \frac{2.7 \times 0.5}{2.77/\text{m}} = 0.487 \, \text{m}$$

Therefore,

$$m = 2.0 \, \text{Kg}$$

Chapter 22

Heterogeneous reactors

Packed bed reactors can be presented in different ways, such as fixed bed, fluid bed, and slurry phase. The reaction is characterized as the presence of a heterogeneous catalyst, which may be fixed or not. The most processes in the petrochemical industry are heterogeneous. There are several examples in petrochemical industry, such as methane reforming; synthesis of ammonia from nitrogen and hydrogen, synthesis of methanol from syngas, and catalytic cracking to produce gasoline.

To evaluate reaction rate of gas–solid systems, one must know the solid surface properties. Moreover, the rate constant is the most important parameter for a correct decision about what type of system will be adopted on the problem.

The solid particles in the bed always have a degree of porosity, regardless of whether the solid is a catalyst or not. The pores contained in each particle may have different characteristics (size, etc.). Thus, the diffusion of molecules on the solid surface is very important, which has various transport mechanisms, such as intraparticle diffusion, Knudsen diffusion, or surface diffusion. The latter, for example, depends on the surface characteristics, such as high or low surface area. Typically, one determines the effective diffusion encompassing both characteristics.

The apparent rate constant, k_a, depends also on the properties of the solid as well as diffusion and/or mass transfer, in addition to temperature and concentration. The rate can be calculated as rate per unit volume, rate per unit area, or rate per unit mass. For example, for a first-order reaction, the rate per unit volume is:

$$r = (\text{mol}/(\text{L s})) = kC_A$$

The ratios of rates for different conditions and for the same reaction order (first order) is:
rate/volume:

$$rV = (\text{mol}/\text{L s})L = kC_A V \text{ with } k \text{ (s}^{-1})$$

rate/mass:

$$r''W = (\text{mol}/\text{L s})L = k''C_A W \text{ with } k''(\text{Lgas}/\text{g s}) \text{ and } W(\text{g}_{solid})$$

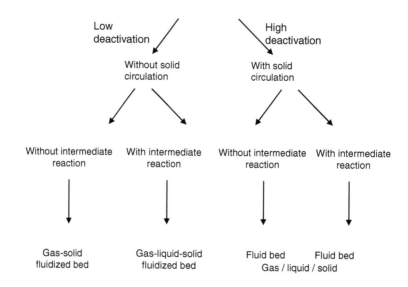

Figure 22.1 Classification of reactors according to rate constant (Adapted from Levenspiel (2001)).

Therefore, for any system:

$$rV = r''W = r'S = r'''V_R$$
$$kV = k''W = k'S = k'''V_R$$
$$s^{-1} = L/gs^{-1} = L/m^2\ s^{-1} = L/m^3_{\text{reactor}}\ s^{-1}$$

The scheme shown in Figure 22.1 allows us to classify reactors according to the magnitude of the rate expression. Assuming a rate that depends on the rate constant k' for a first-order reaction. For reactions such as gas–solid and/or gas–liquid–solid, they can be classified as follows.

For a specific speed k' the next cases [4] have been:

k' small $\rightarrow$ big particles $\rightarrow$ fixed bed
k' big $\rightarrow$ small particles $\rightarrow$ it cannot be fixed bed

   In the solid particles (catalytic or noncatalytic reactions), the main factors that affect the reaction rate are:

Chemical reaction surface
Resistance to pore diffusion
Resistance to mass transfer in the film surrounding the particle
Temperature gradient in the particle
Temperature gradient in the film.

There are many different combinations that may be established, depending on the solid and the fluid, according to the following table (Table 22.1).

Both diffusion coefficients and mass transfer are important, but they depend on the different solids, drainage (flow), and particles porosity. The effective diffusion involves Knudsen and convective diffusion, which depends on the phase of fluid (gas or liquid) and pore size (large or small). These coefficients are characterized by Peclet number (Pe), which depends on the axial or radial dispersion and diffusivity. Depending on the velocity profile, these coefficients can vary radially or axially. The diffusion and dispersion coefficients can also vary due to its dependence on the radial position. If the coefficients vary along the reactor, as in heterogeneous reactors, for example, the velocity is not constant. Thus, the axial dispersion occurs.

There are several correlations that represent how the dispersion varies with Reynolds number in the radial and axial direction, as shown in Figure 22.2.

In this chapter, the Reynolds number is defined in terms of equivalent diameter, i.e.:

$$Pe = \frac{\bar{u}d_p}{\varepsilon_D} \text{ and } Re = \frac{d_e\bar{u}}{\nu}$$

Table 22.1 Different combinations from solid particles

	Porous Solid	Catalyst Anchored	Drops	Living Cells
Chemical reaction surface	Yes	Yes	No	Yes
Pore diffusion	Yes	No	No	Occasionally
Temperature gradient in the particle	Not much	No	No	No
Temperature gradient in the film	Occasionally	Rare	Yes	No
Mass transfer in the film	No	Yes	Yes	Possible

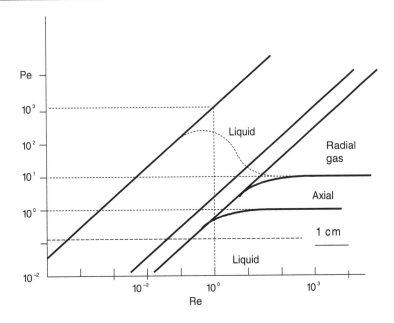

Figure 22.2 Radial and axial dispersion in liquids and gases (Adapted from Levenspiel (2001)).

where:

$$d_e = \frac{d_t}{3/2(d_{t/d_p})(1 - \varepsilon) + 1} \qquad (22.1)$$

$\varepsilon_D =$ dispersion due to diffusion
$\varepsilon =$ bed porosity
d_t and $d_p =$ tube diameter and particle diameter.

For $Re > 10$, it is observed that the Pe (Pe_r or Pe_z) is independent of flow. With Pe_r up to 10 and Pe_z is ~ 1.0 for gases and liquids.

However, for $Re < 1$, Pe decreases linearly for gases, indicating a molecular diffusion. For flows in liquid phase, the behavior is different. Since for low Re, the radial and axial variation is greater.

In particulate systems, the order of magnitude of the radial and axial effects is not uniform. Some hypotheses can be made for simplicity, neglecting the radial and axial effects, approaching them to the ideal conditions.

The first condition is to choose a Reynolds number such that flow be homogeneous and uniform radially and axially, i.e., with a good dispersion. The radial gradients may be insignificant when the ratio of the tube diameter and the particle diameter is large. This makes the difference between the velocity at any position and the velocity of a fluid practically negligible.

It is recommended that:

$$\frac{d_t}{d_p} > (10 \rightarrow 30) \qquad (22.2)$$

By the correlation:

$$Pe_r = \left[19,4 \left(\frac{d_t}{d_p} \right)^2 \right]^{-1} \qquad (22.3)$$

One must calculate the term d_t/d_p for conditions with number of $Pe_r \rightarrow 0$. There is no radial concentration gradient and the radial velocity is uniform in the same direction. However, in adiabatic systems or with a radial temperature variation, the radial concentration range changes. Thus, the radial dispersion cannot be neglected.

For systems with $Re \geq 10$ (i.e., $Pe_z \sim 2$), the axial dispersion cannot be neglected. That dispersion depends not only on flow, but also from bed concentration gradients along certain directions. Under isothermal conditions and steady state, it was established that beyond reason:

$$\frac{d_t}{d_p} > (10 \rightarrow 30)$$

It must satisfy the ratio between length and the catalytic bed diameter (or particle diameter). This ratio must be larger than 50 (Levenspiel, 2001).

Thus:

$$\frac{L}{d} > 50 \tag{22.4}$$

Both axial and radial dispersions can be neglected for differential reactor operating at low conversions.

In summary:

$$\text{For } \frac{d_t}{d_p} > (10 \rightarrow 30)$$

and under isothermal conditions, the radial dispersion is negligible, but the axial dispersion is not. Therefore, the basic equation is:

$$u\frac{dC_A}{dz} = \varepsilon_z^* \frac{d^2 C_A}{dz^2} + r \tag{22.5}$$

where r is reaction rate (mol l^{-1} s^{-1}) while ε_z^* is axial dispersion.

$$\text{For } \frac{L}{d} > 50$$

one can neglect the axial dispersion. Therefore, considering an ideal PFR reactor:

$$u\frac{dC_A}{dz} = -r_A \tag{22.6}$$

22.1   FIXED BED REACTOR

In the design of a fixed bed reactor, it is necessary to know the rate of reaction encompassing mass and diffusion effects. These effects on the reaction rate can be represented by the effectiveness factor η, with pore diffusion, besides the effects of mass. This will be represented by an overall rate r'' (mol mass^{-1} h^{-1}) or r' (mol/area^{-1} h^{-1}).

There are different types of fixed bed reactors, the most known are:

Adiabatic or isothermal reactor without axial or radial dispersion—ideal plug flow reactor (PFR) reactor
Adiabatic reactor with radial dispersion and chemical reaction
Adiabatic reactor with axial dispersion and chemical reaction
Not isothermal reactor with axial dispersion.

In some cases, it is assumed a pseudohomogeneous. That is, considering negligible the radial and axial effects. In other cases, the reaction can occur in the tube wall (monolith).

Besides chemical reaction, it is very important to know the flow even in non-ideal reactors, determining the pressure drop in the reactor, this latter is given by the following equation

$$\frac{dP}{dz} = 2\frac{fu_s^2 \rho_g}{gd_p} \tag{22.7}$$

where:
u_s = superficial velocity
d_p = particle diameter
ρ_g = gas density
f = friction factor.

From the friction factor, one can find several correlations in the literature on this variable. We can use the Ergun's equation (Froment and Bischoff, 1979), which is usually practiced in the literature:

$$f = 6.8\frac{(1 - \varepsilon)^{1.2}}{\varepsilon^3} Re^{-0.2} \tag{22.8}$$

The above equation is valid for $Re/(1 - \varepsilon) < 500$, where ε is the bed porosity.

There are other correlations and they can be found in the literature (Froment and Bischoff, 1979).

Let us consider a special case of adiabatic reactor without dispersion. The reactor can be considered as an ideal nonisothermal reactor. The heat is generated due to chemical reaction, i.e., the process is exothermic.

A reversible first-order reaction involves mass and/or diffusive effects. Thus, the resulting rate for $A \underset{k_{-1}}{\overset{k_1}{\rightleftharpoons}} R$ is:

$$r = k_1 C_A - k_{-1} C_R \tag{22.9}$$

With k_1 representing a specific rate constant for products and k_{-1} for reactants. Rearranging this equation as a function of conversion X_A, one obtains in dimensionless form:

$$r = k_1 C_{A0}\left[1 - \frac{X_A}{X_{Ae}}\right] \tag{22.10}$$

X_{Ae} is the equilibrium conversion. As initial guess, we started with the pure reagent, X, i.e., $R = 0$.

At equilibrium, the resulting rate is null and we can determine the equilibrium conversion from Equations 22.9 and 22.10. And the equilibrium constant at a constant temperature is (see Equation 3.20):

$$K = \frac{k_1}{k_{-1}} = \frac{X_{Ae}}{1 - X_{Ae}} \tag{22.11}$$

From the mass balance for this system, we have:

$$u \frac{dC_A}{dz} = r \tag{22.12}$$

We can also represent the previous equation in dimensionless form:

$$\frac{\partial X_A}{\partial \xi} = \frac{kL}{u*} \left(1 - \frac{X_A}{X_{Ae}} \right) \tag{22.13}$$

whose dimensionless variables are:

$$\xi = \frac{z}{L} \quad \text{and} \quad u* = \frac{u}{\bar{u}}$$

$\bar{u}$ is the average superficial velocity.

With the following boundary conditions:

$$\xi = 0 \quad X_A = 0$$

Using these boundary conditions and integrating Equation 22.13, one obtains:

$$\tau = \int_0^{X_A} \frac{dX_A}{k_1 \left(1 - \frac{X_A}{X_{Ae}} \right)} \tag{22.14}$$

Remembering that:

$$\tau = \frac{L}{4\bar{u}} = \frac{V}{v_0} (\text{min})$$

The rate constant forward (k_1) is related to the reverse (k_{-1}) through the equilibrium constant or by equilibrium conversion, according to Equation 22.11. In adiabatic system, the constant depends on temperature and therefore by the Arrhenius equation.

$$k = k_0\, e^{-(E/RT)} = k_0\, e^{-(\gamma/(T/T_0))} \tag{22.15}$$

where:

$$\gamma = \frac{E}{RT_0} \tag{22.16}$$

is the Arrhenius' parameter.

The energy balance for an adiabatic reactor is:

$$\rho C_P \bar{u} \frac{dT}{dz} = (-\Delta H_R)\, r \tag{22.17}$$

From the condition $z = 0$, the inlet temperature is T_0.

Equation 22.17 can be rearranged to dimensionless form, then integrating it between T_0 and T, we get (see Equation 14.83):

$$\frac{T}{T_0} = 1 + \bar{\beta} X_A \tag{22.18}$$

where:

$$\bar{\beta} = \frac{(-\Delta H_R).C_{A0}}{\rho \cdot C_P \cdot T_0} \tag{22.19}$$

is the energy's parameter.

Thus, it is concluded that the solution depends on the parameters $\bar{\beta}$ and γ, beyond the equilibrium conversion. The influence of these parameters on the conversion and the reactor volume is shown qualitatively in Figure 22.3 (a–d). Figure 22.3a shows how

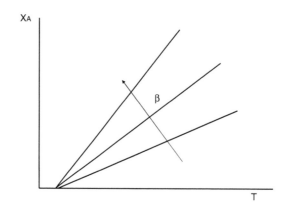

Figure 22.3a Influence of the parameter β on the conversion as a function of temperature (Adapted from Froment and Bischoff (1979)).

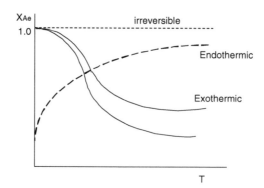

Figure 22.3b Influence of the enthalpy of reaction on the conversion of equilibrium as a function of temperature.

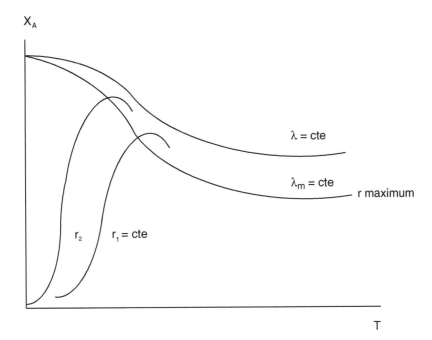

Figure 22.3c Influence of the reaction rate on the conversion as a function of temperature (Adapted from Coulson and Richardson (1971)).

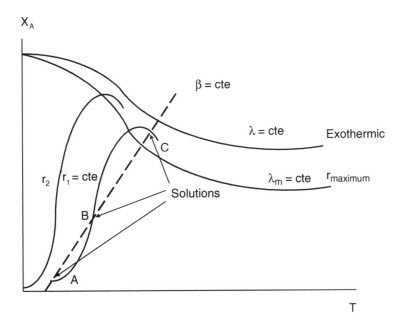

Figure 22.3d Different solutions conversions and temperatures for the adiabatic reactor (Adapted from Coulson and Richardson (1971)).

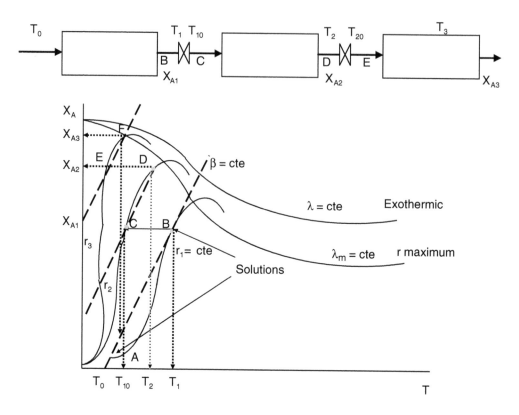

Figure 22.4 Adiabatic reactors in series.

the conversion varies with temperature from the parameter $\bar{\beta}$. Figure 22.3b shows how the conversion of equilibrium changes with temperature. And, finally, Figure 22.3c and d demonstrates how the conversion varies with temperature for different values of rate constants and the solutions to different temperatures, respectively.

22.1.1   Reactors in series

In adiabatic reactors in series, each stage of cooling intermediates can be used from the knowledge of the rate expression as a function of conversion and temperature. The main objective is achieving higher conversions at lower temperatures in the output of the reactor. As an example for a series of three reactors, the temperature T_0 represents the initial value of the first reactor at a rate constant $\bar{\beta}$ (Figure 22.4). At the intersection of straight line (Equation 22.18) and the curve r_1, we can reach "B," temperature T_1 and conversion X_{A1}. At the exit of the first reactor, cooling the mixture to a temperature smaller T_{10} in C. Assuming the same rate β and with the intersection of the line and the curve r_2, one can find a temperature smaller T_2 and a larger conversion D, as can be seen in Figure 22.4.

Thus, one can construct a system with three reactors in series, achieving high conversion efficiencies from smaller values of temperatures.

Example

E22.1 (Coulson and Richardson, 1971) According W.J. Thomas, the production of SO_3 from SO_2 is carried out adiabatically. The composition of the mixture at the inlet of the reactor is: 7% of SO_2, 11% of O_2, and 81% of N_2. The inlet temperature in the first reactor and the molar flow are 400°C and 170 mol/min, respectively. Calculate the mass of catalyst to a conversion of 79% at the outlet of the reactor.
Data:

   Density of catalyst: $2 \, g/cm^3$
   Pressure: 1 atm
   Enthalpy of reaction: $\Delta H_R = -23.290 \, kcal/kmol$

Component	Average Specific Heat (kcal/kg°C)	MW
SO_2	0.190	80
SO_3	0.226	64
O_2	0.246	32
N_2	0.240	28

Solution
Average molecular weight:

$$\overline{M} = 0.07 \times 64 + 0.11 \times 32 + 0.01 \times 80 + 0.81 \times 28 = 31.48$$

Average specific heat:

$$\overline{C_P} = 0.07 \times 0.190 + 0.11 \times 0.246 + 0.01 \times 0.240 + 0.81 \times 0.226$$
$$= 0.239 \, kcal/kg$$

Initial concentration:

$$C_{A0} = \frac{y_{A0} \, P}{RT} = 0.00126 \, kmol/m^3$$

Density:

$$\overline{\rho} = \frac{P\overline{M}}{RT} = 0.570 \, kg/m^3$$

Parameter:

$$\beta = \frac{-\Delta H_R \, C_{A0}}{\rho C_P T_0} = 0.537$$

Therefore, the temperature varies linearly with conversion:

$$\frac{T}{T_0} = 1 + \beta X_A = 1 + 0.537 X_A$$

For $X_A = 0$, we have $T_0 = 400°C$.
The rate is known and it obeys the Langmuir–Hinshelwood equation:

$$r = \frac{k_1 P_o P_{SO_2} \left(1 - \frac{P_{SO_3}}{K_p P_{SO_2} P_{O_2}^{1/2}}\right)}{22 \times 4 \left(1 + K_2 P_{SO_2} + K_3 P_{SO_3}\right)}$$

where:
$k_1 = \exp(12{,}160 - 5{,}473/T)$
$K_2 = \exp(-9{,}953 + 86{,}191/T)$
$K_3 = \exp(-71{,}745 + 52{,}596/T)$
$K_p = \exp(11{,}300 - 10.68/T)$

To calculate the mass of the catalyst, we use the PFR model, assuming a pseudohomogeneous rate and without axial or radial dispersion effects. Thus:

$$W = y_{A0} F_t \int_0^{0.78} \frac{dX_A}{r} = 0.07 \times 0.170 \times 60 \int_0^{0.78} \frac{dX_A}{r}$$

$$W = 0.714 \int_0^{0.78} \frac{dX_A}{f(X_A)}$$

By calculating the rates, we present the results in the table below:

$T(°C)$	X_A	$r(kmol\ kg^{-1}\ h^{-1})$	(r^{-1})
400	0	0.005	200
423	0.1	0.0075	133
435	0.16	0.01	100
450	0.24	0.015	66
465	0.3	0.02	50
500	0.45	0.03	33
545	0.65	0.03	33
560	0.73	0.02	50
565	0.76	0.015	66
570	0.79	0.005	200

From this table, we obtain Figure 22.5.

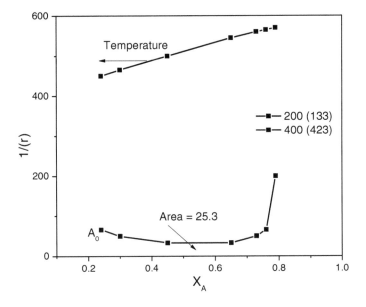

Figure 22.5 Rate and temperature as a function of conversion.

$A_t = 25.3$ kg h/kmol
Therefore:

$W = 0.714 \times 25.3 = 18.0\,\text{kg}$

22.2 FLUIDIZED BED REACTOR

A typical fluidized bed reactor is characterized by an ascending gas (or liquid by flu-idizing agents) flow, which keeps or drags small particles located in the reactor. The reactant gas in contact with the solid particles of the bed facilitates the reaction because of the high flow velocity and short contact time. However, the velocity of flow may vary according to the specifications of each process.

These effects are important and facilitate the transport of heat and mass transfer, in most cases. In general, particles are dragged and they can recirculate within the bed by a process called regeneration. For example, process for catalytic cracking of petroleum.

The advantages of the fluidized bed are:

- Surface area is large, because the particles can be very small, which facilitates heat transfer and mass;
- High reaction rates if compared to fixed bed reactors;
- Increased heat transfer coefficients and mass;

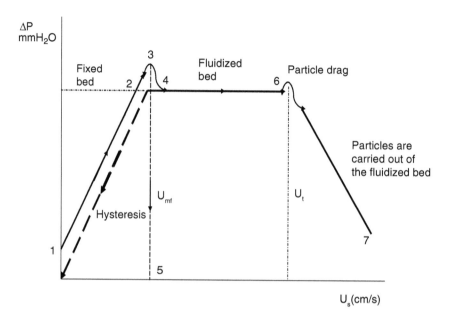

Figure 22.6 Pressure drop versus superficial velocity.

- Easy to control both heat transfer and fluid flow system;
- Easiness in flow in pipelines because the solid particles behave like fluid.

The disadvantages are:

- Backmixing due to particle distribution in dense and dilute phases
- The calculation of the average residence time is too hard, it was not possible to predetermine the position of the particle in the bed
- Possible channeling, slugging, and attrition of catalyst
- Energy consumption due to high pressure drop (requires high fluid velocity)
- Possible agglomeration and sintering of fine particles in the dilute phase under certain conditions (e.g., high temperature).

The starting point to determine the hydrodynamic behavior of the fluidized bed is to define if the particles remain static or dragged by gas flow. To verify and establish the flow regime, one must measure the pressure drop as a function of the superficial velocity, as shown in Figure 22.6.

The pressure drop in the bed is proportional to the superficial velocity (1–2), whose variation is characteristic of the fixed bed and reaches a maximum value in (3). From this point, fluidization of the particles begins. At this instant, the minimum fluidization velocity u_{mf} is identified, maintaining constant pressure drop in the fluidized bed (4–6). Then, particles begin to be dragged out (6) of the catalyst bed, whose pressure drop decreases until it reaches the point (7).

The pressure drop can be determined directly, according to the equation below:

$$\Delta p = L_{mf}(1 - \varepsilon_{mf})(\rho_s - \rho) \tag{22.20}$$

where:
L_{mf} =bed height at minimum fluidization
ε_{mf} =void fraction of the bed at minimum fluidization
ρ_s =solid density
ρ_g =gas density.

It can be observed that it is possible to calculate the height corresponding to the minimum fluidizing condition. From the knowledge of the values of solid density, gas density, and void fraction, one can easily determine the pressure drop.

The minimum fluidization velocity, u_{mf}, can be determined directly from experimental data (3), as shown in Figure 22.6. And, similarly, one can determine the terminal velocity, u_t, or drag velocity, u_t (7). There are other empirical correlations for these velocities (Kunii and Levenspiel, 1991).

$$u_{mf} = 1.118 \times 10^{-13} \frac{d_p^{1.82} (\rho_s - \rho)^{0.94}}{\rho^{0.06} \mu^{0.88}} \tag{22.21}$$

where:
d_p =particle diameter (μm)
μ =dynamic viscosity (N s/m^2).

As the superficial velocity increases, the bed becomes less dense where the particles are dragged and distributed along the bed. Thus, the terminal velocity may be calculated from classical empirical equation:

$$u_t = \sqrt{\frac{4d_p g (\rho_s - \rho)}{3\rho \cdot C_D}} \tag{22.22}$$

C_D is called as drag coefficient. For a laminar flow, we can calculate it from equation below:

$$C_D = \frac{24}{Re}$$

where:

$$Re = \frac{d_p \rho u_t}{\mu} \tag{22.23}$$

To obtain a proper fluidization, the actual fluid velocity, u_{fl}, must be considerably greater than the minimum fluidization velocity, u_{mf}. However, to avoid excessive entrainment, u_{fl} should be less than the terminal velocity, u_t. Thus, the ratio u_t/u_{mf} is a guide to selection of the value of u_{fl}.

However, one must consider that the hydrodynamics of the fluidized bed undergoes some deviations, as described below.

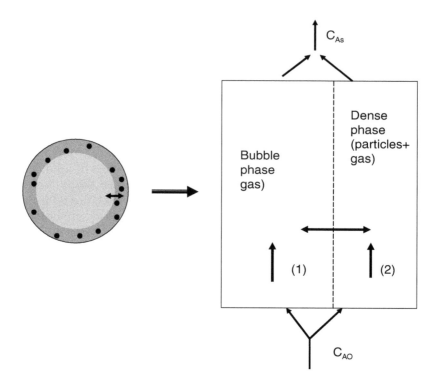

Figure 22.7 Description model of a fluidized bed (Adapted from Froment and Bischoff (1979)).

- When the diameter of the bubble (gas) has a value of the same order of magnitude as the diameter of the tube, the fluidized bed will operate in slugging regime, where the bubble occupies the entire cross section of the bed. Depending on the size of the bubble, we might have preferred paths in bed. These cases do not contribute to a good contact of the gas with the suspended particles in the bed, therefore, they should be avoided. The most suitable solution is the choice of a particular type of gas distributor in the inlet of the reactor.
- In the fluidized bed, the effects of mass transfer and heat are benefited, which may allow a homogenization of the bed temperature and facilitate contact of the reagent with the solid particles.

Then, it can be affirmed that the fluidized bed does not behave as a reactor ideal. That is, their behavior differs of the flow in PFR or continuous stirred-tank reactor (CSTR) ideals. These particulars do not cover the scope of this chapter; hence they will not be studied here.

The flow of gas bubbles in contact with the particulate phase results in a dense phase or phase emulsion. The bubble is surrounded by a "dense cloud" of particles, where there is the exchange of flows with heat and mass transfer and chemical reaction. Figure 22.7 shows this phenomenon.

In the gas phase (bubbles) (1), the velocity is high and there is no chemical reaction. In this case, we can admit a flow identical those used in PFR.

While in "phase emulsion" (2), the flow is not homogeneous. Therefore, their behavior undergoes large deviations from ideal behavior of both PFR and CSTR. Therefore, diffusion effects can occur, as well as preferred paths.

The main problem is the lack of knowledge about the interface between phases (1) and (2) where the exchange phenomenon occurs. Therefore, the interfacial area, the diffusion and mass transfer between the gas and the emulsion phase are unknown.

To understand the behavior of the fluidized bed, one can determine the average residence time or the residence time distribution (RTD) from the tracer technique. For instance, RTD in fluidized bed dryer is usually carried out by means of the stimulus-response technique, in which an impulse of solids marked with some appropriate tracer is fed to the dryer and its time of elution and concentration measured at the exit of the dryer. The material of the tracer has to be such that it can be detected and does not react with the substrate material, and its form of application and response are well known (Levenspiel, 1972).

Besides RTD, other techniques are needed to predict the behavior for the design of a fluidized bed, using mathematical models with parameters to be adjusted based on information from experimental data.

The mass balance of the representative diagram in Figure 22.7 for the gas and emulsion phase, considering continuous and isothermal system, are described below:

Gas phase (bubble)

$$f_b u_b \frac{dC_{Ab}}{dz} + k_i (C_{Ab} - C_{Ae}) + r'' \rho f_b = 0 \tag{22.24}$$

where:
f_b = fraction of the bed occupied by bubbles
k_i = interchange coefficient $(cm^3/cm^3_{bed} \times \sigma)$
u_b = superficial velocity of the bubble (cm/s)
C_{Ab} = concentration of gas in bubble
C_{Ae} = concentration of A in phase emulsion
r'' = reaction rate (mol/g s)
ρ = gas density (g/cm^3).

Emulsion phase (gas + particles)

$$f_e u_e \frac{dC_{Ae}}{dz} + k_i(C_{Ab} - C_{Ae}) - f_e D_e \frac{dC^2_{Ae}}{dz^2} + r'' \rho_e(1 - f_b) = 0 \tag{22.25}$$

where:
ε_e = gas fraction in the emulsion
f_e = fraction of the bed occupied by emulsion
D_e = effective diffusion coefficient (or apparent diffusion coefficient) (cm^2/s)
u_e = superficial velocity of the emulsion (cm/s).

The balance mol/area at the inlet and outlet of the bed, considering the average concentration in the output and output superficial velocity, u_s, is:

$$u_s \overline{C}_A = f_b u_b C_{Ab} + f_e u_e C_{Ae} \tag{22.26}$$

This system of equations can be solved from the following boundary conditions.

For gas phase: $z = 0 \rightarrow C_{Ab} = C_{A0}$ $\qquad$ (22.27)

For emulsion phase: $-D_e \dfrac{dC_{Ae}}{dz} = u_e(C_{A0} - C_{Ae})$

There is an interface between the bubble phase (gas) and the dense phase (emulsion) containing the solid particles. The reaction occurs at the interfacial zone in dense area within a given volume V_i, where it is assumed that the density of the solid phase, ρ_s, must be equal to density of the emulsion phase, ρ_e. This volume was determined empirically by Patridge and Rowe (1965) as follows:

$$\frac{V_i}{V_c} = \frac{1.17}{0.17 + \alpha} \tag{22.28}$$

and

$$\rho_e = \frac{\rho_b}{(1 - f_b)} \tag{22.29}$$

with

where ρ_e corresponds to the fraction of the total volume occupied by the bubble in interface zone (emulsion), disregarding the solid mass of the bubble.

The exchange coefficients at the interface were determined empirically. Usually, the following correlations are used:

$$k_{bc} = 4.5 \times \frac{u_{mf}}{d_b} + 5.85 \left(\frac{D_e^{1/2} \cdot d_t^{1/4}}{d_b^{5/4}} \right) \tag{22.30}$$

$$k_{ce} = 6.78 \left(\frac{\varepsilon_{mf} \cdot D_e . u_b}{d_b^3} \right)^{1/2} \tag{22.31}$$

Equation 22.25 can be solved admitting a first-order reaction. Thus, a solution of type conversion versus spatial time is obtained ($X_A \times W/F_0$), as shown in Figure 22.8, according to various parameters.

Note that there are significant differences when comparing the fluidized bed reactors with the ideal reactors. The most important parameters are effective diffusion coefficients, D_e, and interface transfer coefficients, k_i. These parameters can be determined by Equations 22.30 and 22.31.

The model presented by Kunii and Levenspiel (1991) is intended for a simple system, which comprises a "bubble + interface" phase, called intermediate zone

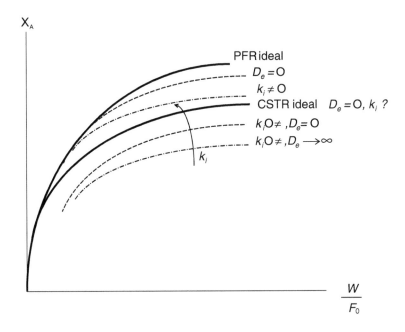

Figure 22.8 Conversion of a reactant to different interfacial and diffusional effects (Adapted from Froment and Bischoff (1979)).

or "cloud." The authors suggest that the superficial velocity must be greater than minimum fluidization velocity:

$$\frac{u_s}{u_{mf}} \geq 6 \text{ to } 1.1$$

Consequently, there is a large volume of gas in the emulsion, forcing them to return to the wall, avoiding the accumulation and neglecting diffusion.

After defining the exchange coefficients for bubble–cloud and cloud–emulsion interfaces (Equations 22.30 and 22.31), and assuming a first-order reaction, we get the following equation:

$$-u_b \frac{dC_{Ab}}{dz} = K_r C_{Ab} \tag{22.32}$$

With the boundary conditions:

$$z = 0 \rightarrow C_{Ab} = 0$$
$$z = L \quad C_{Ab} = \overline{C}_A$$

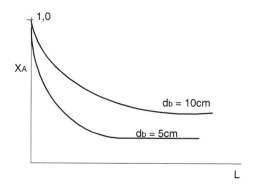

Figure 22.9 Conversion along the reactor considering a first-order reaction.

where the constant k_r is given by:

$$K_r = k \left[\rho + \cfrac{1}{\cfrac{k}{k_{bc}} + \cfrac{1}{\rho_c \frac{V_{iz}}{V_c} + \cfrac{1}{\cfrac{k}{(k_{ce})_b} + \cfrac{1}{\rho_e \frac{(1-\varepsilon_b)}{\varepsilon_b}}}}} \right] \qquad (22.33)$$

$\frac{V_{iz}}{V_c} =$ ratio between the volume of the bubble at the interface and the volume of the cloud is determined empirically, according to the equation below:

$$\frac{V_{iz}}{V_c} = \frac{1.17}{0.17 + \alpha} \qquad (22.34)$$

$$\rho_e = \frac{\rho}{1 - f_e} \qquad (22.35)$$

Solving Equation 22.33 for a first-order reaction, we obtain:

$$\frac{\overline{C_A}}{C_{A0}} = \exp\left(-p\frac{K_r L}{u_b}\right) \qquad (22.36)$$

The solution is shown in Figure 22.9, it shows the conversion as a function of bed length by using as the main parameter the bubble diameter.

Example

E22.2 It is desired to produce 40,400 ton/year of acrylonitrile in a fluidized bed by propene ammoxidation. The fluidized bed reactor operates at 1 atm and 400°C*. The mole fraction of propylene is 0.24. The conversion of propylene obtained was 78% and assuming a first-order reaction, whose specific velocity of reaction is $1.44\,m^3/kg_{cat}\,h$ (at 400°C).

Calculate the length of the fluidized bed.

Additional data: *Adapted from Froment and Bischoff (1979)).

Bed porosity $\rightarrow \varepsilon = 0.24$
Average particle diameter $\rightarrow \bar{d}_p = 51\,\mu m$
Solid density $\rightarrow \rho_s = 2500\,kg/m^3$
Minimum fluidization velocity $\rightarrow u_{mf} = 7.2\,m/h$
Porosity at minimum fluidization $\rightarrow \varepsilon_{mf} = 0.6$
Bubble diameter $\rightarrow d_b = 0.1\,m$
Outer diameter of reactor $\rightarrow d_{out} = 0.14\,m$
Superficial velocity $\rightarrow u_s = 1800\,m/h$
Gas density $\rightarrow \rho_g = 1.0\,kg/m^3$
Effective diffusion coefficient $\rightarrow D_e = 0.14\,m^2/h$
Gas dynamic viscosity $\rightarrow \times\,\mu_g = 0.144\,kg/(m.h)$
Void fraction of gas at minimum fluidization $\rightarrow \varepsilon_b = 0.5$

Solution

To calculate the bed length, we must know the global reaction rate. However, the mass transfer coefficients at the interface must be determined as a function of reaction variables. Furthermore, we need to first determine the velocity of the gas bubble. From the feed stream annual, we can calculate the gas stream daily knowing the gas molecular weight, i.e., $MW = 53\,g/gmol$. Thus:

$$F_0 = \frac{G_0}{hX_A\,WM} = \frac{40,400 \times 10^6}{8000 \times 0.78 \times 53} = 120\,kmol/h$$

Considering the gas void fraction $\rightarrow F_{G0} = \frac{120}{0.24} = 500\,kmol/h$
But, under normal conditions, the volumetric flow is:

$1\,kmol \rightarrow 22.4\,m^3$
$500 \rightarrow x \rightarrow v_0 = 11{,}200\,m^3/h$

The superficial velocity of the bubble can be calculated by determining the relative velocity of a bubble from the following equation:

$$u_{br} = 0.711\sqrt{g \cdot d_b} = 0.711\sqrt{1.27 \times 10^8 \times 0.1} = 2534\,m/h$$
and
$$g = 9.8\,m/s^2 \times 3600^2 = 1.27 \times 10^8\,m/h^2$$

where:

$$u_b = u_s - u_{mf} + u_{br} = 1800 - 7.2 + 2534 = 4326\,m/h$$

Thus:

$$k_{bc} = 4.5 \times \frac{u_{mf}}{d_b} + 5.85 \left(\frac{D_e^{1/2} \cdot d_t^{1/4}}{d_b^{5/4}} \right)^2 \qquad (22.37)$$

$$k_{bc} = 4.45 \times 10^3 \qquad (22.38)$$

From these data, it is possible to calculate the transfer coefficients at the bubble interface with the cloud, k_{bc}, using the following empirical equation (Equation 22.30):

$$k_{ce} = 6.78 \left(\frac{\varepsilon_{mf} \cdot D_e \cdot u_b}{d_b^3} \right)^{1/2} = 4.087 \times 10^3$$

One can calculate the transfer coefficient between the cloud–dense interfaces (Equation 22.31) by:

For the phase emulsion, various parameters are required. They are:

Emulsion density:

$$\rho_e = \rho_s (1 - \varepsilon_{mf}) = 2500(1 - 0.6) = 1000 \, \text{kg/m}^3$$

Fraction of the bed occupied by bubbles:

$$f_b = \frac{u_s - u_{mf}}{u_b} = \frac{1800 - 7.2}{4326} = 0.414$$

Bubble density in the emulsion:

$$\rho_{be} = \rho_e \frac{(1 - f_b)}{f_b} = 1420 \, \text{kg/m}^3$$

Interfacial volume emulsion/gas (Equation 22.28):

$$\frac{V_{iz}}{V_c} = \frac{1.17}{0.17 + \alpha}$$

where:

$$\alpha = \frac{u_b}{\frac{u_{mf}}{\varepsilon_{mf}}} = \frac{4326}{\frac{7.2}{0.6}} = 360$$

Therefore, we get:

$$\rho_e \cdot \frac{V_{iz}}{V_b} = 1000 \frac{1.17}{360} = 3.26 \, \text{kg/m}^3$$

We calculate the density of the solid phase gas considering a solid fraction in the gas equal to 1.5%. Thus:

$$\gamma_b = \frac{(1 - \varepsilon_{mf})(1 - f_b)}{f_b} \cdot 0.015 = 0.01 \frac{m_s^3}{m_b^3} \tag{22.39}$$

$$\rho_b = \gamma_b \rho_s = 0.01 \times 2500 = 25 \, \text{kg/m}_b^3 \tag{22.40}$$

Now, calculate the coefficient K_r from Equation 22.33:

$$K_r = k \left[\rho_b + \cfrac{1}{\cfrac{k}{k_{bc}} + \cfrac{1}{\rho_c \frac{V_{iz}}{V_c} + \cfrac{1}{(k_{ce})_b + \cfrac{1}{\rho_e \frac{(1-f_b)}{f_b}}}}} \right]$$

$$k_r = 1.082 \times 10^3$$

However, the desired conversion is 78%. Thus:

$$\left(1 - \frac{\overline{C}_A}{C_{A0}}\right) = X_A = 0.78$$

Thus,

$$\frac{\overline{C}_A}{C_{A0}} = 0.22$$

Finally, for determining the length of the fluidized bed, we use Equation 22.37:

$$\frac{\overline{C}_A}{C_{A0}} = \exp\left(-\frac{K_r \cdot L}{u_b}\right)$$

But,

$$\exp\left(-\frac{K_r \cdot L}{u_b}\right) = 0.22$$

Finally,

$$L = 1.51 \times 4326/1082 = 6.0 \, \text{m}$$

Chapter 23

Nonideal reactors

23.1 INTRODUCTION

Various types of industrial reactors may occur in different phases as applications and desired properties of the final product, for example, the fixed bed, fluidized bed, slurry bed, and bed phase reactors. In fluidized bed reactors as in slurry bed, the solid (catalyst) is composed of very small particles and moving along the reactor. The fluid flow over these reactors is complex. In these systems, the flow of the fluid phase is not homogeneous and there are large deviations from the ideal behavior of a CSTR or plug flow reactor (PFR), characterizing them in nonideal reactors.

On the other hand, there are exceptional advantages in these reactors, such as improved heat and mass transfer, increased contact between the reactants, and mainly lower contact time reaction.

As in ideal reactors, the kinetics and reaction conditions are similar. However, the distribution of products is quite different and to correlate them with the experiments, it requires a more detailed study of the conditions of nonideality, for example, interfacial and surface phenomena, heat and mass transfer, and flows types. These phenomena characterize the axial and radial dispersion, caused by diffusion and convection.

The most known reactors are shown in the Figure 23.1. In the PFR reactor, the flow can take place in a velocity profile laminar or turbulent. The laminar flow in the PFR reactor is characterized as not ideal because the velocity profile within a cross section is not uniform and, consequently, the contact time between molecules is different in the center and close to the reactor wall. In a turbulent flow, the velocity profile is not uniform, but generally it behaves like uniform flow and closer to the ideal conditions of an ideal reactor because the contact time between the molecules of the reactants is approximately equal.

The flows in catalyst beds are different from those presented in the fixed bed, fluidized bed, or slurry. The flows are random; they depend on empty spaces within a fixed bed, through which the gases or fluids flowing, and to the apparent velocity of the solid within a fluidized bed/slurry. These phenomena are characteristic of the nonideal reactors.

There are two ways that allow us to characterize the nonideal reactors:

1   From the dispersion effects, which cause deviation from ideal behavior. This analysis is performed from the determination of the residence time distribution (RTD) in a system without chemical reaction and determining the effect on the

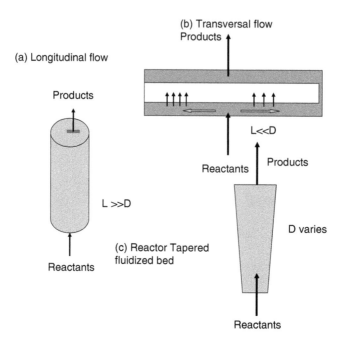

Figure 23.1   Different types of reactors.

reaction, comparing the calculated conversions from the experimental conversions. This method is conducted to analyze the systems separately, segregated models (micromixing) and not segregated (macromixing).

2   From the basic equations of momentum, energy, and mass diffusion effects, and radial and axial convection, with or without chemical reaction. Additionally, we evaluate also the parameters that cause the deviation from ideal behavior and we adopt criteria to estimate the effects of radial and axial dispersion on the reactor.

For the first item, an analysis is made through the segregated and nonsegregated models using experimental results of RTD. In the second item, we will use the balance equations of mass, heat, and momentum.

In studying the ideal reactors, which are limiting cases, no significant influence of other phenomena caused by the flow, the mass, and heat transfer. Therefore, the parameters studied so far were determined in kinetic regime and without considering the effects of transport phenomena. When this happens the actual rate is less than the intrinsic rate constant.

In CSTR reactors, the flow has preferential paths. In continuous tubular reactors, the flow may be laminar, turbulent, and have "dead" volumes. The flow may cause radial and longitudinal diffusion effects and therefore to result temperature gradients and radial/axial concentration. Therefore, the flow may affect the chemical reaction.

The recognition of the reactor can be done by using the so-called "population balance." What does this mean? Consider an irreversible chemical reaction of first order and under isothermal conditions, whose solution is:

$$\tau = -\frac{1}{k} \ln (1 - X_A)$$

where X_A = conversion of A, k = reaction rate constant, and t = reaction time.

The time "t" is the contact time or residence time of molecules in the reactor. In the batch reactor, it is assumed that measured time is equal to the average contact time. However, in a continuous system, this time may or may not be equal to the contact time, because the distribution of molecules or properties (in the reactor inside) may not be uniform (or homogeneous), and it depends on the type of flow. Therefore, it is impossible to determine the kinetic properties without knowing the "true" reaction time.

In fact, the flow of a fluid element formed by a set of molecules is not uniform. Thus, the residence times are different, which characterizes the nonideal reactors.

As we have seen in previous chapters, the flow depends on the geometry of the reactor. The cylindrical tubular reactor is the most used, but there are other ways, such as conical or cylindrical shapes with cross flows.

About the type of reactor to be selected, it depends not only on the stream, but also on several other factors:

(a)   Flow closer to the ideal conditions and residence time equals;
(b)   Lower pressure drop in the reactor;
(c)   Charge of the reactor (catalyst or filling), which aims a better flow and lower pressure drop;
(d)   Heat exchange internal or external, which depends on the type of reaction conducted, that is, if it is exothermic or endothermic and if the process is isothermal or adiabatic.

The flow in the reactor (superficial velocity within the cross section of the reactor) varies according to the operating conditions and geometry of the reactor.

The velocity profile of the laminar flow into a cylindrical tube, for example, may be calculated and to provide a variable velocity distribution in the cross section. This causes a variation in contact time of the molecules along the reactor and therefore cannot be considered an ideal reactor.

In turbulent flow (high Reynolds number), the velocity distribution varies in axial and radial direction with higher velocities close to the inner walls of the reactor and providing greater uniformity in the center of the tube. Thus, the higher the velocity, the greater the uniformity of its profile in the reactor, favoring a uniform contact and an average residence time between the ideal molecules. Therefore, in turbulent flow, there is an approximation to the ideal flow.

The packed bed reactor, particularly catalytic reactors, present a distribution of velocity quite heterogeneous and is very difficult to predict the void space and preferential paths, which depend on the type and shape of the packed column, in addition to

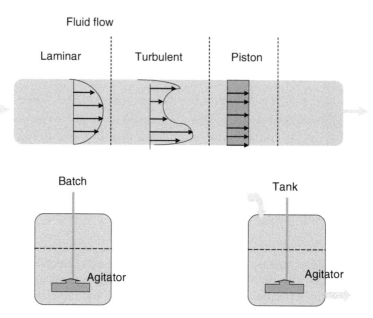

Figure 23.2 Characteristics of fluid flow.

the catalyst particles and their positions inside the reactor. However, the contact of the molecules is higher due to the increased contact area, which facilitates the reaction.

In addition to the flow, one should consider the effect of mass and heat transfer. In reactors without filling, it is possible to predict the heat exchange and to determine the conditions for isothermal or adiabatic operation from the temperature profile in catalytic bed. For uniform flows, the heat transfer depends on the heat capacity. Thus, temperature profile can be uniform, causing large deviations in conditions where heat capacity is very high.

In the packed bed reactor, there is also the influence of heat conduction from particle. As we know, the temperature affects substantially the rate constant and therefore the reaction rate. In parallel, there are the effects of mass transfer by convection and diffusion in the pores of the particles. Therefore, these effects change the kinetics considerably and hence the chemical reaction rate.

The considerations in the previous paragraph are also valid for the flows in tanks and batch reactors. However, the contact between the molecules also depends on the geometry of the reactors. To avoid the undesirable "dead volumes," greater agitation of the system becomes very important, increasing the contact between molecules. Contact is instantaneous and concentration at the outlet of the tank or batch reactor should be as uniform as possible. Thus, the ideal condition is achieved when the reaction mixture is perfect. Figure 23.2 shows some examples.

We will see later that techniques developed to address these issues using concepts of fluid properties distribution in systems based on probability theory. We will also see the concepts of RTD for the reactions and fluid flow, and other properties, such as distribution of solid particles treated by particle population balance.

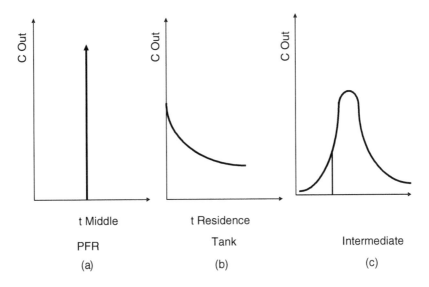

Figure 23.3 Response to pulse tracer at the outlet of the reactor.

23.2 RESIDENCE TIME DISTRIBUTION

The residence time considers the time that each fluid element or group of molecules remains in the reactor; it also depends on the velocity of the molecules within the reactor, and therefore the flow in the reactor. The residence time can be equal to the space time if the velocity is uniform within a cross section of the reaction system, as is the case of an ideal PFR. However, this situation is not the same for tank-type reactor, because the velocity distribution is not uniform. In most nonideal reactors, residence time is not the same for all molecules. This result in variations in concentration along the reactor radial, i.e., its concentration inside and outlet tank reactors are not uniform. This means a need to define the residence time and calculate their distribution for each system.

One can visualize RTD through an experiment with the use of "chemical tracers," introducing them at a particular moment or since the beginning of the reaction. This chemical tracer should necessarily be a compound not reactive to the system under study, by measuring its concentration in the reactor outlet. In general, dye compounds are used, but also other materials can be used with conductive or radioactive material properties that can be measured quantitatively.

Monitoring the tracer at the outlet of the reactor, one can observe different situations for the labeled molecules. For instance, an ideal flow, such as a piston, it will be noted that for an average time all molecules come out in the same instant (Figure 23.3a). In a reactor tank the opposite occurs, the concentration of labeled molecules decreases with time at the outlet of the reactor (Figure 23.3b). Nonideal reactors have an intermediate behavior where the concentration of the labeled molecules at the exit of the reactor will have a Gaussian distribution. At the beginning, a small fraction of labeled

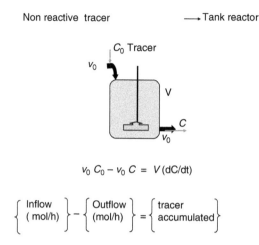

Figure 23.4 Balance molar of a tracer.

molecules are identified in the output of the reactor; most molecules leave the system at a given instant and then, a small fraction of molecules leaves the reactor, as shown in Figure 23.3c.

The "tracer" can be introduced into the reactor in the form of step or pulse. For simplicity, we will use the example of a tank reactor with a volume V and a flow of an inert liquid (water) v_0. At a given instant, a chemical dye is introduced into a stream in the step-form with a concentration C_0, and the concentration C is measured at the outlet of the reactor from the instant $t = 0$. Thus, from the mass balance presented in Figure 23.4, we have:

$$v_0 C_0 - v_0 C = V \left(\frac{dC}{dt} \right) \tag{23.1}$$

From this, on integration between at $t=0$, $C=0$, and at $t \neq 0$, $C_{\text{out}} = C$, and considering τ as space time:

$$\tau = \frac{V}{v_0}(\text{h}) \tag{23.2}$$

We get:

$$\frac{C}{C_0} = 1 - \exp\left(-\frac{t}{\tau}\right) \tag{23.3}$$

The concentration of tracer at the outlet varies exponentially with time, indicating a variation of the distribution in the reactor. Thus, the molecules have different residence times. This is the distribution of concentration within an ideal tank reactor.

Generally, the concentration is related to a function of RTD and assuming that a fraction of molecules having a residence time between the time interval t and $t + dt$. At the instant t, the concentration of tracer at the outlet is C. Thus, we measured a fraction of molecules that remained in the reactor in a time less than t and another fraction that remained in the reactor for a time longer than t. The first fraction is represented by the cumulative distribution function $F(t)$ and the second fraction is represented by the difference $(1 - F(t))$. This last fraction C_0 does not contain at the output of the reactor.

From the mass balance at the output of the reactor, we have:

$$v_0[1 - F(t)]C_0 + v_0\,F(t)C_0 = v_0C \tag{23.4}$$

Thus, relating this equation to Equation 23.3:

$$F(t) = \frac{C}{C_0} = 1 - \exp\left(-\frac{t}{\tau}\right) \tag{23.5}$$

Therefore, the cumulative distribution function of residence times is determined by measuring the concentrations versus time at the output of the reactor. Graphically, we can represent it as given in Figure 23.5:

The average residence time will be when $t = \tau$ or when the area (1) is equal to area (2), as shown in Figure 23.5. This figure indicates that a fraction of molecules in the

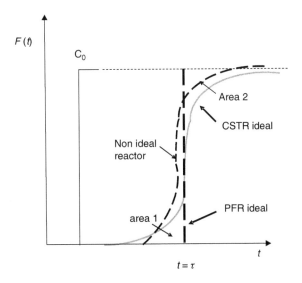

Figure 23.5 Response to pulse tracer (Adapted from Hill, 1979).

area (1) has a short residence time and another fraction (area 2) has a time longer than the average residence time, but the larger fraction has a residence time corresponding to the average value.

In the ideal PFR, the residence time is the same for all molecules, assuming uniform and constant velocity such that the output concentration of the tracer is equal to the concentration of the reactor input. Thus, $F(t) = 1$ or $t = \tau$. The average residence time in the ideal PFR is equal to the space time.

Any other form of RTD between the CSTR and PFR reactors (ideal) is not considered ideal.

To determine the function $F(t)$ from experimental data, we use the G property already defined in Chapter 1. If G is any one property (such as conductivity, ionization, and wavelength), it is proportional to the concentration G_1 at the entrance and G_2 at the outlet. Then, the cumulative residence time fraction which remains in the reactor at an instant less than t will be:

$$F(t) = \frac{G(t) - G_1}{G_2 - G_1} \tag{23.6}$$

E23.1 A tracer is introduced into the reactor feed stream and without chemical reaction at 2 g/m^3. The concentration of this tracer is measured at the output of the reactor, as shown below (Denbigh, 1965).

t (min)	0.1	0.2	1	2	5	10	20	30
C (g/m^3)	1.96	1.93	1.642	1.344	0.736	0.286	0.034	0.004

The reactor volume is 1 m^3 and the flow inlet is given by volumetric flow rate 0.2 m^3/min. Determine function RTD (F) and average residence time.

Solution

Space time

$$\tau = \frac{V}{v_0} = \frac{1}{0.2} = 5\,\text{min}$$

Function $F(t)$

$$F(t) = \frac{G(t) - G_1}{G_2 - G_1} = \frac{C(t) - C_1}{C_2 - C_1} = \frac{C(t) - 2}{-2} = \frac{2 - C(t)}{2}$$

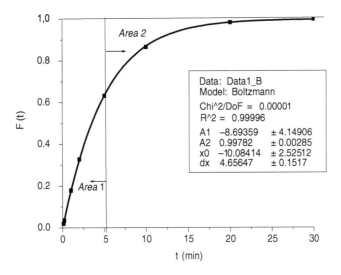

Figure E23.1   Cumulative distribution function.

The average residence time can be determined from Figure E23.1, assuming equal area for A_1 and A_2, according to the dashed line that corresponds at $t = 5$ min.

The cumulative distribution function versus time is shown in the table below:

t (min)	0.1	0.2	1	2	5	10	20	30
F(t)	0.02	0.035	0.179	0.328	0.632	0.866	0.983	0.998

It is observed that longer times indicate that fraction of molecules leaving the reactor is larger than the average residence time, where the inverse is also true.

Trace in shape of pulse

Another way to determine the residence time occurs when the tracer is injected in the form of pulse. The response can be calculated at the outlet of the reactor by concentration, assuming that a fraction of molecules ΔF came out of the reactor in the time interval Δt. This fraction of molecules left the reactor with concentration C_0. Thus, in the time interval Δt, we have the following mass balance:

$$v_0 \Delta F(t) C_0 = v_0 C \tag{23.7}$$

In the limit $\Delta t \to 0$, we get:

$$\lim_{\Delta t \to 0} \frac{F(t)}{\Delta t} = \frac{dF}{dt} = \frac{C}{C_0} \tag{23.8}$$

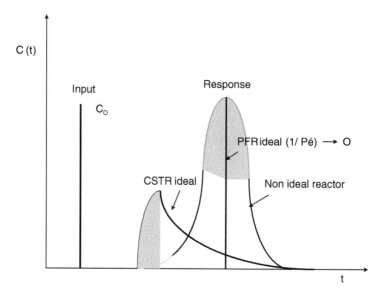

Figure 23.6 Response to pulse tracer.

The variation of the cumulative distribution function of residence times is represented by a Gaussian distribution, according to Figure 23.6.
Integrating it:

$$\int_0^1 dF = \int_0^\infty \left(\frac{C}{C_0}\right) dt$$

But from the curve one may conclude that:

$$\int_0^\infty \left(\frac{C}{C_0}\right) dt = 1 = \frac{1}{\tau C_0} \int_0^t C(t) dt \qquad (23.9)$$

The function $E(t)$ is defined as function of RTD which represents the age distribution function of molecules in the fluid element derived from the cumulative distribution function.

$$\frac{dF}{dt} = E(t)$$

Or:

$$\frac{dF(\theta)}{d\theta} = \tau E(t) = \frac{C(t)}{C_0} \tag{23.10}$$

Thus:

$$\int_0^F dF = \int_0^\infty E(t)dt$$

Therefore, the mean residence time is:

$$\bar{t} = \int_0^\infty tE(t)dt \tag{23.11}$$

where $\bar{t}$ = mean residence time and $\tau = \frac{V}{v_0}$ (h) = space time.

The dimensionless time θ is conveniently represented with the corresponding function $E(\theta)$. This relation is possible because both represent the same physical phenomenon, i.e., the fraction fluid with a time t at the outlet:

$$E(t)dt = E(\theta)d\theta$$

Thus:

$$E(\theta) = \tau E(t) \tag{23.12}$$

23.2.1 Ideal cases

In PFR ideal, the response is immediate in shaped-pulse. For ideal CSTR, the ideal response will be a distribution of molecules not instantaneous (Figure 23.6).

To determine the RTD of a tracer in shaped-pulse at a volume V and a volumetric flow v_0, it is assumed that N tracer units are introduced. The global balance gives us the total time retained in this volume.

Thus:

$$N = \int_0^\infty v_0 C(t)dt \tag{23.13}$$

Or assuming constant volumetric flow:

$$N = v_0 \int_0^\infty C(t)dt \tag{23.14}$$

E23.2 Determine the average residence time of the CSTR from the tracer using the following data (Denbigh, 1965).

t (min)	0	5	10	15	20	25	30	35
C (mol/m³)	0	84.9	141.5	141.5	113.3	56.6	28.3	0

The concentration of the tracer was measured at the outlet of the reactor. The volume is 2 m³ and the flow at the outlet is 7.2 m³/h.

Solution

From the data, we can determine the concentration profile at the outlet of the reactor (Figure E23.2.1).

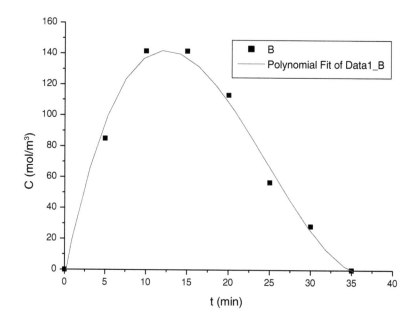

Figure E23.2.1 Concentration versus time.

On integrating over the area from that figure, we obtain τC_0:

$$\tau C_0 = 2,830$$

Therefore, by Equations 23.10 and

$$E(t) = \frac{C(t)}{\tau C_0} = \frac{C(t)}{2830} \tag{23.15}$$

Calculate the values of $E(t)$ as shown below:

Calculation of $E(t)$ and $t \times E(t)$			
t (min)	C (mol/m³)	$E(t)$	$t \times E(t)$
0	0	0	0
5	84.9	0.03	0.15
10	141.5	0.05	0.5
15	141.5	0.04	0.75
20	113.2	0.02	0.8
25	56.6	0.01	0.5
30	28.3	0	0.3
35	0	0	0

The average residence time is determined by Equation 23.11:

$$\bar{t} = \int_0^\infty tE(t)dt$$

Figure E23.2.2 represents $t \times E(t)$ versus t.

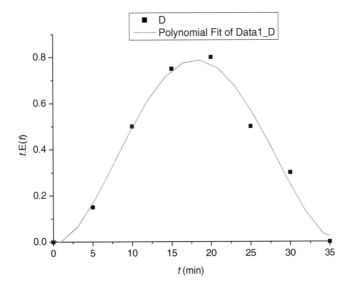

Figure E23.2.2 $t \times E(t)$ versus t.

On integrating over the area from this figure using Equation 23.11, we obtain the average residence time:

$$\bar{t} = 15 \, \text{min}$$

Finally, we can calculate the space time:

$$\tau = \frac{V}{v_0} = \frac{2}{7.2/60} = 16.6 \, \text{min}$$

$$\tau = 16.6 \, \text{min}$$

E23.3 The data below were obtained using a radioactive tracer in the fluidized bed reactor as shaped-pulse from the reactor. The space time was 3.25 min. Calculate the average residence time.

t (s)	0	0.5	1	1.5	2	2.5	3	3.5	4	4.5	5	5.5	6	6.5	7
C (mol/m³)	0	5	22	27	26	22	19	15	10	7	4	3	3	1	0

The data are presented in Figure E23.3.1.

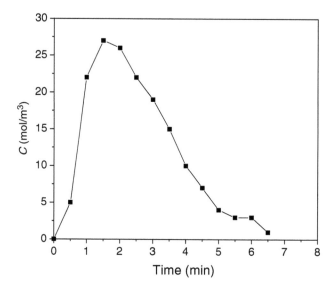

Figure E23.3.1 Concentration distribution versus t.

Integrating the curve in the above figure, we obtain a total area of 81.75×10^3 c/min.

Therefore, from Equation 23.10, we can calculate $E(t)$:

$$E(t) = \frac{C(t)}{\tau C_0} = \frac{C(t)}{81.75 \times 10^3} \qquad (23.16)$$

In the table below, we have the results of the functions of RTD $E(t)$ and $E(\theta)$ (by Equation 23.12).

t (min)	Cont/ (min × 10³)	$E(t)$	$t \times E(t)$	θ	$E(\theta)$	θ^{-2}	$\theta^{-2}E(\theta)$
0	0	0	0	0	0	0	0
0.5	5	0.0611	0.0305	0.197628	0.15474	0.039057	0.006044
1	22	0.269	0.269	0.395257	0.680856	0.156228	0.106369
1.5	27	0.330	0.495	0.592885	0.835596	0.351513	0.293723
2	26	0.318	0.636	0.790514	0.804648	0.624912	0.502834
2.5	22	0.269	0.672	0.988142	0.680856	0.976425	0.664805
3	19	0.232	0.697	1.185771	0.588012	1.406052	0.826776
3.5	15	0.183	0.642	1.383399	0.46422	1.913793	0.888422
4	10	0.122	0.489	1.581028	0.30948	2.499648	0.773592
4.5	7	0.085	0.385	1.778656	0.216636	3.163618	0.685354
5	4	0.049	0.244	1.976285	0.123792	3.905701	0.483495
5.5	3	0.0366	0.201	2.173913	0.092844	4.725898	0.438771
6	3	0.0366	0.220	2.371542	0.092844	5.624209	0.522174
6.5	1	0.0122	0.0795	2.56917	0.030948	6.600634	0.204277
7	0	0	0	2.766798	0	7.655173	0

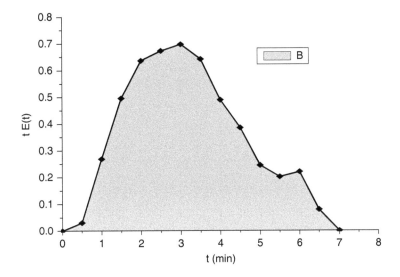

Figure E23.3.2 Distribution $t \times E(t)$.

Figure E23.3.2 shows $t \times E(t)$
From the area calculation (figure above), we find the average residence time:

$\bar{t} = 2.35$ min

The space time is 3.25 min. Thus we get a standard deviation of 22%.

23.2.2 Variance

The average residence time is the average time of contact between the molecules within a fluid element. The RTD function, either by $F(\theta)$ or $E(\theta)$, reports a range of periods of contact. Therefore, there is a time variation around the average residence time or variance of RTD.

Analyzing the distribution curves, it can be inferred that the width of the curve represents the variance. In the case of PFR, this width is null. While the ideal CSTR, it is relatively large. Therefore, the width of the peak may indicate deviation of behavior between the two ideal cases.

Therefore, variance is defined as an RTD around the average value and by probability theory, it can be represented by:

$$\sigma^2 = \int_0^\infty (t-\bar{t})^2 E(t)\, \mathrm{d}(t) \tag{23.17}$$

where:

$$\theta = \frac{t}{\bar{t}}$$

and

$$\bar{t}E(t) = E(\theta)$$

Therefore:

$$\sigma^2 = \int_0^\infty \theta^2 E(\theta)\mathrm{d}(\theta) - 1 \tag{23.18}$$

For an ideal CSTR, one obtains a *variance* $\sigma^2 = \bar{t}^2$. For any nonideal reactor, one obtains a variance for the peak width and thereby, deviation from ideal behavior. This method can also be used to determine the RTD function $E(t)$.

E23.4 From the previous example and by using Equation 23.18, calculate the variance.

$$\sigma^2 = \left(\int_0^\infty (\theta^2 E(\theta)\mathrm{d}(\theta) \right) - 1 = 1.26416 - 1 = 0.264$$

$$\sigma^2 = 0.264$$

Or:

$$\bar{t}^2\sigma^2 = \int_0^\infty \bar{t}^2 E(t)dt - \bar{t}^2$$

$$\bar{t}^2 \cdot \sigma^2 = 8.09174 - 6.40 = 1.689$$

$$\sigma^2 = 0.264$$

Therefore, by calculating the variance, it can be seen the deviation from ideal behavior of the reactor using axial dispersion by equation that relates the variance with Peclet number (Equation 15.25):

$$\sigma^2 = \frac{2}{P\acute{e}} = 0.264 \tag{23.19}$$

$$P\acute{e} = 7.6$$

When the Peclet number is large ($P\acute{e} \to \infty$), we will have a behavior close to ideal PFR and when it is too small ($P\acute{e} \to 0$), it approaches an ideal CSTR.

In this example, we conclude that the reactor has a nonideal behavior.

23.3 NONIDEAL COMPARTMENTAL REACTOR MODELS

In the previous sections, different ideal reactor configurations were described and analyzed. As a matter of fact, the use of ideal reactors for interpretation of kinetic problems can indeed be important, because it can allow for capture of the most important features of typical reaction vessels and better comprehension about the distinguishing characteristics of each particular reaction process. Nevertheless, when compared with the proposed ideal benchmarks, real industrial reactors can present distinct behaviors for many reasons, including imperfect mixing (due to high viscosity, as in some polymerization processes, and existence of geometrical constraints for design of agitated tanks), development of radial velocity profiles (due to small diameters of tubular reactors)among others, as described in the next chapters. For this reason, there are many incentives to use simple mathematical tools to represent more complex reactor vessels with the help of the simple ideal benchmarks discussed in the previous sections. One of these tools involves the combination of ideal reactors to represent different compartments of the equipment and to form more complex reaction flowsheets – the compartmental reactor models.

Compartmental reactor models can be defined broadly as reactor models that are obtained after combination in series or parallel of simpler ideal reactors that are used to represent smaller sections of the reaction vessel, in the presence or absence of recycling streams. This type of modeling procedure has been used extensively in the literature to represent the behavior of actual reactors with help of the simpler and well-behaved ideal reactor benchmark models. Some examples have been provided by Mattos Neto *et al.* (2005) and Oechsler *et al.* (2018, 2019) to represent the complex mixing behavior of real stirred tank polymerization reactors. In this case, the resulting process flow-sheets do not represent real physical industrial arrangements, as discussed in detail in Chapter 14, but realizations of complex mixing patterns through combination of ideal benchmark models. However, it is not possible to present general rules for building of such compartmental models, although some useful examples are presented in the following examples for illustrative purposes.

E23.3.1 The occurrence of shortcuts (reactor bypasses) and dead volumes (stagnant zones) inside real agitated reaction vessels can be modeled with help of the compartmental procedure, as illustrated in Figure 23.3.1. In this case, the shortcut and deadzone sections can be described with help of standard ideal stirred tank models, leading to more complex reactor configurations in which the distinct reactor sections exchange mass (and energy when the reaction is not performed isothermally). This sort of arrangement emphasizes the fact that a reaction vessel can present distinct mixing zones and that mixing may not be perfect.

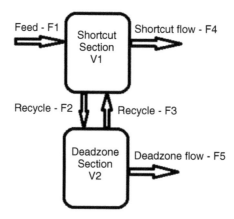

Figure E23.3.1 Example of reactor configuration with shortcut and dead zone.

Assuming that a first-order reaction rate model can be used to represent the reaction:

$$A \xrightarrow{k} B$$

then two mass balance equations must be written in the form:

$$\frac{dC_A^I}{dt} = \frac{F_1}{V_1}C_A^e - \frac{F_2}{V_1}C_A^I + \frac{F_3}{V_1}C_A^{II} - \frac{F_4}{V_1}C_A^I - kC_A^I$$

$$\frac{dC_A^{II}}{dt} = \frac{F_2}{V_2}C_A^I - \frac{F_3}{V_2}C_A^{II} - \frac{F_5}{V_2}C_A^{II} - kC_A^{II}$$

where it is assumed that the volumes of the two reactor sections remain constant during the operation. As one can see, the behavior of a reactor vessel that presents different mixing zones can be different from the one discussed in Section 14.3.2 for an ideal stirred tank. As a matter of fact, the technique can be extended to accommodate multiple mixing zones, as often performed when the reactor vessel is equipped with a mixer that is composed of multiple blades (Embiruçu et al., 2000, 2008a, 2008b).

In the analyzed case, at steady-state conditions, it is possible to write:

$$0 = \frac{F_1}{V_1}C_A^e - \frac{F_2}{V_1}C_A^I + \frac{F_3}{V_1}C_A^{II} - \frac{F_4}{V_1}C_A^I - kC_A^I$$

$$0 = \frac{F_2}{V_2}C_A^I - \frac{F_3}{V_2}C_A^{II} - \frac{F_5}{V_2}C_A^{II} - kC_A^{II}$$

and

$$\left(\frac{F_3}{V_2} + \frac{F_5}{V_2} + k\right) C_A^{II} = \frac{F_2}{V_2}C_A^I$$

$$\frac{F_1}{V_1}C_A^e = \left(\frac{F_2}{V_1} + \frac{F_4}{V_1} + k\right) C_A^I - \frac{\frac{F_3 F_2}{V_1 V_2}}{\frac{F_3}{V_2} + \frac{F_5}{V_2} + k}C_A^I$$

so that

$$F_1 C_A^e = \left(F_2 + F_4 + kV_1\right) C_A^I - \frac{F_3 F_2}{\left(F_3 + F_5 + kV_2\right)}$$

$$C_A^I = \left[\frac{\left(F_2 + F_4 + kV_1\right)\left(F_3 + F_5 + kV_2\right) - F_3 F_2}{\left(F_3 + F_5 + kV_2\right)}\right] C_A^I$$

and finally

$$C_A^I = \left[\frac{F_1\left(F_3 + F_5 + kV_2\right)}{\left(F_2 + F_4 + kV_1\right)\left(F_3 + F_5 + kV_2\right) - F_3 F_2}\right] C_A^e$$

$$C_A^{II} = \left[\frac{F_2 F_1}{\left(F_2 + F_4 + kV_1\right)\left(F_3 + F_5 + kV_2\right) - F_3 F_2}\right] C_A^e$$

so that

$$C_A^{out} = \frac{F_4 C_A^I + F_5 C_A^{II}}{F_4 + F_5}$$

One must observe that the behavior of the ideal reactor depends essentially on two parameters, as discussed in Section 14.3.2: the residence time, $\tau = V_1/F_1$, and the kinetic rate constant (k). On the other hand, the proposed compartmental model depends on a much larger number of parameters: the residence time of the shortcut section, $\tau_1 = V_1/(F_1 + F_3)$, the residence time of the deadzone section, $\tau_2 = V_2/F_2$, the kinetic rate constant, k and the recycle ratio, $\phi = F_2/F_1$. Therefore, the behavior of nonideal reactors can be far more complex than their ideal counterparts. It is also interesting to observe that some additional constraints must be satisfied:

$$V = V_1 + V_2$$

$$F_1 + F_3 = F_2 + F_4$$

$$F_2 = F_3 + F_5$$

Therefore, when $F_5 = 0$:

$$F_2 = F_3 = \phi F_1$$

$$F_1 = F_4$$

$$\tau = \tau_1 (1 + \phi) + \tau_2 \phi$$

and it becomes possible to specify τ, τ_1, ϕ and k. Particularly,

$$\frac{V_1}{V_1 + V_2} = \frac{\tau_1 (1 + \phi)}{\tau} = \alpha$$

which clearly indicates that global residence time of the reactor, τ, is related to recycle ratio, ϕ and the relative volumes of the two compartment sections (that define α and τ_1). Assuming $V_1 = 1$ m^3, $F_1 = 1$ m^3/h ($\tau = 1$ h), $F_5 = 0$ m^3/h, $C_A^e = 1000$ gmol/m^3 and $k = 1$ h^{-1}, Figures 23.3.2 to 23.3.3 can be used to illustrate the effects of α and ϕ on the concentrations of each compartment and of the output stream ($C_A^{out} = C_A^I$), described in the form:

$$C_A^I = \left[\frac{1 + k\tau_2}{(1 + \phi + k\alpha\tau)(1 + k\tau_2) - \phi} \right] C_A^e$$

$$C_A^{II} = \frac{1}{(1 + \tau_2 k)} C_A^I$$

As one can see in Figures 23.3.2 and 23.3.3, concentrations of the main reactant A are always smaller in the stagnant section of the reactor, due to the higher effective residence time and lower feed concentration, when compared to the shortcut section. One can also observe the asymptotic convergence of concentrations to the reference value of 500 gmol/m^3 in both sections as recycle ratio and volume fraction of the shortcut section increase. This is because the compartmental model approaches the ideal stirred tank reactor model when mixing is enhanced and volume segregation is diminished.

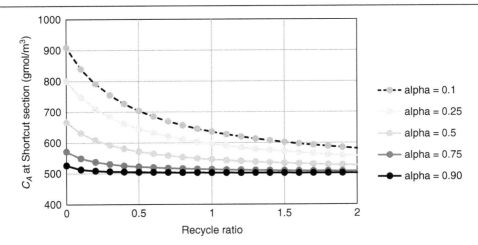

Figure 23.3.2 Concentration in the shortcut section and output stream as a function of the recycle ratio and relative volume of the shortcut section.

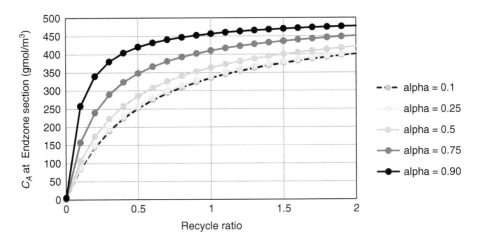

Figure 23.3.3 Concentration in the deadzone section as a function of the recycle ratio and relative volume of the shortcut section.

If the mass balance equation in the shortcut section is differentiated in respect to time, one gets:

$$\frac{d^2 C_A^I}{dt^2} = -\left(\frac{F_2}{V_1} + \frac{F_4}{V_1} + k\right)\frac{dC_A^I}{dt} + \frac{F_3}{V_1}\frac{dC_A^{II}}{dt}$$

If the mass balance equation in the deadzone section is inserted into the previous equation:

$$\frac{d^2 C_A^I}{dt^2} = -\left(\frac{F_2}{V_1} + \frac{F_4}{V_1} + k\right)\frac{dC_A^I}{dt} + \frac{F_3}{V_1}\left[\frac{F_2}{V_2}C_A^I - \left(\frac{F_3}{V_2} + \frac{F_5}{V_2} + k\right)C_A^{II}\right]$$

Finally, if the mass balance equation in the shortcut section is used to remove variable C_A^{II} from the previous equation:

$$\frac{d^2 C_A^I}{dt^2} = -\left(\frac{F_2}{V_1} + \frac{F_4}{V_1} + k\right)\frac{dC_A^I}{dt}$$

$$+\frac{F_3}{V_1}\left\{\frac{F_2}{V_2}C_A^I - \frac{V_1}{F_3}\left(\frac{F_3}{V_2} + \frac{F_5}{V_2} + k\right)\left[\frac{dC_A^I}{dt} - \frac{F_1}{V_1}C_A^e + \left(\frac{F_2}{V_1} + \frac{F_4}{V_1} + k\right)C_A^I\right]\right\}$$

Then

$$\frac{d^2 C_A^I}{dt^2} + \left(\frac{F_2}{V_1} + \frac{F_4}{V_1} + \frac{F_3}{V_2} + \frac{F_5}{V_2} + 2k\right)\frac{dC_A^I}{dt}$$

$$+\left[\left(\frac{F_3}{V_2} + \frac{F_5}{V_2} + k\right)\left(\frac{F_2}{V_1} + \frac{F_4}{V_1} + k\right) - \frac{F_3}{V_1}\frac{F_2}{V_2}\right]C_A^I$$

$$= \frac{F_1}{V_1}\left(\frac{F_3}{V_2} + \frac{F_5}{V_2} + k\right)C_A^e$$

which is a second-order linear differential equation, that leads to distinct dynamical responses, when compared to the ideal stirred tank reactor. This can be very interesting for practical purposes, as the ideal reactor is characterized by a single time constant (that depends on k and τ), while the nonideal reactor is characterized by as many time constants as the number of compartments that are needed to represent the mixing pattern. For instance, using the variables defined previously, the equation becomes:

$$\frac{d^2 C_A^I}{dt^2} + \left(\frac{1}{\tau_1} + \frac{1}{\tau_2} + 2k\right)\frac{dC_A^I}{dt} + \left[\left(\frac{1}{\tau_2} + k\right)\left(\frac{1}{\tau_1} + k\right) - \left(\frac{\phi}{\alpha\tau}\right)\frac{1}{\tau_2}\right]$$

$$C_A^I = \left(\frac{1}{\alpha\tau}\right)\left(\frac{1}{\tau_2} + k\right)C_A^e$$

Using the previously defined numerical data, for $\phi = 1$ and $\alpha = 0.5$, $\tau_1 = 0.25$ h and $\tau_2 = 0.5$ h. Then the equation becomes:

$$\frac{d^2 C_A^I}{dt^2} + 8\frac{dC_A^I}{dt} + 11 C_A^I = 6000$$

while for the ideal stirred tank reactor the mass balance equation is equal to:

$$\frac{dC_A^I}{dt} + 2C_A^I = 1000$$

Assuming the initial concentrations are $C_A^I(0) = C_A^{II}(0) = 0$ the analytical solutions are, respectively:

$$C_A^I(t) = 545.45 - 232.07e^{-6.24t} - 313.38e^{-1.76t}$$

$$C_A^{II} = 0.5 \left[\frac{dC_A^I}{dt} - 2C_A^e + 5C_A^I \right]$$

$$C_A(t) = 500 - 500e^{-2t}$$

as illustrated in Figures 23.3.4 and 23.3.5. As one can see, the nonideal compartmental model contains two time constants, while the ideal reactor is described by a single time constant. Therefore, obtained dynamic responses are different in both cases. As shown in Figure 23.3.4, the variation of the reactant concentration in the output stream is faster in the nonideal reactor, because of the shortcut zone. As shown in Figure 23.3.5, the variation of the reactant concentration in the stagnant zone is slower in the nonideal reactor, because of the second-order process, as feeding of the stagnant section depends on the previous increase of concentrations in the shortcut section. Therefore, very complex dynamic patterns can evolve if multiple mixing zones are present in the reacting system.

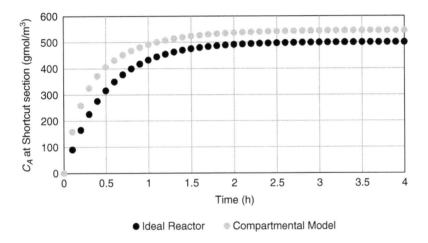

Figure 23.3.4 Dynamic reactant concentration responses in the shortcut section of the reactor.

E23.3.1 Example 23.3.2. Occurrence of transport delays can also affect concentration responses in reaction systems. For instance, reaction can take place in feed and output lines, so it can be interesting to investigate the compartmental model illustrated in Figure 23.36. In this case, the feed and output sections can be described with standard plug flow reactor models, leading to reactor configurations in which the distinct reactor sections exchange mass (and energy, when the reaction is not performed isothermally). Delays associated with transportation of mass and occurrence of undesired reactions outside the borders of the reaction equipment are not unusual in industrial sites.

Assuming a first-order reaction rate model can be used to represent the reaction:

$$A \xrightarrow{k} B$$

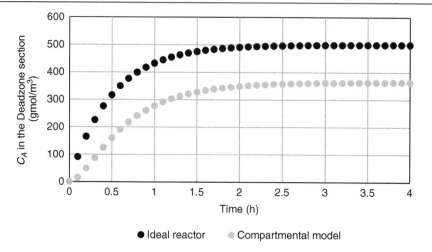

Figure 23.3.4 Dynamic reactant concentration responses in the deadzone section of the reactor.

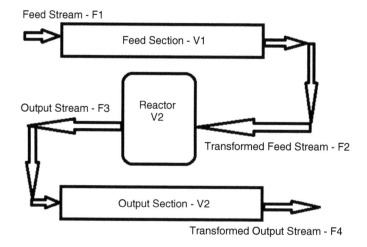

Figure 23.3.4 Example of reactor configuration with feed and output transformation effects.

then three mass balance equations must be written of the form:

$$\frac{\partial C_A^I}{\partial t} + \frac{F}{A_1}\frac{\partial C_A^I}{\partial z} = -kC_A^I$$

$$\frac{dC_A^{II}}{dt} = \frac{F}{V_2}C_A^I(L_1) - \frac{F}{V_2}C_A^{II} - kC_A^{II}$$

$$\frac{\partial C_A^{III}}{\partial t} + \frac{F}{A_3}\frac{\partial C_A^{III}}{\partial z} = -kC_A^{III}$$

where it was assumed that the volumes of the three reactor sections remain constant during the operation. L_1 and L_3 represent the lengths of the tubular reactors, while A_1 and A_3 represent the sectional flow areas of both tubular reactors. As one can see, the final reactor behavior is different from the ones discussed in Sections 14.3 and 14.4. For example, at steady-state conditions, it is possible to write:

$$\frac{dC_A^I}{dz} = -\frac{kA_1}{F}C_A^I$$

$$0 = \frac{F}{V_2}C_A^I(L_1) - \frac{F}{V_2}C_A^{II} - kC_A^{II}$$

$$\frac{dC_A^{III}}{dz} = -\frac{kA_3}{F}C_A^{III}$$

Therefore,

$$C_A^I(L_1) = C_A^e e^{-\frac{kA_1 L_1}{F}} = C_A^e e^{-k\tau_1}$$

$$C_A^{II} = \frac{C_A^I(L_1)}{\left(1 + k\frac{V_2}{F}\right)} = \frac{C_A^e}{1 + k\tau_2} e^{-k\tau_1}$$

$$C_A^{III}(L_3) = C_A^{II} e^{-\frac{kA_3 L_3}{F}} = \frac{C_A^e}{1 + k\tau_2} e^{-k(\tau_1 + \tau_3)}$$

The interesting point regarding this compartmental configuration is that the output responses become delayed because of the flow through the tubular sections, in the form:

$$C_A^I(L_1, t) = C_A^e\left(t - \frac{A_1 L_1}{F}\right) e^{-\frac{kA_1 L_1}{F}} = C_A^e(t - \tau_1) e^{-k\tau_1}$$

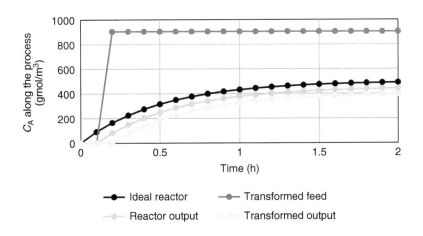

Figure 23.3.7 Dynamic reactant concentration responses in the different sections of the reactor.

$$\frac{dC_A^{II}}{dt} = \frac{C_A^e \, (t - \tau_1)}{\tau_2} e^{-k\tau_1} - \left(\frac{1}{\tau_2} + k\right) C_A^{II}$$

$$C_A^{III} \, (L_3) = C_A^{II} \left(t - \frac{A_3 L_3}{F}\right) e^{-\frac{kA_3 L_3}{F}} = C_A^{II} \, (t - \tau_3) \, e^{-k\tau_3}$$

Assuming $V_1 = 0.1 \text{ m}^3$, $F = 1 \text{ m}^3/\text{h}$ ($\tau_1 = 0.1$ h), $V_2 = 1 \text{ m}^3$ ($\tau_2 = 1$ h), $V_3 = 0.1 \text{ m}^3$ ($\tau_3 = 0.1$ h), $C_A^e = 1000 \text{ gmol/m}^3$ and $k = 1 \text{ h}^{-1}$, Figure 23.3.7 can be used to illustrate the dynamic response of output reactant concentrations when $C_A(t= 0) = 0$ throughout the reacting system. As one can see, the output response is delayed 0.2 h ($\tau_1 + \tau_3$) is converges to a lower concentration value (when compared to the ideal reactor) because of the reaction in the feed and output lines.

Part VI

Catalysis

Chapter 24

Catalysis: Analyzing variables influencing the catalytic properties

24.1 INTRODUCTION

Catalysts are very important for many applications and different processes, in particular in the chemical and petrochemical industries, also in energetic and environmental processes, and nowadays in material developments. Generically, catalytic reactions are classified into *homogeneous* and *heterogeneous or heterophase systems*.

Although most catalytic processes have been solved, there are new processes and catalyst developments aiming higher efficiency in different industrial applications in homogeneous and heterogeneous processes. Older processes or catalysts can be optimized with respect to the performance and stability in view of many variables affecting activity, provoking deactivation, or poisoning. In particular, new and alternative catalysts have been studied and modified for environmental processes, fine chemistry, hydrotreating of heavy oils, hydrogen generation, C_1 chemistry, biomass processes, and new nanometric materials.

24.2 SELECTION OF CATALYSTS

Reactions are classified into groups or families and catalysts in different solid materials. For each family of reactions, there are one or more solid groups, like metals, oxides having particular specific activities. The reactions can be classified by reactivity or similar functionality and so are materials with specific properties.

In general, empirical developments are limited, focusing global kinetic parameters of different reaction family groups for specific material groups (metals or oxides) and, if possible, correlating the activity with material properties.

Table 24.1 shows a general classification of reaction families and materials.

24.3 ACTIVITY PATTERNS

24.3.1 Model reactions

The activity and selectivity of catalysts can be evaluated using model reactions. However, it is advisable to remember some important characteristics of these reactions.

Table 24.1 Classification of reactions and materials.

Solid Groups	Reaction Families (Groups)	Suggested Catalysts
Conductors (metals)	Hydrogenation Hydrogenolysis Dehydrogenation Oxidation	Fe, Co, Ni Ir, Pt Ru, Rh, Pd Ag, Cu, Zn
Semiconductors (oxides and sulfates)	Oxidation Reduction Hydrogenation	NiO, CuO, ZnO Co_2O_3, Cr_2O_3 WS_2, MoS_2
Insulators (zeolites and acids	Oxidation Redution	NiO, CuO, ZnO Co_2O_3, Cr_2O_3
	Hydrogenation	WS_2, MoS_2
Polymerization		

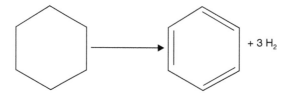

Figure 24.1 Cyclohexane dehydrogenation.

First, is the reaction structure sensitive or insensitive? According to Boudart,[4] supported catalysts have dispersed metallic particles of different sizes d. With increasing or decreasing particle sizes, the concentration of accessible atoms or sites varies significantly, indicating surface structural changes. Structure sensitive reactions (SSRs) are those reactions where the intrinsic reaction rate relative to the accessibility of surface sites or the turnover frequency (TOF) varies with particle sizes or diameters, but it does not vary for structure insensitive reactions (SIRs). It means, for SSRs, the TOF depends on particle sizes, or dispersion of particles or on the accessible surface sites of the catalyst. On the other hand, for SIRs, the activity or TOF is independent of the particle sizes.

Dehydrogenation reactions are known as SIRs, while hydrogenolysis reactions are SSRs. Some reactions include several steps simultaneously with different reactions, which are structure sensitive or insensitive.

24.3.2 Cyclohexane dehydrogenation

This is an SIR. The reaction rate is directly proportional to the surface active sites, and the intrinsic activity or TOF does not depend on particle diameters. The cyclohexane dehydrogenation (CHD) reaction forms only benzene and hydrogen as products, according to Figure 24.1, and occurs under atmospheric pressure and temperatures varying between 250°C and 300°C.

Table 24.2 shows the results of CHD for a Pt/Al_2O_3 catalyst for low concentrations. The Pt dispersion on alumina is almost 100%.

Table 24.2 Cyclohexane dehydrogenation on Pt/Al$_2$O$_3$.

Catalysts	Initial rate at 543 K (10^3 mol/h/g$_c$)	Dispersion H/Pt^a	TOF^b (s^{-1})	E$_a^c$ (kcal/mol)
0.7% Pt/Al$_2$O$_3$	179.5	1.1	1.2	27
0.9% Pt/Al$_2$O$_3$	161.5	1.0	1.0	24

Note: $P = 1$ atm; H$_2$/C$_6$H$_{12} = 13.2$.
^{a}Dispersion of Pt.
^{b}Activity.
^{c}Apparent activation energy.

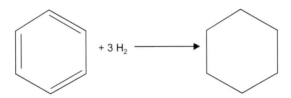

Figure 24.2 Hydrogenation of benzene.

Benzene is the final product for all tests. The activity or TOF values remain constant around $1\,\text{s}^{-1}$, suggesting that the rate is proportional to accessible active sites. From the Arrhenius plot, one determines the activation energy which is also constant.

24.3.3 Benzene hydrogenation

The hydrogenation of benzene is a model reaction for characterization of metallic surface sites which is an SSR. According to Figure 24.2, the final product is only cyclohexane.

Experiments were performed using palladium Pd^0 supported on carbon and were conducted isothermally, without interference of mass or heat transfer. Table 24.3 presents the results of Pd^0 dispersion and the specific activities in terms of TOF.

The activity or TOF values were calculated from the rate consumption of benzene at 373 K and from CO chemisorption measurements. TOF values increase with decreasing dispersions.

Table 24.3 Results of dispersion and activity.

Catalisador	Área BET (m^2/g$_{cat}$)	Dispersão (%)*	TOF^a (s^{-1})
10% Pd/C$_V$	309	20	0.64
9% Pd/C$_C$	876	6	2.84

Note: Carbon (C$_V$), graphite (C$_C$).
* CO chemisorption at 308 K

24.4 CONVENTIONAL PREPARATION METHODS OF CATALYSTS

The most important challenge is to confer desired characteristics to the solid, aiming to design specific catalysts. In most cases, there are conventional direct procedures

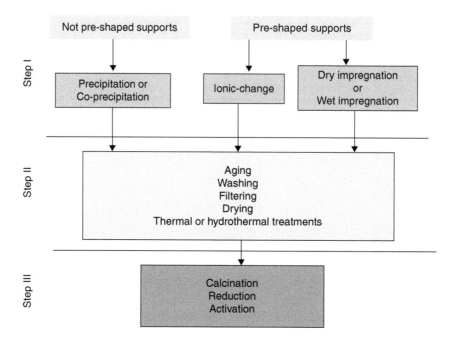

Figure 24.3 Scheme of catalysts preparation.

that confer desired properties; however, there are other new and innovative advanced methodologies in study, searching for more specific properties.

The formulation of a correct catalyst is a compromise between the fluid flowing system of the reactor and the activity or stability. The relative importance of these factors depends on the reaction system, the reactor design, reaction conditions, and economic factors.

Flow distribution and low pressure drop can be achieved selecting particles with appropriate sizes and formats and good mechanical resistance. The formulation or fabrication of a catalyst must attend a specific industrial process. The more severe are the conditions (charge, temperature, and high pressures or space velocities) the more difficult are the design parameters of the catalyst.

In general, catalyst can be classified into three categories:

1. Bulk catalyst
2. Impregnation on preshaped supports
3. Mixed/sintered materials

The preparation method depends on which application it is used in. In general, there are different steps as shown in Figure 24.3. The catalysts are formed by:

- Active phases
- Supports
- Promoters

Table 24.4 Examples of oxide supports.

Tipo	Oxides	Melting Point (°C)
Acids	γ-Al$_2$O$_3$	2318
	SiO$_2$	1973
	SiO$_2$–Al$_2$O$_3$	1818

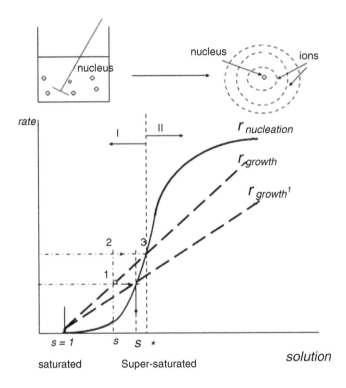

Figure 24.4 Rate of nucleation and growth.

Active phases

The active sites or phases are the most important components and responsible for reaction occurring. The first choice to design a new catalyst is the selection of the active component.

Supports

The supported catalyst may have different attributes but the most important are the specific surface areas and the stability of the catalyst. A support stabilizes the active phases and is thermally stable. Materials must support high temperatures, in particular for exothermic reactions. Therefore, these materials must have high melting points (Table 24.4).

The ideal and unique function of the support is to improve the dispersion of the active phases, without reactivity or inert for undesirable side reactions. Indeed, such characteristics can be achieved with oxides having high melting points.

The simultaneous presence of acid sites may favor reactions involving free radicals. In fact, with γ-Al_2O_3, there are undesirable site reactions, such as cracking, isomerization, and coke formations, and the later one favoring the deactivation of the catalyst.

However, there are cases where the acidity of the support influences positively the main reaction. In this case, the catalyst is bifunctional, and the best example is the reforming of hydrocarbons. The main objective of the reforming is the conversion of naphtha compounds with low octane number (paraffins and naphtenes) into branched or aromatic compounds with high octane index.

Promoters

The promoter is the third component of the catalyst that when added (in general small amounts) makes the catalyst more active, selective, and more stable. The aim of the promoter is to improve the stability of the support and of the active phases. The most used support is the gamma alumina (γ-Al_2O_3).

24.4.1 Precipitation/coprecipitation methods

In this section, we present a summary of the principal preparation methods, whose details can be found and described in several other references.

It is important to remember thermodynamic properties or in particular the solubility constants, besides the energy involved and verify if the process is thermodynamically possible. Very important is the precipitation kinetics and how far it is from the equilibrium which influences the precipitation rate.

There are two kinds of materials resulting from the precipitation: amorphous and crystalline or mixed materials, with different properties. For crystallites, the precipitation occurs in two steps (:

1. *Nucleation*: germination in a homogenous solution phase. The interaction of ions and molecules occurs in the solution with the formation of germs and clusters beginning the crystallization process. If the precipitation occurs in the presence of a support, it is a heterogeneous process where germination occurs due to the interaction between ions of the solution and the surface occurring nucleation and formation of crystallites.
2. *Growth*: it is a physical–chemical process where crystallite nucleus grows at the solid–solution interface; the rate growth depends on the kinetic rate.

Cerium oxide is a good example and one of the most significant materials for supports or catalyst in the reforming and partial oxidation processes. It is special, having high oxygen storage capacity, promoting the partial or total oxidation reactions. It is also a good support for metals, like platinum, palladium, and gold. The cerium oxide was prepared starting with a solution of 0.20 mol/L of cerium nitrate ($Ce(NO_3)_3 \times 6H_2O$ 99.0%, Vetec). Then, cerium hydroxide is precipitated after slow addition of 1.0 mol/L of ammonium hydroxide (NH_4OH 99.0%, Vetec) at room temperature. The precipitate was filtered under vacuum and washed with deionized water

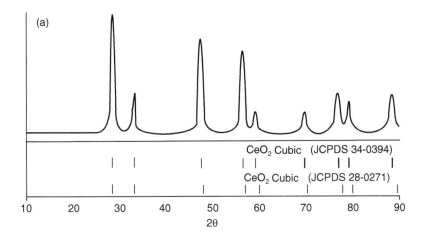

Figure 24.5 Diffraction pattern of CeO₂.

in a Buchner until pH 7. Then, it is dried at 110°C in a muffle and calcined at 500°C under air flux for 10 h. Figure 24.5 shows the diffraction pattern. The specific surface area was 71 m²/g and Figure 24.6 shows the microanalysis.

24.4.2 Impregnation of metals on supports

This is the most important method for preparation of industrial catalysts, and there are different methodologies, however, the most important is the impregnation over frame-shaped solids, like pellet or extruded. The solution with specified metal concentration, varying time, temperature, and pH transfers the active phase to the support, and after drying and calcination, it is fixed and stable. The support is in general an inert material or can be partially active; however, the main properties must be well defined. In this case, the volume of solution must be sufficient to fill all pores, known as dry impregnation. When the support is in powder form, the volume of solution needed is higher than the pore volume of the support, and the impregnation is known as wet impregnation. Thus, the impregnation method can be divided into two types:

- Without interaction
- With interaction

In the first case, the solution contains the active substance and the support is inert or partially active. The preparation can be dry or wet impregnation (Figure 24.7). In this method, the solution drops slowly in the Becker containing the solid.

$V_{\text{pores}} = V_{\text{solution}} \Rightarrow$ volume for dry impregnation
$V_{\text{solution}} > V_{\text{pores}} \Rightarrow$ volume for wet impregnation

Figure 24.6 Photo of CeO$_2$ sample prepared by precipitation. (NUCAT/PEQ/COPPE, FEG sample, Neto).

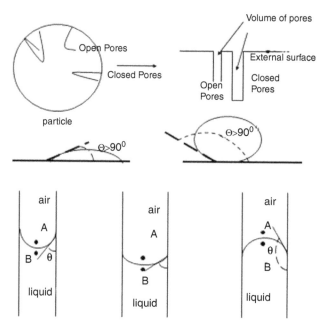

Figure 24.7 Scheme of pores.

The impregnation occurs by diffusion or wetting. For diffusion, the time estimated is:

$$t_{diffusion} = t_d = \frac{x^2}{D} \tag{24.1}$$

where x is the distance in pores and D is the diffusion coefficient (cm^2/s).

The impregnation time must be greater than the diffusion time:

$$t_{impregnation} > t_d$$

The impregnation time can also be estimated as follows:

$$t = \frac{4\mu.x^2}{\sigma.r} \tag{24.2}$$

where μ is the viscosity.

In the impregnation with interaction with the solid, there are covalent and ionic bindings. The pH is the most important parameter and indicates superficial electric charges. If the resultant electric charge is zero, it is known as zero electric charge and identified as isoelectric point (IP), where the electric charges are compensated. In this state, neither the solution nor the colloidal system does transport electric charges.

If pH > IP, then the surface or the solution has positive electric charges attracting anions, annulling total electric charges. In the colloidal solution, it forms amorphous aggregates. However, when pH < IP, then the negative superficial charge or the solution attracts cations which are compensated. In the colloidal solution occurs precipitation. Each oxide has its IP, or zero electric charge, for example:

Al_2O_3 PI pcz 8.0
TiO_2 PI pcz 5.0 – 6.0
SiO_2 PI pcz 2.0

Summarizing:

$$[M^+ \cdots O^{2-}] + H^+ \leftrightarrow M^+ - OH^- \leftrightarrow [M^+ \cdots OH_2]^+ + OH^- \tag{24.3}$$

If acid → surface (+) → exchange anions
If surface (−) ← basic → exchange cations

Thus, proton H^+ can exchange cations of type Ni^{2+}, Pd^{2+}, Pt^{2+}, and Au^{3+}. Hydroxyls OH can exchange complex anions.

24.4.3   Ion exchange

The ion exchange occurs between ions A of solid MO with ions B of solution S, according to the following reaction:

$$A_{MO} + B_S \leftrightarrow B_{MO} + A_S \tag{24.4}$$

The well-known example for ion exchange is Pt on zeolites. The exchange occurs between ions of solution and ions of zeolite as follows:

$$[Pt(NH_3)_4^{2+}]_s + 2[NH_4^+]_z \leftrightarrow [Pt(NH_3)_4^{2+}]_z + 2[NH_4^+]_s \qquad (24.5)$$

More details can be seen in the literature [Le Page].

Example

For example, we present here the monolithic support (Figure 24.8) which is an inert material ($2MgO \times 5SiO_2 \times 2Al_2O_3$). It is highly resistant and has been used in different processes. The main advantage is the low residence time or contact time of the order of milliseconds. It presents 400 cells in^{-2} ($D = 12$ mm, $L = 8$ mm) as shown in Figure 24.8.

First, γ-alumina was deposited by washcoating and then impregnated with the active phase. The γ-alumina was prepared with a solution which contains urea and deposited by dipcoating of small pieces. until reaching 10% concentration of Al_2O_3. The active phase was impregnated with a solution of $Ni(NO_3)_2 \times 6H_2O$ (Vetec) and concentration of 291 g/L, then dried and calcined and successively impregnated until reaching 6% concentration of nickel. The metal oxide content on the $NiO/Al2O3$ coated cordierite monolith catalysts was 5.3 wt. %, based on the total mass of the monolith support, after three cycles of immersion and heating treatment. The adherence was qualitatively measured by ultrasonic vibration test. The weight losses were 1.3% and 0.3% for /Al2O3 and NiO, respectively, after exposure to ultrasonic vibration for 30 min, which are extremely good when compared to reported values in the literature, about 4.0%. These results confirm that the urea method is efficient for washcoating Al_2O_3 phase and nickel oxide over a cordierite monolith.

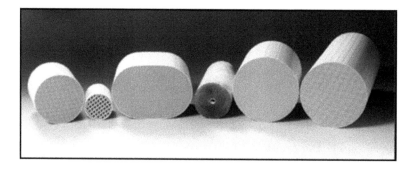

Figure 24.8 Monoliths of different forms and sizes.

The microscopic images (FEG) are shown in Figure 24.9. Figure 24.9B and C shows details of particles and fissures at the surface, respectively. Images indicate that the surface was completely covered by nickel particles.

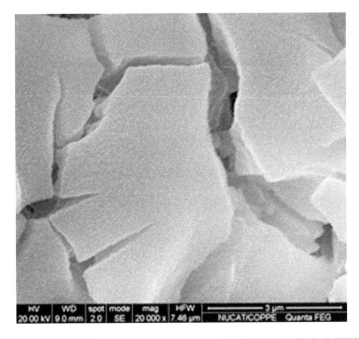

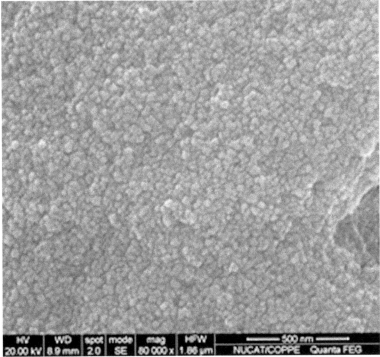

Figure 24.9 (Continued).

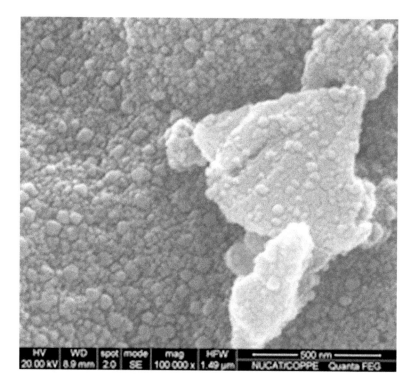

Figure 24.9 SEM microscopy images relative to NiO/Al$_2$O$_3$/cordierite: (A) 20 k, (B) 80 k, and (C) 100 k.

24.5 ANALYSES OF VARIABLES INFLUENCING FINAL PROPERTIES OF CATALYSTS

Variables influencing the preparation by precipitation, coprecipitation, precipitation–deposition, and impregnation transform chemical and physical properties approaching the equilibrium conditions. The main variables are:

- pH
- Aging time
- Temperature
- Precursors

24.5.1 Influence of pH

The pH values influence significantly the formation of intermediate complexes during the preparation by precipitation or precipitation–deposition. For example, gold catalysts are prepared with a gold precursor (HAuCl$_4$ solution), and pH influences the formation of gold complexes, the maximum content, and the presence of chlorine. The ions [AuCl$_4$]$^-$ form different complexes, as shown in Figure 24.10. Complexes

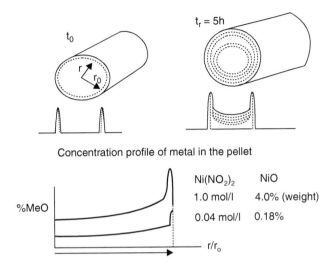

Figure 24.10 Gold complexes from a solution of 2.5 × 10^{-3} M HAuCl$_4$ (Lee and Gavriilidis, 2002) (Adapted).

like $[AuCl_{4-x}(OH)_x]^-$ ($x = 1-3$) are adsorbed at the surface with the formation of $Au(OH)_3$ species which are the precursors of gold nanoparticles.

24.5.2 Autoclaving

Autoclaving is performed in a batch reactor under high pressure and different temperatures. There are thermal and hydrothermal transformations in the presence of steam water, resulting in structural changes. In the presence of a solvent, the precipitated solid suffers structural and textural modifications, but in principle the main steps are:

1. *Dissolution*: part of the solid in the presence of a solvent or water undergoes solvation of the metallic ions with the rupture of bindings and formation of new bonds with the solid.
2. *Diffusion*: transfer of the solvated ion from the solid to the final solid.
3. *Dissolvation*: the inverse phenomena of the first step, where the ion goes to the solid and then is integrated in the new solid structure occurring again as precipitation.

The transformations occurring are:

Small crystallites → large crystallites
Small amorphous particles → large amorphous particles
Amorphous solid → crystalline solids
Crystallites (1) → crystallites (2)

24.5.3 Influence of time, concentration, and impregnation cycles

There are maximum concentrations that can be deposited on the surface. The difficulty is to impregnate high metallic concentrations in one step. When the support is preshaped, such as cylinders or spheres, the impregnation time is important. For short times, the metallic distribution is concentrated at the external surface. Long impregnation times improve the metal distribution inside pores of the pellet or extruded, as shown in Figure 24.11.

For higher metal concentrations, the successive impregnation is employed for several cycles until the desired concentration is reached. The metal distribution must be homogeneous. In practice, higher concentrations can be reached with 3–5 cycles (Figure 24.12).

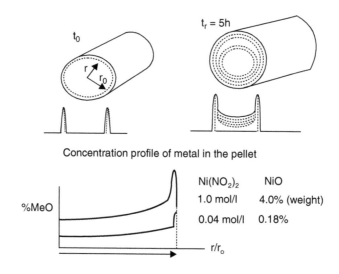

Concentration profile of metal in the pellet

	Ni(NO$_2$)$_2$	NiO
	1.0 mol/l	4.0% (weight)
	0.04 mol/l	0.18%

%MeO

r/r$_o$

Figure 24.11 Influence of time and concentration on pellets/extruded.

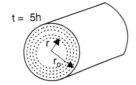

Concentration profile of Ni in the cylindrical pellet

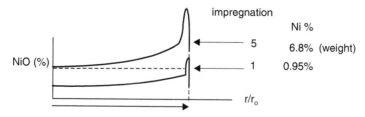

impregnation

	Ni %
5	6.8% (weight)
1	0.95%

NiO (%)

r/r$_o$

Figure 24.12 Impregnation cycles—metal concentration.

24.6 THERMAL TREATMENTS

24.6.1 Drying

During the drying process, the residues and solvents are eliminated, transforming gels into xerogels. The mass loss depends on the heating rate, and involves heat and mass transfer phenomena.

$$\dot{m} = a k_m (p_s - p_g) \tag{24.6}$$

$$\dot{q} = a' h (T_g - T_s) \tag{24.7}$$

where a and a' are the interface areas; k_m, h the mass and heat transfer coefficients, respectively; and p, T pressure and temperature of the gas (g) and solid (s), respectively. Figure 24.13 shows the mass loss and drying rates and the temperature of the different transformation phases.

The rate increases up to the equilibrium temperature at the surface, and then it remains constant. During this step, the temperature increases. With time, drying proceeds to the inner cake, and the rate as well as rate mass transfer decreases significantly with increasing temperature till the final temperature. The heat transfer rate is important to obtain a homogenous drying. In the drying processes, usually the rate is of the order of 10°C/min.

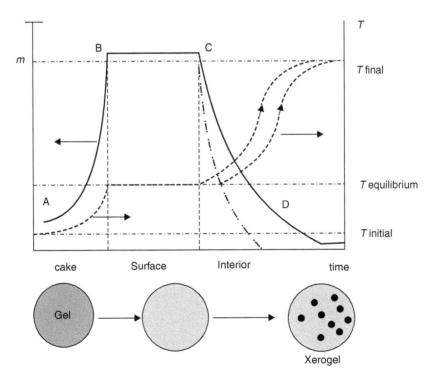

Figure 24.13 Drying of cake.

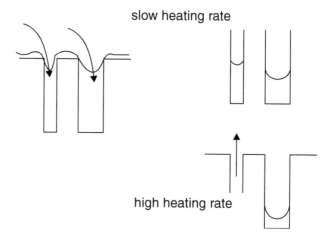

Figure 24.14 Drying of pores.

The drying process is also very important for wet impregnation and for the elimination of solvents. The higher the heating rate the lesser the contact of the gas with the surface, which favors the formation of smaller particles and in opposite, the slower the heating rate the higher the contact time of the drying gas with surface, and consequently bigger particles are favored, as shown in Figure 24.14.

24.6.2 Calcination

During the calcination with or without gases, there are decompositions of hydroxides, carbonates (gel, xerogel) transforming into oxides which are chemical reactions, transforming one compound to other compounds. During thermal treatments, there are following cases:

(a)   Decomposition

solid 1 → solid 2

$$2Al(OH)_3 \xrightarrow{\Delta} Al_2O_3 + 3H_2O$$

$$Ni(HCOO)_2 \xrightarrow{\Delta} NiO + CO_2 + H_2O \tag{24.8}$$

(b)   Presence of gas

solid + gas → solid 2 + gas

$$Ni(HCOO)_2 + O_2 \xrightarrow{\Delta} NiO + 2CO_2 + H_2O \tag{24.9}$$

$$MoO_2 + 2H_2S \xrightarrow{\Delta} MoS_2 + 2H_2O \tag{24.10}$$

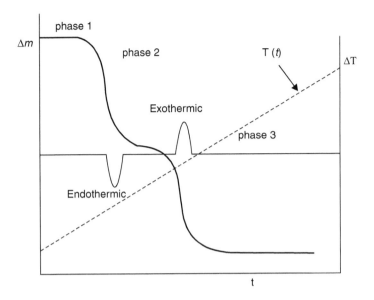

Figure 24.15 Thermal treatments—TGA and TDA.

(c)   Phase transformation

$$\gamma - Al_2O_3 \xrightarrow{\Delta} \theta - Al_2O_3 \xrightarrow{\Delta} \alpha - Al_2O_3 \tag{24.11}$$

Phase transformation can be followed by thermogravimetry, measuring the mass loss and the temperature difference with increasing temperature, as shown in Figure 24.15.

The effects of calcination are significant in modifying textural and morphological properties, affecting the surface area, pore volumes, and structures. There are also differences between bulk and supported materials. In the first case, one observes from Figure 24.16 that how these parameters change with increasing calcination temperature.

It shows that the specific surface area decreases with increasing temperature of calcination.

Starting from precursors, like nitrates or sulfates, the specific surface area increases drastically reaching high values and then decreases due to structural crystalline rearrangement. Starting from crystals, like alumina or silica-alumina, with increasing calcination temperature, the specific surface area decreases due to sintering or agglomeration of crystallites. The classical case is the alumina. Starting from γ-Al_2O_3 with a surface area of around 200 m^2/g, it is transformed into α- Al_2O_3 which presents a surface area of the order of 20 m^2/g.

On supported catalysts, the effect of calcination is significant on particle sizes, dispersions, and crystallite sizes. Measurements of particle sizes or metal dispersions after calcination and reduction of supported catalysts indicate different situations and there are three cases that are illustrated below.

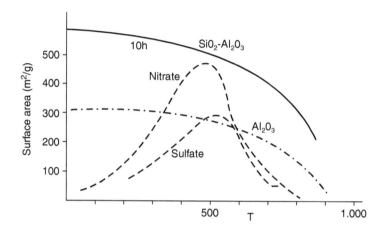

Figure 24.16 Specific surface area as a function of the calcination temperature adapted from reference Le Page.

Table 24.6 Influence of calcination temperature and digestion time.

Ni (%)	$T_{calcination}$ (°C)	$t_{digestion}$ (h)	$S_{metallic}$ (m²/g$_{Ni}$)	Reduction (%)
9	400	1	4.3	7
9	Without calcination	–	10	16
15	400	4	5.2	5
15	Without calcination		14.6	14

24.6.2.1 Effect of calcination and digestion over reduction and metallic surface area

The first case illustrates how the calcination temperature and digestion time affect metallic area and degree of reduction of the alumina-supported NiO for different Ni contents.

Results indicate that the reduction without calcination presented higher reduction of NiO and higher metallic area for different Ni contents. After calcination, both reduction and metallic area decreased significantly, independent of the digestion time, which suggests agglomeration of particles or interaction with the support during the calcination step. The digestion time did not affect the metallic area and the reduction.

Figure 24.17 displays the effect of calcination and reduction temperature on the metallic Ni⁰ area. The precursors were calcined separately at 300°C and 500°C and reduced with H₂. When calcined at 500°C, the metallic area increases with increasing reduction temperature allowing the reduction of big and small particles at the surface. However, when calcined at 300°C, the metallic area indicates two different situations.

In the initial situation between range A and B, there are big NiO particles which are reduced to metallic Ni⁰:NiO → Ni⁰.

In the range B and C, the metallic area decreases because NiO interacts with the support and the reduction of these particles is more difficult.

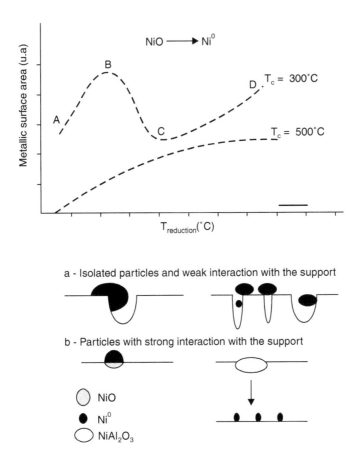

Figure 24.17 Effect of the calcination temperature over the metallic area adapted from Le Page et al. 1978.

In between C and D, the metallic area increases due to the reduction of nickel particles with strong interaction with the support and the reduction of nickel aluminate $NiAl_2O_3 \rightarrow Ni^0 + Al_2O_3$. These particles are not easily reduced to metallic Ni^0.

Therefore, there are following situations as shown in Figure 24.17.

24.6.2.2   Influence of calcination temperature over dispersion and particle sizes

The effect of calcination temperature on particle diameters of a Pt/SiO_2 catalyst after reduction is displayed in Figure 24.18. The precursor was calcined and present as PtO_2 or PtO. After reduction, it is transformed into metallic Pt^0. Figure 24.18 shows the effect of calcination temperature on dispersion and particle sizes.

The dispersion decreases drastically whereas particle diameters increase with increasing temperature. It is assigned to crystal growth or strong interaction with the support or sintering and therefore results in the formation of silicates which are

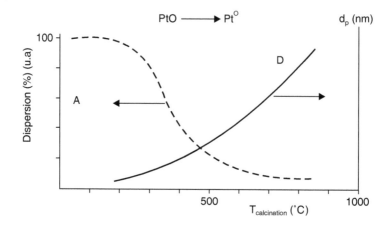

Figure 24.18 Effect of calcination temperature on particle diameters adapted from Le Page et al. 1978.

difficult to reduce. With increasing temperature, there are migrations of particles or growth from smaller to bigger particles.

24.7   EFFECT OF REDUCTION TEMPERATURE ON INTERACTION AND SINTERING

The influence of the reduction temperature on the metal support interaction and particle agglomeration occurs frequently on supported catalysts and will be illustrated for nickel-supported catalysts. It depends on the support, the metal concentration, the calcination temperature, the reducing agent, the H_2 concentration, and the presence of water. When promoted with potassium, lanthanum, or any other promoter, the reducibility and the metallic distribution improve. After calcination, nickel is converted as oxide (NiO), also as suboxide (Ni_2O_3) and as aluminate form.

$$NiO + Al_2O_3 \rightarrow NiAl_2O_4 \tag{24.12}$$

The reduction of nickel oxide after calcination occurs as follows:

$$NiO + H_2 \rightarrow Ni^0 + H_2O, \quad \Delta H = 0.4\,\text{kcal / moles} \tag{24.13}$$

If the reduction is direct of the precursor nitrate, then the following reactions may occur:

$$Ni(NO_3)_2 + 2H_2 \rightarrow Ni^0 + H_2O + 2NO_2, \quad \Delta H = 11.9\,\text{kcal / moles} \tag{24.14}$$

$$Ni(NO_3)_2 + 4H_2 \rightarrow Ni^0 + 4H_2O + 2NO_2, \quad \Delta H = -86.5\,\text{kcal / moles} \tag{24.15}$$

$$Ni(NO_3)_2 + 9H_2 \rightarrow Ni^0 + 6H_2O + 2NH_3, \quad \Delta H = -261.5\,\text{kcal / moles} \tag{24.16}$$

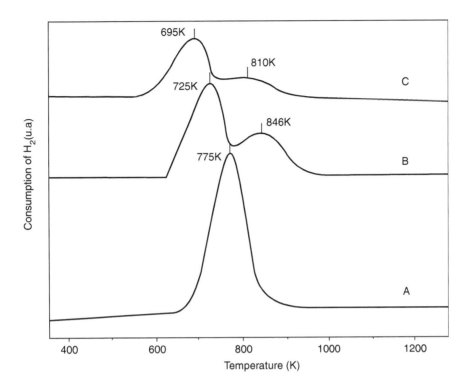

Figure 24.19 TPR : (A) α-Al$_2$O$_3$, (B) 8%Ni/α-Al$_2$O$_3$, and (C) 16%Ni/α-Al$_2$O$_3$.

24.8 INFLUENCE OF THE SUPPORT AND METAL CONCENTRATION OVER THE REDUCTION

The supports γ-Al$_2$O$_3$ and α-Al$_2$O$_3$ exhibit different structures and have specific surface area of around 200 m^2/g and <10 m^2/g, respectively. Typical reduction profiles of supported NiO, with 8% and 16% content are shown in Figures 24.19 and 24.20.

These figures show two reduction peaks of nickel species those Alberton et al. attributed to the reduction of NiO without interaction with the support for lower temperatures, while for higher temperature there are reduction of NiO species linked to Al^{3+} species of the alumina support formed during the impregnation step.

The α-Al$_2$O$_3$ sample shows shifting of the maximum reduction temperature with increasing metal content. Comparing the first reduction peak of the supported samples with the reduction of bulk NiO, it turns out that with increasing metal content, the reduction is facilitated. Results show the same behavior for the second peak. According to Li and Chen, the reduction rate of NiO depends, besides other factors from the nucleation of metallic Ni. Indeed, higher metal concentrations favored the reduction of Ni species and so the reduction rate.

The degrees of reduction of the 8% and 16%Ni/α-Al$_2$O$_3$ were 59.4% and 22.0%, respectively. In fact, there is a significant amount of Ni species that is not reduced,

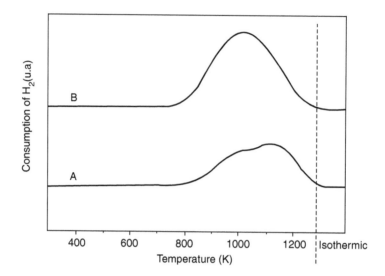

Figure 24.20 TPR analyses of bulk NiO and supported catalysts: (A) 8%Ni/γ-Al$_2$O$_3$ and (B) 16%Ni/γ-Al$_2$O$_3$.

although under H$_2$ atmosphere and temperatures up to 1000°C. It seems that the catalyst with higher metal content has lower reducible Ni sites.

The absence of NiO on the supported catalysts γ-Al$_2$O$_3$ is probably due to the high-specific surface area of the support. It suggests high dispersion of Ni species which under calcination at 550°C provoked the formation of NiAl$_2$O$_4$, as detected by XRD. Indeed, H$_2$ chemisorption measurements do not allow determining dispersions of Ni particle sizes. TEM or XRD *in situ* reduction analyses succeeded. Figure 24.21 shows the diffraction of 8%Ni/α-Al$_2$O$_3$. The NiO lines were not more observed above 500°C, but typical lines of metallic Ni are observable. Thus, only metallic Ni was detected by XRD above 653°C. From the diffraction line (1 1 1) at 44.5°, 0.232 rad, we calculated the crystallite sizes (d_{Ni}) 45 Å and thus the dispersion (D_{Ni}) 1.9% of Ni.

24.9 INFLUENCE OF THE HEATING RATE

The influence of the heating rate of reduction after calcination and of direct reduction on the metallic area and dispersion of the Ni/Al$_2$O$_3$ catalyst is illustrated in Table 24.7.

As seen, heating rates do not affect the dispersion or metallic area of the calcined sample, but influence these variables on catalysts after direct reduction, evidencing higher dispersion with lower heating rates.

24.10 INFLUENCE OF VAPOR

The effect of the presence of H$_2$O vapor on the reduction of a 10% nickel-supported catalysts is shown in Figure 24.22. The reduction was performed with a mixture of

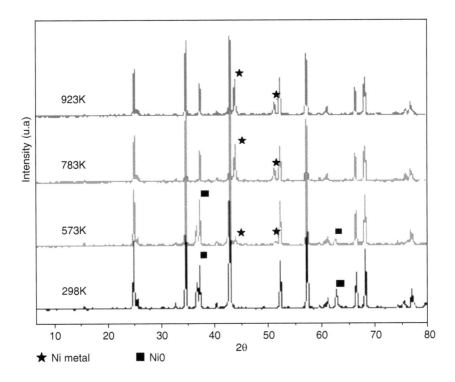

Intensity (u.a)

923K

783K

573K

298K

10   20   30   40   50   60   70   80

2θ

★ Ni metal   ■ NiO

Figure 24.21 XRD *in situ* of reduced 8%Ni/α-Al$_2$O$_3$.

Table 24.7 Effect of heating rate.

$T_{calcination}$ (°C)	Heating Rate (°C)/min	Metallic Area, $S_{metallic}$ (m^2/g)	Dispersion (%)
400	15	3.2	5.1
400	5	4.3	6.9
S	15	6.5	6.1
S	5	14.6	13.8

10%H$_2$/Ar and without (a) or in the presence of vapor (b). Profiles evidence that vapor influences the reduction temperature shifting the maximum peak to higher temperatures.

The first peak is associated to the reduction of NiO with low interaction with the support, while the second and third peaks, which are partially superimposed, are associated to the reduction of NiO interacting with the support, thus, to the reduction of nickel aluminates. The reduction degree was 100% and the particle sizes of 200 Å. XRD data confirmed the existence of metallic Ni, aluminates and lanthanum aluminates in this sample after reduction.

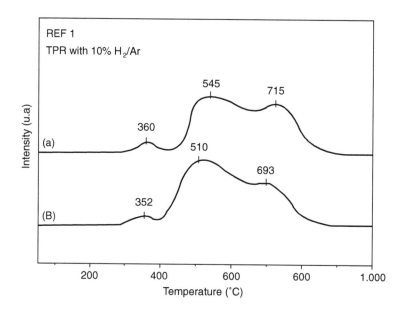

Figure 24.22 TPR profiles of a nickel sample and with 10%H_2/Ar, without (a) and with water vapor (b).

24.11   EFFECT OF TEMPERATURE AND REACTION TIME

The reaction temperature, mainly for exothermic reactions, affects the structure and textural properties of the catalysts which cause deactivation. The consequences are structural modifications or sintering of the catalyst. The most important identification of sintering is the analysis of crystal sizes or particle sizes before and after reaction.

Frusteri et al. measured the particle sizes by TEM analysis of a 21%Ni/MgO sample for ethanol reforming at 650°C with H_2O/EtOH after 20 h reaction, as well as, the influence of alkaline metals. Figure 24.23 displays the particle size distribution before and after reaction, showing the presence of big particles evidencing sintering. When doped with Li and K, the deactivation is less affected.

24.12   STRONG METAL SUPPORT INTERACTION

Transition metal oxides are easily reduced promoting interaction with metals of group VIII. This phenomenon was designated by Tauster as strong metal support interaction (SMSI). The main characteristic of the interaction is the lower metal surface exposition, drastically decreasing the chemisorption capacity of H_2 above 500°C. The TiO_2 is an n-type semiconductor, with defects in the lattice conferring special conditions for interactions and hence geometrical or electronic effects.

Horsley assigned the SMSI effect of Pt/TiO_2 to charge electric transfer from titanium to platinum atoms in the covalent bonding Pt–Ti that becomes stronger than the bonding between Pt–Pt atoms. When reduced with H_2, it forms oxygen vacancies producing Ti^{3+} ions which attract platinum atoms, modifying the electronic density.

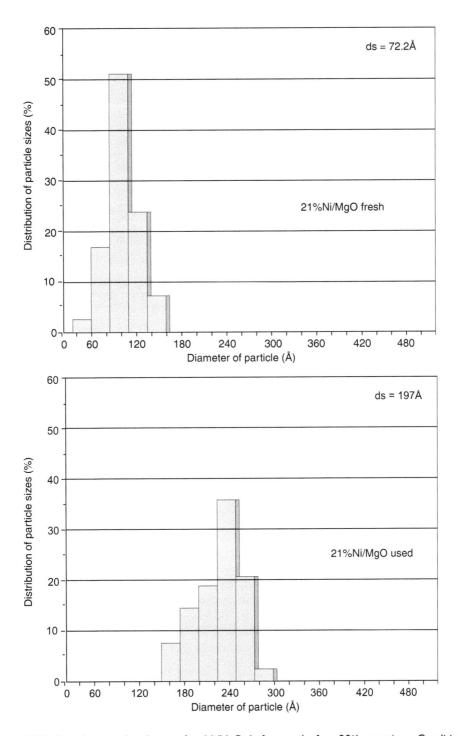

Figure 24.23 Particle size distribution for Ni/MgO before and after 20°h reaction. Conditions: $GHSV_V = 667\,mL_{EtOH}/(min\,mL_{cat})$, $T = 923\,K$, and ratio $H_2O/EtOH = 3$ (Frusteri et al., 2004a).

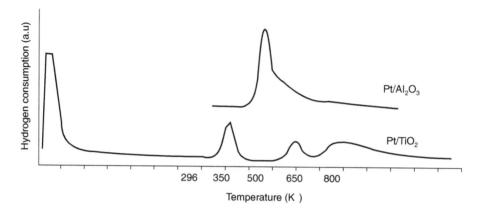

Figure 24.24 TPR profiles of 1% of platinum supported on Al_2O_3 and TiO_2.

Table 24.8 H_2 consumption.

Catalyst, % Pt° (weight)		TPR, μmol of H_2/g_{cat}	Chemisorption Dynamic,[b] μmol of H_2/g_{cat}	H/M
Pt/Al$_2$O$_3$	1.01	84.70	11.3	0.54[c]
Pt/TiO$_2$	1.03	182.90	4.4	0.17

Note: [a]After atomic absorption.
[b]At 300 K.
[c]82% reduction.

The reduction profiles of platinum-supported catalyst (1%Pt w/w) are displayed in Figure 24.24. The Pt/Al$_2$O$_3$ catalyst exhibits a maximum peak at 250°C with a shoulder at 370°C, in agreement with the literature. The hydrogen consumption is presented in Table 24.10 and corresponds to the reduction of Pt^{4+} to Pt^{0+}.

However, the Pt/TiO$_2$ catalyst exhibits reduction at room temperature and at 112°C, 372°C, and above 500°C, besides peak shift to lower temperatures due to the reduction of PtO. Partial reduction of TiO$_2$ to TiO$_{2-x}$ occurs at higher temperatures which are attributed to SMSI. The presence of Pt atoms promoted the reduction of titanium which in turn facilitates the reduction of platinum at the surface.

The calculated dispersion of Pt atoms at the surface of alumina was 54%, considering the reduction degree of 82% of platinum oxide (Table 24.8). On the other hand, the Pt/TiO$_2$ catalyst has two times more hydrogen consumption than the alumina-supported catalyst, indicating reduction of TiO$_2$ and thus interaction with metallic Pt^0. The H$_2$ chemisorption is three times less (17%) than the Pt/Al$_2$O$_3$ catalyst and confirms SMSI effect after reduction at 500°C.

Jiang and collegues claim that the Ti^{3+} cations may adsorb hydrogen and then diffuse through the support, decreasing the adsorption rate of hydrogen dissociation.

Table 24.19 presents the mean particle sizes of metallic Pt^0 from CO chemisorption and TEM analyses of these samples. The results are in good agreement and in accordance with XRD analyses, suggesting mean particles sizes less than 5 nm.

Table 24.9 Mean particle sizes (d_p) from CO chemisorption and microscopic analyses (TEM)[31].

		573 K[a]		773 K		773 K
	Pt (%peso)	$Sg_{Pt} \times 10^{-2}$ (m^2/g)	$d_{p\ CO}*$ (nm)	$Sg_{Pt} \times 10^{-2}$ (m^2/g)	$d_{p\ CO}*$ (nm)	$d_{p\ MET}$ (nm)
Pt/Al$_2$O$_3$	1.01	0.94	3.0	0.82	3.4	2.0–4.0
Pt/TiO$_2$	1.03	1.95	1.4	0.60	4.7	1.8–4.0

[a]After reduction, $*d_p = 6/\rho Pt.SgPt18$, $\rho Pt = 21.45$ g/cm^3.

TPR and chemisorption results confirm the SMSI effect after reduction with H$_2$ above 500°C. Indeed, oxygen vacancies are formed, producing Ti^{3+} ions, attracting Pt atoms. This effect is caused by charge transfer of TiO$_2$ atoms to platinum atoms through the covalent bonding Pt–Ti.

24.13 EXPERIMENTAL DESIGN—INFLUENCE OF PARAMETERS ON THE CATALYTIC PERFORMANCE

The main goal is to design minimum number of experiments needed for the selection of variables influencing catalytic properties. The effect of experimental conditions on the parameters used for evaluating the performance using the minimum number of experiments (runs) was proposed using a full factorial design of experiments.

Experimental design is an important tool to determine the dependence of variable and to analyze statistically and coherently the experimental results. The classical method is the full factorial design.

The idea is to design the experiments that provide exact information about the variables and thus defining the main objectives and selecting the appropriate techniques. For each target, there are associated techniques. Knowing the most important variables and the effect of these variables on the system and the interactions between them, then the full factorial design may predict minimum experiments.

In general, the system can be represented by a functional equation containing correlated factors or variables (entrance) and responses (exit). In particular, the full factorial design requires specifying the upper and lower limits of each factor. This method suggests two limits which are maximum value represented by +1 and minimum value represented by −1.

Besides, it is important to estimate errors for evaluation of these effects. Therefore, deviation of different factors indicate how experiments are reliable, assuming the same error for similar experiments. These calculations allow determining the main effects and the interaction effects.

The main effect (b_1) of a factor (x_1) is defined as the difference between the average response when the maximum value is ($y(x_{1+})$) and the minimum value is ($y(x_{1-})$), according to equation:

$$b_1 = y(x_{1+}) - y(x_{1-})$$

Without interaction between the factors ($x_1, x_2, ..., x_n$), the effect of a specific factor (x_j) is equivalent to the effects in the lowest level, when the other factors are in highest level. Therefore, the difference between these effects can be used as a measure of the interaction x_j with the other factors. In fact, consistent with the definition of the main effect, the *interaction effect* is defined as the difference between two averages, and thus, the tests must be separated into two groups, according to the levels of each factor ($+1$ or -1). For example, the interaction between factors x_1 and x_j can be calculated by doing separate tests where both factors assume the maximum (x_{1+} and x_{j+}) and minimum (x_{1-} and x_{j-}) levels and where the factors assume these levels alternatively (x_{1+} and x_{j-} or x_{1-} and x_{j+}). Therefore, the interaction effect between x_1 and x_j, or b_{1j}, can be calculated by the difference between averaged responses of each testing groups, according to equation:

$$b_{1j} = [y(x_{1+}, x_{j+}) + y(x_{1-}, x_{j-})]/2 - [y(x_{1+}, x_{j-}) + y(x_{1-}, x_{j+})]/2$$

where $y(x_{1+}, x_{j+})$ is the response when the factors assume maximum values, $y(x_{1-}, x_{j-})$ the responses when the factors assume minimum values, and analogous $y(x_{1+}, x_{j-})$ and $y(x_1, x_{j+})$ the responses when these factors assume maximum and minimum values alternatively.

In fact, the levels are represented as $+1$ for the highest and -1 for the lowest level. For calculation of the interaction effect between two or more factors, tests are grouped and those factors which are significant assume levels that multiplied result in $+1$ or -1.

For example, in the selection of a catalyst for soot combustion, the evaluation of the combustion temperature (T_C), the selectivity of CO_2 (S_{CO_2}), and possible interaction between these factors can be calculated using inlet variables. The main factors are the ratio of catalyst and particulates (cat:PM), heating rate, and gas inlet flow rate. From the experimental oxidation profiles and the amount of CO and CO_2, we determined the combustion temperature T_C, as the temperature for maximum CO_2 formation, or selectivity (S_{CO_2}), after complete conversion of particulates (PM).

Two levels, maximum ($+1$) and minimum (-1) values, were attributed for these three factors and according to the full factorial design resulted in eight tests.

1. Ratio cat:PM: 2:1 and 95:1
2. Heating rate: 2 K/min and 20 K/min
3. Flow rate O_2/He: 5 mL/min and 115 mL/min

Therefore, the average values were calculated and following values were obtained: Ratio cat:PM: 49:1, heating rate: 11 K/min, and flow rate O_2/He: 60 mL/min.

Table 24.10 presents the values of combustion temperature (T_C) and selectivity of CO_2 (S_{CO_2}) according to the experimental design and the averages, deviation values of T_C and S_{CO_2}.

Table 24.10 presents also tests with corresponding deviation values and excellent reproducibility.

The flow rate affected the CO_2 profiles as shown in Figure 24.25. As observed, there are interactions between the rate cat:PM and heating rates when the flow is low, or for high heating rates, elevated temperatures, while for high cat:PM ratios, the reaction is not limited and reaction occurs at lower temperature. The oppositeoccurs

Table 24.10 Combustion temperature (T_C), selectivity of CO_2 (S_{CO_2}), and deviation values.

Test	Factors			Responses		
	Cat:PM 2:1–95:1	R_{aq} 2–20 K/min	Flow 5–115 mL/min	S_{CO_2} (%)	T_c (K)	
1	−1	−1	−1	98	856	
2	+1	−1	−1	97	787	
3	−1	+1	−1	97	>923	
4	+1	+1	−1	99	797	
5	−1	−1	+1	56	825	
6	+1	−1	+1	75	821	
7	−1	+1	+1	79	883	
8	+1	+1	+1	93	879	
PC	0	0	0	82	863	Average ± deviation
PC	0	0	0	90	860	S_{CO_2} T_c
PC	0	0	0	83	863	85 ± 4% 859 ± 5 K

Note: Symbols +1, −1, or 0 indicate, maximum and minimum values for each factor.

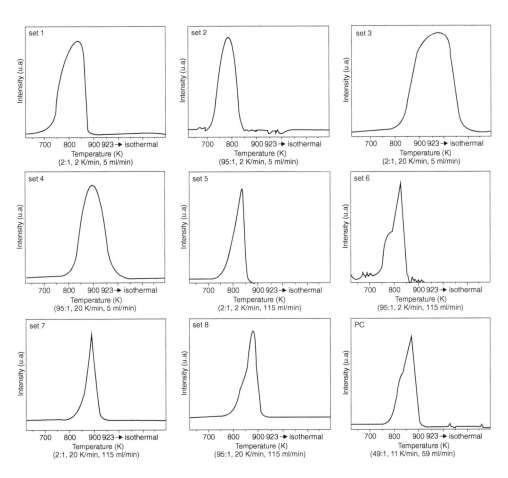

Figure 24.25 TPO analysis of the combustion of particulates (PM mixed with catalysts (cat). Profiles of CO_2 varying the ratio cat:PM, heating rate and gas flow rate [34].

for low heating rates and cat:PM ratios. These results are in good agreement with the expected interaction factors cat:PM and heating rate parameters.

From these results, it turns out that the combustion temperature is an important parameter for testing and comparison of the catalytic performance with certain limitations, since it can be strongly influenced not only by activity tests but also by experimental measurements. Moreover, it depends on the maximum peak temperature and precise evaluations.

24.14 CONCLUSION

The main goal of this chapter is to present the catalytic properties under the influence of different variables before and after treatments, after a brief introduction of the catalyst preparation, with specific examples showing affected differences and not characterization methods which were reproduced partially from my book on heterogeneous catalysis, published before. It shows how industrial catalysts are affected and modified during pre- and posttreatment or reaction conditions.

Part VII

Practices

Chapter 25

Experimental practices

In this final chapter, we present some laboratory practices to apply the theory learned in previous chapters and also demonstrate some practical examples of kinetics and reactors. The practical examples aimed to determine the kinetic rates and activation energies. Whereas reactor studies demonstrate to determine the performance in relation to contact time and yield of different types of reactors presented in this book.

25.1 REACTIONS IN HOMOGENEOUS PHASE

25.1.1 Free radical polymerization of styrene

A polymer may be formed by hundreds, thousands, or even tens of thousands of monomeric units. This material can also be present in natural form, for example, cellulose and rubber. The characteristics of these macromolecules depend on the monomer, chain length, and composition of the mixture. The mechanism and kinetics have been presented. We present the free radical polymerization of styrene (solution polymerization) as an example.

This type of polymerization is quite simple. We used as an example, an experiment from an organic mixture of benzene and styrene with azobisisobutyronitrile (AIBN). AIBN is the reaction initiator, which undergoes a thermal dissociation to form free radicals.

Problem: Determine the conversion of polymerization reaction solution (styrene/benzene) to different concentrations of initiator ($[I] = 0.07$; 0.03; 0.01 mol/L) with styrene concentration $[M] = 4$ mol/L. Determine also the conversion from three different concentrations of monomer, $[M] = 6$; 3; 1 mol/L with $[I] = 0.03$ mol/L. Use the temperature at 50°C.

Material: See Figure 25.1. Two glass flasks—round-bottom flasks (K) with a standard ground glass stopper (S); 1 magnetic stirrer (R) or an agitated stirrer (M); 2 vases (W); 1 thermometer (1/10°C); pipette, styrene, benzene, and AIBN.

25.1.1.1　Experimental system

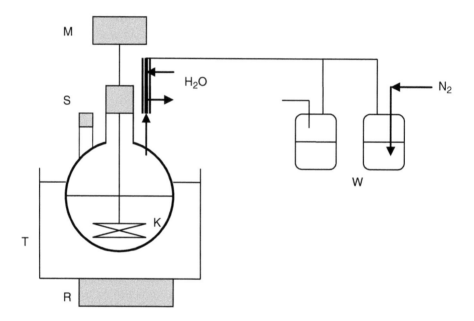

Figure 25.1 Experimental layout of a phase homogeneous reaction: polymerization system.

25.1.1.2　Procedure

Styrene was dried over $CaCl_2$ and purified by two successive vacuum distillations (50 mbar) over Na. The same procedure was performed with benzene. AIBN was purified with methanol recrystallization before use and drying in vacuum at room temperature. The round-bottom flasks must be clean and dried repeatedly. Place the flask in a water bath (or another heating fluid) at 50°C. The system was deaerated with nitrogen gas flow before reaction (to remove O_2). The reagents (benzene and styrene) are introduced into the flask. After the system reached the reaction temperature, the initiator was introduced. Small samples ($\sim$10 mL) are removed from the system every 2 h for determination of the kinetics. These samples are put into a container which contains an inhibitor (hydroquinone) and then weighed (by gravimetric method). During the experiment, the reaction solution must be maintained with strong agitation. After the end of reaction, the material is weighed and then sent to a vacuum oven.

The product is "washed" with methanol and dried at 60°C for 4 h (or longer). The influence of temperature on the density can be neglected.

Plot a graph X and t and determine the reaction rates for different conditions in the proposed experiment. Use Equation 25.1:

$$(-r_M) = -\frac{\Delta[M]}{\Delta t} = [M_0]\frac{\Delta X_M}{\Delta t} \qquad (25.1)$$

25.1.2 Polymerization of isobutylene

It is intended to study the polymerization reaction of isobutylene. This reaction is known as one of the most important processes for cationic polymerization, and also called catalytic reaction Friedel-Crafts. The high reactivity of the polyisobutene is obtaining a product with high concentration of double bonds.

The aim of this example is to determine a mode of production of polybutene to obtain a product similar to standard UV-10. This means, to produce a material with a greater concentration of terminal double bonds from the process variables and a commercial catalyst.

Initially, the experimental system was designed for adjustment to process, as shown in Figure 25.2. RAF1 and butane are added continuously to the reactor. After reaching the reaction volume, the system was maintained under pressure for about 1 h. The reaction was initiated from the addition of co-catalyst and a HCl gas stream. At the end of reaction, the product was "washed" in a solution of NaOH (1.0 N), separated, and distilled. Some tests with temperature variation were applied to the system (0 to T_{amb}). The reagents used were isobutene pure (AGA 99.0%) and RAF1.

Figure 25.2 Polymerization system equipped with a stainless steel reactor.

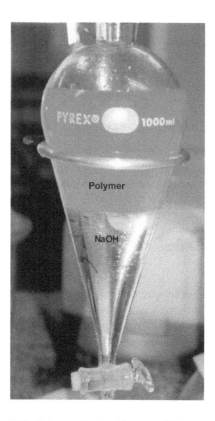

Figure 25.3 Polymer obtained by using BF$_3$ as catalyst.

The variables investigated during the process were the temperature and the concentration of co-catalyst (HCl/Cat $= 3$ and HCl/Cat $= 1$). The products were characterized by infrared (FT-IR). The infrared analyzes were conducted at room temperature in a Perkin-Elmer 2000 FT-IR at a resolution of 4 cm^{-1}.

Due to the formation of HF during the procedure, it was adopted to use a Teflon reactor. The process is highly exothermic (as well as most of the processes of polymerization). After using the Teflon reactor in the first reactions, it was replaced by a stainless steel reactor with the aim of increasing the safety of the process. Other variables chosen were the largest volume and a good cooling system, where it was possible to control the temperature in a better way. For a better understanding one of the process steps, see Figure 25.3.

One of the conditions used for polymerization reactions is described in Table 25.1. The pure isobutene and its mixture with RAF1 (from the polybutene) were used as initial reactants in the process. The addition of reagent is conducted slowly, avoiding a significant increase in temperature. After the end of the addition of isobutylene, the temperature remained stable at around 7°C. In an initial volume (reactant + ethanol in dichloromethane) of 1000 mL, a yield of 600 mL was obtained, approximately.

Table 25.1 Reaction Conditions for the Polymerization

Temperature (°C)	5
Volumetric ratio ethanol/dichloromethane	1:100
Molar ratio BF$_3$/ethanol	1:1
Volume of solution ethanol/dichloromethane (mL)	101
Total reaction volume (mL)	1.000
Reaction time (min)	45

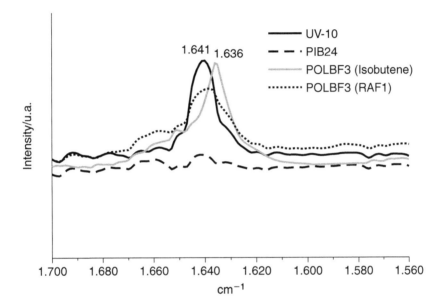

Figure 25.4 IR spectra between 1700 and 1560 cm^{-1}.

The stretching frequencies 1640–1642 cm^{-1} and 885–895 cm^{-1} are characteristic of the vinylidene group (double bond), while in 1662 and 820–830 cm^{-1} it belongs to trisubstituted olefins. Figure 25.4 shows the IR spectra between 1700 and 1560 cm^{-1}.

25.2 REACTIONS IN HETEROGENEOUS PHASE

25.2.1 Experimental system

Figure 25.5 shows the flowchart of the experimental testing unit. It is used in a "U"-shaped quartz tube without bulb, the temperature is controlled by a thermocouple with a K-Type sensor associated to a controller/programmer and coupled to an electric resistance furnace.

The experimental system also has a mass flow controller for gases with four channels. Four-way valves that allow these currents are diverted from saturators (by-pass). A three-way valve allows to replace the gas (He) through the reaction mixture.

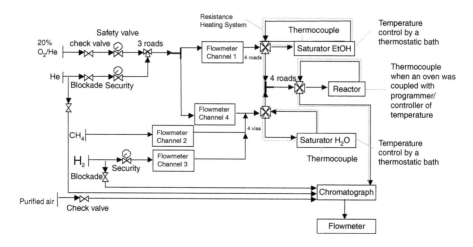

Figure 25.5 The experimental testing system.

The determination of the volumetric flow rates is easily performed by a flowmeter and a stopwatch (by soap-film method). It is necessary to consider the pressure drop when the gas bubbling through the saturator to yield calibration curves as a function of the soap-film burette method.

To prevent condensation of liquids, gases are maintained at a temperature of 75°C by electric resistance heating system, in this case, a VR115 voltage regulator.

25.2.1.1   Chromatographic analysis

Analysis of the results is conducted by a gas chromatograph in series with the experimental unit, in which the chromatograph is equipped with an active phase column (Hayesep D), thermal conductivity detector (TCD), and flame ionization detector (FID). The carrier gas used was helium at a flow rate of about 12 mL/min at 20°C.

The calibration system is made by injection of mixtures-certified standards with known concentrations (H_2, CO, CH_4, CO_2, and H_2O). According to the calibration, one can calculate the respective conversion factors obtained from the areas of the chromatographic peaks and subsequently a mathematical treatment is performed to normalize the data.

Figure 25.6 shows one typical chromatogram obtained, which is observed a low sensitivity to H_2. In Table 25.2 presents the retention times and the calibration factors for each of the components.

25.2.1.2   Determination of mole fractions and partial pressures

The calculation of partial pressures for the reactor effluent from the chromatographic areas is conducted from an external standard (analytical chemistry) and carbon balances. Before the beginning of the kinetics tests, the mixture is bubbled into the

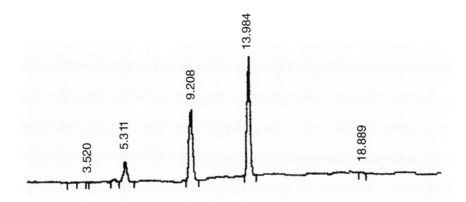

Figure 25.6 Typical chromatogram obtained during reaction ($CH_4 + CO_2$).

Table 25.2   Retention Times and Calibration Factors of Peak Areas from the Chromatograms

Gas	Retention Time (min)	Calibration Factor
H_2	3.5	4.275
CO	5.3	0.67
CH_4	9.2	0.45
CO_2	13.9	0.915
H_2O	18.9	0.55

saturator. From the knowledge of the ratio, for example, O_2/He (0.2/0.8) and chromatographic areas, it is possible to determine the molar fraction that corresponds to a particular area "A," according to the Equation 25.2:

$$y_{Ei} = \frac{0.2 \left(\dfrac{A_{Ei}}{A_{O_2}} \right)}{1 + 0.2 \left(\dfrac{A_{Ei}}{A_{O_2}} \right)} \tag{25.2}$$

where E_i is a compound, y_{Ei} is the molar fraction, A_{Ei} and A_{O_2} are corrected areas of the compound and oxygen, respectively.

To analyze the feed to the reactor, helium gas is bubbled into saturators containing the liquid for at least 40 min at desired temperatures. The mole fraction corresponding to the area of a given component (reactor feed) can be determined from the external standard, from a simple calculation as shown in Equation 25.3 below:

$$y_{Ei}^0 = \frac{A_{Ei}^0}{A_{Ei}} \cdot y_{Ei} \tag{25.3}$$

where y_{Ei}^0 is the feed molar fraction (ethanol, for example); A_{Ei}^0 is the area corresponding to the mole fraction of component in the feed.

For instance, the mole fraction of water in the reactor feed (y_{H_2O}) can be determined from the ratio of the corrected areas (A_{H_2O}) and also the component (A_{Ei}), as shown below:

$$y_{H_2O}^0 = \frac{A_{H_2O}^0}{A_{H_2O}} \cdot y_{Ei}^0 \tag{25.4}$$

25.2.1.3 Verification of the kinetic regime

To verify the kinetic regime, two tests are performed, in which the volumetric flow (v_0) is changed and keeping the ratio constant (m_{cat}/v_0).

For the system of kinetic regime, the conversion remains unchanged in experiments that have the same ratio (m_{cat}/v_0). If the external mass transfer determines the reaction rate, an increase in volumetric flow tends to decrease the boundary layer between gas and catalyst surface, increasing the concentration of gaseous species on the surface and consequently the reaction rate.

25.2.1.4 Catalytic tests

Tests are conducted at atmospheric pressure. The partial pressures of liquids are determined by the thermostatic baths (temperature control) and volumetric flow rates passing through saturators. The methodology of these calculations is described in detail in Patat and Kirchner (1975).

In general, the catalysts are reduced under flow v_0 (mL/min) from a mixture 10% H_2/He and heated to a temperature T (°C) at a specific heating rate (10°C/min). For the adjustment of the reactants partial pressures, the gas flow is stopped. After the system reached the desired conditions in saturators, several samples from the feed (containing liquid and water) are analyzed in the chromatograph. Since conditions have stabilized input, the reactor temperature is raised to the reaction conditions (T_r) and then, the reaction starts.

To analyze the external and internal mass transfer, the space velocity must be increased and keeping constant the ratio m_{cat}/F_{Ei} (catalyst mass/molar flow of liquid). These tests are made on the temperature and partial pressure of liquid in reactor feed. For example, the catalyst mass and volumetric flow rate of the gas used in the first test are, respectively, 25 mg and 100 mL/min ($m_{cat}/v_0 = 25 \times 10^{-5}$ $g_{cat} \times mL^{-1} \times min^{-1}$). For the second test, the mass and volumetric flow rate are reduced to half to obtain the same ratio, thus, the conversion is evaluated.

The space velocity is calculated (GHSV) as follows:

$$GHSV = y_{Ei}^0 \cdot \frac{v_0}{m_{cat}} \tag{25.5}$$

where y_{Ei}^0 is the molar fraction of liquid in the reactor inlet, v_0 is the volumetric flow (mL min^{-1}); and m_{cat} is the mass of catalyst (g) used in the test.

Table 25.3 Dehydrogenation of Cyclohexane over Pt/Al₂O₃ Catalyst ($P = 1$ atm $H_2/C_6H_{12} = 13.2$)

Catalyst	H/Pt[b]	Initial Rate at 543 K (10^{-3} mol/h/g_{cat})	TOF (s^{-1})	E_a (kcal/mol)
0.9% Pt/Al₂O₃	1.0	161.5	1.0	24

25.2.2 Determination of activation energy: dehydrogenation of cyclohexane

This example shows a reaction conducted in a microreactor at 1 atm, from a quantity of 10 mg of catalyst previously dried and under flowing of N_2 (30 cm³/min) at 393 K in 30 min. Then, the catalyst is reduced with 1.5 H_2/N_2 and the same flow at 500°C in 30 min. After reduction and cooling was introduced, a stream of H_2 passes through the saturator containing cyclohexane (99.9%) at 15°C ($H_2/C_6H_{12} = 13.2$). The space velocity (GHSV) used was 170 h⁻¹ to a temperature variation of 250 at 300°C. The conversion was kept below 10%. The products were analyzed by a chromatograph FID. The dispersion was determined by chemisorption of H_2 and the TOF was calculated by the dispersion of the Pt/Al₂O₃ catalysts.

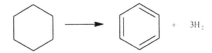

Benzene was the only product found and the catalysts were not deactivated after 4 h of reaction. The alumina showed no activity. The results are shown in Table 25.3. Figure 25.7 shows the TOF with the inverse of temperature. The activation energy found was 24 ± 2 kcal/mol.

25.2.3 Kinetic study—methane reforming with CO₂—heterogeneous reaction

In this practical experimental reports how can we determine the catalytic activity and reaction kinetics in methane reforming with CO_2. A great effort has been directed to the development of active and stable catalysts for this reaction.

Among the several catalysts tested in methane reforming with CO_2, the Pt/ZrO₂ catalyst yields good results of activity and stability. The objective is to determine the catalytic activity of the Pt/ZrO₂ systems in different experimental conditions.

25.2.3.1 Experimental part

The tests are conducted on the multipurpose unit, as shown in Figure 25.5. The pretreatment of the catalysts consisted of drying up to 150°C, under a flow of helium gas for 30 min, followed by reduction at a mixture 10% of H_2/Ar (30 mL/min) until the temperature of 500°C (10°C/min). Then, the catalysts are cooled to room temperature

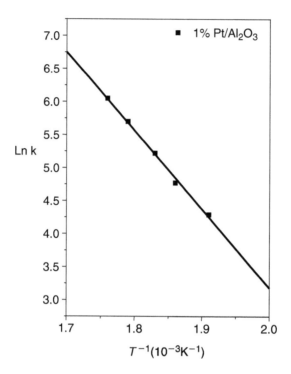

Figure 25.7 Dehydrogenation of cyclohexane over Pt/Al_2O_3 catalyst, $pH_2/pC_6H_{12} = 13.2$ ($P = 1$ atm).

(25°C) under He flow. Gaseous mixtures available for the reaction are 10% CH_4/He and 10% CO_2/He.

The catalytic tests with chromatographic analysis can be performed at atmospheric pressure in a fixed-bed quartz microreactor. The mass of catalyst used was 20 mg and a total flow of the reagents of 200 mL/min, with a ratio CH_4:CO_2:He of 1:1:18 (GHSV = 600,000 cm^3/h g_{cat}). The gas flow of gases is monitored by the mass flow controllers with transducers valves connected to a four channel control panel (MKS), which allows an accurate dilution of the mixture reagents.

25.2.3.2 Activity of the catalysts

The results for different catalysts (Pt/Al_2O_3 and Pt/ZrO_2) are shown in Figure 25.8. The equilibrium curves were obtained from commercial process package Aspen Plus[TM] (Aspen, 2006).

25.2.3.3 Determination of rates

In order to determine the kinetics in a particular model, one determines initially the effect of variation of partial pressure of CH_4 and CO_2 on the rate of consumption CH_4 for PtZr catalyst at 550°C, whose results are shown in Figures 25.9 and 25.10. Each

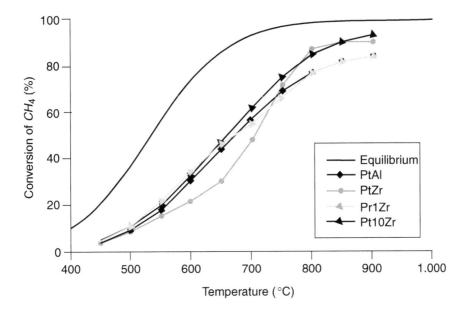

Figure 25.8 Conversion as a function of temperature.

experimental measurement was obtained after 30 min of reaction. When the pressure of CH_4 is kept fixed at 19 Torr, the consumption rate of CH_4 shows two different levels, $P_{CO_2} < P_{CH_4}$ and $P_{CO_2} \geq P_{CH_4}$. When the pressure of CO_2 is kept fixed at 19 Torr, the consumption rate of CH_4 increases with increasing pressure until 27 Torr; for pressures above this value, the rate is nearly constant.

25.2.3.4   Kinetic model

From the analysis of the data available in the literature and our experimental results, the following bifunctional mechanism has been proposed for the reaction CH_4–CO_2:

$$\{A\}\ CH_4 + (p) \leftrightarrow CH_{x^-}(p) + \frac{4-x}{2}H_2 \tag{25.6}$$

$$\{B\}\ CO_2 + (z) \leftrightarrow CO_2(z) \tag{25.7}$$

$$\{C\}\ H_2 + 2(p) \leftrightarrow 2H^+(p) \tag{25.8}$$

$$\{D\}\ CO_2(z) + 2H^+(p) \leftrightarrow CO + H_2O \tag{25.9}$$

$$\{E\}\ CO_2(z) + CH_{x^-}(p) \xrightarrow{k} 2CO_2 + \frac{x}{2}H_2 + (z) + (p) \tag{25.10}$$

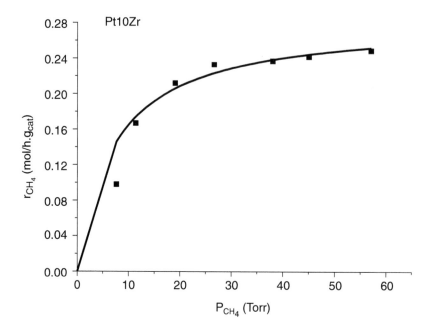

Figure 25.9 Reaction rate as a function of the partial pressures of CH$_4$ (at constant pressure and temperature, $T = 550°C$).

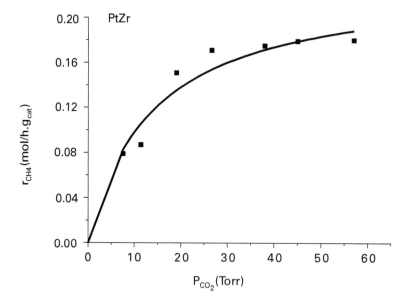

Figure 25.10 Reaction rate as a function of the partial pressures of CO$_2$ versus CH$_4$ (at constant pressure and temperature, $T = 550°C$).

(In this mechanism, p represents the platinum site and z the support site.)

The slow steps of this model are decomposition of CH_4 {A} and the reaction of the CH_x fragments with adsorbed species of CO_2 {E}. For catalysts that inhibit the formation of residual carbon (kinetically), the surface concentration of carbon in the metal surface (as CH_x) remains constant, that is, the decomposition rate of CH_4 is equal to the reaction rate of CH_x with CO_2:

$$r_{CH_4} = k_{CH_4} \cdot P_{CH_4} \cdot (p) - k_{CH_4}^{-1} \cdot [CH_{x^-}(p)] \cdot P_{H_2}^{(4-x)/2} = k \cdot [CO_2(z)] \times [CH_{x^-}(p)] \tag{25.11}$$

Assuming step {B} in equilibrium, it has been:

$$K_{CO_2} \cdot P_{CO_2} \cdot [z] = [CO_2 - z] \tag{25.12}$$

Therefore, two equations are formulated at independent sites. Assuming the surface concentration of H_2 negligible:

$$(p) + CH_{x^-}(p) = 1 \tag{25.13}$$

$$(z) + CO_2(z) = 1 \tag{25.14}$$

and substituting Equations 25.12, 25.13, and 25.14 into 25.10, we find the following equation for the reaction rate of methane:

$$r_{CH_4} = \frac{k \cdot K_{CO_2} \cdot k_{CH_4} \cdot P_{CO_2} \cdot P_{CH_4}}{k \cdot K_{CO_2} \cdot P_{CO_2} + \left(1 + K_{CO_2} \cdot P_{CO_2}\right)\left(k_{CH_4} \cdot P_{CH_4} + k_{CH_4}^{-1} \cdot P_{H_2}^{(4-x)/2}\right)} \tag{25.15}$$

Kinetic constants were calculated using the following procedure:

(a)   When the partial pressure of H_2 is null and with CH_4 at constant pressure, the reaction rate of methane can be expressed by:

$$r_{CH_4} = \frac{a \cdot P_{CO_2}}{b \cdot P_{CO_2} + c} \tag{25.16}$$

where a, b, and c are as follows:

$$a = k \cdot K_{CO_2} \cdot k_{CH_4} \cdot P_{CH_4} \tag{25.17}$$

$$b = k \cdot K_{CO_2} + K_{CO_2} \cdot k_{CH_4} \cdot P_{CH_4} \tag{25.18}$$

$$c = k_{CH_4} \cdot P_{CH_4} \tag{25.19}$$

Table 25.4 Metric Parameters

Constant	PtZr
$k_{CH4}^{(a)}$	0.030
$k_{CO2}^{(b)}$	0.0148
$k^{(c)}$	1.138
Correlation	0.983

(a) Unit: mol/hg$_{cat}$ Torr.
(b) Unit: Torr^{-1}.
(c) Unit: mol/hg$_{cat}$.

This equation can be linearized as follows:

$$\frac{P_{CO_2}}{r_{CH_4}} = A \cdot P_{CO_2} + B \tag{25.20}$$

where $A = \frac{1}{k_{CH_4} \cdot P_{CH_4}} + \frac{1}{k}$ and $B = \frac{1}{k \cdot K_{CO_2}}$

Hence, by plotting P_{CH_4}/r_{CH_4} versus P_{CH_4}, we obtain the constants A and B.

(b) When the partial pressure of H_2 is null and with CO_2 at constant pressure, the reaction rate of methane can be expressed by:

$$r_{CH_4} = \frac{a' \cdot P_{CH_4}}{b' \cdot P_{CH_4} + c'} \tag{25.21}$$

where a', b' and c' are as follows

$$a' = k \cdot K_{CO_2} \cdot k_{CH_4} \cdot P_{CO_2} \tag{25.22}$$

$$b' = k_{CH_4}(1 + K_{CO_2} \cdot P_{CO_2}) \tag{25.23}$$

$$c' = k \cdot K_{CO_2} \cdot P_{CO_2} \tag{25.24}$$

This equation can be linearized as follows:

$$\frac{P_{CH_4}}{r_{CH_4}} = A' \cdot P_{CH_4} + B' \tag{25.25}$$

where $A' = \frac{1 + K_{CO_2} \cdot P_{CO_2}}{k \cdot K_{CO_2} \cdot P_{CO_2}}$ and $B' = \frac{1}{k_{CH_4}}$

Hence, by plotting P_{CH_4}/r_{CH_4} versus P_{CH_4}, we obtain the constants A' and B'.

(c) From the constants A, B, A', and B', we can calculate k_{CH4}, K_{CO2}, and k.

The kinetics parameters calculated from this procedure are shown in Table 25.4, as well as the correlation coefficients obtained for the PtZr catalyst at 55°C.

The proposed model fit well to the data obtained by kinetic tests for the catalysts containing zirconia. However, the results show that the kinetics can be simplified. Thus, considering:

$$1 \gg K_{CO_2} \cdot P_{CO_2}$$

Therefore:

$$r_{CH_4} = k_{CH_4} \cdot P_{CH_4} = 0.03 \cdot P_{CH_4} \tag{25.26}$$

25.3 PERFORMANCE OF REACTORS

In this section, some practices in batch and continuous reactors are presented in order to determine the performance in terms of conversion, selectivity, and yield. These examples were conducted in partnership with some Brazilian industries.

25.3.1 Batch reactor–hydrogenation of sucrose

The evaluation of catalytic hydrogenation was carried out in batch reactor of 450 mL and under pressure (Parr autoclave). To eliminate the diffusion effects, the reactor was equipped with double-helix stirring device at specific positions: near the bottom and on the surface of the reactant (height of the liquid mixture). To visualization of the system, Figure 25.11 presents the reactor used for this example.

Figure 25.12 shows a schematic representation of the interior of the reactor. We can see that the system is provided with an inner cooling coil (in case there is a thermal runaway), one valve for the gas flow, one pressure relief valve, and one for sampling. From that figure, one can observed the agitation system, input reagent (gas), and thermocouple.

The analysis of the operating conditions for the hydrogenation of sucrose was based on industrial conditions to obtain a conversion of approximately 90% to 65 min of reaction.

The introduction of hydrogen was controlled to maintain the reactor pressure at 40 bar (580 psig), and agitation rate used was 1200 rpm. The heating procedure used is described in Table 25.5. The concentration of the sucrose solution was fixed at 50% (w/w) of sucrose diluted in distilled and deionized water.

This procedure was adopted in the production of various experiments by varying the reaction time to obtain levels of conversion of around 60%, and then compare the various catalysts used. Thus, we worked with a reaction time of 50 min to reach a temperature of 130°C. These conditions are summarized below:

Reactor volume $= 450 \, \text{mL}$
Mass of the Raney–Nickel catalyst with 6% $(p_{newcat.}/p_{solid}) = 10.92 \, \text{g}$
Reaction temperature $= 130°C$
Total pressure $= 40 \, \text{bar} \, (580 \, \text{psig})$

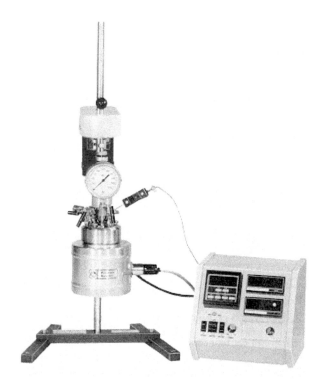

Figure 25.11 Schematic representation of the reactor used in catalytic tests, Parr autoclave (with permission).

Table 25.5 Temperature × Pressure

Time (min)	Temperature (°C)	Pressure (psig)
0	T_{amb}	atm (14.7 psig)
5	56	Start feeding hydrogen
25	90	450
30	100	580
45	120	580
65	140	580

Sample concentration HD to be hydrogenated:

- Sucrose: 1.14%
- Glucose: 51.72%
- Fructose: 47.14%

Agitation rate = 1200 rpm
Reaction time = 50 min
Hot filtration (80°C) to prevent the polymerization of the products formed.

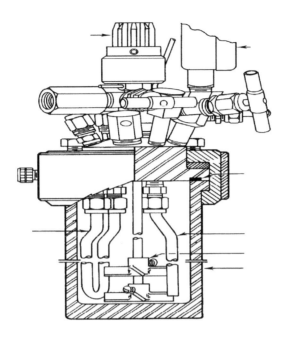

Figure 25.12 Schematic representation of the reactor used in catalytic tests, Parr autoclave (with permission).

The analysis of the reactant and reaction products were performed on liquid chromatography (HPLC) equipped with a column (Shodex SC101) with aqueous mobile phase.

The results of the hydrogenation reactions at 50 min and 130°C of the leached catalysts at 100°C (except commercial) are described in Figure 25.13. The Raney–Nickel commercial catalyst showed the highest conversion (~60%) followed slightly below of the Cr–Mo promoted catalyst. The other additive catalysts showed a conversion level of about 35%, which was similar to that obtained by the Ni–Raney (100) (leached) catalyst (manufactured in NUCAT lab) at 100°C. The order of activity of the catalysts tested is shown below:

Ni–Al (commercial) > Ni–Al–Cr ≈ Ni–Al–Mo > Ni–*Raney*(100)

≈ Ni–Al–Fe ≈ Ni–Al–Co

Reaction at 130°C (leached cat. 100°C)

25.3.2 Integral continuous flow reactor (tubular)—isomerization of xylenes

The experimental unit for isomerization of xylenes has the function of increasing the content of ortho-xylene in the xylenes stream from the ethylbenzene conversion.

Ethylbenzene → *o*-xylene   (25.27)

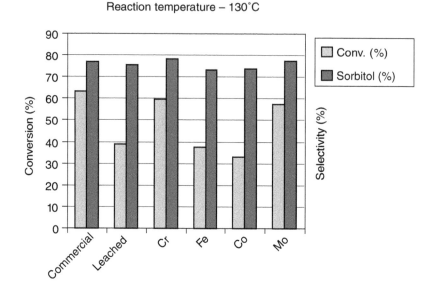

Figure 25.13 Conversion and selectivity in the hydrogenation reactions at 130°C.

During the isomerization process, ethylbenzene can react in different ways:
O-xylene isomerization;
Disproportionation to benzene and diethylbenzene;
Dealkylation.

Bifunctional catalysts (acid hydrogenation over supported metal) are required for the isomerization reaction. Under these conditions, reactions of hydrogenation of aromatics (reactants and products) and of alkene reactions occur simultaneously as well as secondary reactions, such as hydrocracking of naphthenic and hydrogenolysis. Due to the acidic character of the support, the transalkylation reactions, disproportionation, and isomerization of xylenes also occur in bifunctional catalysts.

$$o\text{-xylene} \leftrightarrow m\text{-xylene} \leftrightarrow p\text{-xylene} \tag{25.28}$$

We emphasize that conversion of meta-xylene to ortho-xylene is undesired.
Studies already published in the literature report that a higher activity for platinum catalysts supported on zeolites compared with γ-Al_2O_3 supported catalysts (Katzer, 1977). In addition, the reactions of ethylbenzene through routes of isomerization, transalkylation, and disproportionation are reported as preferred.
In this way, an evaluation of commercial catalysts requires a methodology for simultaneous monitoring of activity and selectivity, as well as evaluating the catalytic stability under operating conditions.
The objective is to evaluate commercial catalysts based on platinum (Pt1, Pt2, and Pt3), comparing them on the activity and selectivity for use in an industrial isomerization of xylenes.

Table 25.6 Operating Conditions of the Catalytic Tests

	Nominal	Pt1	Pt2	Pt3
LHSV (h^{-1})	3.5	3.49	3.30	3.13
H_2/HC8 (mol/mol)	4	4.01	–	–
$P_{reactor}$ (bar)	12	12	12	12/9
V_{bed} (cm^3)	26.9	28.4	26.9	28.4
D_{bed} (cm)	1.90	1.90	1.90	1.90
L_{bed} (cm)	8.58	9.3	9.5	9.3
M_{cat} (g)	18.32	18.29	16.69	15.57
Q_{C8} (cm^3/min)	1.57	1.69	1.5	1.48
Q_{H2} (cm^3/min)	1.150	1.240	1.150	1.150

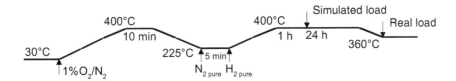

Figure 25.14 Pretreatment and reaction of the catalytic tests.

25.3.3 Goals

1 Evaluation of the activity of catalysts based on platinum for the conversion of ethylbenzene.
2 Determination of selectivity of platinum catalysts for the conversion of ethylbenzene to xylenes.
3 Performance evaluation of catalysts subjected to forced deactivation.

The catalysts were tested for performance under operational conditions, as shown in the Table 25.6. The steps of drying, activation, reaction, deactivation, and revaluation were conducted as shown in Figure 25.14.

25.3.3.1 Drying

The catalyst is dried in a N_2 gas stream (24 L/h) with 1% of O_2 (1.2 L/h of air) at 400°C for 10 min (400°C/h). Then, the catalyst is cooled to 225°C under the same gas. At 225°C, a N_2 gas stream (pure) for 5 min is used.

25.3.3.2 Activation

N_2 is replaced by H_2 and changing the flow rate to 8.3 L/h g_{cat}, simultaneously, the reactor is heated to 400°C and maintained for 1 h.

25.3.3.3 Reaction

After the complete reduction, reactor is pressurized and the reactants are introduced. After introduction of feed, the reactor is heated to 425°C and kept at this value for 24 h. Thus, the temperature is reduced to 360°C and the load is introduced.

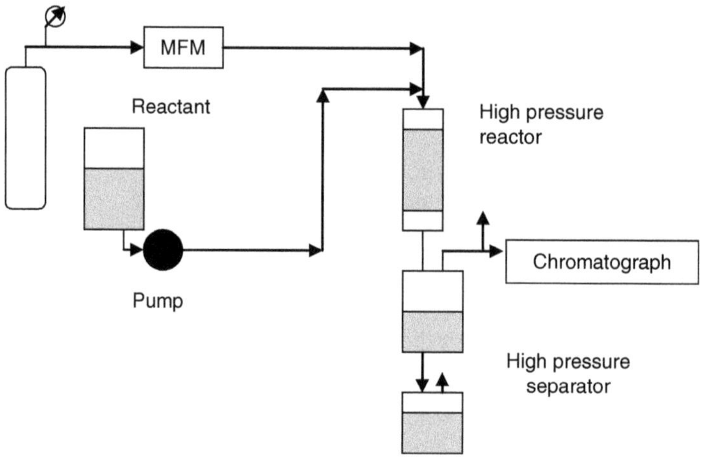

Figure 25.15   Bench unit for catalytic tests from NUCAT lab.

25.3.3.4   Deactivation

After withdrawal of the last sample at 360°C, feeding is interrupted and the reactor is cooled to room temperature, under H_2 flow. H_2 is replaced by N_2 at 600 mL/min, increasing the temperature to 450°C at a rate of 5°C/min for 3 h. We introduced a liquid load (1 mL/min). The system is cooled to 385°C and the load is interrupted, N_2 is replaced by H_2. And again, the liquid load (2 mL/min) is fed to the system.

The tests were carried out in bench unit, as shown in Figure 25.15.

Samples were removed at 360°C, 370°C, and 385°C and again at 360°C. The catalytic deactivation step was performed under nitrogen flow at 450°C for 3 h. And again, the experimental points at 385°C and 360°C were evaluated.

The load and the liquid phase of the reactor effluent were analyzed by gas chromatography and were identified at least for the following compounds: cyclohexane and carbon chains of up to C7 (nonaromatic), styrene, benzene, toluene, ethylbenzene, meta-, ortho-, and para-xylenes.

The tests were performed using a specific load industry. The results of load composition and products at 360°C for the three catalysts are shown in Table 25.7.

For Pt1 and Pt2 catalysts, a large decrease in the products of xylenes at 360°C was observed, which were converted into lighter compounds (nonaromatic). The catalyst Pt3, although it has the highest percentage of ethylbenzene in product composition, did not show significant change in the stream of xylenes.

25.3.3.5   Activity

The determination of the activity of three platinum catalysts in the isomerization reaction of xylenes was by converting ethylbenzene. As well as the formation of o-xylene, benzene and nonaromatic and m-xylene the same procedure was used:

$$X_{EB} = \frac{EB_{load} - EB_{product}}{EB_{load}} \times 100 \qquad (25.29)$$

Table 25.7 Load and Products Composition for Activity Tests at 360°C

Compounds	Composition (% mol)			
		Products (360°C)		
	Load/Catalyst	Pt2	Pt1	Pt3
Nonaromatic	0.67	44.73	37.16	0.68
Benzene	0.02	0.05	0.02	0.18
Toluene	0.04	0.28	0.22	0.52
Ethylbenzene	27.64	07.34	08.35	16.79
$\sum$ xylene	71.63	47.07	53.99	70.06
p-xylene	20.6	12.51	14.10	17.72
m-xylene	46.0	26.07	29.66	36.91
o-xylene	5.1	8.50	10.23	15.44

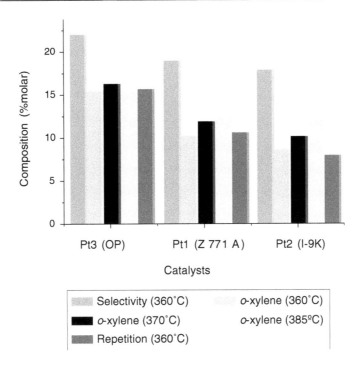

Figure 25.16 Effect of temperature on the selectivity of *o*-xylene.

25.3.3.6 Selectivity

The relationship between the selectivity of *o*-xylene to total xylenes amount (%-$\sum$xylene) was calculated from the Equation 25.30 and shown in the Figure 25.16. The production (%) of *o*-xylene is also presented:

$$S_{\text{o-xylene}} = \frac{\text{Xylene} - \text{Xylene}_{\text{product}}}{\sum \text{Xylene}_{\text{product}}} \times 100 \qquad (25.30)$$

These results show that o-xylene selectivity decreases with increasing temperature for all the catalysts. The selectivity was significantly higher in the presence of the Pt3 catalyst at low temperatures, although at a lower conversion.

The results showed that the Pt3 catalyst had the lowest activity when compared to the conversion of ethylbenzene and better performance for the selectivity of o-xylene, further lower formation of nonaromatic compounds. It was observed that this catalyst remained away from the equilibrium conversion when compared to other catalysts tested; it also showed increased coke formation for samples that have undergone deactivation forced.

The Pt1 and Pt2 catalysts presented a similar behavior of activity and selectivity; they are more active in the conversion of ethylbenzene and less selective for the formation of o-xylene with high formation of nonaromatic in conversions near to equilibrium.

Under the conditions of this example, the Pt1 catalyst showed to be more resistant to deactivation processes, maintaining a reasonable selectivity for o-xylene.

References

Alberton, A.L., Schwaab, M., Schmal, M. & Pinto, J.C. (2009) *Chem. Eng. J.*, 155, 816–823.

Alberton, A.L., Schwaab, M., Lobão, M.W.N. and Pinto, J.C. (2013) "Experimental Design for the Joint Model Discrimination and Precise Parameter Estimation through Information Measures", *Chem. Eng. Sci.*, 75, 130–131.

Alberton, A.L., Souza, M.M.V.M. & Schmal, M. (2007) *Catalysis Today*, 123, 257.

Alexopoulos, A.H. and Kiparissides, C. (2007) "On the Prediction of Internal Particle Morphology in Suspension Polymerization of Vinyl Chloride. Part I: The Effect of Primary Particle Size Distribution", *Chemical Engineering Science*, 62, 3970–3983.

Alves, P.S. (1999) *M.Sc. Thesis*. COPPE-Univeristy Federal of Rio de Janeiro.

Anderson, R.B. (1968) *Experimental Methods in Catalytic Research*. Academic Press.

Antal Jr., M., Croiset, E., Dai, X., De Almeida, C., Mok, W.S. & Norberg, N. (1996) *Energy Fuels*, 10 (3), 652–658.

Araujo, L.R.R. & Schmal, M. (2000) *Appl. Catal. A.*, 203 (2), 275.

Aris, R. (1969) *Elementary Chemical Reactor Analysis*. New Jersey, Prentice Hall.

Aspen Plus help manual (2006) Aspen Tech.

Bailey, J.E. & Ollis, D.F. (1986) *Biochemical Engineering Fundamentals*. New York, McGraw-Hill.

Bard, Y. (1974) Nonlinear Parameter Estimation, Academic Press, New York.

Beckman, D. & Elliott, D.C. (1985) Chem. Can. J. Eng., 63, 99.

Bhering, D.L., Nele, M., Pinto, J.C. and Salim, V.M.M. (2002) "Preparation of High Loading Silica Supported Nickel Catalyst: Analysis of the Reduction Step", *Appl. Catal. A: Gen.*, 234, 55–64.

Billmeyer, F. (1984) *Textbook of Polymer Science*. New York, John Wiley & Sons.

Bond, C.C. (1974) *Heterogeneous Catalysis and Applications*. Oxford, Clarendon Press.

Boudart, M. (1968) *Kinetics of Chemical Process*. New Jersey, Prentice Hall.

Boudart, M. & Djega-Mariadassou, G. (1984) *Kinetics of Heterogeneous Catalytic Reactions*. New Jersey: Princeton University Press.

Box, G.E.P. (1960) Fitting Empirical Data, *Annals of the New York Academy of Sciences*, 86, 792–816.

Box, G.E.P., Hunter, W.G. and Hunter, J.S. (1978) *"Statistics for Experimenters - An Introduction to Design, Data Analysis, and Model Building"*, John Wiley & Sons, New York.

Braido, R., Borges, L.E.P. and Pinto, J.C. (2018) "Chemical Recycling of Crosslinked Poly(Methyl Methacrylate) and Characterization of Polymers Produced with the Recycled Monomer", *Journal of Analytical and Applied Pyrolysis*, 132, 47–55.

Bridgwater, A.V. (2012) *Biomass and Bioenergy*, 38, 68–94.

Cardoso, D. (1987) *Introdução a Catálise Heterogênea*. U.F. São Carlos.

Carothers, W.H. (1929) "Studies on Polymerization and Ring Formation. I. An Introduction to the General Theory of Condensation Polymers", *Journal of the American Chemical Society*. 51, 2548–2559.

Cesar, D.V., Perez, C.A., Vera, M.M.S. & Schmal, M. (1999) *Appl. Catal. A: General*, 176, 2055.

Choudhary, T.V. & Phillips, C.B. (2011) *Appl. Catal. A: General*, 397 (1–2), 30, 1–12.

Ciola, R. (1981) *Fundamentos em Catálise*. Editora Moderna, S.Paulo.

Colman, R.C., Baldanza, M.A.S. & Schmal, M. (2010), *J. Phys. Chem. C*, 114, 18501–18508.

Coulson, J.M. & Richardson, J.F. (1971) *Chemical Engineering*, v. 3. England, Pergamon Press.

Czernik, S. & Bridgwater, A.V. (2004) *Energy & Fuels*, 18, 590–598.

Da Ros, S., Jones, M.D., Mattia, D., Schwaab, M., Barbosa-Coutinho, E., Rabelo-Neto, R.C., Noronha, F.B., Pinto, J.C. (2017) *Chem. Eng. Jl.*, 308, 988–1000.

Denbigh, K.G. (1965) *Chemical Reactor Theory*. Cambridge University Press.

Dinesh Mohan, Charles U. Pittman, Jr. & Philip H. Steele (2006) *Energy & Fuels*, 20, 848–889.

Dowding, P. J. and Vincent, B., (2000) "Suspension Polymerisation to Form Polymer Beads", *Colloids and Surfaces A: Physicochemical and Engineering Aspects*, 161, 259–269.

Edgar, T.F. and Himmelblau, D.M. (1988) *"Optimization of Chemical Processes"*, McGraw-Hill, New York.

Embiruçu, M., Lima, E.L. and Pinto, J.C. (2000) Continuous Soluble Ziegler-Natta Ethylene Polymerizations in Reactor Trains– I. Mathematical Modeling, *Journal of Applied Polymer Science*, 77, 1574–1590.

Embiruçu, M., Lima, E.L. and Pinto, J.C. (2008b) Continuous Soluble Ziegler-Natta Ethylene Polymerizations in Reactor Trains – III. Influence of Operation Conditions upon Process Performance, *Macromolecular Reaction Engineering*, 2, 2, 161–175.

Embiruçu, M., Prata, D.M., Lima, E.L. and Pinto, J.C. (2008a) Continuous Soluble Ziegler-Natta Ethylene Polymerizations in Reactor Trains – II. Estimation of Kinetic Parameters from Industrial Data, *Macromolecular Reaction Engineering*, 2, 2, 142–160.

Espie, D.M. and Macchietto, S. (1988) Nonlinear Transformations for Parameter Estimation, *Industrial and Engineering Chemistry Research*, 27, 2175–2179.

Exercícios resolvidos de cinética e de reatores (1980) Divulgação interna.

Faria, W.L.S., Dieguez, L.C. & Schmal, M. (2008) *Appl. Catal. B: Environ.*, 85, 77–85.

Figueiredo, J.L. & Ramôa Ribeiro, F. (1987) *Catálise Heterogênea*. Lisboa, Fundação Calouste Gulbekian.

Flory, P.J. (1936) "Molecular Size Distribution in Linear Condensation Polymers1", *Journal of the American Chemical Society*, 58, 1877–1885.

Fogler, H.S. (2000) *Elements of Chemical Reaction Engineering*. 2nd edition. Upper Saddle River, NJ, Prentice Hall, 2000.

Fonseca, S.C.M. (2006) *M.Sc. Thesis*. Porto, Portugal, Faculdade de Ciências da Universidade do Porto.

Freitas Filho, I.P., Biscaia Jr., E.C. and Pinto, J.C. (1994) "Steady-State Multiplicity of Continuous Bulk Polymerization Reactors - A General Approach", *Chemical Engineering Science*, 49, 3745–3755.

Froment, G.F. & Bischoff, K.B. (1979) *Chemical Reactor Analysis and Design*. New York, John Wiley & Sons.

Froment, G.F., Bischoff, K.B. and De Wilde, J. (2011) "Chemical Reactor Analysis and Design", 3rd Ed.; John Wiley & Sons, New York.

Frusteri, F., Freni, S., Chiodo, V., Spadaro, L., Bonura, G. & Cavallaro, S. (2004a) *Appl. Catal.*, 270 (1), 1.

Furminsky, E. (2000) *Appl. Catal. A: General*, 199, 147–190.

Gallezot, P., Alarcon-Diaz, A. & Dalmon, J.A. (1975) *J. Catal.*, 39, 334.

Goldberg, D.E. (1989) Genetic Algorithms in Search, Optimization and Machine Learning, Addison Wesley Longman Inc., Boston.

Gonzáles-Velasco, J.R., Gonzáles-Marcos, J.A. and Romero, A. (1991) "Sequential Design of Experiments for Optimal Model Discrimination and Parameter Estimation in Isopropanol Dehydration", *Chem. Eng. Sci.*, 46, 2161–2166.

Guo, B., Chang, L. and Xie, K. (2006) Adsorption of Carbon Dioxide on Activated Carbon, *Journal of Natural Gas Chemistry*, 15, 223–229.

Haller, G.D.E.R. (1989) *Adv. Cat.*, 36, 173.

Haruta, M.J. (2004) *New Mater. Electrochem. Syst.*, 7, 163.

Heuts, J.P.A. and Russell, G.T. (2006) "The Nature of the Chain-Length Dependence of the Propagation Rate Coefficient and its Effect on the Kinetics of Free-Radical Polymerization. 1. Small-Molecule Studies", *European Polymer Journal*, 42, 3–20.

Hill, C.G. (1977) *An Introduction to Chemical Engineering Kinetics and Reactor Design*. New York, John Wiley & Sons, 1977.

Himmelblau, D.M. (1970) Process Analysis by Statistical Methods, John Wiley & Sons, New York.

Hui, A.W. and Hamielec, A.E. (1972) "Thermal Polymerization of Styrene at High Conversions and Temperatures. An Experimental Study", *Journal of Applied Polymer Science*, 16, 749–769.

Horsley, J.A. (1979) *Am. J. Chem. Soc.*, 101 (11), 2870.

Hougen, O.A. & Watson, K.M. (1959) *Chemical Process Principles. Part 3. Kinetics and Catalysis*. New York, John Wiley & Sons.

Jiang, X-Z, Hayden, T.F.& Dumesic, J.A. (1983) *J. Catal.*, 83, 168.

Katzer, J.R. (1977) *Molecular Sieves – II. American Chemical Society*, Volume 40.

Knözinger, H., Kochloefl, K. and Meye, W.J. (1973) "Kinetics of the Bimolecular Ether Formation from Alcohols over Alumina", *J. Catal.*, 28, 69–75.

Kennedy, J., Eberhart, R. (1995) In *Neural Networks*, Proceedings of the IEEE International Conference, Perth, Australia, 4, 1942–1948.

Kirkpatrick, S., Gelatt, C.D., Vecchi, M.P. (1983) Science, 220, 671–680.

Kreyszig, E. (1978) Introductory Functional Analysis with Applications, John Wiley & Sons, New York.

Larentis, A.L., Bentes Jr., A.M.P., Resende, N.S., Salim, V.M.M. and Pinto, J.C. (2003) "Analysis of Experimental Errors in Catalytic Tests for Production of Synthesis Gas", *Appl. Catal. A: Gen.*, 242, 2, 365–379.

Larentis, A.L., Resende, N.S., Salim, V.M.M. and Pinto, J.C. (2001) "Modeling and Optimization of the Combined Carbon Dioxide Reforming and Partial Oxidation of Natural Gas", *Appl. Catal. A: Gen.*, 215, 1–2, 211–224.

Le Page, J.F., Cosyns, J., Courty, P., Freud, E., Franck, J.P., Joaquin, Y., Jugin, B., Marcelly, G., Martino, G. et al. (1978) *Catalyse de Contact*. Technip.

Lédé, J. (2000) *Ind. Eng. Chem. Res.*, 39 (4), 893–903.

Lédé, J., Panagopoulos, J. & Villermaux, H.Z.L.I.J. (1985), *Fuel*, 64, 1514–1520.

Lee, S-J & Gavriilidis, E.A. (2002) *J. Catal.*, 206, 305.

Leocadio, I.C.L., Braun, S. & Schmal, M. (2004) *J. Catal.*, 223 (1), 114–121.

Levenspiel, O. (2001) *Chemical Reaction Engineering*. 3rd Edition, New York, John Wiley & Sons.

Li, C. & Chen, Y.W. (1995), *Thermochim. Acta*, 256, 457.

Matos, V., Mattos Neto, A.G. and Pinto, J.C. (2001) "Method for Quantitative Evaluation of Kinetic Constants in Olefin Polymerizations. I. Kinetic Study of a Conventional Ziegler–Natta Catalyst Used for Propylene Polymerizations", *J. Appl. Polym. Sci.*, 79, 2076–2108.

Mattos Neto, A.G., Freitas, M.F., Nele, M. and Pinto, J.C. (2005) Modeling Ethylene/1-Butene Copolymerizations in Industrial Reactors, *Industrial Engineering Chemistry Research*, **44**, 2697–2715.

Masel, I.R. (1996) *Principles of Adsorption and Reaction on Solid Surfaces*. New York, John Wiley & Sons.

Mercader, F.M., Groeneveld, M.J., Kersten, S.R.A., Geantet, C., Toussaint, G., Way, N.W.J., Schaverien, C.J. & Hogendoorn, K.J.A. (2011), *Energy Environ. Sci.*, 4, 985–997.

Monteiro, A.D., Delgado, J.J.S. and Pinto, J.C. (2019) *"Reciclagem Química de Resíduos Plásticos: Tecnologias e Impactos"*, E-Papers, Rio de Janeiro.

Monteiro, R.S., Dieguez, L.C., & Schmal, M. (2001) *Catal. Today*, 65, 77.

Montgomery, D.C. and Runger, G.C., *"Applied Statistics and Probability for Engineers"*, 3rd Ed., John Wiley & Sons, New York, 2002.

Moujijn, J.A., Van Leeuwen, P.W.N.M. & Van Santen, R.A. (1993) *Catalysis – Studies Science and Catalysis*, Volume 79. Elsevier, Scientific Publishing Company.

Moulijn, J.A., Makke, M. & Van Diepen, A. (2001) *Chemical Process Technology*. New York, John Wiley & Sons.

Naik, S.N., Goud, V.V., Rout, P.K. & Dalai, A.K. (2010) *Renewable and Sustainable Energy Reviews*, 14, 578–597.

Nele, M., Vidal, A., Bhering, D.L., Pinto, J.C. and Salim, V.M.M. (1999) "Preparation of High Loading Silica Supported Nickel Catalyst: Simultaneous Analysis of the Precipitation and Aging Steps", *Appl. Cat. A: Gen.*, 178, 177–189.

Nogueira, E.S., Borges, C.P. and Pinto, J.C. (2005) "In-Line Monitoring and Control of Conversion and Weight-Average Molecular Weight of Polyurethanes in Solution Step-Growth Polymerization Based on Near Infrared Spectroscopy and Torquemetry", *Macromolecular Materials Engineering*, **290**, 272–282.

Oasmaa, A. & Peacocke, G.V.C. (2010) VTT. Technical Research Centre of Finland, Espoo, Finland. Properties and fuel use of biomass derived fast pyrolysis liquids, vol. 731. VTT, Publication.

Oechsler, B.F., Poblete, I.B.S., Melo, P.A. and Pinto, J.C. (2018) Mixing Effects in Continuous Free-Radical Solution Copolymerization Tank Reactors: I – Characterization of Residence Time Distributions, *Macromolecular Reaction Engineering*, **12**, 6, 1800037.

Oechsler, B.F., Poblete, I.B.S., Melo, P.A. and Pinto, J.C. (2019) Mixing Effects in Continuous Free-Radical Solution Copolymerization Tank Reactors: II – Investigation of Micromixing Effects, *Macromolecular Reaction Engineering*, **13**, 5, 1900018.

Odian, G. (2004) "Principles of Polymerization", New York: Wiley–Interscience.

Orfao, J.J.M., Antunes, F.J.A. & Figueiredo, J.L. (1999) *Fuel*, 78, 349–358.

Pacheco, H., Thiengo, F., Schmal, M. and Pinto, J.C. (2018) A Family of Kinetic Distributions for Interpretation of Experimental Fluctuations in Kinetic Problems, *Chemical Engineering Journal*, **332**, 303–311.

Passos, F.B., Schmal, M. & Vannice, M.A. (1996) *J. Catal.*, 160, 106.

Patat, F. & Kirchner, K. (1975) *Praktikum der teschnishen chemie*, 3. ed. Berlin, De Gruyter.

Petersen, E.E. (1965) *Chemical Reaction Analysis*. New Jersey, Prentice Hall.

Pichler, H. & Schulz, H. (1970) *Chem.-Ing. Techn.*, 42, 1162.

Pinto, J.C. (1995) "The Dynamic Behavior of Continuous Solution Polymerization Reactors - A Full Bifurcation Analysis of a Full Scale Copolymerization Reactor", *Chemical Engineering Science*, 50, 3455–3475.

Pinto, J.C., Lobão, M.W. and Monteiro, J.L., (1990) "Sequential Experimental Design for Parameter Estimation: A Different Approach", *Chem. Eng. Sci.*, 45, 883–892.

Pinto, J.C., Lobão, M.W. and Monteiro, J.L. (1991) "Sequential Experimental Design for Parameter Estimation: Analysis of Relative Deviations", *Chem. Eng. Sci.*, 46, 3139–3138.

Pinto, J.C. and Ray, W.H., (1995a) "The Dynamic Behavior of Continuous Solution Polymerization Reactors - VII. Experimental Study of a Copolymerization Reactor", *Chemical Engineering Science*, 50, 715–736.

Pinto, J.C. and Ray, W.H. (1995b) "The Dynamic Behavior of Continuous Solution Polymerization Reactors - VIII. A Full Bifurcation Analysis of a Lab - Scale Copolymerization Reactor", *Chemical Engineering Science*, 50, 1041–1056.

Pinto, J.C. and Ray, W.H. (1996) "The Dynamic Behavior of Continuous Solution Polymerization Reactors IX - Effects of Inhibition", *Chemical Engineering Science*, 51, 63–79.

Pollaco, G., Semino, D. and Palla, M. (1996) "Temperature Profiles in Batch Methyl Methacrylate Polymerization in Gelled Suspension", *Polymer Engineering Science*, 36, 2088–2100.

Pritchard, D.J. and Bacon, D.W. (1975) Statistical Assessment of Chemical Kinetic Models, *Chemical Engineering Science*, 30, 567–574.

Project PADCT-GETEC. n. 301139895. COPPE/UFRJ (1996).

Project PEQ2470/COPPE/UFRJ (2002).

Project PEQ 496/ COPPE/UFRJ – Polybutene Project (2003).

Ralph, J., Brunow, G., Boerjan, W. (2007) *Lignins, Encyclopedia of Life Sciences*. John Wiley & Sons.

Resende, N.S.De, Eon, J.G. & Schmal, M. (1999), *J. Catal.*, 183, 6.

Richardson, J.T., Lei, M., Thrk, B., Forester, K. & Twigg, M.V. (1994) *Appl. Catal. A*, 110, 217.

Rodrigues, C.P., Da Silva, V.T. & Schmal, M. (2009), *Catal. Comm.*, 10, 1967.

Sachtler, W.M. & Zhang, Z., *Adv. Catal.* 39 (1993) 129.

Santos, D.C.R.M., Lisboa, J.S., Passos, F.B. & Noronha, F.B. (2004) *Braz. J. Chem. Eng.*, 21, 203.

Saracco, G. & Montanaro, L. (1995) *Ind. Eng. Chem. Res.*, 34, 1471–1479.

Satterfield, C.N. (1970) *Mass Transfer in Heterogeneous Catalysis*. Cambridge, MA, MIT Press.

Satterfield, C.N. (1991) *Heterogeneous Catalysis in Industrial Practice*. 2nd edition. New York, McGraw-Hill.

Schmal, M. (1982) *Cinética homogênea aplicada e cálculo de reatores*. 2. ed. Rio de Janeiro, Guanabara.

Schmal, M. (2010) *Cinética e Reatores, Aplicação na Engenharia Química*. Synergia.

Schwaab, M., Biscaia, E.C., Monteiro, J.L., Pinto, J.C. (2008) Chem. Eng. Sci., 63, 1542–1552.

Schwaab, M., Lemos, L.P., Pinto, J.C. (2007) *Chem. Eng. Sci.*, 62, 2750–2764.

Schwaab, M., Monteiro, J.L. and Pinto, J.C. (2008) "Sequential Experimental Design for Model Discrimination: Taking into Account the Posterior Covariance Matrix of Differences between Model Predictions", *Chem. Eng. Sci.*, 63, 2408–2419.

Schwaab, M. and Pinto, J.C. (2011) *"Análise de Dados Experimentais II – Planejamento de Experimentos"*, E-papers, Rio de Janeiro.

Schwaab, M. and Pinto, J.C. (2007) Optimum Reference Temperature for Reparameterization of the Arrhenius Equation. Part 1: Problems Involving One Kinetic Constant, *Chemical Engineering Science*, 62, 2750–2764.

Schwaab, M., Pinto, J.C. (2007) *"Análise de Dados Experimentais I – Fundamentos de Estatística e Estimação de Parâmetros"*, E-Papers, Rio de Janeiro.

Schwaab, M.; Pinto, J.C. (2007b) *Chem. Eng. Sci.*, 62, 2750–2764.

Schwaab, M., Silva, F.M., Queipo, C.A., Barreto Jr., A.G., Nele, M. and Pinto, J.C. (2006) " A New Approach for Sequential Experimental Design for Model Discrimination", *Chem. Eng. Sci.*, 61, 5791–5806.

Schmal, M., Souza, M.M.V.M., D^a. Aranda, G. & Perez, C.A.C. (2001) *Studies in Surface Science and Catalysis*, 132, 695.

Schmal, M., Toscani, A. & Castelan, L. (1983) *Ind. Eng. Chem. Develop.*, 22 (4), 563–570.

Schwaab, M. and Pinto, J.C. (2007) Análise de Dados Experimentais I: Fundamentos de Estatística e Estimação de Parâmetros, E-Papers, Rio de Janeiro.

Schwaab, M. and Pinto, J.C. (2008) Optimum Reparameterization of Power Function Models, *Chemical Engineering Science*, 63, 4631–4635.

Schwaab, M. and Pinto, J.C. (2007) Optimum Reference Temperature for Reparameterization of the Arrhenius Equation. Part 1: Problems Involving One Kinetic Constant, *Chemical Engineering Science*, 62, 2750–2764.

Scott Fogler (2000) *Elements of Chemical Reaction Engineering*. 2nd edition.

Senol, O.I., Ryymin, E.M., Viljava, T.R. & Krause, A.O.I. (2007) *J. Mol. Catal. A: Chem.*, 277 (1–2), 16, 107–112

Sexton, B.A., Hughes, A.E. & Foger, K. (1982) *J. Catal.*, 77, 85.

Smith, J.M. (1981) *Chemical Engineering Kinetics*, 3rd edition. New York, McGraw-Hill.

Smith, J.M., Van Ness, H.C. & Abbott, M.M. (2005) *Introduction to Chemical Engineering Thermodynamics*, 7th edition. McGraw-Hill Higher Education.

Smith, G.B., Russell, G.T., Yin, M. and Heuts, J.P.A. (2005) "The Effects of Chain Length Dependent Propagation and Termination on the Kinetics of Free-Radical Polymerization at Low Chain Lengths", European Polymer Journal, 41, 225–230.

Software HYSYS®version 3.1. Gibbs reactor module with package thermodynamic Peng-Robinson.

Somorjai, G.A. (1994) *Introduction to Surface Chemistry and Catalysis*. New York, John Wiley & Sons.

Souza, A.G.F. (2010) *D.Sc. Thesis*. COPPE-Univeristy Federal of Rio de Janeiro.

Souza, M.M.V.M. & Schmal, M. (2005) *Appl. Catal. A: General*, 281 (1–2), 19–24.

Souza, M.M.V.M. & Schmal, M. (2008) *Otimização, avaliação e aumento de escala de preparo de catalisadores a base de níquel para a reforma a vapor do metano*. Internal Report COPPETEC.

Souza, M.M.V.M., Aranda, D.A.G. & Schmal, M. (2001) *J. Catal.*, 204 (2), 498.

Souza, M.M.V.M., Clave, L., Dubois, V., Perez, C.A.C. & Schmal, M. (2004) *Appl. Catal. A*, 272, 133.

Storn, R., Price, K.V. (1997) *J. Glob. Optim*, 12, 341–359.

Tauster, S.J. & Fung, S.C. (1978) *J. Catal.*, 55, 29.

Toft, A.J. (1998) *Ph.D. Thesis*, Aston University.

Toniolo, F.S, Magalhães, R.N.S.H., Perez, C.A.C. & Schmal, M. (2012) *Appl. Catal. B: Environ.*, 117–118, 156–166.

Trimm, D.L. (1980) *Design of Industrial Catalysis*. Elsevier, Scientific Publishing Company.

Van Santen, R.A. & Niemanstverdriet, J.W. (1995) *Chemical Kinetics and Catalysis – Fundamental and Applied Catalysis*. New York, Plenum Press.

Vannice, M.A., Benson, J.E. & Boudart, M. (1970) *J. Catal.*, 16, 348.

Vassilev, S.V., Baxter, D., Andersen, L.K. & Vassileva, C.G. (2010) *Fuel*, 89 (5), 913–993.

Veglio, F., Trifoni, M., Pagnanelli, F. and Toro, L. (2001) Shrinking Core Model with Variable Activation Energy: A Kinetic Model of Manganiferous Ore Leaching with Sulphuric Acid and Lactose, *Hydrometallurgy*, 60, 167–179.

Viljava, T.R., Kromulainen, R.S. & Krause, A.O.I. (2000) *Catal. Today*, 60, 83–92.

Watts, D.G. (1994) Estimating Parameters in Nonlinear Rate Equations, *Canadian Journal of Chemical Engineering*, 72, 701–710.

Weigold, H. (1982) *Fuel*, 61, 1021–1026.

White, J.E., James Catallo, W., Legendre, B.L. (2011) *J. Anal. Appl. Pyrolysis*, 91, 1–33.

Wingard, L.B. (1972) *Enzyme Engineering*. New York, John Wiley & Sons.

Wolf, A. & Schüth, E.F. (2002) *Appl. Catal. A: General*, 226, 1.

York, A.P.E., Xiao, T. & Green, M.L.H. (2003) *Topics Catal.*, 22 (3–4), 345–358.

Subject index

Milton Keynes UK
Ingram Content Group UK Ltd.
UKHW052016071024
449327UK00027B/2297